in your class each day. With WileyPLUS you can:

Track Student Progress

Keep track of your students' progress via an instructor's gradebook, which allows you to analyze individual and overall class results. This gives you an accurate and realistic assessment of your students' progress and level of understanding.

Now Available with WebCT and eCollege!

Now you can seamlessly integrate all of the rich content and resources available with *WileyPLUS* with the power and convenience of your WebCT or eCollege course. You and your students get the best of both worlds with single sign-on, an integrated gradebook, list of assignments and roster, and more. If your campus is using another course management system, contact your local Wiley Representative.

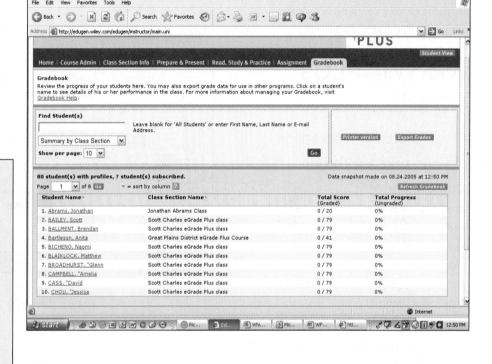

"I studied more for this class than I would have without *WileyPLUS*."

Melissa Lawler, *Western Washington Univ.*

For more information on what *WileyPLUS* can do to help your students reach their potential, please visit

www.wileyplus.com/experience

84% of students said they would recommend *WileyPLUS* to their other instructors *

You have the potential to make a difference!

WileyPLUS is a powerful online system packed with features to help you make the most of your learning potential, and get the best grade you can!

With Wiley**PLUS** you get:

A complete online version of your text and other study resources

Study more effectively and get instant feedback when you practice on your own. Resources like self-assessment quizzes, tutorials, and animations bring the subject matter to life, and help you master the material.

Problem-solving help, instant grading, and feedback on your homework and quizzes

You can keep all of your assigned work in one location, making it easy for you to stay on task. Plus, many homework problems contain direct links to the relevant portion of your text to help you deal with problem-solving obstacles at the moment they come up.

The ability to track your progress and grades throughout the term.

A personal gradebook allows you to monitor your results from past assignments at any time. You'll always know exactly where you stand.

If your instructor uses *WileyPLUS*, you will receive a URL for your class. If not, your instructor can get more information about *WileyPLUS* by visiting www.wileyplus.com

---

"It has been a great help, and I believe it has helped me to achieve a better grade."

Michael Morris, *Columbia Basin College*

74% of students surveyed said it helped them get a better grade.*

Essential Environmental Science

THE WILEY BICENTENNIAL–KNOWLEDGE FOR GENERATIONS

$\mathcal{E}$ach generation has its unique needs and aspirations. When Charles Wiley first opened his small printing shop in lower Manhattan in 1807, it was a generation of boundless potential searching for an identity. And we were there, helping to define a new American literary tradition. Over half a century later, in the midst of the Second Industrial Revolution, it was a generation focused on building the future. Once again, we were there, supplying the critical scientific, technical, and engineering knowledge that helped frame the world. Throughout the 20th Century, and into the new millennium, nations began to reach out beyond their own borders and a new international community was born. Wiley was there, expanding its operations around the world to enable a global exchange of ideas, opinions, and know-how.

For 200 years, Wiley has been an integral part of each generation's journey, enabling the flow of information and understanding necessary to meet their needs and fulfill their aspirations. Today, bold new technologies are changing the way we live and learn. Wiley will be there, providing you the must-have knowledge you need to imagine new worlds, new possibilities, and new opportunities.

Generations come and go, but you can always count on Wiley to provide you the knowledge you need, when and where you need it!

WILLIAM J. PESCE
PRESIDENT AND CHIEF EXECUTIVE OFFICER

PETER BOOTH WILEY
CHAIRMAN OF THE BOARD

Essential Environmental Science

Edward A. Keller

Professor of Environmental Studies and Earth Science
University of California, Santa Barbara

Daniel B. Botkin

Professor Emeritus
Department of Ecology, Evolution, and Marine Biology
University of California, Santa Barbara

President
The Center for the Study of the Environment
Santa Barbara, California

BICENTENNIAL
BICENTENNIAL
1807
WILEY
2007
BICENTENNIAL
BICENTENNIAL

John Wiley & Sons, Inc.

ACQUISITIONS EDITOR   *Rachel Falk*
DEVELOPMENT EDITOR   *Ellen Ford*
MARKETING MANAGER   *Clay Stone*
DEVELOPMENT ASSISTANT   *Justin Bow*
SENIOR MEDIA EDITOR   *Linda Muriello*
ASSOCIATE EDITOR   *Merillat Staat*
SENIOR PRODUCTION EDITOR   *Lisa Wojcik*
SENIOR DESIGNER   *Kevin Murphy*
SENIOR PHOTO EDITOR   Elle *Wagner*
COVER PHOTO   *Dr. Melissa Songer, Conservation GIS Lab, Smithsonian National Zoological Park*
COVER DESIGN   *David Levy*
SENIOR ILLUSTRATION EDITOR   *Anna Melhorn*
ANNIVERSARY LOGO DESIGN   *Richard J. Pacifico*

This book was set in Quark by Prepare and printed and bound by Courier Kendallville.

This book is printed on acid free paper. ⊚

To order books or for customer service please, call 1-800-CALL WILEY (225-5945).

ISBN-13 978-0-471-70411-9

Printed in the United States of America

10   9   8   7   6   5   4   3   2

DEDICATION

For Valery Rivera
who contributed so much to this book and
is a fountain of inspiration in our work and lives.

Ed Keller

For my sister, Dorothey B. Rosenthal
who has been a source of inspiration, support, ideas,
and books to read, and is one of my harshest and
best critics.

Dan Botkin

About the Authors

Edward A. Keller was chair of the Environmental Studies and Hydrologic Sciences Programs from 1993 to 1997 and is Professor of Earth Science at the University of California, Santa Barbara, where he teaches earth surface processes, environmental geology, environmental science, river processes, and engineering geology. Prior to joining the faculty at Santa Barbara, he taught geomorphology, environmental studies, and earth science at the University of North Carolina, Charlotte. He was the 1982–1983 Hartley Visiting Professor at the University of Southampton, a Visiting Fellow in 2000 at Emmanuel College of Cambridge University, England, and receipent of the Easterbrook Distinguished Scientist award from the Geological Society of America in 2004. Professor Keller has focused his research efforts into three areas: studies of Quaternary stratigraphy and tectonics as they relate to earthquakes, active folding, and mountain building processes; hydrologic process and wildfire in the chaparral environment of Southern California; and physical habitat requirements for the endangered Southern California steelhead trout. He is the recipient of various Water Resources Research Center grants to study fluvial processes and U.S. Geological Survey and Southern California Earthquake Center grants to study earthquake hazards.

Professor Keller has published numerous papers and is the author of the textbooks *Environmental Geology, Introduction to Environmental Geology* and (with Nicholas Pinter) *Active Tectonics* (Prentice-Hall). He holds bachelor's degrees in both geology and mathematics from California State University, Fresno; an M.S. in geology from the University of California; and a Ph.D. in geology from Purdue University.

Daniel B. Botkin is President of The Center for the Study of Environment, and Professor Emeritus of Ecology, Evolution, and Marine Biology, University of California, Santa Barbara, where he has been on the faculty since 1978, serving as Chairman of the Environmental Studies Program from 1978 to 1985. For more than three decades, Professor Botkin has been active in the application of ecological science to environmental management.

Trained in physics and biology, Professor Botkin is a leader in the application of advanced technology to the study of the environment. He was one of the pioneers in doing research on possible ecological effects of global warming, starting this work in the late 1960s, and continuing to the present.

The originator of widely used forest gap-models, he has conducted research on endangered species, characteristics of natural wilderness areas. His recent research includes studies of the bowhead whales, an endangered species hunted actively by Yankee whalers in the 19th century, important since ancient times to the Eskimos, and one of the longest-lived species, with individuals known to live 120 years.

During his career, Professor Botkin has advised the World Bank about tropical forests, biological diversity, and sustainability; the Rockefeller Foundation about global environmental issues; the government of Taiwan about approaches to solving environmental problems; and the state of California on the environmental effects of water diversion on Mono Lake. He served as the primary advisor to the National Geographic Society for its centennial edition map on "The Endangered Earth." He directed a study for the states of Oregon and California concerning salmon and their forested habitats. He has published many articles and books about environmental issues. His books include: *Beyond the Stoney Mountains: Nature in the American West from Lewis and Clark to Today* (Oxford University Press), *Strange Encounters: Adventures of a Renegade Naturalist* (Penguin/Tarcher), *The Blue Planet* (Wiley), *Our Natural History: The Lessons of Lewis and Clark* (Oxford University Press),

Discordant Harmonies: A New Ecology for the 21st Century (Oxford University Press), and *Forest Dynamics: An Ecological Model* (Oxford University Press).

Professor Botkin was on the faculty of the Yale School of Forestry and Environmental Studies (1968–1974) and was a member of the staff of the Ecosystems Center at the Marine Biological Laboratory, Woods Hole, MA (1975–1977). He received a B.A. from the University of Rochester, an M.A. from the University of Wisconsin, and a Ph.D. from Rutgers University.

He is the winner of the Mitchell International Prize for Sustainable Development and the Fernow Prize for International Forestry, and he has been elected to the California Environmental Hall of Fame. Recently he was awarded the Astor Lectureship of Oxford University, Great Britain.

Preface

You can't pick up a paper or turn on the TV without coming across some item about the environment. Everybody is talking about global warming. Even computer-animated movies seem to focus on endangered species like penguins, and extinct animals like dinosaurs. When leaders of nations get together these days, they usually spend some of their time talking about one or another environmental concern. And the issues seem to go on and on. One day we hear that honeybees are in trouble, another day we learn that genetically engineered crops are being banned in some nations because of concerns about environmental effects. There is continual discussion about future sources of energy and where we can turn as fossil fuels become scarce and continue to pollute the atmosphere. We hear that more and more species may be in trouble, threatened with extinction. Our biological natural resources—forests, fisheries—are suffering, we are told, from overexploitation and from failure to understand how forests and oceans work as complex ecological systems.

The debate is so constant and intense that it is getting hard to know what is the truth and what is simply talk. But one thing is clear: Each of us, every citizen of every nation, has to come to terms with environmental questions and understand them, at least enough to know what is reliable information and what is not; which are the important issues and which are not.

Science is at the center of the environmental debate for two reasons: our modern technology, an outgrowth of modern science, seems to be causing many new environmental problems, and only through science can we understand how our environment works. That basic understanding of environmental science is the purpose of this book. More to the point, the primary goal of this book is to help the student learn how to think about environmental issues, especially the science behind them. We present lots of facts, but our goal is not to tell you what to think, only how to interpret the information and arrive at your own conclusions.

We have been at this a long time. Our first environmental science textbook was published more than 20 years ago! But times are changing fast, and we believe there is a need for a change of direction in the way to present all the complexities of environmental science to the beginning student. Although this book builds on our long experience in teaching and writing about the environment, it does so in a fresh way, which we hope is clearer, easier, and more fun to read. In addition, our textbook is connected to the internet through *WileyPLUS*, so we can help you use some of the modern computer-based tools that are so important now in environmental science.

Goals of This Book

Modern science and technology have given us the power to change the environment, but before we start changing things, we need to understand how the environment works — what to change and what is best left alone. As two active scientists with decades of research experience, we know what is needed to solve environmental problems. And with this background, we have made sure that *Essential Environmental Science* provides students with an up-to-date introduction to the study of the environment. Again, we do not try to tell them *what* to think; we provide the information, knowledge, and understanding necessary for clear thinking.

Most students today are interested in their environment, but many do not have either the background in, or appreciation of, the underlying science needed to understand it. They often underestimate the rigorous nature of the Environmental Science course, and are then overwhelmed by the variety of topics, complexity of the issues and scientific terminology. If students are not prepared to read or think critically about the underlying science, they can leave the course without an appreciation of core concepts that will help them be environmentally responsible citizens.

Essential Environmental Science is a learning tool to help educate the next generation of environmentally aware college students by introducing them to some of the most important questions in Environmental Science today. By building scientific confidence and critical thinking skills, students will be better able to understand the natural world around them, how environmental issues affect them and their future, and how to act responsibly in a global environment.

We take a positive approach to identifying the most significant environmental issues and questions basic to the study of environmental science. By linking these fundamental principles to the underlying science, presenting key themes (population, urbanization, sustainability, globalization and values), encouraging critical thinking, and offering approaches to finding acceptable solutions to environmental problems, we provide the motivation to maintain students' interest in, and understanding of, the world around them.

The Book's Organization

Environmental science is necessarily interdisciplinary, which makes the presentation of it to the beginning student challenging, perhaps uniquely so among academic subjects. One way to approach it is to start with the broadest, most general and most integrating discussions, typically within social, political, and economic contexts, and then move downward to more and more specific topics. This has the advantage of showing the big picture, with all of its multifaceted contexts. But at an introductory level, this approach generally fails, because it ends up providing superficial and confusing presentations of extraordinarily complex ideas.

Another common approach is to work from the bottom up. This has the advantage of discussing each topic at a level that beginners can understand, but it can leave students without a clear view of the overall context for each environmental problem.

We have chosen to use the best combination of both the top-down and bottom-up approaches. Standard chapter topics are covered, but written, condensed and presented at a level intended for students in business, social sciences and the humanities.

For most of the book, we build up from the basic to the integrating, except for the first section, which presents a brief overview of environmental science and introduction to the scientific method (Chapter 1); an introduction to the human population problem, in some ways *the* underlying environmental problem and therefore presented early (Chapter 2); and fundamentals of complex systems — an integrating scientific approach (Chapter 3).

The second section of the book focuses on living resources, beginning with the fundamentals of how ecosystems work (Chapter 4), of biological diversity (Chapter 5), and of restoration ecology (Chapter 6). Then this section deals with commercially and economically important biological resources: forests and wildlife (Chapter 7), health effects of pollution (Chapter 8), and agriculture and the environment (Chapter 9).

The book's third section deals with physical and chemical aspects of the global environment, including energy (Chapter 10), freshwater (Chapter 11), oceans (Chapter 12), atmosphere, climate, and global warming (Chapter 13). The section then focuses on air and water pollution (Chapter 14), our mineral resources and their environmental effects (Chapter 15), waste management (Chapter 16), and — unusual in this kind of book but included because of modern concerns — an entire chapter on natural hazards, disasters and catastrophes (Chapter 17).

The book's final section consists of two chapters that broaden the discussion to societal aspects of environment, bringing the reader back to the big picture of the first section. The first of the two discusses environmental economics (Chapter 18); the second, how we might plan for a sustainable future (Chapter 19).

Chapter Format

In *Essential Environmental Science*, we focus on the science behind the stories and emphasize the need to think critically about major topics such as human population growth, sustainability, biogeochemical cycles, ecosystems processes, ecosystems management, global warming, air pollution, water pollution, waste management, and environmental planning. We do this by creating a framework around a set of *Big Questions*. Each chapter opens with a *Big Question*. These questions are designed to focus students on the main theme of each chapter, so they are not distracted or overwhelmed by the numerous facts and terminology. The question selected is certainly not the only question concerning the subject matter of the chapter, but is one that we believe significant. At the end of the chapter, after discussing the main points, we return to the *Big Question* and attempt to answer it. Often there is no single answer but the information in the chapter will assist students in formulating their own ideas as to what are the big questions and how they might be approached.

All the chapters are organized in a similar pattern, designed to focus students on the central questions and underlying science issues. After the *Big Question* is presented, we begin the chapter with a set of *Learning Objectives* for that chapter. This brief overview and set of objectives serve to focus students' attention on the key topics to come so that they understand right from the start where they are going. A short *Case Study* follows that is intended to highlight a particular issue or story of environmental significance to the chapter and position the ideas presented as a real-world application. At the end of the chapter, we return to the *Big Question* and then provide a *Summary* and a list of *Key Terms* to help students organize their studying. Finally, we provide three types of study questions. The first group, which we call *Getting It Straight*, reviews the basic factual information. The second, called *What Do You Think?*, encourages critical thinking about issues. The last group consists of broader questions that involve *Putting It All Together* as a way of integrating knowledge.

We have also carefully considered the kinds of ancillary matter to make available. Some is included at the end of the book — an extensive list of references cited in each chapter (our book is unusual among texts in this feature, which we consider essential for a scientific text), and several appendices to help the student solve problems and think through the chapter material.

Other ancillary material is available on the web at the book companion site and through *WileyPLUS*. Any teacher and student these days is aware of the great transition taking place in the presentation of knowledge, the big move to computer- and internet-based presentations. While everyone seems to agree that books continue to play a fundamental role in higher education, no one has yet worked out the perfect combination of printed and electronic media. We see this as a work in progress, for which we attempt to do the best for our students. We have tried to avoid presentations that simply make use of the "wow!" factor, and instead have focused on those that are truly advantageous to education.

Supplementary Materials

A full line of teaching and learning resources has been developed to help professors create more dynamic and innovative learning about the environment. For students we offer tools to build their ability to think clearly and critically. As a convenience for both professors and students, we provide these teaching and learning tools on the companion book website (*www.wiley.com/college/keller*).

For the Student

The student website is completely new, redesigned and content-rich. The activities and resources for students feature: reviews of learning objectives, online quizzing, virtual field trips, interactive environmental debates, as well as a map of regional case studies, critical-thinking readings, a glossary, flash cards, web links to important data and research in the field of environmental studies and video and animations covering a wide array of specific topics.

For the Instructor

Instructor Resources on the book companion site include a Biology Visual Library containing all the line illustrations in the textbook in jpeg format, as well as access to numerous other life science illustrations; the Instructor's Manual, Test Bank, PowerPoint Presentations, and select flash animations. Also included is the **Environmental Science Community Resource Library**. Instructor Resources are password protected.

- **Test Bank** by Christy Bazan, Illinois State University. Containing approximately 60 multiple choice and essay test items per chapter, this test bank offers assessment of both basic understanding and conceptual applications. The *Essential Environmental Science Test Bank* is offered in two formats: MS Word files and a Computerized Test Bank. The easy-to-use test-generation program fully supports graphics, print tests, student answer sheets, and answer keys. The software's advanced features allow you to create an exam to your exact specifications.

- **Instructor's Manual** prepared by Sandy Buczynski, San Diego University and Anthony Gaudin, Ivy Tech Community College, provides over 70 creative ideas for in-class activities, lecture outlines and answers to all end-of-chapter questions in the text.

- **All Line Illustrations and Photos** from *Essential Environmental Science*, in jpeg files and PowerPoint format.

- **PowerPoint Presentations** by Michael Freake, Lee University. Tailored to *Essential Environmental Science's* topical coverage and learning objectives, these presentations are designed to cover key text concepts, illustrated by embedded text art.

- **Animations.** Select text concepts and topics are illustrated using flash animation.

- **Personal Response System** questions by Raymond Beiersdorfer, Youngstown State University, are specifically designed to foster student discussion and debate in class.

- **Environmental Science Community Resource Library** is an exciting new resource that enables instructors from across the country or even across the world to share resources and ideas that have worked for them in teaching environmental science. The Resource Library includes resources such as activities, video, or animation all contributed by your fellow environmental scientists. You may access these resources at *www.wiley.com/college/keller*. If you are interested in contributing to the Library, please contact your Wiley sales representative for assistance.

WileyPLUS

WileyPLUS provides an integrated suite of teaching and learning resources, including an online version of the text, in one easy-to-use website. Organized around the essential activities you perform in class, *WileyPLUS* helps you:

- **Prepare and Present**. Create class presentations using a wealth of Wiley-provided resources, including an online version of the textbook, PowerPoint slides, animations, and more—making your preparation time more efficient. You may easily adapt, customize, and add to this content to meet the needs of your course.

- **Create Assignments**. Automate the assigning and grading of homework or quizzes by using Wiley-provided question banks or by writing your own. Student results will be automatically graded and recorded in your gradebook. *WileyPLUS* can link homework problems to the relevant section of the online text, providing students with context-sensitive help.

- **Track Student Progress**. Keep track of your students' progress via an instructor's gradebook, which allows you to analyze individual and overall class results to determine student progress and level of understanding.

- **Administer Your Course**. *Wiley PLUS* can easily be integrated with other course management systems, gradebooks, or other resources you are using in your class, providing you with the flexibility to build your course in your own way.

Acknowledgments

This book owes much to the cooperation and work of many people. To everyone who provided advice and encouragement, we offer our sincere appreciation. Diana Perez was the developmental editor. She also reviewed the manuscript to improve its clarity and provided much of the writing style to help make the book much more readable for introductory students. We wish to acknowledge the support of Rachel Falk, Senior Acquisitions Editor; Ellen Ford, Senior Developmental Editor; Merillat Staat, Associate Editor; Linda Muriello, Senior Media Editor; Lisa Wojcik, Senior Production Editor; Kevin Murphy, Senior Designer; Elle Wagner, Photo Editor; and Anna Melhorn, Senior Illustration editor.

It is particularly important to obtain reviews of the book chapter by chapter. To those reviewers listed below we offer our gratitude and thanks:

Mark Anderson, University of Maine – Orono
John Baldwin, Florida Atlantic University
Christy Bazan, Illinois State University
Edward Becker, Palm Beach Community College
Cheryl Berg, Gateway Community College – Phoenix
Shannon Bliss, Cerro Coso Community College
Rene Borgella, Ithaca College
Hunting Brown, Wright State University – Dayton
J. Christopher Brown, University of Kansas
Catherine Carter, Georgia Perimeter College
Robert Cordero, Holy Family University
Dagmar Cronn, Oakland University
Michael Dann, Pennsylvania State University – University Park
Stephanie Dockstader, Monroe Community College
James Dunn, Grand Valley State University
Robert East, Washington and Jefferson College
Brad Fiero, Pima County Community College
Todd Fritch, Northeastern University
Marcia Gillette, Indiana University – Kokomo
Joseph Goy, Harding University
Linda Ingram, Texas A&M University
Dianne Jedlicka, Art Institute of Chicago
Robert Keesee, SUNY – Albany

Dawn Keller, Hawkeye Community College
Gregory Kientop, Devry University – Tinley Park of Nevada – Reno
Barry Vroeginday, DeVry University North Brunswick
Karen Wellner, Arizona State University
James White, University of Colorado
Jennifer Withington, SUNY – Plattsburgh
Catherine Koning, Franklin Pierce College
Paul Kramer, SUNY – Farmingdale
Peter Lortz, North Seattle Community College
Tim Lyon, Ball State University
Anthony Marcattilio, St. Cloud State University
Lynn McCartney, Simpson College
Paige Mettler-Cherry, Lindenwood University
Woody Moses, Highline Community College
Jay Odaffer, Manatee Community College
Barry Perlmutter, Community College of Southern Nevada – Las Vegas
Roy Sofield, Chattanooga State Technical Community College
Bo Sosnicki, Piedmont Virginia Community College
Julie Stoughton, University

Thanks go out also to those colleagues who participated in focus groups on the book and plan. Their time and insights were particularly helpful and appreciated.

Jennifer Anderson, Johnson County Community College
Steve Browder, Franklin College
Dagmar Cronn, Oakland University
Robert East, Washington & Jefferson College
Robert Harrison, University of Washington
Diane Jedlicka, Art Institute of Chicago
Greg Kientop, DeVry University
Paul Kramer, Farmingdale State University of NY
Meredith Lassiter, Winona State University
Lynn McCartney, Simpson College
Natalie Osterhoudt, Broward Community College
Charles Shorten, West Chester University
Richard Stevens, Monroe Community College
Arlene Westhoven, Ferris State University

Brief Table of Contents

Table of Contents

6

Restoration Ecology   96

Forests and Wildlife   114

8

Environmental Health, Pollution and Toxicology   134

Agriculture and Environment 154

Energy and Environment 178

11

Water and Environment   210

? **Big Question:** Can We Maintain Our Water Resources for Future Generations?   210

Case Study: The Colorado River: Water Resources Management, Water Pollution, and the Environment   211

Oceans and Environment   247

Earth's Atmosphere
and Climate   267

Air Pollution and Environment   290

Minerals and Environment   318

Waste Management   332

Natural Hazards 351

18

Environmental Economics 374

19

Planning for a Sustainable Future   389

Image Source/Getty Images

1

Fundamental Issues in Environmental Science

Big Question

Why Is Science Necessary to Solve Environmental Problems?

?

Learning Objectives

Certain issues are basic to the study of environmental science. After reading this chapter, you should understand . . .

- why rapid growth of the human population is the fundamental environmental issue;

- that we must learn to use our environmental resources in a way that assures that they will be available in the future;

- how people affect the environment of the entire planet and why we must take a global perspective on environmental problems;

- that urban environmental issues and the effects of urban areas on other environments should be a primary focus of our attention;

- that people and nature are intimately linked;

- that developing solutions to environmental problems requires us to make value judgments based on knowledge of scientific facts;

- that solutions to many environmental problems involve an understanding of systems and rates of change;

- why solving environmental problems is often complex and difficult;

- what the Precautionary Principle is, and when to apply it.

Case Study

Easter Island

The story of Easter Island spans a period of approximately 1,500 years and illustrates both the importance of science and the sometimes irreversible consequences of rapid human population growth, accompanied by depletion of resources necessary for survival. Evidence of the island's history is based on detailed studies by earth scientists and social scientists who investigated the anthropological record left in the soil where humans lived and sediment in ponds where pollen from plants that lived at different times was deposited. The goals of the studies were to estimate the number of people, their diet, and their use of resources. This was linked to studies of changes in vegetation, soils, and land productivity.

Easter Island lies several thousand kilometers west of South America, and when Polynesian people first reached it about 1,500 years ago they colonized a green island covered with rich soils and forest. The small group of settlers grew rapidly, and by the 16th century over 10,000 people had established a complex society. They were spread among a number of small villages that raised crops and chickens, supplementing their diet of fish from the sea. They used the island's trees to build their homes, and to build boats. They also carved massive 8-meter-high statues from volcanic rock and moved them into place at various parts of the island using tree trunks as rollers (Figure 1.1).

When Europeans first reached Easter Island in the 17th century, the only symbols of the once-robust society were the statues. A study suggested that the island's population had collapsed in just a few decades to about 2,000 people because they had used up (degraded) the isolated island's limited resource base.

At first there were abundant resources, and the human population grew fast. To support their growing population, they cleared more and more land for agriculture, and cut more trees for fuel, homes, and boats, and for moving the statues into place.

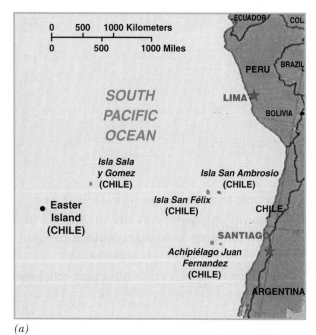

(a)   (b)   Michael Wozniak/iStockp▪

▪ **FIGURE 1.1**
Easter Island, collapse of a society.
(a) Location of Easter Island in the Pacific Ocean several thousand kilometers west of South America; (b) large statues carved from volcanic rock before the collapse of a society of about 10,000 people.

Some of the food plants they brought to the island didn't survive, possibly because the voyage to the island was too long, or because the climate was not suitable for them. In particular, they did not have the breadfruit tree, a nutritious starchy food source, and so they relied more heavily on other crops, which required clearing more land for planting. The island was also relatively dry, so it is likely that fires for clearing land got out of control sometimes and destroyed even more forest than intended.

The cards were stacked against the settlers to some extent—but they didn't know this until too late. Other islands of similar size that the Polynesians had settled did not suffer forest depletion and fall into ruin.[1, 2] This isolated island, however, was more sensitive to change. As the forests were cut down, the soils, no longer protected by forest cover, were lost to erosion. Loss of the soils reduced agricultural productivity, but the biggest loss was the trees. Without wood to build homes and boats, the people were forced to live in caves and could no longer venture out into the ocean for fish.

These changes did not occur immediately—it took over 1,000 years for the increasing population to deplete their resources. The loss of the forest was irreversible: Because the loss of trees led to loss of soil, new trees could not grow to replace the forests. As resources grew scarcer, wars between the villages became common, as did slavery and even cannibalism.

Easter Island is small, but its story is a dark one that suggests what can happen when people use up the resources of an isolated area. We note that some aspects of the history of Easter Island stated above have recently been challenged as being only part of the story. Deforestation certainly played a role in the loss of the trees, and rats that arrived with the Polynesians were evidently responsible for eating seeds of the palm trees, not allowing regeneration. The alternative explanation is that the Polynesian people on Easter Island at the time of European contact in 1722 numbered about 3,000. This population may have been close to the maximum reached around the year 1350. Contact with Europeans introduced new diseases and enslavement, which reduced the population to about 100 by the late 1870s.[3]

As more of the story of Easter Island emerges from scientific and social studies, the effects of human resource exploitation, invasive rats, and European contact will become clearer, and the environmental lessons of the collapse will lead to a better understanding of how we can sustain our global human culture. However, the primary lesson is that *limited resources can support only a limited human population.*

Like Easter Island, our planet Earth is isolated in our solar system and universe and has limited resources. As a result, the world's growing population is facing the problem of how to conserve those resources. We know it takes a while before environmental damage begins to show, and we know that some environmental damage may be irreversible. We are striving to develop plans to ensure that our natural resources, as well as the other living things we share our planet with, will not be damaged beyond recovery.[4]

1.1 Fundamental Principles

The environment is complex and multifaceted. As a result, this book must cover a wide range of topics, from environmental ethics to the chemistry of the ozone layer in the atmosphere. However, all of these topics are linked together by several concepts. The purpose of this chapter is to acquaint you with concepts that will be repeatedly used throughout your exploration of environmental science.

1.2 Human Population: The Basic Environmental Problem

The most dramatic increase in the history of the human population occurred in the last part of the 20th century. In merely the past 35 years, the number of people in the world more than doubled, from 2.5 billion to over 6.6 billion. This rapid explosion of the human population is sometimes referred to as the "population bomb."[5]

Our rapidly increasing population underlies all environmental problems because most environmental damage results from the very large number of people on Earth. Ultimately, then, we cannot solve our environmental problems unless we can learn to limit our population to a number that Earth can sustain.

We discuss human population in detail in Chapter 2.

1.3 Sustainability

Scientists are still debating the definition of **sustainability.** The concept as we use it in this book can be viewed this way:

Use it, but don't use it *up.* When we harvest a product from an ecosystem, we need to be concerned about the *sustainability of the product, the sustainability of our harvest, and, most important, the sustainability of the ecosystem itself.* We want to be able to continue to harvest regularly as long as possible, and so we want production to continue as long as possible so. Therefore the ecosystem must be sustainable at least in maintaining whatever it takes to produce the amount we are harvesting.

(a)   Mira/Alamy Images

(b)   Thinkstock/Jupiter Images

▪ **FIGURE 1.2**
How many people do we want on Earth?
(a) Streets of Calcutta; (b) Davis, CA.

Other uses of the term *sustainability*. People often refer to the goal of a *sustainable society*—meaning that we seek ways to ensure that Earth's resources will remain adequate to support future generations. This leads to two additional types of sustainability, involving economics and development. A *sustainable economy* can maintain its level of activity over time despite its uses of environmental resources, and at least doesn't cause irreversible harm to the environment. *Sustainable development* typically means that a society is able to maintain the environment while continuing to develop its economy and social institutions for an indefinite time. In other words sustainable development is development that is economically viable, socially just, and doesn't harm the environment. Just recently we heard a discussion about *sustainable architecture*—that is, designing buildings that do not waste energy and are environmentally friendly in other ways as well.

Earth's Carrying Capacity

How many people can Earth sustain? **Carrying capacity** is usually defined as the maximum number of individuals of a species that can be sustained by an environment without lessening the environment's ability to sustain that same number of individuals in the future. When we ask how many people Earth can sustain, we are asking about the Earth's carrying capacity, and we are also asking about sustainability.

It's partly a matter of values. Earth's human carrying capacity depends in part on how we want to live, and how we want those who follow us to live. Will we and, even more so, our children live short lives in crowded surroundings (Figure 1.2) without a chance to enjoy Earth's scenic beauty and diversity of life? Or do we hope that we and our descendants will enjoy a high quality of life and good health? Once we choose a goal for the quality of life, we can use scientific information to understand what the carrying capacity might be and how we might achieve it.

1.4 A Global Perspective

Today our actions are experienced worldwide. Because human actions have begun to change the environment all over the world, the next generation will have to take a global perspective on environmental issues (Figure 1.3). The recognition that civilization can change the entire planet's environment is relatively recent. As we discuss in detail in later chapters, scientists now believe that emissions of modern chemicals are changing the ozone layer high in the atmosphere. Scientists also believe that the burning of fossil fuels increases the concentration of greenhouse gases in the atmosphere, which may change Earth's climate. These atmospheric changes suggest that the actions of many groups of people at many locations affect the environment the world over.[6,7]

Another new idea being explored is that nonhuman life affects the environment of our planet at a global level and has changed it over the course of several billion years. These two new ideas have profoundly affected our approach to environmental issues.

■ **FIGURE 1.3**
Earth from space.
Isolated from other planets, Earth is "home," the only habitat we have.

Life makes Earth's environment unlike that of other planets. Awareness of the global interactions between life and the environment has led to the **Gaia hypothesis,** named for the Greek goddess Mother Earth by British chemist James Lovelock and promoted by Lovelock and American biologist Lynn Margulis.[8] The Gaia hypothesis proposes that the environment at a global level has been profoundly changed by life throughout the history of life on Earth, and that these changes have tended to improve the chances that life on Earth will continue.

Because life affects the environment at a global level, the environment of our planet is different from that of a lifeless one. The Gaia hypothesis states that life manipulates the environment for the maintenance of life. For example, some scientists believe that algae floating near the surface of the ocean influence rainfall and the carbon dioxide content of the atmosphere, thereby significantly affecting the global climate. This suggests that the planet Earth is capable of regulating itself.

The idea of a living Earth is probably as old as humanity. James Hutton, who viewed the present as a key to the past, stated in 1785 that he believed Earth to be a superorganism whose cycling of nutrients from soils and rocks in streams and rivers could be compared to the circulation of blood in an animal.[8] In this analogy, the rivers are the arteries and veins, the forests are the lungs, and the oceans are the heart of Earth.

What we do now could determine our planet's future. The Gaia hypothesis is really a series of hypotheses. Few scientists disagree with the first hypothesis—that life, since it began, has greatly affected the planetary environment. And certainly there is some evidence that life has altered Earth's environment, particularly the climate, in ways that have allowed life to persist. Although few scientists accept the idea that life deliberately (consciously) controls the global environment, we have become conscious of our own effects on our planet. Thus, the idea that we can consciously make a difference in the future of our planet is not as extreme as would once have been thought. The future status of our environment may depend in part on actions we take now and in coming years.

1.5 Cities Affect the Environment

In part as a result of the rapid growth of the human population and in part as a result of changes in technology, we are becoming an urban species, and our effects on the environment are more and more the effects of urban life.[9] With economic development comes urbanization—people move from farms to cities and then perhaps to suburbs. Cities and towns get bigger, and because they are commonly located near rivers and along coastlines, urban sprawl often overtakes the good agricultural land of river floodplains, as well as the coastal wetlands, an important habitat for many rare and endangered species. As urban areas expand, wetlands are filled in, forests are cut, and soils are covered over.

In developed countries, about 75% of the people live in urban areas and 25% live in rural areas. In developing countries, by contrast, about 40% of the people are city dwellers.[10] Today nearly one-half of people on Earth live in urban regions and this trend will increase. For example, a massive movement of people to cities is occurring in China today. It is estimated that by 2025 almost two-thirds of the world's population—5 billion people—will live in cities.

We must look more closely at the effects of urbanization. Environmental organizations have often focused on wilderness, endangered species, and natural resources, including forests, fisheries, and wildlife. Although these will remain important, in the future we must place more emphasis on urban environments and their effects on the rest of the planet.

1.6 People and Nature

People and nature are intimately linked through the "principle of environmental unity," which holds that everything affects everything else. People and nature affect each other in ways ranging from simple to complex.

▪ **FIGURE 1.4**
People and nature.
We feel safe around a campfire—a legacy from our Pleistocene ancestors?

For example, we depend on nature for many natural "service functions," such as providing us with soil, water, and air. As long as people have had tools, including fire, they have changed nature, often in ways that we like and consider "natural." In fact, it can be argued that it is natural for organisms to change their environment. Elephants clear trees, changing forest to grassland, and people clear forest to plant food crops. Who can say which is more natural? Today, environmental sciences are showing us how people and nature connect, and in what ways this is mutually beneficial.

We are becoming more Earth-centered, and spend more time in nature for recreation and spiritual activities. We have evolved on and with Earth and are not separate from it. We are genetically very similar if not identical to people who lived on Earth over 10,000 years ago and who were by necessity Earth-centered in their hunting and gathering. Do you ever wonder why we like to go camping, sitting around a fire at night roasting marshmallows and exchanging scary stories about bears and mountain lions (Figure 1.4)? We should celebrate our union with nature as we work toward sustainability.

1.7 Science and Values

We use both to solve environmental problems. Before we decide what kind of environment we want, we need to know what is possible. That requires scientific data. Once we know our options, we make choices based on our values. An example of an environmental value judg-

ment is the choice between one's desire to have many children and the need to limit the human population worldwide.

Science is a process of discovery. Often, the fact that scientific ideas change over time or differ at the same time seems frustrating. Why can't scientists agree on what is the best diet for people? Why do scientists consider a chemical dangerous to the environment for a while and then decide that it isn't? Why do scientists in one decade believe that fire in nature is an undesirable disturbance and in a later decade decide that it is important and natural? Why can't they tell us whether there is going to be significant global warming or not? Can't scientists just find out the truth for each of these questions once and for all, and agree on it?

Rather than looking to science for answers to such questions, think of science as a continuing adventure with increasingly better knowledge about how the world works. Sometimes changes in ideas are small, and the major context remains the same. Sometimes a science undergoes a fundamental revolution in ideas.

Science is one way of looking at the world. It begins with observations about the natural world. From these observations, scientists formulate hypotheses that can be tested. Modern science does not deal with things that cannot be tested by observation, such as the ultimate purpose of life or the existence of a supernatural being. Thus, science does not deal with values, such as standards of beauty or issues of good and evil. Our criterion for deciding whether a statement is in the realm of science is whether it is possible to disprove the state-

☐ **TABLE 1.1   SELECTED INTELLECTUAL STANDARDS**

- *Clarity:* If a statement is unclear, you can't tell whether it is relevant or accurate.
- *Accuracy:* Is a statement true? Can it be checked? To what extent does a measurement agree with the accepted value?
- *Precision:* The degree of exactness to which something is measured. Can a statement be more specific, detailed, and exact?
- *Relevance:* How well is a statement connected to the problem at hand?
- *Depth:* Did you deal with the complexities of a question?
- *Breadth:* Did you consider other points of view or look at it from a different perspective?
- *Logic:* Does a conclusion make sense and follow from the evidence?
- *Significance:* Is the problem an important one? Why?
- *Fairness:* Are there any vested interests, and have other points of view received attention?

Modified after Paul, R., and L. Elder. 2003. Critical thinking. Dillon Beach, CA: The Foundation for Critical Thinking.

ment. If so, it is a scientific statement. For example, if you say there is life elsewhere in the universe, that is not a scientific statement, because it's impossible to prove it's not true.

Scientists rely on critical thinking. Critical scientific thinking is disciplined thinking using intellectual standards, effective communication, clarity, and commitment to developing scientific knowledge and skills (Table 1.1). It leads to conclusions, generalizations, and sometimes scientific theories and even scientific laws. Taken together, these comprise a body of beliefs that, at the present time, account for all known observations about a particular phenomenon.

What is environmental science? Environmental science is a group of sciences that attempt to explain how life on Earth is sustained, what leads to environmental problems, and how these problems can be solved. Environmental science includes ecology (the part of biology that deals with the relationships among living things and their environment), geology, hydrology, climatology, meteorology, oceanography, and soil science.

We depend on our environment. People can live only in an environment that has certain characteristics, certain resources, and certain ranges of availability of those resources. Because modern science and technology give us the power to affect the environment, we have to understand how the environment works so that we don't affect it in ways that will harm it and us and other living things.

Environmental science is different from other sciences for two reasons:

- It includes sciences, but also is often linked with nonscientific fields that have to do with how we value the environment, such as environmental ethics.
- It deals with many topics that have great emotional effects on people, often stirring up political debate and strong feelings that sometimes override scientific information.

Placing a value on the environment. How do we place a value on any aspect of our environment? How do we choose between two different concerns? We may justify our choice in four different ways: utilitarian, ecological, aesthetic, and moral.

A *utilitarian justification* views some aspect of the environment as valuable because it is useful—it provides economic benefits or is directly necessary to people's survival. For example, fishermen earn a living from the ocean and need a continued supply of fish so they can continue to earn a living.

An *ecological justification* places a value on some factor that is essential to larger life-support functions, even though it may not directly benefit an individual.

Aesthetic justification has to do with the value we place on beauty. For example, many people find wilderness scenery beautiful and would rather live in a world with wilderness than without it.

Moral justification is based on one's view of right and wrong. One example is the belief that certain aspects of the environment have a right to exist and that it is our moral obligation to allow them or help them to persist. Moral arguments have been extended to many nonhuman organisms, to entire ecosystems, and even to inanimate objects.

1.8 Solving Many Environmental Problems Involves Systems and Rates of Change

What is a system? A **system** is a set of parts that function together to act as a whole. A single organism, such as your body, is a system, as is a sewage-treatment plant, a city (Figure 1.5), a river (Figure 1.6), and your dorm room. On a much larger scale, the entire Earth is a system.

Systems respond to inputs and have outputs. If you are hiking and see a grizzly bear, the sight of the

Steve Geer/iStockphoto

▪ **FIGURE 1.5**
Urban system.
Lake Michigan, Lincoln Park, and the city of Chicago make up a large, complex urban system that includes air, water, and land resources. Urban systems are particularly important in environmental science because more and more people are living in urban areas.

Courtesy Ed Keller

▪ **FIGURE 1.6**
River system.
The Owens River, on the east side of the Sierra Nevada in California, is a system that includes water, sediment, vegetation, and animals such as fish and insects, all of which function together as a whole.

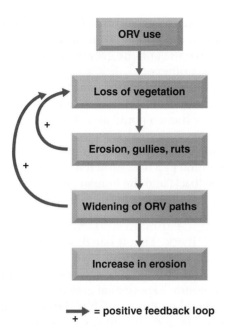

■ **FIGURE 1.7**
Positive feedback.
How the use of off-road vehicles (ORVs) produces positive feedback that increases erosion.

bear is an input. Your body reacts to that input: The adrenaline level in your blood goes up, your heart rate increases, and the hair on your head and arms may rise. Your response—perhaps moving slowly away from the bear—is an output. When you move away, you feel safer and your heart rate slows as adrenaline decreases. This is **feedback**—that is, *the output of the system also serves as input and leads to further changes in the system.*

Negative and positive feedback. If you go out in the sun and get hot, your higher temperature affects your sensory perceptions (input). If you stay in the sun, your body responds physiologically: Your pores open, you sweat, and you are cooled by the evaporating water.

In our example, an increase in temperature is followed by a response that leads to a decrease in temperature. This is an example of *negative feedback*. Negative feedback is self-regulating, or stabilizing; it usually keeps a system in a relatively constant condition.

Positive feedback occurs when an increase in output leads to a further increase in the output. Consider a fire starting in a forest. The wood may be slightly damp at the beginning and not burn well. But once a fire starts, the flame dries out nearby wood, which begins to burn and in turn dries out even more wood, leading to an even larger fire. The larger the fire, the drier the wood and the faster the fire spreads. Positive feedback is sometimes called "a vicious cycle," when it is destabilizing.

Positive feedback can cause problems. Environmental damage can result when people's use of the environment leads to positive feedback. For example, the churning tires of off-road vehicles may loosen the soil and uproot plants, increasing the rate of erosion (Figure 1.7). As more soil is eroded, running water carves ruts and gullies. To avoid the ruts and gullies, drivers move onto adjacent sections, which exposes an even greater area to erosion. The gullies themselves worsen erosion because they concentrate runoff and tend to get longer, wider, and deeper (Figure 1.8). Thus, from positive feedback, areas frequently used by off-road vehicles may become wastelands.

Some situations involve both positive and negative feedback. Changes in human population in large cities offer an example, as shown in Figure 1.9. Positive feedback, which increases the population in cities, may occur when people perceive greater opportunities and

Ed Keller

■ **FIGURE 1.8**
Off-road-vehicle damage to rare plants living on coastal dunes near San Luis Obispo, California. Note the tire tracks extending into the dune field.

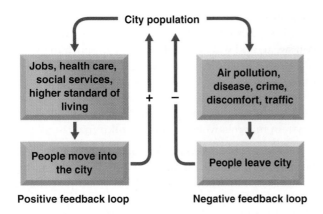

▪ **FIGURE 1.9**

Potential positive and negative feedback loops
for changes in human populations of large cities. The
left side shows that as jobs, health care, and the
standard of living improve, the city population
increases. The right side shows that increases in air
pollution, disease, crime, discomforts, or traffic tend
to reduce city populations. [*Source:* Modified from M.
Maruyama. The second cybernetics: Deviation-
amplifying mutual causal processes, *American
Scientist* 51:164–670, 1963. Reprinted by permission
of *American Scientist,* magazine of Sigma Xi, The
Scientific Research Society.]

a higher standard of living in cities. Negative feedback
may result if air and water pollution, disease, crime, and
overcrowding cause some people to migrate from the
cities to more rural suburban areas.

Which kind of feedback is more desirable?
Feedback naturally results in change. Whether we consider the change useful, desirable, or undesirable reflects our values. Sometimes positive feedback has
desired effects. For example, suppose we are interested
in restoring the ecology of Yellowstone National Park by
reintroducing wolves. We will expect positive feedback
in the population for a while as the number of wolves
grows. In this case, positive feedback for a period of time
is desirable because it produces a change we want.

By contrast, consider a system that is in a stable but
undesirable state—for example, a polluted stream in an
urban environment. Urban runoff carrying pollutants,
such as oil and other chemicals from streets, into the
stream's system may, through negative feedback mechanisms, reach a stable state between the water and the
pollutants in the stream. However, most people would
consider this system undesirable. If they began a channel restoration project, controlling pollutants by collecting and treating them before they entered the
stream, the stream might reach a new state that is ecologically more desirable.

Whether we view positive or negative feedback as desirable depends on the system and the potential
changes. Nevertheless, some of the major environmen-

tal problems we face today result from positive feedback
mechanisms that are out of control. These problems include resource use and the growth of the human population, among others.

Interestingly, throughout most of human history,
strong negative feedback cycles resulted in very slow
growth of the human population. Disease and a limited
capacity to produce food kept growth low. However, in
the past hundred years, modern medicine, sanitation,
and agricultural practices turned negative feedback into
positive feedback, and the human population increased
rapidly.

Exponential growth is an important outcome of
positive feedback. Simply stated, something is growing
exponentially when it increases by a constant percentage
per time period—let's say per year—rather than by a
constant amount. Exponential growth is also characterized by a **doubling time,** which is the time it takes for
what is growing to double in size or number. The doubling time depends on the rate of growth. Due to the
mathematics of exponential growth, the doubling time
is approximately 70 divided by the growth rate. For example, if the growth rate is 7% per year, the doubling
time is 10 years.

A population growing exponentially increases by a
greater and greater amount each year. Think of a bank
deposit earning a fixed interest rate each year. The first
year you earn, say, 3% on your money. The second year
you earn that same percentage on your original deposit
but also on the 3% that was added the first year. The
third year you earn 3% on your original deposit and also
on the two previous years' interest. And so on. Thus,
each year (assuming you don't take any money out),
your money grows by a greater amount than it did the
year before. At an annual growth of 3% how soon will
your bank account double?

In the same way, an increase in a population's output (growth) may lead to a further increase in the output (more growth; see Figure 1.10). As you will see in
Chapters 2 and 4, exponential growth can occur with
populations of organisms, at least for short periods.

Exponential growth is incompatible with sustainability. In fact, the term *sustainable growth* is an oxy-
moron—a self-contradiction. Even at modest growth
rates, the number of whatever is growing will eventually
reach levels that are impossible to maintain.[11,12]

Environmental Unity

It's (mostly) all connected. Our discussion of positive and negative feedback sets the stage for another
fundamental concept in environmental science: **environmental unity.** Simply stated, the term means that it is
impossible to change only one thing; everything affects
everything else. Of course, this is something of an over-

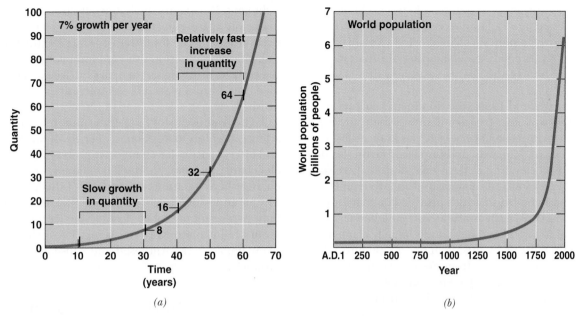

■ **FIGURE 1.10**

Exponential growth.
(a) Idealized curve illustrating exponential growth. The growth rate is constant at 7%, and the time necessary to double the quantity is constant at 10 years. Notice that growth is slow at first and much faster after several doubling times. For example, the quantity changes from 2 to 4 (absolute increase of 2) for the time from 10 to 20 years. It increases from 32 to 64 (absolute increase of 32) during the time from 50 to 60 years. (b) Human population increase for the last 2000 years. [*Source:* Data from U.S. Department of State.]

statement and is not absolutely true; the extinction of a species of snails in North America, for instance, is hardly likely to change the flow of the Amazon River. However, many aspects of the natural environment are closely linked. Changes in one part of a system often have secondary and tertiary effects within the system and effects on adjacent systems. Earth and its ecosystems are complex entities in which, as we saw with the case history of Easter Island, any action may have several or many effects. With this in mind, consider the following story.

Deforestation in Amboseli National Reserve. Environmental change is often caused by a complex web of interactions (environmental unity). Therefore, the most obvious answer to the question of what caused a particular change may not be the right answer. Amboseli National Reserve (Figure 1.11) is a case in point. In just a

■ **FIGURE 1.11**

Geologic and landform map of Amboseli National Reserve in southern Kenya, Africa. Mount Kilimanjaro is a volcano composed of alternating layers of volcanic rock and ash. Rainfall on the slopes of the mountain infiltrates and becomes groundwater that moves slowly down the mountain to an ancient lakebed. There it emerges as springs in the swampy, seasonally flooded Lake Amboseli. [*Source:* After T. Dunn and L. B. Leopold. *Water in environmental planning.* San Francisco: W. H. Freeman, 1978.]

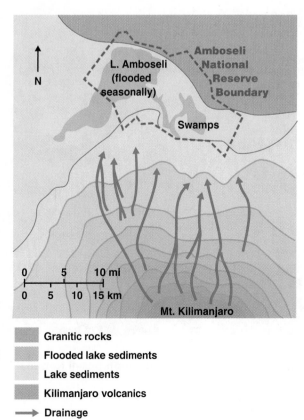

few decades, this park, in southern Kenya at the foot of Mount Kilimanjaro (a large volcano), underwent a significant environmental change. To explain what happened required an understanding of physical, biological, and human-use factors—and how these factors are linked.

Before the mid-1950s, the dominant vegetation in the Reserve was fever tree woodlands, which provided habitat for mammalian species such as kudu, baboons, vervet monkeys, leopards, and impalas. Starting in the 1950s, and accelerating in the 1960s, despite increased rainfall, these woodlands disappeared and were replaced by short grass and brush, which provided habitat for typical plains animals such as zebras and wildebeest.

Loss of the woodland habitat was initially blamed on overgrazing of cattle by the Masai people (Figure 1.12) and damage to the trees from elephants. In the end, however, careful study by environmental scientists showed that neither people nor elephants but changes in rainfall and soils were the primary culprits.[13,14] Research on rainfall, groundwater history, and soils suggested that the Amboseli National Reserve area is very sensitive to changing amounts of rainfall. When the groundwater flowing within the slopes of Mount Kilimanjaro enters the lake beds, it becomes salty from small amounts of salt that accumulate slowly in the lake

sediments. During dry spells, the salty groundwater sinks lower into the earth, and soil near the surface has relatively little salt. The fever trees grow well in the nonsalty soil. During wet periods, the groundwater rises close to the surface, bringing salt that kills the trees. As the trees died, they were replaced by salt-tolerant grasses and low brush.[13,14,15]

The Amboseli story illustrates how environmental science attempts to work out sequences of events before and after a particular change. At Amboseli, rainfall cycles change hydrology and soil conditions, which in turn change the vegetation and animals of the area. This is why an understanding of environmental unity is important. As a result of global warming the ice and snow on Mount Kilimanjaro is fast retreating. The melt water will likely impact Lake Amboseli and cause a rise in salty groundwater.

Changes and Equilibrium in Systems

When the input to a system is equal to the output (Figure 1.13a), there is no net change in the size of the storage compartment (the amount of whatever is being measured), and the system is said to be in a **steady state.** The steady state is a **dynamic equilibrium,** because it's not just sitting still; something is happening—

M. Renaudeau/HOA-QUI

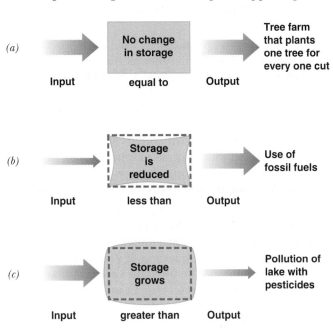

■ **FIGURE 1.13**

Major ways that the amount of material in a storage compartment of a system can change.
Row (a) represents steady-state conditions; rows (b) and (c) are examples of decrease and increase in storage. [*Source:* Modified from P. R. Ehrlich, A. H. Ehrlich, and J. P. Holvren. *Ecoscience: population, resources, environment,* 3rd ed. San Francisco: W. H. Freeman, 1977.]

■ **FIGURE 1.12**

Masai people grazing cattle in Amboseli National Reserve, Kenya.
Grazing was prematurely blamed for the loss of fever tree woodlands.

material or energy is entering and leaving the system in equal amounts. An approximate steady state may occur on a global scale, such as in the balance between incoming solar radiation and outgoing radiation from Earth, or on the smaller scale of a university, where new freshmen begin their studies and seniors graduate at about the same rate.

When the input is less than the output (Figure 1.13b), the amount of material in the storage compartment shrinks. For example, if a resource, such as groundwater, is consumed faster than it can be replaced by nature or by people, it may be used up.

When input exceeds output (Figure 1.13c), the amount of material in the storage compartment will expand. Two examples of this are the buildup of heavy metals in lakes and the pollution of groundwater.

Residence time, and why it's important. By using rates of change or input-output analysis of systems, we can calculate an *average residence time* for material moving through a system—that is, how long it takes for that particular material to be cycled through the system. Average residence time has important implications for environmental systems. In a system such as a small lake with an inlet and an outlet, water has a short residence time. This makes the lake especially vulnerable to change—if, for example, a pollutant is introduced. On the other hand, the pollutant soon leaves the lake. In large systems, such as oceans, water has a long residence time. This makes them much less vulnerable to quick change. However, once polluted, they are difficult to clean up.

The "balance of nature." Once you understand input and output, you have a framework for interpreting some of the changes that may affect systems. An idea that has been used and defended in the study of our natural environment is that if natural systems are not disturbed by people, they will tend to persist in a steady state, sometimes called the "balance of nature." However, if we examine natural systems in detail and evaluate them over a variety of time frames, we find that a steady state is seldom established or maintained for very long. Most systems are disturbed not only by people but also by natural events, such as floods and wildfires, which cause systems to change over time. In fact, studies of such diverse systems as forests, rivers, and coral reefs suggest that disturbances due to natural events such as storms, floods, and fires are necessary for the maintenance of those systems.

The lesson is that systems change naturally. If we are going to manage systems for the betterment of the environment, we need to gain a better understanding of the following:[15,16]

- types of disturbances and changes that are likely to occur;
- time periods over which changes occur;
- the importance of each change to the long-term productivity of the system.

These concepts are at the heart of understanding the principles of environmental unity and sustainability.

Biota: Biosphere and Sustaining Life

Earth as a planet has been profoundly altered by the life that inhabits it (recall our discussion of the Gaia hypothesis, Section 1.4). Earth's air, oceans, soils, and sedimentary rocks are very different from what they would be on a lifeless planet. In many ways, life helps control the makeup of the air, oceans, and sediments.

Biota is a general term used to refer to all living things (animals and plants, including microorganisms) within a given area—from an aquarium to a continent to Earth as a whole.

The biosphere is the region of Earth where life exists. It extends from the depths of the oceans to the summits of mountains. The biosphere includes all life as well as the lower atmosphere and the oceans, rivers, lakes, soils, and solid sediments that actively interchange materials with living things. All living things require energy and materials. In the biosphere, energy is received from the sun and from Earth's interior and is used and given off as materials are recycled.

What is needed to sustain life? To answer this question, first consider another question: How small a part of the biosphere could be isolated from the rest and still sustain life? If you put parts of the biosphere into a glass container and seal it, which parts would you choose in order to sustain life in the jar? If you placed a single green plant in the jar along with air, water, and some soil, the plant could make sugars from water and from carbon dioxide in the air. It could also make many organic compounds, including proteins and woody tissue, from sugars and from inorganic compounds in the soil. But no green plant can decompose its own products and recycle the materials. Eventually, your green plant would die.

It takes several species. We know of no single organism, population, or species that both produces all its own food and completely recycles all its own metabolic products. For life to persist, there must be several species within a system that includes air and water to transport materials and energy. Such a system is called an **ecosystem,** our next important topic of discussion.

Ecosystems

What is an ecosystem? In brief, it is a community of organisms and that community's local nonliving environment in which matter (chemical elements) cycles and energy flows. It is a fundamental principle that sustained life on Earth is a characteristic of ecosystems, not of individual organisms or populations or single species.

Ecosystems vary greatly in size, from the smallest puddle of water to a large forest or the entire global biosphere. Ecosystems also vary greatly in composition—that is, in the number and kinds of species and the kinds and relative proportions of nonbiological components. Sometimes the borders of an ecosystem are well defined—for example, the transition from the ocean to a rocky coast or from a pond to the surrounding woods. Sometimes the borders are vague, as in the gradual shift from forest to prairie in Minnesota and the Dakotas, or from grasslands to savannas or forests in East Africa. What all ecosystems have in common is not their physical structure—size, shape, variations of borders—but the processes we have mentioned: the flow of energy and the cycling of chemical elements.

Characteristics of Environmental Systems that Make Solving Environmental Problems Harder

Three main problems. Environmental systems—whether the entire planet or the hydrosphere, lithosphere, biosphere, or human population—are open systems characterized by poorly defined boundaries and the transfer of material and energy. They are inherently difficult to work with. Global environmental problems are particularly difficult because of three characteristics:[12] (1) exponential growth and the positive feedback that accompanies growth; (2) long lag times (a decade or more) between input and responses to input; and (3) the potential for some events to produce irreversible changes. Let's look at each of these in turn.

Exponential growth. As we noted earlier, the consequences of exponential growth and its accompanying positive feedback can be dramatic, leading to incredible increases in whatever is being evaluated or measured.

Lag time. Lag time is the time between a stimulus and a system's response to that stimulus. If the lag time is very short, the response is easier to identify. For example, the release of a toxic gas from a chemical plant may have almost immediate consequences to people living near the plant. If there is a long delay between stimulus and response, then the resulting changes are much more difficult to recognize. When we are dealing with biological resources, a long delay in seeing the results of a population's exponential growth may lead to **overshoot**, when the carrying capacity is exceeded.[11,12,17] As

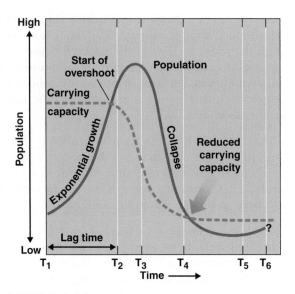

▪ **FIGURE 1.14**

The concept of overshoot.
Exponential growth of a population may eventually exceed the carrying capacity, resulting in overshoot and a decline (collapse) in the population to a lower carrying capacity. The time of growth to overshoot is lag time. [*Source:* Modified after D. H. Meadows, D. L. Meadows, and J. Randers. *Beyond the limits: Confronting global collapse; envisioning a sustainable future.* Post Mills, VT: Chelsea Green, 1992.]

an example, see Figure 1.14, which shows the relationship between the population of a species (perhaps a fish being harvested or people living on an isolated island) and the carrying capacity for that species. The carrying capacity starts out being much higher than the population, but as the population grows exponentially, it eventually overshoots the carrying capacity. Ultimately, this causes the population to decline or even collapse.

Irreversible consequences. Not all adverse effects of environmental change lead to irreversible consequences. But some do (remember Easter Island), and these cause particular problems. When we talk about irreversible consequences, we mean consequences that may not be easily rectified even within a few hundred years. A good example is soil erosion, or the harvesting of old-growth forest. There may be a long lag time before the soil no longer has enough nutrients to produce a crop, and at that point it may take hundreds or thousands of years for new soil to form.[11] Similarly, it may take hundreds of years to restore old-growth forests after logging. Lag time may be even longer if the soils have been damaged or eroded due to timber harvesting.

In summary, we see that exponential growth, long lag time, and the possibility of irreversible consequences are serious environmental concerns. For example, an

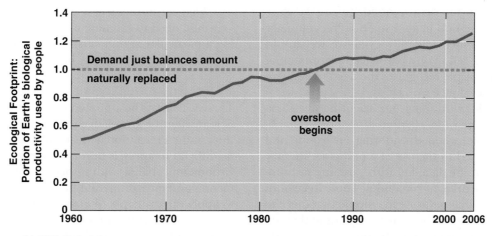

■ **FIGURE 1.15**

Are we already overshooting?
Since the mid-1980s, we may be using some of Earth's renewable resources (such as biological productivity) faster than they can be naturally replaced. [Data from World Wildlife Fund. Living planet report. Gland, Switzerland, 2004, p. 42.]

important question related to sustainability is, Are we consuming natural resources faster than Earth can replace them? There is no easy answer to this question. See the discussion below.

Today we may be using more than Earth can replace. One way to view our **ecological footprint** (the total area each person requires based on the resources we use and the waste we produce) is to compare the consumption of renewable natural resources with their replacement. Figure 1.15, which shows trends in human use of Earth's biological productivity for the past 40 years, suggests that in the mid-1980s we passed the point where people's demand for biological resources just balanced the amount that could be naturally replaced. At that point, after a long lag time of exponential growth in the use of biological resources, overshoot began.

The current overshoot is estimated at about 20%— that is, we may be consuming about 20% more of Earth's biological productivity than is replaced each year. This is possible because Earth's total biological resources are much greater than what is annually produced. But if our estimate is correct, then today we are dipping into Earth's natural capital of renewable biological resources. This is like taking more money from your bank account than you are putting back in. Over time, your bank account will become smaller and smaller.[18]

Thus, finding solutions to environmental problems won't be easy. Coupling exponential growth with long lag times and irreversible consequences makes us realize that solutions to environmental problems may be complex. Remember, the principle of environmental unity states that one activity or change often leads to a sequence of changes, some of which may be difficult to

recognize. Environmental scientists are trained to recognize lag time and the potential for irreversible consequences. However, even with good scientific research, science is incomplete. So how do we make decisions on important environmental problems? This important subject is discussed next as the *Precautionary Principle*.

1.9 The Precautionary Principle: When in Doubt, Play It Safe

Even with careful scientific research, it can be difficult to prove with certainty how relationships between human activities and other physical and biological processes lead to local and global environmental problems, such as global warming, depletion of ozone in the upper atmosphere, loss of biodiversity, endangered species, and declining forest and fishery resources. For this reason, in 1992 the Rio Earth Summit on Sustainable Development listed as one of its principles what we now define as the **Precautionary Principle.** Basically, it says that when there is a threat of serious, perhaps even irreversible, environmental damage, we should not wait for scientific proof before taking cost effective, precautionary steps to prevent potential harm to the environment.

The Precautionary Principle thus requires critical thinking about a variety of environmental concerns, such as the manufacture and use of chemicals such as pesticides, herbicides, and drugs; the use of fossil fuels and nuclear energy; the conversion of land from one use to another (for example, from rural to urban); and the management of wildlife, fisheries, and forests.[19]

How much proof do we need before acting? One important question in applying the Precautionary Principle is how much scientific evidence we should have before taking action on a particular environmental problem. The principle recognizes the need to evaluate all the scientific evidence we have and draw provisional conclusions while continuing our scientific investigation, which may provide additional or more reliable data. For example, when considering environmental health issues related to the use of a pesticide, we may have a lot of scientific data, but with gaps, inconsistencies, and other scientific uncertainties. Those in favor of continuing to use that pesticide may argue that there isn't enough proof to ban it. Others may argue that absolute proof of safety is necessary before a new pesticide is used. Those advocating the Precautionary Principle would argue that we should continue to investigate but, to be on the safe side, should not wait to take cost-effective precautionary measures to prevent environmental damage or health problems.

What constitutes a cost-effective measure? Certainly we would need to examine the benefits and costs of taking a particular action versus taking no action. Other economic analyses may also be appropriate.[19,20]

The Precautionary Principle as a proactive tool. There will always be arguments over what constitutes sufficient scientific knowledge for decision-making. Nevertheless, the Precautionary Principle, even though it may be difficult to apply, is becoming a common part of environmental analysis with respect to environmental protection and environmental health issues. It requires us to apply the principle of environmental unity and predict potential consequences before they occur. In a sense, the Precautionary Principle is a *proactive*, rather than *reactive*, tool—that is, we use it when we see real trouble coming, rather than in reaction to big trouble that has already arisen.

Return to the Big Question

Why is science necessary to solve environmental problems?

Without the scientific method, we would have no way of testing hypotheses to identify potential environmental problems and solutions. For example, environmental scientists can test whether exponential growth is going on, whether a particular action would have undesirable consequences, and whether there is likely to be a lag time between an action and its consequences. We can also construct mathematical models based on scientific principles and assumptions to predict the environmental consequences of certain actions.

Without the scientific method, we would have no way of knowing whether a proposed solution to an environmental problem was likely to work or not. However, which environmental actions we ultimately choose to incorporate into our environmental policy depends not just on science but also on our values.

Summary

▪ Important ideas that run through this text include the urgency of the population issue, the importance of urban environments, the need for sustainability of resources, the importance of a global perspective, how people relate to nature, the role of science and values in the decisions we face, and the importance of systems and rates of change.

▪ In the 20th century, the human population grew at a rate unprecedented in history. Population growth is the underlying environmental problem.

▪ Sustainability is a long-term concept that, with respect to human society, refers to finding ways to ensure that future generations have equal access to Earth's resources, and that what we do today won't cause irreversible harm to the environment.

▪ The Gaia hypothesis states that life on Earth, through a complex system of positive and negative feedback, regulates Earth's environment to help sustain life.

▪ Awareness of how our actions at a local level affect the environment globally gives credence to the Gaia hypothesis. Future generations will need a global perspective on environmental issues.

▨ Placing a value on various aspects of the environment requires understanding of the science involved but also depends on our judgments concerning the uses and aesthetics of the environment and on our moral commitments to other living things and to future generations.

▨ Science has a tradition of critical thinking about the natural world. Its goal is to understand how nature works. Decisions on environmental issues must begin with an examination of the scientific evidence but also require careful analysis of economic, social, and political consequences. Solutions will reflect religious, aesthetic, and ethical values as well.

▨ Science begins with careful observations of the natural world, from which scientists form hypotheses. Whenever possible, scientists test hypotheses with controlled experiments.

▨ A system is a set of components that function together as a whole. Environmental studies deal with complex systems at every level, and solutions to environmental problems often involve understanding systems and rates of change.

▨ Systems respond to inputs and have outputs. Feedback is a special kind of system response. Positive feedback can be destabilizing, whereas negative feedback tends to stabilize, encouraging more constant conditions in a system.

▨ The principle of environmental unity, simply stated, holds that everything affects everything else. It emphasizes linkages among parts of systems.

▨ Two important aspects of exponential growth, in which the increase per time period is a constant rate, are the growth rate as a percentage and the doubling time.

▨ An ecosystem is a community of different species and their local nonliving environment in which energy flows and chemicals cycle.

▨ The so-called balance of nature, a persistent long-term state without change, is largely a figment of our imagination. Disturbance and change are the norm.

▨ Overshoot occurs when the size of a population exceeds its carrying capacity.

▨ Solving environmental problems may be particularly difficult as a result of exponential growth, lag times, and the possibility of irreversible consequences.

▨ The Precautionary Principle is emerging as a proactive tool in addressing potential environmental problems.

Key Terms

biosphere
biota
carrying capacity
doubling time
dynamic equilibrium
ecological footprint
ecosystem
environmental unity

exponential growth
feedback
Gaia hypothesis
overshoot
Precautionary Principle
steady state
sustainability
system

Getting It Straight

1. Which of the following are global environmental problems? Why?
   a. The growth of the human population.
   b. The furbish lousewort, a small flowering plant found in the state of Maine. It is so rare that it has been seen by few people and is considered endangered.
   c. The blue whale, an endangered species under the U.S. Marine Mammal Protection Act.
   d. A car that has air-conditioning.
   e. Seriously polluted harbors and coastlines in major ocean ports.

2. Is it possible that all the land on Earth will someday become one big city? If not, why not? To what extent does the answer depend on the following:
   a. Global environmental considerations
   b. Scientific information
   c. Values

3. Which of the following are scientific statements and which are not? What is the basis for your decision in each case?
   a. The amount of carbon dioxide in the atmosphere is increasing.
   b. Picasso was the greatest painter of the 20th century.
   c. Helping terminally ill people to die is wrong.
   d. The Earth is flat.
   e. The good will reap their reward in the afterlife.

4. a. Identify a technological advance that resulted from a scientific discovery.
   b. Identify a scientific discovery that resulted from a technological advance.
   c. What technological device have you used today? What scientific discoveries were necessary before the device could be developed?

5. What is the difference between positive and negative feedback in systems? Provide an example of each.

6. What are the impacts of overpopulation on Earth's sustainability?

7. What is environmental science?

8. Define the following terms. Where do you live in terms of these definitions?
   Biota   Ecosystem   Biosphere

9. What are the three main problems of environmental systems?

10. What does our ecological footprint tell us about our life on Earth?

What Do You Think?

1. Find a newspaper article on a controversial topic. Identify some loaded words in the article—that is, words that convey an emotional reaction or a value judgment.

2. Discuss some of the intellectual standards that help critical thinking. Which are most significant? Why?

3. How does the Amboseli National Reserve illustrate environmental unity?

4. What is the main point in exponential growth? Is exponential growth good or bad?

5. Why is the idea of equilibrium in environmental systems somewhat misleading? Is a balance of nature possible?

6. Why is the concept of the ecosystem so important in the study of environmental science? Should we be worried about disturbing ecosystems? Under what circumstances should we worry or not worry?

7. Is the Gaia hypothesis a true statement of how nature works? Explain.

8. What does the term *sustainability* mean? How can we ensure sustainability for our futures?

9. Discuss your rationale for the involvement of science in solving environmental problems.

Pulling It All Together

1. In what ways do the environmental effects of a resident of a large city differ from the environmental effects of someone living on a farm? In what ways are the effects similar?

2. Programs were established to supply food from Western nations to feed starving people in Africa. Some people argue that such food programs, which may have short-term benefits, actually increase the threat of starvation in the future. What are the pros and cons of international food relief programs?

3. What are the values involved in deciding whether to create an international food relief program? What are five kinds of information required to determine the long-term effects of such programs?

Mario Tama/Getty Images

Big Question

Why is Human Population Growth the Underlying Environmental Problem?

?

Learning Objectives

The human population today represents something unprecedented in the history of the world: Never before has one species had such a great impact on the environment in such a short time and continued to increase at such a rapid rate. These qualities make the human population the underlying environmental issue. After reading this chapter, you should understand that . . .

■ ultimately, there can be no long-term solutions to environmental problems unless the human population stops increasing;

■ two major questions about the human population are what controls its rate of growth and how many people Earth can sustain;

■ the rapid increase in the human population occurred with little or no change in the maximum lifetime of an individual;

■ modern medical practices, improved sanitation, better control of organisms that spread disease, and improved access to human necessities have lowered death rates and accelerated human population growth;

■ birth rates have declined faster in countries that have a high standard of living than in countries with a low standard of living.

Case Study

How the Great Tsunami of 2004 Affected the Human Population

On the day after Christmas, 2004, one of the greatest earthquakes and tidal waves ever known caused terrible damage in Southeast Asia, including the nations of Indonesia, Thailand, Sri Lanka, India, and many island nations. No one can doubt the suffering of the survivors who lived along the coasts and saw family and friends swept away forever by the enormous tidal wave, called a *tsunami*. In this great human population tragedy, more than 150,000 people died—most within a few hours across a wide region—and more than 5 million watched it happen, experienced it, and suffered from the loss of family, friends, and colleagues (Figure 2.1a).

The sheer quantity of human suffering stretched our ability to respond despite our advanced technologies and best intentions. Years of sadness and misery will track the survivors, but death and suffering are only the first tragedy of the tsunami of 2004.

Coupled with that tragedy is a second: At the world's present human population growth rate of 1.3%, it took less than a day for the 6.3 billion people in the world to replace the more than 150,000 who died. Each day the world's human population grows by more than 224,000—that's almost a quarter of a million people a day! In Indonesia alone, where the population of 218,700,000 is growing

1.41 % per year, the approximately 100,000 people killed there by the tsunami were replaced in just two weeks (Figure 2.1b).[1]

That country is hardly unique. The United States has a 0.4% growth rate per year, and our population, which reached the milestone of 300 million in 2006, increases by more than 200,000 every week—that is, the U.S. population replaced all those who died in the tsunami in less than one week! Even Sweden, which estimates that it lost more than 6,000 of its citizens who were visiting or living in Southeast Asia, will replace that number in 3 weeks—and that is at Sweden's amazingly low population growth rate of 0.1% per year.

What we learn from this story is that as our human population continues to grow, human suffering all over the world will increase; tragedies will become greater in terms of sheer numbers, and the need for resources to aid people will stretch our ability to respond. Lulled by modern technologies, many tourists had no thought of danger as they swam in the Pacific Ocean on that fateful day. We had come to believe that we could overcome all environmental problems no matter how many of us there were. But as *this terrible event shows us, there can be no long-term solution to environmental issues as long*

(a)   Choo Youn-Kong/AFP/Getty Images

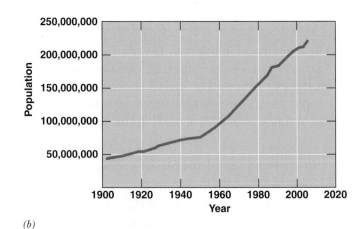

(b)

▪ **FIGURE 2.1**
The Great Tsunami of 2004.
(a) This photograph shows some of the disaster. (b) Indonesia's population of more than 209 million would show hardly a ripple in its growth from the terrible tragedy of the 2004 tsunami. This illustrates the power of the human population growth.

as the human population continues its momentous rise. If our population continues to grow rapidly, it will ultimately overwhelm the environment. That is why human population growth is a major theme of this book.

2.1 How Populations Change Over Time: Basic Concepts of Population Dynamics

The rapid regrowth of Southeast Asia's population following the tsunami of 2004 vividly illustrates the human population's great capacity to multiply and suggests the problems that this poses for the environment. The central question is, can we limit the growth of our population so that it neither increases nor decreases (sometimes referred to as **zero population growth**, or ZPG)? To put this another way: Can the human population become *sustainable?* A related question is, can we forecast how a population will change and what effects it will have on its environment?

These questions can be answered only if we understand basic ideas about any population, and that is the purpose of this chapter.

The Prophecy of Malthus

More than 200 years ago, Thomas Malthus, an English economist and demographer (population expert), eloquently stated the human population problem. He based it on three simple facts. First, that food is necessary for people to survive. Second, that "passion between the sexes is necessary and will remain nearly in its present state," so children will continue to be born. And third, that "the power of population growth" is "greater than the power of Earth to produce subsistence." Malthus argued that it will be impossible to maintain a rapidly multiplying human population on Earth's limited resource base, and that our species can survive only if our population stops growing exponentially. If we did not limit our growth voluntarily, he said, then our numbers would be reduced by our own vices (such as wars) and by epidemics and famines.

Some terms you need to know. A *population* is a group of individuals of the same species living in the same area or interbreeding and sharing genetic information. A **species** is all individuals that are capable of *interbreeding* (producing young together). A species is made up of populations. *Five key properties of any population are (1) abundance* (the size of a population—now, in the past, and in the future); *(2) birth rates, (3) death rates, (4) growth rates,* and *(5) age structure.* Defined most simply, the **birth rate** is the number of individuals born during a specified time interval, such as a year. The **death rate** is the number of individuals who die during that same time interval. The **growth rate** is the difference between the birth rate and death rate. That is, the growth rate is the net change in the size of the population.

Populations change over time and over space. The general study of population change is called **population dynamics.** How rapidly the size of a population changes depends on the growth rate, which is the difference between the birth rate and the death rate.

2.2 Population Growth

The first impressive fact about biological populations is their great capacity for growth, recognized since ancient times—the Greek philosopher Aristotle wrote more than 2,000 years ago about a pregnant female mouse that was shut up in a jar filled with millet seed, and "after a short while," when the jar was opened, "120 mice came to light."[2] Such rapid growth carries its own problems (Figure 2.2).

■ **FIGURE 2.2**
People scavenge in a garbage dump in Rio de Janeiro.
Rapid population growth promotes poverty and a way of life that opens people to disease and malnutrition, as suggested in this photograph.

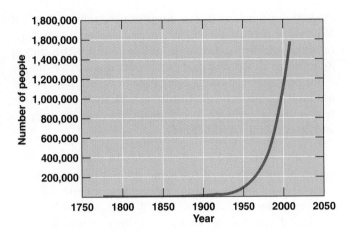

■ **FIGURE 2.3**
Exponential growth.
In exponential growth, a population (or anything, including invested money) increases at a constant percentage per time period.

Exponential growth: a widely used population forecast method. Discussions about the human population problem generally involve forecasts of how big our population will be at various times in the future. What is the basis for these forecasts? The simplest forecasting method is to assume that the population undergoes **exponential growth**. Growth is exponential when it occurs at a constant rate per time period (let's say a year) rather than by a constant amount (see Chapter 1). For instance, suppose you have $1,000 in the bank, and it grows at 10% per year. The first year, $100 in interest is added to your account. The second year, you earn more, because you earn 10% not just on $1,000 but now on $1,100. The greater the stored amount, the greater the amount earned, so the money (or the population, or some other quantity) increases by larger and larger amounts. When we show exponential growth on a graph, we end up with a curve that is said to be **J**-shaped. It looks like a skateboard ramp, starting out nearly flat and then rising steeply (Figure 2.3).

The human population has mostly grown exponentially. It has not actually maintained a constant percentage increase indefinitely (see Figure 2.4, World Population since A.D. 1000). During the mid-20th century (1965–1970), growth of the world's human population peaked at 2.1%. Since then, the growth rate has declined, to about 1.4% today.[1, 3]

We can view the history of the human population in terms of four major periods:

1. **The early period of hunters and gatherers.** At that time there were probably less than a few million people in the world.

2. **The rise of agriculture.** This second period marked the first major increase in the human population,

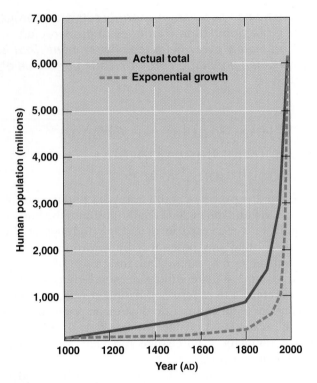

■ **FIGURE 2.4**
World population growth since A.D. 1000.
The world's human population growth has generally followed an exponential growth curve during this period, but has at times grown faster than an exponential.

because farming provided a steadier and more plentiful food supply than did hunting and gathering.

3. **The Industrial Revolution.** This period brought even greater improvements in the food supply and in health care, both of which led to a rapid increase in the human population.

4. **Where we stand today.** Population growth has slowed in wealthy, industrialized nations but continues to increase rapidly in many poorer, less developed nations (Figure 2.5).

How Many People Have Lived on Earth?

Long ago, no one was counting. Before written history, there was no census. The first estimates of population in Western civilization were attempted in the Roman era. Later, during the Middle Ages and the Renaissance, scholars occasionally tried to estimate the number of people.

The first modern census was taken in 1655 in the Canadian colonies by the French and British. Sweden began a series of regular censuses in 1750, and the United States has taken a census every decade since 1790. Most countries began much later. The first Russian census, for example, was taken in 1870. Even today,

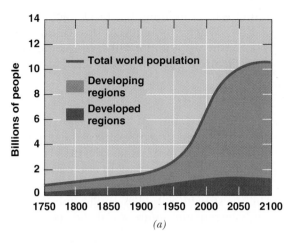
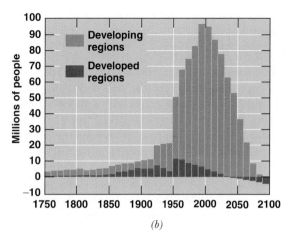

Population growth in developed and developing nations, 1750 to the present.
(a) Billions of people and (b) the net increase in the number (millions) of people per decade.
Since the second half of the 20th century, most of the world's human population growth has
been occurring in the developing nations. The numbers shown after year 2000 are projected
based on a logistic growth curve (see discussion in the text). This means that the projections
have to appear to lead to a stable population.

many countries do not take a census or do not do so regularly. The population of China has only recently begun to be known with any accuracy.

Still, we can get a rough idea. By studying modern primitive peoples and applying principles of ecology, we can arrive at a rough estimate of the total number of people who may have lived on Earth. Adding up all the values, including those since the beginning of written history, we estimate that about 50 billion people have lived on Earth. If so, then, to our surprise, the 6 billion people alive today represent more than 10% of all of the people who have ever lived.

2.3 The Logistic Growth Curve

Exponential population growth cannot go on forever. An exponentially growing population theoretically increases forever, but since Earth itself is not growing, people would eventually run out of food and space. A population of 100 increasing at 5% per year, for example, would grow to 1 billion in less than 325 years. If the human population had increased at this rate since the beginning of recorded history, it would now exceed all the known matter in the universe! So if we use the exponential to forecast the growth of our population, the answer we get is that it will need all the matter in the universe in a relatively short time.

What are the alternatives? If a population cannot increase forever, what changes in the population can we expect over time? The simplest alternative is that a popu-

lation's growth will decline gradually. The idea here is that each individual in a population uses enough resources to affect every other individual, and the effect of every individual is the same. If this is true, then the birth rate should decline slowly and the death rate should rise slowly, so that the growth rate gradually slows to zero. Over time, the population would follow a smooth **S**-shaped curve known as the **logistic growth curve** (Figure 2.5a).

The logistic curve. The logistic curve has been widely used for long-term population forecasts. In logistic growth, the population begins to increase exponentially when it is still very small, so the **S**-shaped curve first rises steeply. But after a while the rate of growth begins to gradually decline, until an upper population limit, called the **logistic carrying capacity**, is reached. Once the logistic carrying capacity is reached, the population remains at that number.

Is the logistic growth curve a realistic forecasting method? The point at which the curve begins to decline is the *inflection point*. Until a population has reached the inflection point, we cannot project its final size. The problem is, *the human population has not yet made the bend around the inflection point.* Many forecasters have dealt with this problem by assuming that today's population is just reaching the inflection point. This leads them to greatly underestimate the maximum population. For example, one of the first projections of the upper limit of the U.S. population, made in the 1930s, assumed that the inflection point had occurred then. That assumption resulted in an estimate that the final population of the United States would be about 200 million. The U.S. population has long since exceeded that number and in fact has passed 300 million.[4]

Why doesn't the logistic growth curve work? The logistic growth curve was first suggested in 1838 by a European scientist, P. F. Verhulst, as a theory for the growth of animal populations. It has been applied widely in forecasting the growth of many animal populations, including those important in wildlife management, endangered species, and those in fisheries, as we will see in later chapters. Unfortunately, there is little evidence that human populations—or any animal populations, for that matter—actually follow this growth curve. This is because the theory relies on assumptions that are unrealistic for humans and for other mammals. These assumptions include a constant environment, a constant carrying capacity, and a homogeneous population, in which all individuals are identical in their effects on each other.

The logistic curve is especially unlikely to occur if death rates continue to decline from ongoing improvements in health care and food supplies. After benefiting from these improvements, the population must pass through what has become known as the **demographic transition** (see below) to achieve zero population growth, which leads to a stabilized population.

Still, the logistic curve has been useful for a simplified forecast, sometimes referred to as a *first-order approximation forecast* or, informally, as a "back-of-the-envelope" forecast, since you can do the math on a piece of scrap paper. The World Bank, an international organization that makes loans and provides technical assistance to developing countries, has made a series of forecasts based on current birth rates and death rates and assumptions about how these rates will change. These projections (which form the basis for the logistic curves presented in Figure 2.5) assume that there will be no major worldwide catastrophe and that the world population would reach equilibrium at 10–13 billion people, which is about double the number of people alive today.[5] Developed countries would experience population growth from 1.2 billion today to 1.9 billion, but populations in developing countries would increase from 4.5 billion to 9.6 billion. Bangladesh (an area the size of Wisconsin) would reach 257 million; Nigeria, 453 million; and India, 1.86 billion. In these projections, the developing countries contribute 95% of the increase.[5]

2.4 Other Clues to How Our Population May Change

If these fairly simple forecasting methods, the exponential and the logistic, are not reliable, how else can we gain insight into how our population may change? The answer is by examining specific characteristics of a population. When it comes to forecasting population growth, the most important factor is the **population's age structure**, so we will look at this first.

Age Structure

The proportion of the population in each age group. The age structure affects the population's current and future birth rates, death rates, and growth rates. It gives us clearer insight into how that population will use resources and how it will affect its environment. The age structure also has implications for the population's current and future social and economic status.

Four general types of population age structure. Picture the age structure as a pile of blocks, one for each age group, with the youngest group at the bottom and with the size of each block reflecting the number of people in that age group. Although age structures can take many shapes, four general types are most important to our discussion: a *pyramid*, a *column*, an *inverted pyramid* (top-heavy), and a *column with a bulge*. The pyramid age structure occurs in a population that has many young people (the block at the bottom is big) and a high death rate at each age (the blocks get smaller as they go up)— and therefore a short average lifetime. A column shape occurs where the birth rate and death rate are low and a high percentage of the population is elderly. A bulge occurs if some event in the past caused a high birth rate or death rate for some age group but not others.

Age structure varies considerably by nation (Figure 2.6). Kenya's pyramid-shaped age structure indicates a rapidly growing population with a great many young people. About a third of the populations in developing countries today are under 15 years of age. Such an age structure indicates that the population will grow very rapidly in the future, when the young get old enough to marry and have children. It suggests that in the future, such a nation will require more jobs for the young, and it has many other social implications that go beyond the scope of this book.

The age structure of the United States is more like a column, showing a population with slow growth. Italy's age structure is a slightly top-heavy pyramid, indicating a nation with declining growth. Elderly people make up a small percentage of Kenya's population but a much larger percentage of the populations of the United States and Italy.[1]

Age structure can offer insights into a population's history, current status, and likely future. For example, a "baby boom" occurred after World War II in the United States (a great increase in births from 1946 through 1964). This surge in the birth rate can be seen

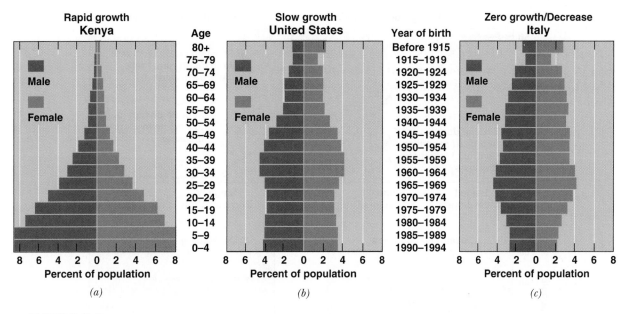

■ **FIGURE 2.6**
Age structures.
(a) Kenya); (b) the United States; and (c) Italy in 1995. [*Sources:* U.S. Bureau of the Census. U.S. population estimates by age, sex, and race: 1990 to 1995, PPL-41, February 14, 1996; Council of Europe, Recent demographic developments in Europe 1997, Table 1–1; United Nations. The sex and age distribution of the world populations—The 1996 revision, 500–501.]

as a bulge in the population's age structure, especially among those aged 40–50 in 2000. A secondary, smaller bulge resulted from offspring of the first baby boom, which can be seen as a slight increase in 5- to 15-year-olds. This second peak shows that the baby-boom "pulse" is moving through the age structure. Each baby boom increases demand for social and economic resources. For example, schools became crowded when the baby boomers reached school age.

One economic implication of age structure involves care of the elderly. In preindustrial and nonindustrial societies, average lifetimes are short, children care for their parents, and therefore it benefits parents to have many children. In modern technological societies, family size is smaller, and care for older people is distributed throughout the society through taxes, so that those who work provide funds to care for those who cannot. Parents tend to benefit when their children are well educated and have high-paying jobs. Rather than relying on a large family in which each child has fewer resources, parents tend to have fewer children and invest more in each. This makes zero population growth possible. However, a shift from a youthful age structure (like Kenya's) to an elderly age structure (like Italy's) means that a smaller percentage of the population works—and less tax money is available for elder care.

A population weighted toward the elderly poses problems for a nation. The heated debate about the

Social Security program in the United States today is one example of this. In general, older people no longer hold jobs and thus have lower incomes and pay less income tax, while at the same time having greater health-care needs. The easiest way to increase tax income is to increase the percentage of young people. In this way, a nation's short-term economic pressures can lead to political policies supporting rapid population growth, which is not in the long-term best interest of the nation.

The Demographic Transition

This is an important factor in population growth. In brief, a three-stage pattern of change in birth rates and death rates occurred during the industrial and economic development of Western nations, leading to a decline in population growth.

In Stage I, the birth rate declines (Figure 2.7)[3]. As we have noted, in a nonindustrial country, birth rates and death rates tend to be high and the population growth rate low.[5] With industrialization, health and sanitation improve, causing the death rate to drop rapidly. The birth rate remains high, however, and the population enters Stage II.

Stage II is a period with a high growth rate. Most European nations passed through this period in the 18th and 19th centuries. In this stage, more people have access to education, the standard of living

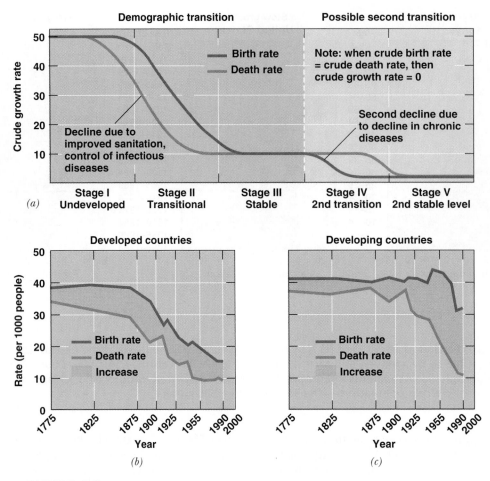

■ **FIGURE 2.7**
The demographic transition.
(a) Theoretical, including possible fourth and fifth stages that might take place in the future; (b) as has taken place for developed countries since 1775; and (c) as may be occurring in developing nations since 1775. [*Source:* M. M. Kent and K. A. Crews. *World population: Fundamentals of growth.* Washington, DC: Population Reference Bureau, 1990. © 1990 by the Population Reference Bureau, Inc. Reprinted by permission.]

increases, and family planning becomes more widely used. As a result, the population reaches Stage III.

In Stage III, the birth rate drops toward the death rate, so growth declines, eventually to a low or zero growth rate. However, the birth rate declines only if families believe there is a direct connection between their future economic well-being and funds spent on the education and care of their young. Such families have few children and put all their resources toward nurturing and educating those few.

Some nations are slow to move from Stage II to Stage III. Historically, parents have preferred to have large families. Without other means of support, parents can depend on children for a kind of "social security" in their old age, and even young children help with many kinds of hunting, gathering, and low-technology farming. Unless a change in attitude occurs among parents—unless they see more benefits from a few well-educated children than from many poorer children—nations face a problem in making the transition from Stage II to Stage III (see Figure 2.7).

Some developed countries are approaching Stage III, but it is an open question whether some of the developing nations will make the transition before a serious population crash occurs. *The key point here is that the demographic transition will take place only if parents come to believe that having a small family is to their benefit.*

Here we see again the connection between science and values. Scientific analysis can show the value of small families, but this knowledge must become part of cultural values to have an effect.

Medical advances can affect the demographic transition. Although the demographic transition is traditionally defined as consisting of three stages, advances in treating chronic health problems, such as heart disease, can lead a Stage III country to a second decline in the death rate. This could bring about a second transitional phase of population growth (Stage IV), in which the birth rate would remain the same while the death rate fell. A second stable phase of low or zero growth (Stage V) would be achieved only when the birth rate declined to match the decline in the death rate. Thus, there is danger of a new growth spurt even in industrialized nations that have passed through the standard demographic transition.

What's good for the individual may not always benefit the group. Advances in our understanding of aging, along with new biotechnology, are lengthening the average longevity and maximum lifetime of human beings. However, something positive from each individual's point of view—a longer, healthier, and more active life—could have negative effects on the environment and therefore on the society as a whole.

Thus, we may ultimately need to make some difficult choices: Stop trying to increase life expectancy even further and stop trying to lower the birth rate, or do neither and simply wait for Malthus's projections to come true—for famine, environmental catastrophes, and epidemics to periodically eliminate sizable portions of the population. The first choice—ceasing our efforts to help older people live even longer—seems inhumane and will likely seem downright unacceptable when we ourselves are older. But limiting family size by birth control is highly controversial. For the people of the world, this is one of the most important issues, concerning science and values and people and nature.

Human Death Rates and the Rise of Industrial Societies

Death rates differ widely between developed and developing countries. We touched on this earlier in discussing the first stage in the demographic transition. We will illustrate it now by comparing a modern industrialized country, such as Switzerland, which has a crude (unadjusted) death rate of 9 per 1,000, with a developing nation, such as Sierra Leone, which has a crude death rate of 25 per thousand.[1] The low death rates in countries such as Switzerland are due largely to modern medicine, which has greatly reduced death rates from disease, particularly from *acute* or *epidemic diseases.*

An acute or epidemic disease is one that appears rapidly in the population, affects a large percentage of the people, and then declines or almost disappears for a while, only to reappear later. Epidemic diseases typically are rare but have occasional outbreaks—influenza (flu), plague, measles, mumps, and cholera are examples of epidemic diseases. A *chronic disease,* in contrast, is always present in a relatively small but constant percentage of the population. Heart disease, cancer, and stroke are examples.

The great decrease in deaths due to acute or epidemic diseases that has occurred with the Industrial Revolution can be seen in a comparison of causes of deaths in Ecuador in 1987 and the United States in 1900, 1987, and 1998 (Figure 2.8).[6] In Ecuador, a developing nation, acute diseases and those listed as "all others" accounted for about 60% of mortality in 1987. In the United States in 1987, these accounted for only 20% of mortality, while chronic diseases accounted for about 70% of mortality in the modern United States. In contrast, chronic diseases accounted for less than 20% of the deaths in the United States in 1900 and about 33% in Ecuador in 1987. Death rates in Ecuador in 1987, then, resembled those in the United States of 1900 more than it resembled the United States of either 1987 or 1998.

There is concern about new diseases and new strains of old diseases. Although outbreaks of the well-known traditional epidemic diseases are much rarer today in industrialized nations, there is some concern that outbreaks of new diseases may occur, due to several factors. One is that as the human population grows, people live in new habitats where they are exposed for the first time to previously unknown diseases. Another concern is the appearance of new strains of diseases we thought we had wiped out; the new strains have developed a resistance to antibiotics and other modern methods of control.

We are a tempting target. A broader view of why diseases are likely to increase comes from an ecological and evolutionary perspective (which we will explain in later chapters). Stated simply, 6 billion people on Earth constitute a huge and easily accessible host for other species, and it would be foolish to think that other species will not take advantage of the opportunity we present. From this perspective, the future holds more diseases, rather than fewer. This is a new perspective. In the mid-20th century, it was easy to believe that modern medicine would eventually cure all diseases and that most people would live the maximum human life span.

Diseases can travel by jet . . . Modern transportation can lead to the rapid spread of epidemic diseases. One recent example is the sudden occurrence of severe acute respiratory syndrome (SARS), a new disease, in February 2003. The disease began in China, perhaps spread from some wild animal to human beings. By late spring 2003, SARS had spread to two dozen countries, in part because China has become much more open to

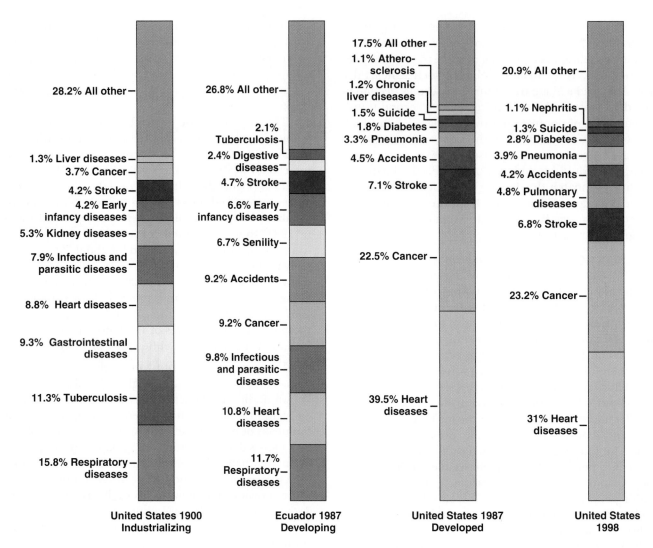

28.2% All other

1.3% Liver diseases
3.7% Cancer
4.2% Stroke
4.2% Early infancy diseases
5.3% Kidney diseases
7.9% Infectious and parasitic diseases
8.8% Heart diseases
9.3% Gastrointestinal diseases
11.3% Tuberculosis
15.8% Respiratory diseases

**United States 1900
Industrializing**

26.8% All other

2.1% Tuberculosis
2.4% Digestive diseases
4.7% Stroke
6.6% Early infancy diseases
6.7% Senility
9.2% Accidents
9.2% Cancer
9.8% Infectious and parasitic diseases
10.8% Heart diseases
11.7% Respiratory diseases

**Ecuador 1987
Developing**

17.5% All other
1.1% Athero-sclerosis
1.2% Chronic liver diseases
1.5% Suicide
1.8% Diabetes
3.3% Pneumonia
4.5% Accidents
7.1% Stroke
22.5% Cancer
39.5% Heart diseases

**United States 1987
Developed**

20.9% All other

1.1% Nephritis
1.3% Suicide
2.8% Diabetes
3.9% Pneumonia
4.2% Accidents
4.8% Pulmonary diseases
6.8% Stroke
23.2% Cancer
31% Heart diseases

United States 1998

■ **FIGURE 2.8**

Causes of mortality in industrializing, developing, and industrialized nations.
[*Sources:* U. S. 1900, Ecuador 1987, and U.S. 1987 data from M. M. Kent and K. A. Crews. *World population: Fundamentals of growth.* Washington, DC: Population Reference Bureau, 1990. © 1990 by the Population Reference Bureau, Inc. Reprinted by permission. *National Vital Statistics Report* 48(11): July 24, 2000.]

foreign travelers, with more than 90 million visitors in a recent year.[7] Quick action led by the World Health Organization (WHO) contained the disease, which as of this writing appears well under control.[8]

. . . and by other things that fly. West Nile virus is another example of how rapidly and widely diseases can spread. Before 1999, West Nile virus occurred in Africa, West Asia, and the Middle East, but not in the New World. Related to encephalitis, West Nile virus infects birds. It is spread to people when a mosquito bites an infected bird, becomes infected itself, and then bites a person. It reached the Western Hemisphere through infected birds and has now been found in more than 25 species of birds native to the United States, including

crows, the bald eagle, and the black-capped chickadee— a common visitor to bird feeders in the northeastern U.S. Fortunately, about 80% of people so infected show no symptoms and most of the rest are ill for only a few days. But about one person in 150 suffers severe effects, including neurological ones, which can be permanent.[9,10]

Crowding and poverty are factors in TB and AIDS. Increasingly dense populations of poor people also make the spread of diseases more likely. In China more than 550 million people are infected with tuberculosis, and 120,000–205,000 die each year from this disease. TB is one disease that is worrying health experts because of its persistence and its capacity to increase worldwide.[11] AIDS (acquired immunodeficiency syndrome) is a factor in this, because individuals who have AIDS lack resistance to diseases. Some 95% of the 33.6 million people who either have full-blown AIDS or are infected with HIV, the virus that causes AIDS, live in the developing world. Of these 33.6 million, 70% live in Sub-Saharan Africa, which is home to 10% of the world's population.[12] However, because birth rates in African countries continue to be high, their populations will continue to grow.[5]

Longevity and Its Effect on Population Growth

The **maximum lifetime** (**longevity**) is the genetically determined maximum possible age to which an individual of a particular species can live. **Life expectancy** is the average number of years an individual can expect to actually live given his or her present age, health, and other factors. Most often, however, the term is used, without qualification, to mean the life expectancy of a newborn.

The human population has grown despite little or no change in longevity. A surprising aspect of the second and third periods in the history of human population is that population growth occurred with little or no change in the maximum lifetime. What changed were birth rates, death rates, population growth rates, age structure, and average life expectancy.

Life Expectancy

Life expectancy is affected by many factors. Life expectancy, as stated earlier, is the number of years a person of a specific age can expect to live. Each age group within a population therefore has its own life expectancy—the younger groups naturally expecting to live longer than the older groups. However, in comparing the life expectancies of different populations, we refer to the number of years newborns in those populations can be expected to live. This differs by nation and by sex, age, and other factors. The life expectancy in a hunter-gatherer society is short—for example, among the !Kung bushmen of Botswana, life expectancy at birth is 39 years.[13,14]

Studies suggest that death rates were much higher among young people in ancient Rome than in 20th-century England. In ancient Rome, the life expectancy of a one-year-old was about 22 years, whereas in 20th-century England it was about 50 years. Life expectancy in 20th-century England was greater than in ancient Rome for all ages until about age 55; after that, the life expectancy appears to have been higher for ancient Romans than for a 20th-century Briton. In fact, surprisingly, ages at death carved on tombstones tell us that the chances of a person 75 years old living to be 90 were greater in an-

cient Rome than they were at the close of the 20th century in the United States (Figure 2.9). This suggests that the aged may be more susceptible to many of the hazards of modern life, such as diseases caused by pollution.

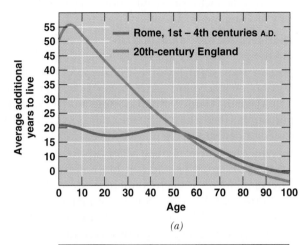

(a)

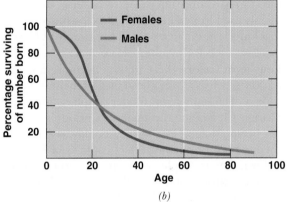

(b)

■ FIGURE 2.9

Life expectancy in ancient Rome and 20th-century England.
(a) This graph shows the average additional number of years one could expect to live after reaching a given age. For example, a 10-year-old in England could expect to live about 55 more years; a 10-year-old in Rome could expect to live about 20 more years. Among the young, life expectancy was greater in 20th-century England than in ancient Rome. However, the graphs cross at about age 55. An 80-year-old Roman could expect to live longer than an 80-year-old Briton. The graph for Romans is reconstructed from ages given on tombstones. (b) Approximate survivorship curve for Rome for the first four centuries A.D. The percentage surviving drops rapidly in the early years, reflecting the high mortality rates for children in ancient Rome. Females had a slightly higher survivorship rate until age 20, after which males had a slightly higher rate. [*Source:* Modified from G. E. Hutchinson. *An introduction to population ecology.* New Haven, CT: Yale University Press, 1978. © 1978 by Yale University Press. Used by permission.]

2.5 Limiting Factors

The world population cannot keep on growing. Because our planet and its resources cannot expand as our population expands, human populations will eventually be limited by some factor or combination of factors. We can classify limiting factors into three groups:

▪ **Short-term factors:** those that affect a population during the year in which they become limiting. One important short-term factor is the disruption of food distribution in a country, often caused by drought or by a shortage of energy for transporting food. Other important short-term factors are the outbreak of a new disease or a new strain of a previously controlled disease.

▪ **Intermediate-term factors:** those whose effects become apparent after one year but before ten years. Intermediate-term factors include desertification (land turning desertlike due to mismanagement or drought—see Chapter 11); pollutants, such as toxic metals, entering waters and fisheries; disruption in the supply of nonrenewable resources, such as rare metals used in making steel alloys for transportation machinery; and a decrease in the supply of firewood or other fuels for heating and cooking.

▪ **Long-term factors:** those whose effects are not apparent for ten years. Long-term factors include soil erosion, a decline in groundwater supplies, and climate change. Changes in the amount of resources available per person suggest that we may already have exceeded the long-term human carrying capacity of Earth (see below). For example, wood production peaked in 1967, fish production in 1970, beef in 1977, mutton in 1972, wool in 1960, and cereal crops in 1977.[15] Before these peaks were reached, per-capita production of each resource had grown rapidly.

Some factors fit into more than one category, having, say, both short-term and intermediate-term effects.

The Quality of Life and the Human Carrying Capacity of Earth

What is the human carrying capacity of Earth—that is, how many people can live on Earth at the same time? The answer depends on what quality of life people desire and are willing to accept.

The logistic curve is one way of estimating Earth's human carrying capacity. Drawing conclusions based on past growth, this approach, as discussed earlier, assumes that the population will follow an S-shaped logistic growth curve and will gradually level off at a population size that we can calculate from a knowledge of birth rates and death rates and an inflection point in the curve (Figure 2.3).

Or we can view the problem simply in terms of how many people Earth has space for. This is some- times termed the "packing problem" approach—it estimates how many people can be packed onto our planet, without taking into account Earth's ability to supply their needs, such as the need for food, water, energy, homes, and scenic beauty, and the need to maintain biological diversity. It could in fact be called the "standing-room-only approach" and obviously leads to very high estimates of the total number of people that might occupy Earth—as many as 50 billion.

Deep ecology takes the opposite approach. Recently, a philosophical movement has developed at the other extreme. Known as "deep ecology," this philosophy makes sustaining the biosphere the primary moral imperative. Its proponents argue that the whole Earth is necessary to sustain life. Therefore, everything else must be sacrificed to the goal of sustaining the biosphere. People are considered destructive of the biosphere, and therefore their number should be greatly reduced. Estimates of the desirable number of people vary greatly, starting as low as a few million.

Is there a "happy medium"? Between standing room only and the relative handful of people advocated by some deep ecologists, there are a number of options. But setting goals between these two extremes involves both value judgments and science, again reminding us of the theme of science and values. What constitutes a desirable quality of life is a value judgment. What kind of life is possible is affected by technology, which in turn is affected by science. And scientific understanding tells us what is required to meet each quality-of-life level.

The higher the quality of life, the lower the Earth's carrying capacity. If all the people of the world were to live at the same level as those in the United States, with our high resource use, then the carrying capacity would be comparatively low. If all the people of the world were to live at the level of those in Bangladesh, Earth's carrying capacity would be much higher.

In summary, an acceptable carrying capacity is not simply a scientific issue; it is an issue combining science and values. Science plays two roles. First, by leading to new knowledge that leads to new technology, it makes possible a higher density of people. Second, scientific methods can be used to forecast a probable carrying capacity once people decide on a goal for the average quality of life, in terms of their values. Science can tell us the implications of our value judgments, but it cannot provide those value judgments.

2.6 How Can We Achieve Zero Population Growth?

Raise the age of first childbearing. One of the simplest and most effective ways to slow population growth is to delay childbearing.[16] This delay comes about naturally as education levels and standards of living increase, and as more women enter the workforce and seek to establish careers outside the home. That's why lowering the rate of human population growth is strongly linked to education.

Social pressures that delay marriage can also be effective. In countries with high population growth rates, people are expected to marry early. In South Asia and in Sub-Saharan Africa, about 50% of women marry between the ages of 15 and 19. In Bangladesh, women marry at age 16 on average, whereas in Sri Lanka the average age for marriage is 25. The World Bank estimates that if Bangladesh adopted Sri Lanka's marriage pattern, families could average 2.2 fewer children.[17] Simply raising the marriage age could account for 40–50% of the decline in fertility required to achieve zero population growth for many countries.

The average age at marriage has risen in some countries, especially in Asia. For example, in Korea, the average marriage age rose from 17 in 1925 to 24 in 1975. China passed laws fixing minimum marriage ages, first at 18 for women and 20 for men in 1950, then at 20 for women and 22 for men in 1980.[6] Between 1972 and 1985, China's birth rate dropped from 32 to 18 per thousand people, and the average fertility rate went from 5.7 to 2.1 children.

Birth control: biological and societal. One simple means of lowering birth rates is breast-feeding, which can delay resumption of ovulation after childbirth.[18] Women in a number of countries use breast-feeding as a birth-control method. In fact, according to the World Bank, in the mid-1970s in developing countries the practice of breast-feeding provided more protection against conception than did family-planning programs.[4]

Family planning is widely emphasized.[9] Traditional methods range from sexual abstinence to attempts to induce sterility with natural agents, such as some vegetation products. Modern methods include the birth-control pill (which prevents ovulation through control of hormone levels), surgical techniques to make men and women permanently sterile, and mechanical devices. Contraceptive devices are used widely in many parts of the world, especially in East Asia, where data indicate that 78% of women use them. In Africa, only 18% of women use them; in Central and South America, the numbers are 53% and 62%, respectively.[1]

Abortion is also widespread. Although it is now medically safe in most cases, abortion is one of the most controversial methods from a moral perspective. Ironically, it is one of the most important birth-control methods in terms of its effects on birth rates—approximately 46 million abortions are performed each year.[19]

National Programs to Reduce Birth Rates

Reducing birth rates requires a change in attitude, knowledge of the various methods of birth control, and the ability to afford these methods. As we have seen, a change in attitude can occur simply with an increase in the standard of living. In many countries, however, it has been necessary to provide formal family-planning programs to explain the problems caused by rapid population growth and how individuals will benefit from reducing population growth. These programs also provide information about birth-control methods and provide access to these methods.[20] The choice of population-control methods involves social, moral, and religious beliefs, which vary from country to country.

The first country to adopt an official population policy was India in 1952. Few developing countries had official family-planning programs before 1965. Since 1965, many such programs have been introduced, and the World Bank has lent $4.2 billion to more than 80 countries to support "reproductive" health projects.[18]

Many countries now have some kind of family-planning program, but effectiveness varies greatly. The approaches used range from simply providing more information, to promoting and providing birth-control methods, to offering rewards and imposing penalties. Penalties usually take the form of taxes. Ghana, Malaysia, Pakistan, Singapore, and the Philippines have used a combination of methods, including limits on tax allowances for children and limits on maternity benefits. Tanzania has restricted paid maternity leave to once in three years. Singapore does not take family size into account in allocating government-built housing, so larger families find themselves more crowded. Singapore also gives higher priority in school admission to children from smaller families.[4] Some countries—including Bangladesh, India, and Sri Lanka—have paid people to be voluntarily sterilized. In Sri Lanka, this practice has applied only to families with two children, and parents sign a statement of voluntary consent.

China has one of the oldest and most effective family-planning programs. In 1978, China adopted an official policy to reduce its population growth from 1.2% in that year to zero by the year 2000. Although the growth rate did not drop to zero by 2000, it did slow to 1.0%, which shows significant progress in curbing the country's rapid population growth. The Chinese program encourages couples to have only one child. The government has used education, a network of family planning that provides information and means for birth control, and a system of rewards and penalties. Women receive paid leave for abortions and for surgical sterilization. Families with a single child have received benefits, including financial subsidies in some areas. In some parts of China, families that have a second child have had to return the bonuses received for the first. Other rewards and penalties vary from province to province.[4]

Should governments force people to limit family size? The Chinese program has raised questions relating to several larger issues involving science and values, and people and nature. In the past, the Chinese government mandated certain practices, raising the question: How can we allow people freedom of choice and still emphasize the need to reduce the birth rate? The government of China responded to this concern in 2002 with a new law that prevents coercion.[21]

2.7 How Many People Can Earth Support?

In mid-1992 the world population reached 5.5 billion. Today it exceeds 6 billion. Estimates of how many people the planet can support range from 2.5 billion to 40 billion. Why do the estimates vary so widely?

An estimate of 2.5 billion assumes that we will maintain current levels of food production and that everyone eats as well as Americans do now—that is, 30–40% more calories than we need. The estimate of 40 billion assumes that all the remaining flat land of the world can be used to produce food, although in fact most of it is too cold or too dry to farm. What is a realistic carrying capacity? What factors need to be considered to answer this question?

The food supply. World grain production has apparently leveled off since reaching its highest levels in the mid-1980s. From 1984 to 1994, the production of grain remained at approximately 1.7 billion tons, up from 631 million tons in 1950. The remarkable increase in productivity after 1950 resulted from the development of high-yielding varieties, use of chemical fertiliz-

ers, application of pesticides, and doubling of cropland acreage. If the present harvest were distributed evenly and everyone ate a vegetarian diet, it could support 6 billion people. As the world population has continued to grow, the per-capita allotment of grain has been falling. It stood at 346 kg per person in 1984, but had fallen to just 311 kg per person by 1994.

Land and soil resources. Almost all the usable agricultural land—approximately 1.5 billion hectares (3.7 billion acres)—is already being cultivated. An increase of 13% in agricultural lands is possible but would be costly. The amount of land devoted to raising crops has dropped since 1950 to 1.7 ha (4.2 acres) per person and will likely continue to drop, to approximately 1 ha (2.5 acres) per person by 2025 if present population predictions prove accurate. Each year more soil is lost to erosion than is formed.

Water resources. Less than 3% of all the water on Earth is suitable for drinking and irrigation. Underground reservoirs are being depleted by several feet per year but are being replaced in inches or even fractions of inches per year. Per-capita water consumption varies: It averages 350–1,000 liters (371–1,060 quarts) a day in the developed countries and 2–5 liters (2.1–5.3 quarts) a day in rural areas, where people may obtain water directly from streams or primitive wells.

Population density. Population density varies greatly, from 3,076 people per square kilometer (km^2) on the tiny island of Malta to 66 people/km^2 in Africa as a whole. Bangladesh has 2,261 people/km^2; the Netherlands, 1,002/km^2; and Japan, 869/km^2.

Technology. Earth's carrying capacity is not merely a matter of numbers of people. It also involves the impact they have on the world's resources—most critically on energy resources. Multiplying population by per-capita energy consumption gives a relative measure of the impact people have on the environment. By that measure, each American has the impact of 35 people in India or 140 people in Bangladesh.

Return to the Big Question

Why is human population growth the underlying problem?

Growth of the human population is said to be the underlying environmental problem for several reasons. First, because there are already so many of us. Second, because more of us are living longer, and our modern technology allows each one of us to have an increasingly large effect on our environment. Then, too, as the only species with civilizations, cultures, and written histories, we think of ourselves as somehow outside of nature, and therefore our increasing numbers are seen as an even greater environmental threat. The first two points are factual. The third is a matter of interpretation, and for this reason has led to much environmental controversy.

Since this is such a complex and controversial idea, we suggest you make your own evaluation. To begin your evaluation, you may want to review the prophecy of Malthus and decide whether you agree with the general thesis that over the long term it is impossible to maintain a rapidly growing population on a limited resource base. You will also want to consider the possibility of a potential Stage IV and V in the demographic transition.

Summary

- The human population is the underlying environmental issue, because most environmental damage today stems from the very high number of people on Earth and their great power to change the environment.

- Throughout most of our history, the human population and its average growth rate were small. The growth of the human population can be divided into four major phases. Although the population increased in each phase, the current situation is unprecedented.

- Countries whose birth rates have declined have experienced a demographic transition marked by a decline in death rates followed by a decline in birth rates. Many developing nations, however, have experienced a great decrease in their death rates but still have very high birth rates. It remains an open question whether some of these nations will be able to achieve a lower birth rate before reaching disastrously high population levels.

- The maximum population Earth can sustain and how large the human population will ultimately grow are controversial questions.

- How the human population might stabilize, or be stabilized, raises questions concerning science and values, and people and nature.

- One of the most effective ways to lower a population's growth rate is to raise the age of first childbearing. This is also fairly uncontroversial, because it involves relatively few societal and value issues.

Key Terms

birth rate
carrying capacity
death rate
demographic transition
exponential growth
growth rate
life expectancy
logistic carrying capacity

logistic growth curve
longevity
maximum lifetime
population
population age structure
population dynamics
species
zero population growth

Getting It Straight

1. What are the principal reasons that the human population grew so rapidly in the 20th century?

2. Based on the history of human populations in Brazil, France and the United States, how would you expect the following to change as per-capita income increased: (a) birth rates, (b) death rates, (c) average family size, and (d) age structure of the population? Explain.

3. What is population dynamics? What most influences population change?

4. Why has human population mostly grown exponentially?

5. What is the most important characteristic of a population to consider when forecasting population growth? Explain.

6. What is demographic transition? What can affect demographic transition within a population?

7. Why do death rates vary so much between developed and developing countries?

8. How might you convince a family to consider demographic transition in their reproductive plan?

9. What is the difference between longevity and life expectancy?

10. What are the long- and short-term effects of limiting factors on population growth?

11. Can birth control and family planning alone cure the current exponential population growth?

12. What environmental issues are affected most by population growth? Explain

What Do You Think?

1. The population of Japan is maturing, hence a greater percentage of the population is elderly. How might this affect (a) the average hourly pay received by gardeners; (b) the kinds of automobiles that sell best in that nation?

2. In 2100 the United States sends 100 people to establish a new colony on Mars. Create a scenario for how this population will change (if at all) by 2200, assuming that it does not become extinct.

3. A recent newspaper article warned that there may be infestations of rats on some of the bigger airliners. Apparently, they are feeding on leftover food from the airlines' meals and snacks. Discuss how such a population would change over time on a single aircraft. Discuss changes in birth, death, and population growth rates. Propose a solution.

4. Oysters produce huge numbers of fertilized eggs and provide no parenting of their offspring. The mortality rate among the very young is high, but once established, adults can live a long time. Draw the age structure for an oyster population.

5. Which of the following has contributed to the great increase in the human population since the beginning of the Industrial Revolution: changes in human (a) birth rates, (b) death rates, (c) longevity? Explain.

Pulling It All Together

1. What might be the effects of the Great Tsunami of 2004 on the age structure of a seaside town's population in (a) the first year after the tsunami; (b) in 20 years? Explain in terms of the population's death rates, birth rates, and growth rates.

2. This chapter discusses how the logistic growth curve has been used to forecast the future of the human population, especially to estimate the maximum number of people that will live on Earth in the future. How would a major bird flu epidemic alter this forecast? Take into account that past flu epidemics caused especially high mortality among young adults, and this is expected to happen with bird flu.

3. Suppose a cure for cancer is discovered in 2010. Would this lead to a new demographic transition? If so, what would it be like? See Figure 2.6 and draw a new version of that figure showing the effects of this cancer cure.

4. White-tailed deer have undergone a population explosion throughout much of the United States, especially in the Northeast, in part because suburban areas are not good places for people to hunt deer but provide many of the kinds of food that these deer like.

   a. Do you think the future of this population is likely to be forecast accurately by the logistic growth curve? Why or why not?

   b. Assuming that the population began to increase rapidly 20 years ago, draw a diagram of its age structure. (Assume that the deer can live 15 years—not necessarily an accurate estimate of their longevity but useful for this example.)

Further Reading

Brown, L. R., G. Gardner, and B. Halueil. 1999. *Beyond Malthus: Nineteen dimensions of the population challenge.* New York: W. W. Norton.—A discussion of recent changes in human population trends and their implications.

Cohen, J. E. 1995. *How many people can the Earth support?* New York: W. W. Norton.—A detailed discussion of world population growth, Earth's human carrying capacity, and factors affecting both.

Haupt, A., and T. T. Kane. 2004. *Population reference handbook.* Washington, DC: Population Reference Bureau.— The basic handbook about human populations.

Malthus, T. 1990. *An essay on the principle of population.* Edited with an introduction and notes by Geoffrey Gilbert. New York: Oxford University Press.—The classic essay that began modern understanding of the human population problems and also played an important role in Charles Darwin's work.

World Population Data Sheet. 2004. Washington, DC: Population Reference Bureau.—An easy-to-use source of world population information. Also available as a computer .PDF file.

World Resources 1996–97. 1996. Database diskette. New York: Oxford University Press.—A joint publication by World Resources Institute, the United Nations Environment Programme, the United Nations Development Programme, and the World Bank. A basic reference of facts about population and environment.

3

Biogeochemical Cycles

Taxi/Space Frontiers/Dara/Getty Images, Inc.

Big Question

Why Are Biogeochemical Cycles Essential to Long-Term Life on Earth?

Learning Objectives

Living things are made up of many chemical elements, and these elements must be present in specific amounts, specific concentrations, and specific ratios to one another. The study of the chemicals necessary for life and of **biogeochemical cycles**—the movement of these chemicals through land, water, air, and living things—is important in understanding and solving many environmental problems. After reading this chapter, you should understand . . .

■ what the major biogeochemical cycles are;

■ what the major factors and processes that control biogeochemical cycles are;

■ why some chemical elements cycle quickly and some slowly;

■ how each major component of Earth's global system (the atmosphere, waters, solid surfaces, and

life) are involved and linked with biogeochemical cycles;

■ how the biogeochemical cycles most important to life, especially the carbon cycle, generally operate;

■ how humans affect biogeochemical cycles.

Case Study

Lake Washington

People may harm the environment by unknowingly altering the natural cycling of chemicals. Science can suggest potential solutions, but we may or may not decide to pursue these solutions, depending on our values. The following story about Lake Washington shows that we can sometimes reduce the adverse effects of changes in chemical cycling if we decide to try.

The city of Seattle, Washington, lies between two major bodies of water—saltwater Puget Sound to the west and freshwater Lake Washington to the east (Figure 3.1). In the 1930s, people began to use the freshwater lake for disposal of sewage. By 1959, 11 sewage-treatment facilities had been constructed along the lake. These plants removed disease-causing organisms and much of the organic matter that used to enter the lake. Each day, the sewage-treatment plants released this treated wastewater into the lake. This was supposed to be a good strategy, but it didn't turn out that way. The introduction of treated wastewater caused a major bloom of undesirable algae and photosynthetic bacteria, which turned the water cloudy and made the lake less attractive.[1]

Seattle's mayor appointed an advisory committee to determine what might be done. Scientific research showed that phosphorus (a nutrient to plants) in the treated wastewater had stimulated the growth of algae and bacteria. Although the treatment process had eliminated disease-causing organisms, it had not removed phosphorus, which came from common laundry detergent and other urban sources. Phosphorus can be reduced in wastewater if people use low-phosphorus detergents, but removing the nutrient from all urban sources is not possible, and when it entered the lake, it acted as a potent fertilizer, posing a chemical problem for the lake's ecosystem.

Puget Sound, on the other hand, is a large body of ocean water with a rapid rate of exchange with the Pacific Ocean. Because of this rapid exchange, the phosphorus would be quickly diluted to a very low concentration. Therefore, the committee advised the city to discharge the treated wastewater into Puget Sound instead of into the lake.

The change was made by 1968, and the lake improved rapidly. Diverting the phosphorus-rich wastewater into Puget Sound greatly reduced the amount of phosphorus cycling through the lake's ecosystem. After just a year, the unpleasant algae had decreased, and surface waters had become much clearer than they had been five years before.[1]

Urbanization around Lake Washington remains a concern because urban runoff is a potential source of water pollution that may degrade the quality of streams or lakes it enters. Nevertheless, the water quality of Lake Washington in the early 21st century is still good for a large urban lake with heavy recreational use (Figure 3.2).

TerraNova International/Photo Researchers

■ **FIGURE 3.1**

The city of Seattle, Washington, is between Puget Sound and Lake Washington. The view is to the south. The city and region have experienced rapid urban development in recent decades, expanding around the shore of Puget Sound and Lake Washington.

■ **FIGURE 3.2**

Lake Washington in the summer of 1998 at a popular recreational beach on the eastern shoreline.

Ed Keller

The story of Lake Washington illustrates the importance of understanding how chemicals cycle in ecosystems. As the population in Seattle increased, so did its sewage. Because the people of Seattle valued Lake Washington, they initiated a scientific study that led to the discovery of the role of phosphorus and the development of a plan to sustain the water quality of their urban lake. In acting locally to solve a water-pollution problem, they provided a positive example of pollution abatement.

3.1 How Chemicals Cycle

Earth is a particularly good planet for life from a chemical point of view. Earth's atmosphere contains plenty of water and oxygen, which we and other animals need to breathe. In many places, soils are fertile, containing the chemical elements necessary for plants to grow; and Earth's bedrock contains valuable metals and fuels. Of course, some parts of Earth's surface are not perfect for life—deserts with little water, "chemical deserts" (such as the middle regions of the oceans) where nutrients necessary for life are not abundant, and certain soils that lack some of the chemical elements required for life or contain some others that are toxic to life.[2] If we place a value on a quality environment, we need to ask ourselves two scientific questions:

▪ What kinds of chemical processes benefit or harm the environment, ourselves, and other life-forms?

▪ How can we manage chemicals in the environment to improve and sustain ecosystems, both locally and at the global level?

To answer these questions, we need to know how chemical elements cycle, and this is therefore our starting point.

Biogeochemical Cycles

The path a chemical takes through Earth's system. The term *chemical* refers here to an element, such as carbon (C) or phosphorus (P), or to a compound, such as water (H_2O). (You will find a short review of chemistry and energy in Appendix A.) A biogeochemical cycle is the complete path a chemical takes through the four major components, or reservoirs, of Earth's system: (1) atmosphere, (2) hydrosphere (oceans, rivers, lakes, groundwaters, and glaciers), (3) lithosphere (rocks and soils), and (4) biosphere (plants and animals). We term this a **biogeochemical cycle** because *bio*- pertains to life, *geo*- pertains to Earth (atmosphere, water, rocks, and soils), and it is chemicals that are cycled.

Consider as an example an atom of carbon (C) in carbon dioxide (CO_2) emitted from burning coal (which is made up of fossilized plants hundreds of millions of years old). The carbon atom is released into the atmosphere and then taken up by a plant and incorporated into a seed. The seed is eaten by a mouse. The mouse is eaten by a coyote, and the carbon atom is expelled as scat following digestion. Decomposition of the scat allows our carbon atom to enter the atmosphere again. It may also enter another organism, such as an insect, which uses the scat as a resource.

Chemical Reactions

How new chemicals are formed. In our discussion of how chemical cycles work, it is important to acknowledge that the emphasis is on chemistry. Many chemical reactions occur within and between the living and nonliving portions of ecosystems. A **chemical reaction** is a process in which new chemicals are formed from elements and compounds that undergo a chemical change. For example, a simple reaction between rainwater (H_2O) and carbon dioxide (CO_2) in the atmosphere produces weak carbonic acid (H_2CO_3).

This weak acid reacts with Earth materials, such as rock and soil, to release chemicals into the environment. The released chemicals include calcium, sodium, magnesium, and sulfur, as well as smaller amounts of heavy metals, such as lead, mercury, and arsenic.

Making chemicals usable by living things. Many other chemical reactions determine whether chemicals are available to living things. For example, photosynthesis is a series of chemical reactions by which living green plants, using sunlight as an energy source, convert carbon dioxide and water into sugar. Oxygen is a byproduct of this process, and that is why we have free oxygen in our atmosphere. (We return to the topic of photosynthesis later in this chapter.)

After considering the two chemical reactions described above and applying critical thinking, you may recognize that both reactions combine water and carbon dioxide, but the products are very different: carbonic acid in one combination and a sugar in the other. How can this be so? The answer lies in an important difference between the simple reaction in the atmosphere to produce carbonic acid and, in the case of photosynthesis, a series of reactions that produce sugar and oxygen. Green plants absorb energy from the sun through the chemical chlorophyll and thus convert active solar energy into stored chemical energy in sugar.

All biogeochemical cycles involve four main parts of the Earth system. Perhaps the simplest way to think of a biogeochemical cycle is as a "box-and-arrow" diagram, which shows where a chemical is stored and the pathways along which it is transferred from one storage place to another (Figure 3.3). The boxes represent *storage compartments* (places where a chemical is stored).

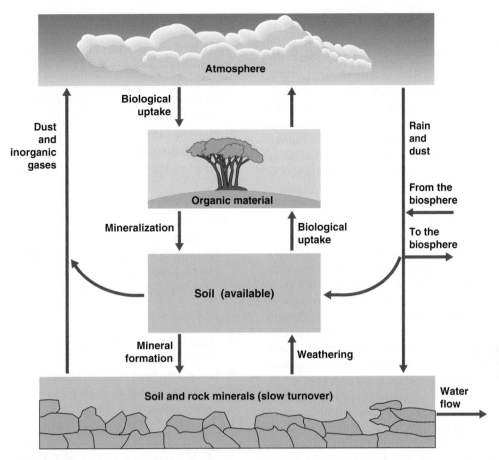

■ FIGURE 3.3

Biogeochemical cycle: pathways for cycling of chemicals in an ecosystem.
Chemical elements cycle within an ecosystem or exchange between an ecosystem
and the biosphere. Organisms exchange elements with the nonliving environment;
some elements are taken up from and released to the atmosphere, and others are
exchanged with water and soil or sediments. The parts of an ecosystem can be
thought of as storage compartments for chemicals. The chemicals move among
these compartments at different rates and remain in them for different average
lengths of time. For example, the soil in a forest has an active part, which rapidly
exchanges elements with living organisms, and a less active part, which exchanges
elements slowly (as shown in the lower part of the diagram). Generally, life
benefits if needed chemicals are kept within the ecosystem and are not lost
through geologic processes, such as erosion, that remove them from the ecosystem.

Storage compartments may also be called *reservoirs*, and
you may see the term *reservoir* in reading about biogeochemical cycles. The arrows represent pathways of transfer. The rate of transfer, or **flux,** is the amount of a
chemical that enters or leaves a storage compartment
per unit of time.

What are source and sink and what is flux? A
source is a storage compartment that releases a chemical to another location. For example, fossil fuels, when
burned, are a source of carbon, which is released into
the atmosphere. When a chemical enters a storage compartment from another compartment, we call the re-
ceiving compartment a **sink.** For example, the forests of

the world (which are a storage compartment for carbon) are a sink for carbon from the atmosphere, storing
it in wood, leaves, and roots. The amount of carbon
transferred from the atmosphere to the forest on a
global basis is the *flux*, which can be measured in units,
such as billions of metric tons of carbon per year. A sink
is called a *net sink* if input exceeds output. It is called a
net source if output exceeds input. For example, the carbon stored in fossil fuels, such as coal, is a net source of
carbon. The atmosphere where some of the carbon
from burning coal goes is a net sink.

In each compartment, we can identify an average
length of time that a given part (atom, grain, cubic centimeter, etc.) is stored before it is transferred. This is

called the **average residence time** and is calculated by dividing the total amount of material in the storage compartment by the rate of transfer (flux) through the compartment (see Chapter 1).

Certain factors and processes control the flow between compartments. Another crucial aspect of a biogeochemical cycle is the set of factors or processes that control the flow from one compartment to another. To understand a biogeochemical cycle, we need to quantify (measure the amounts) and understand these factors and processes. For example, understanding how air temperature and wind velocity vary across a lake is important in calculating the rate of evaporation of water from the lake.

We can consider biogeochemical cycles on any spatial scale of interest to us, from a single ecosystem to the whole Earth. It is often useful to consider such a cycle from a global perspective—for example, the problem of potential global warming calls for an understanding of how carbon cycles into and out of Earth's atmosphere. Sometimes, though, it is more useful to consider a cycle at a local level, as in the case study of Lake Washington. The link that unifies all these cycles is that they all involve the four principal components of the Earth system: atmosphere, hydrosphere, lithosphere, and biosphere (all of which are systems in their own right). Chemicals in these four major components have different average times of storage, or *residence time*. In general, chemicals reside for a longer time in rocks; average residence time is short in the atmosphere and intermediate in the hydrosphere and biosphere.

3.2 Environmental Questions and Biogeochemical Cycles

Now that you have a general idea of how chemicals cycle, let us consider some of the environmental questions that the science of biogeochemical cycles can help answer.

Biological Questions

▪ What factors, including chemicals necessary for life, place limits on the abundance and growth of organisms and their ecosystems?

▪ What toxic chemicals may be present that adversely affect the abundance and growth of organisms and their ecosystems?

▪ How can we improve the production of a desired biological resource?

▪ What are the sources of chemicals required for life, and how might we make these more readily available?

▪ What problems occur when a chemical is too abundant, as was the case with phosphorus in Lake Washington?

Geologic Questions

▪ What physical and chemical processes control the movement and storage of chemical elements in the environment?

▪ How are chemical elements transferred from solid earth to the water, the atmosphere, and living things?

▪ How does the long-term storage of chemical elements (for thousands of years or longer) in rocks and soils affect ecosystems locally and globally?

Atmospheric Questions

▪ What determines the concentrations of elements and compounds in the atmosphere?

▪ Where the atmosphere is polluted as the result of human activities, how might we alter a biogeochemical cycle to lower the pollution?

Hydrologic Questions

▪ What determines whether a body of water will be biologically productive?

▪ When a body of water becomes polluted, how can we alter biogeochemical cycles to reduce the pollution and its effects?

3.3 Biogeochemical Cycles and Life: Limiting Factors

The first of our "Biological Questions" concerns chemicals necessary for life, and the limits these chemicals may impose on the growth and abundance of living things and ecosystems.

Living things need just 24 elements. All living things are made up of chemical elements, but of the known elements, only 24 are required for life processes (Figure 3.4). These 24 are divided into two categories: *macronutrients*, elements required in large amounts by all life; and *micronutrients*, elements required either in small amounts by all life or in moderate amounts by some forms of life and not at all by others.

The Big Six. The macronutrients include the "big six"—elements that form the fundamental building blocks of life. These are carbon, hydrogen, nitrogen, oxygen, phosphorus, and sulfur. Each of these elements plays a special role in organisms. Carbon is the basic building block of organic compounds. Along with oxygen and hydrogen, carbon forms carbohydrates. Nitrogen, along with these other three, makes proteins. Phosphorus is the "energy element"; it occurs in compounds that are important in the transfer and use of energy within cells.

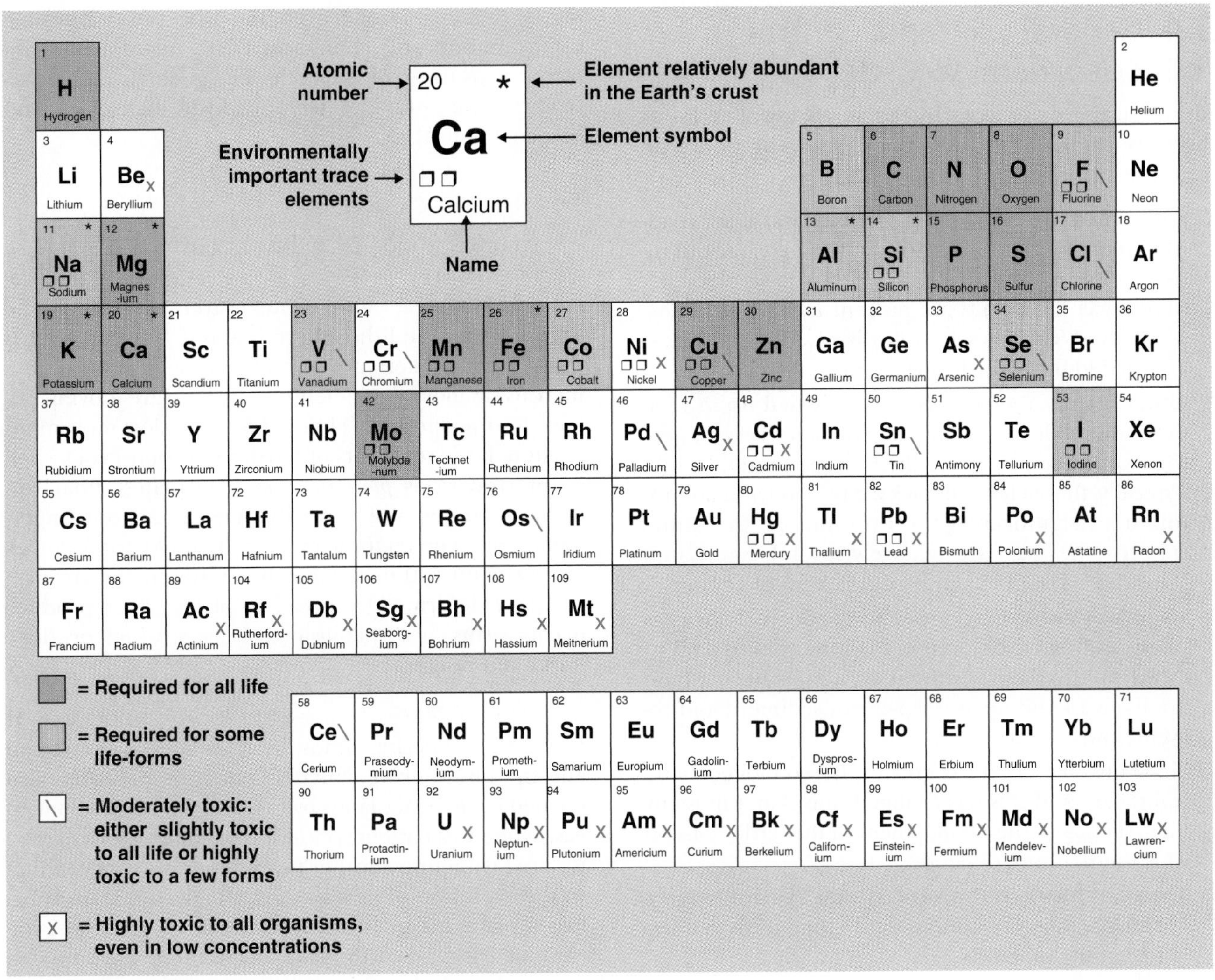

■ FIGURE 3.4
Periodic table of the elements.
The macronutrients are carbon, hydrogen, nitrogen, oxygen, phosphorus, and sulfur.
Others such as calcium, magnesium, sodium, and iron are also very important to life.

In addition to the "big six," other macronutrients also play important roles. Calcium, for example, is the structure element, occurring in bones of vertebrates, shells of shellfish, and wood-forming cell walls of vegetation. Sodium and potassium are important to nerve-signal transmission. Many of the metals required by living things are necessary for specific enzymes. (An enzyme is a complex organic compound that acts as a catalyst—it causes or speeds up chemical reactions, such as digestion.)

Chemicals needed for life may become limiting factors. For any form of life to persist, chemical elements must be available at the right times, in the right amounts, and in the right concentrations relative to each other. When all of this does not happen, a chemical can become a limiting factor, preventing instead of helping the growth of an individual, a population, or a species, or even causing its local extinction. (Limiting factors were discussed in Chapter 2.) Chemical elements may also be toxic to some life-forms and ecosystems. Mercury, for example, is toxic even in low concentrations. Copper and some other elements are required in low concentrations for life processes but are toxic in high concentrations.

3.4 General Concepts Central to Biogeochemical Cycles

Although there are as many biogeochemical cycles as there are chemicals, certain general concepts hold true for these cycles.

- Some chemical elements cycle quickly and are readily regenerated for biological activity. Oxygen and nitrogen are among these. Typically, these elements have a gas phase and are present in Earth's atmosphere, and/or they are easily dissolved in water and are carried by the hydrologic cycle.

- Other chemical elements are easily tied up in relatively immobile forms and are returned slowly, by geologic processes, to where they can be reused by life. Typically, these elements lack a gas phase and are not found in significant concentrations in the atmosphere. They also are relatively insoluble in water. Phosphorus is an example of this kind of chemical.

- Chemicals whose biogeochemical cycles include a gas phase and that are stored in the atmosphere tend to cycle rapidly. Those without an atmospheric phase are likely to end up as deep-ocean sediment and recycle slowly.

- Since life evolved, it has greatly altered biogeochemical cycles, and this has changed our planet in many ways, such as in the development of the fertile soils on which agriculture depends.

- The continuation of processes that control biogeochemical cycles is essential to the long-term maintenance of life on Earth.

- Through modern technology, we have begun to transfer chemical elements among air, water, and soil at rates comparable to natural processes. These transfers can benefit society, as when they improve crop production, but they can also pose environmental dangers, as illustrated by the opening case study. To live wisely with our environment, we must recognize the positive and negative consequences of altering biogeochemical cycles. Then we must attempt to accentuate the positive and minimize the negative.

Discussion of biogeochemical cycles beyond the general concepts listed above requires an understanding of geologic and hydrologic cycles. Of particular importance are geologic processes linked to the cycling of chemicals in the biosphere.

3.5 The Geologic Cycle

Throughout the 4.6 billion years of Earth's history, rocks and soils have been continuously created, maintained, changed, and destroyed by physical, chemical, and biological processes. Collectively, the processes responsible for formation and change of Earth materials are referred to as the **geologic cycle**. The geologic cycle is actually a group of cycles: tectonic, hydrologic, rock, and biogeochemical.

The Tectonic Cycle

Plate tectonics has important effects on the environment. The *tectonic cycle* involves creation and destruction of the solid outer layer of Earth, the lithosphere. The lithosphere is about 100 km (60 mi) thick and is broken into several large segments and numerous smaller ones called *plates,* which are moving relative to one another (Figure 3.5). The slow movement of these large segments of Earth's outermost rock shell is referred to as *plate tectonics.* The plates "float" on denser material and move about as fast as your fingernails grow. The tectonic cycle is driven by forces originating deep within the Earth. Closer to the surface, rocks are deformed by spreading plates, which produce ocean basins, and by collisions of plates, which produce mountain ranges.

Moving plates change the location and size of continents, altering atmospheric and ocean circulation and thereby altering climate. Plate movement has also created ecological islands by breaking up continental areas. When this happens, closely related life-forms may be isolated from one another for millions of years, leading to the evolution of new species. Finally, boundaries between plates are geologically active areas, and most volcanic activity and earthquakes occur there. Earthquakes occur when rocks of the brittle upper lithosphere rupture along *faults* (a fault is a fracture along which movement occurs). Movement of several meters between plates can occur within a few seconds during earthquakes, in contrast to the slow, deeper plate movement described above.

Three types of plate boundaries occur: divergent, convergent, and transform faults.

A divergent plate boundary occurs at a spreading ocean ridge, where plates are moving away from one another and new lithosphere is produced. This process, known as seafloor spreading, produces ocean basins.

A convergent plate boundary occurs when plates collide. Sometimes a plate composed of relatively heavy ocean-basin rocks slips beneath the leading edge of a plate composed of lighter continental rocks. Such a convergence may produce linear coastal mountain ranges, such as the Andes in South America. When two plates composed of lighter continental rocks collide, a continental mountain range may form, such as the Himalayas in Asia.[3, 4]

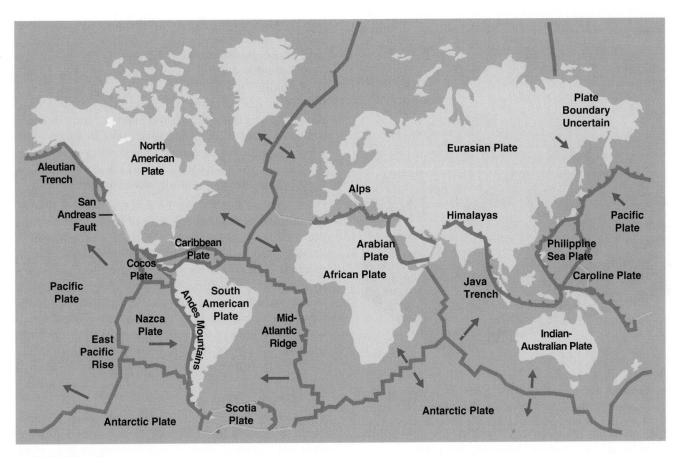

■ FIGURE 3.5

Plate tectonics.
Divergent plate boundaries are shown as heavy lines (for example, the Mid-Atlantic
Ridge). Convergent boundaries are shown as barbed lines (for example, the Aleutian
Trench). Transform fault boundaries are shown as yellow, thinner lines (for example, the
San Andreas Fault). Arrows indicate direction of relative plate movements. [*Source:*
Modified from B. C. Birchfiel, R. J. Foster, E. A. Keller, et al. *Physical geology: The
structures and processes of the Earth.* Columbus, OH: Merrill, 1982.]

*A transform fault boundary occurs where one plate slides
past another.* An example is the San Andreas Fault in
California, which is the boundary between the North
American and Pacific plates. The Pacific plate is moving
north relative to the North American plate at about 5
cm (2 in.) per year. As a result, Los Angeles is moving
slowly toward San Francisco, about 500 km (300 mi)
north. If this continues, in about 20 million years San
Francisco will be a suburb of Los Angeles.

The Hydrologic Cycle

The **hydrologic cycle** (Figure 3.6) is the transfer of water from the oceans to the atmosphere to the land and
back to the oceans. The processes involved include evaporation of water from the oceans; precipitation on land;
evaporation from land; and runoff from streams, rivers,
and subsurface groundwater. The hydrologic cycle is

driven by solar energy, which evaporates water from
oceans, freshwater bodies, soils, and vegetation.

Storage compartments of water. The "storage
compartments" of water are the ocean, glaciers and ice
caps, shallow groundwater, lakes, soil, the atmosphere,
and rivers. Of the total water on Earth, about 97% is in
oceans and about 2% is in glaciers and ice caps. The rest
is freshwater on land and in the atmosphere. Although
the water on land is only a small fraction of the water on
Earth, it is important in moving chemicals, sculpting the
landscape, weathering rocks, transporting sediments,
and providing our water resources. The water in the atmosphere—only 0.001% of the total on Earth—cycles
quickly to produce rain and runoff for our water
resources.

Rates of transfer among storage compartments.
Annual rates of transfer from the storage compartments

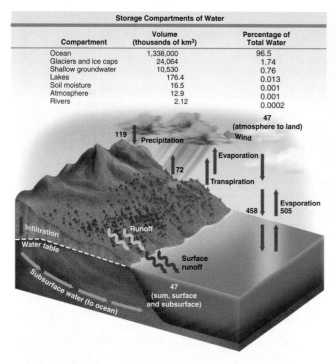

Storage Compartments of Water		
Compartment	Volume (thousands of km³)	Percentage of Total Water
Ocean	1,338,000	96.5
Glaciers and ice caps	24,064	1.74
Shallow groundwater	10,530	0.76
Lakes	176.4	0.013
Soil moisture	16.5	0.001
Atmosphere	12.9	0.001
Rivers	2.12	0.0002

▪ **FIGURE 3.6**

Hydrologic cycle, showing the transfer of water from the oceans to the atmosphere to the continents and back to the oceans again. Units are thousands of cubic kilometers of water. [*Source:* P. H. Gleick. Water in crisis. New York: Oxford University Press, 1993.]

in the hydrologic cycle are shown in Figure 3.6. These rates of transfer define a global "water balance." For example, if we sum the arrows going up in the figure, and then sum the arrows going down, we find that the two sums are the same. Similarly, precipitation on land is balanced by evaporation from land plus surface and subsurface runoff.

Especially important from an environmental perspective is that rates of transfer on land are small compared with what's happening in the ocean. Most of the water that evaporates from the ocean falls into the ocean again as precipitation. On land, however, most of the water that falls as precipitation comes from evaporation of water from land. This means that regional changes in land use, such as the building of large dams and reservoirs, can change the amount of water evaporated into the atmosphere and change the location and amount of precipitation on land—water we depend on to raise our crops and supply water for our urban environments. Furthermore, as we pave over large areas of land in cities, storm water runs off more quickly and in greater volume, increasing flood hazards. Bringing water into semi-arid cities by pumping groundwater or transporting water from distant mountains through aqueducts may increase evaporation, thereby increasing humidity and precipitation in a region.

We can see from Figure 3.6 that approximately two-thirds of water that falls by precipitation on land each year evaporates to the atmosphere. A smaller component (about one-third) returns to the ocean by subsurface runoff. This small annual transfer provides water for rivers and for urban and agricultural lands. Unfortunately, water is not distributed in equal amounts everywhere on land. Water shortages occur in some areas, and as the human population increases, water shortages will become more frequent in arid and semi-arid regions, where water is naturally not abundant.

Drainage basins. At the regional and local level, the fundamental hydrologic unit of the landscape is the drainage basin, also called a watershed or catchment. A *drainage basin* is the area that contributes surface runoff to a particular stream or river. The term *drainage basin* is usually used in evaluating the hydrology of an area, such as the stream flow or runoff from slopes. Drainage basins vary greatly in size, from less than a hectare (2.5 acres) to millions of square kilometers. A drainage basin is usually named for its main stream or river—for example, the Mississippi River drainage basin.

The Rock Cycle

The rock cycle consists of numerous processes that produce rocks and soils. The rock cycle depends on the tectonic cycle for energy and on the hydrologic cycle for water.

Three types of rock: igneous, sedimentary, and metamorphic (Figure 3.7) These types of rock are involved in a global recycling process.

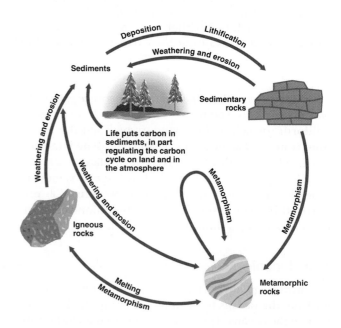

▪ **FIGURE 3.7**

Rock cycle, showing major paths of material transfer as modified by life.

Igneous rock. Internal heat from the tectonic cycle produces molten material called magma that crystallizes to form igneous rocks such as granite deep in the earth. Near the surface, magma may be extruded as lava or explosively ejected as volcanic ash from volcanoes. Lava crystallizes to form volcanic rock such as basalt. These new rocks weather when exposed at the surface. For example, when water gets into cracks in the rocks and freezes, it expands, breaking the rocks apart. This kind of physical weathering makes smaller particles of rocks from bigger ones, producing sediment, such as gravel, sand, and silt. Chemical weathering occurs when the weak acids in water dissolve chemicals from rocks. The sediments and dissolved chemicals are then transported by water, wind, or ice (glaciers).

Sedimentary rock. Weathered materials accumulate in *depositional basins,* such as the oceans. The weight of the layers of sediments deposited in these basins compacts and cements the particles into sedimentary rocks.

Metamorphic rock. When igneous or sedimentary rocks are buried deep enough (usually tens to hundreds of kilometers), heat, pressure, or chemically active fluids may transform them into metamorphic rocks. Sometimes these deeply buried rocks are brought to the surface by an uplift due to plate tectonics. Once they are on the surface, they become subject to weathering, and the cycle continues.

The role of life processes and rock cycle You can see in Figure 3.7 that life processes play an important role in the rock cycle by adding organic carbon to rocks. The addition of organic carbon helps to produce rocks such as limestone, which is mostly calcium carbonate (the material of seashells and bones), as well as fossil fuels, such as coal.

Kim Heacox/Peter Arnold, Inc.

■ **FIGURE 3.8**
Grand Canyon.
In response to slow tectonic uplift of the region, the Colorado River has eroded through the sedimentary rocks of the Colorado Plateau to produce the spectacular scenery of the Grand Canyon.

Geologic processes that lift rocks or cause them to sink down, along with erosion, produce Earth's varied topography. The spectacular Grand Canyon of the Colorado River in Arizona (Figure 3.8), sculpted from mostly sedimentary rocks, is one example. Another is the beautiful Tower Karst in China (Figure 3.9); these resistant blocks of limestone have survived chemical weathering and erosion that removed the surrounding rocks.

Our discussion of geologic cycles has emphasized tectonic, hydrologic, and rock-forming processes and their linkages to life. We can now begin to integrate biogeochemical processes into the picture.

Manfred Gottschalk/Tom Stack & Associates

■ **FIGURE 3.9**
Tower Karst.
This landscape in the People's Republic of China features Tower Karst, steep-sided hills or pinnacles made of limestone. The rock has been slowly dissolving through chemical weathering. The pinnacles and hills are remnants of the weathering and erosion processes.

3.6 Biogeochemical Cycling in Ecosystems

When we want to know what chemicals might limit the abundance of a specific organism, population, or species, we look for the answer first at the *ecosystem* level. An *ecosystem* is a community of species and their nonliving environment in which energy flows and chemicals cycle. The boundaries that we choose for our investigation of limiting factors may be somewhat arbitrary, selected for convenience of measurement and analysis. On land, we often evaluate biogeochemical cycles in a fundamental element of the landscape, usually a drainage basin. Freshwater bodies—lakes, ponds, and bogs—are also convenient for analyzing ecosystems and biogeochemical cycling (see Chapter 1).

Chemical inputs to ecosystems. Chemical cycling in an ecosystem begins with inputs from outside the system. On land, chemical inputs to an ecosystem come from the atmosphere via rain, dust transported by wind (called dry fallout), and volcanic ash from eruptions and from adjoining land via stream flow from flooding and groundwater from springs. Ocean and freshwater ecosystems have the same atmospheric and land inputs (including large, nearshore submarine springs). Ocean ecosystems have additional inputs from ocean currents and submarine vents (hot springs) at divergent plate boundaries.

Chemical cycling within an ecosystem. Chemicals cycle within an ecosystem through air, water, rocks, soil, and food chains by way of physical transport and chemical reactions. When individual organisms die, chemical reactions cause them to decompose, returning their chemical elements to other parts of the ecosystem. In addition, living organisms release some chemical elements directly into an ecosystem. Defecation by animals and the dropping of fruit by plants are examples.

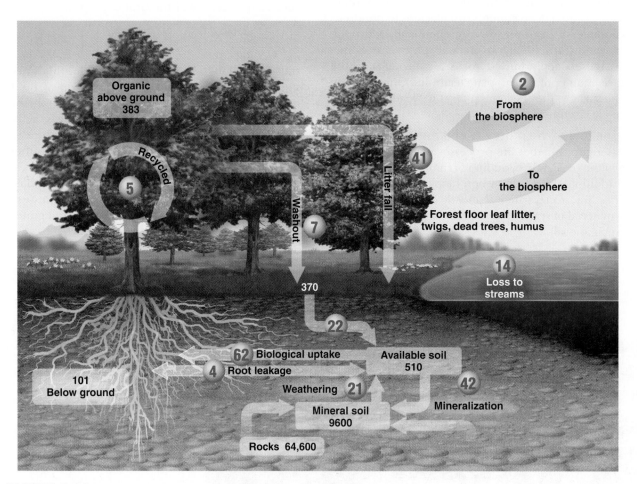

■ **FIGURE 3.10**

The annual calcium cycle in a forest ecosystem.
The circled numbers are flux rates in kilograms per hectare per year. The other numbers are the amounts stored in kilograms per hectare. Unlike sulfur, calcium does not have a gaseous phase, although it does occur in compounds as part of dust particles, transported by wind. Calcium is highly soluble in water in its inorganic form and is readily lost from land ecosystems in water transport. The information in this diagram was obtained from Hubbard Brook Ecosystem. [*Source:* G. E. Likens, F. H. Bormann, R. S. Pierce, et al. *The biogeochemistry of a forested ecosystem*, 2nd ed. New York: Springer-Verlag, 1995.]

Ecosystems can lose chemicals. An ecosystem can lose chemical elements to other ecosystems. For example, rivers transport chemicals from the land to the sea. An ecosystem that loses its chemical elements very slowly can function in its current condition longer than can a more "leaky" ecosystem that loses chemical elements rapidly. All ecosystems, however, lose chemicals to some extent. Therefore, all ecosystems require some external inputs of chemicals.

Ecosystem Cycles of a Metal and a Nonmetal

Within an ecosystem, different chemical elements may have very different pathways, as illustrated in Figure 3.10 for calcium and in Figure 3.11 for sulfur. The calcium cycle is typical of a metallic element, and the sulfur cycle is typical of a nonmetallic element.

Chemical cycles that include a gaseous state are faster. An important difference between these cycles is that calcium, like most metals, does not form a gas on the Earth's surface. Therefore, calcium is not present as a gas in the atmosphere. In contrast, sulfur forms several gases, including sulfur dioxide, a major air pollutant and component of acid rain (see Chapter 14), and hydrogen sulfide (swamp gas or rotten-egg gas, usually produced biologically). Because sulfur has gas forms, it can be returned to an ecosystem more rapidly than can calcium. The annual input of sulfur from the atmosphere to a forest ecosystem has been measured at ten times that of calcium. For this reason, calcium and other elements without a gas phase are more likely to become limiting factors.

Chemical Cycling and the "Balance of Nature"

For life to be sustained indefinitely within an ecosystem, energy must be continuously added, and the store of essential chemicals must not decline. There is a common belief that without human interference, life would be

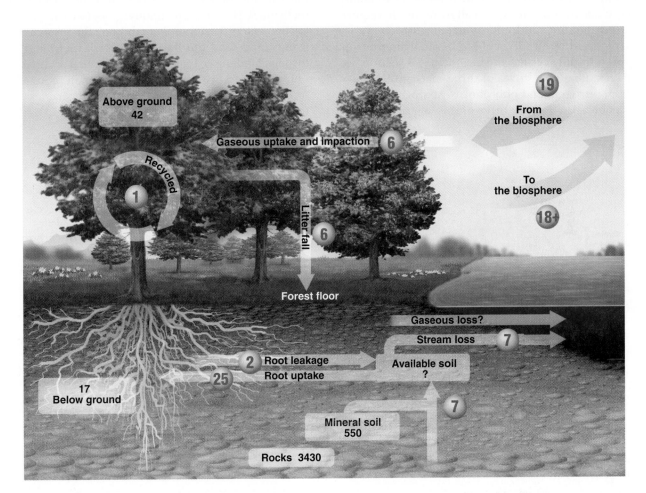

■ **FIGURE 3.11**
The annual sulfur cycle in a forest ecosystem.
The circled numbers are the flux rates in kilograms per hectare per year. The other numbers are the amounts stored, in kilograms per hectare. Sulfur has a gaseous phase as H_2S and SO_2. The information in this diagram was obtained from Hubbard Brook Ecosystem. [*Source:* G. E. Likens, F. H. Bormann, R. S. Pierce, et al. *The biogeochemistry of a forested ecosystem,* 2nd ed. New York: Springer-Verlag, 1995.]

sustained indefinitely in a steady state, or "balance of nature." Another common belief is that life tends to function to preserve an environment that is beneficial to itself. Both beliefs presume that chemical elements within an ecosystem will remain constant over time. However, as you have seen, some fraction of the chemical elements stored in an ecosystem are inevitably lost and must be replaced. In short, studies indicate that ecosystems are never in a dynamic steady state, because rates of chemical input and output do not balance, and concentrations of some chemicals decrease over time.[5]

3.7 Some Major Global Chemical Cycles

Earlier we asked what chemical elements limit the abundance of life. We pointed out that the chemical elements required by life are divided into two major groups: macronutrients, which all forms of life require in large amounts; and micronutrients, which are either required by all forms of life in small amounts or required by only certain life-forms. In this section, we consider the global cycles of three macronutrients—carbon, nitrogen, and phosphorus. We focus on these in part because they are three of the "big six"—the el-

ements that are the basic building blocks of life. Each is also related to important environmental problems that have attracted attention in the past and will continue to do so in the future.

The Carbon Cycle

Carbon is vital for life but is not highly abundant. Carbon is the element that anchors all organic substances, from coal and oil to DNA (deoxyribonucleic acid), the compound that carries genetic information. Although carbon is of central importance to life, it is not one of the most abundant elements in Earth's crust. It contributes only 0.032% of the weight of the crust.[6, 7]

Carbon enters the atmosphere in several ways. The major pathways and storage compartments of the carbon cycle are shown in Figure 3.12. Notice that carbon has a gaseous phase as part of its cycle. It occurs in the Earth's atmosphere as carbon dioxide (CO_2) and methane (CH_4), both greenhouse gases (see Chapter 13). Carbon enters the atmosphere through the respiration of living things and the burning of fossil fuels, and also through land-use changes, including natural wildfires and human-caused fires that burn organic material, such as trees, brush, grass, agricultural plants after harvest, and tree branches left after timber harvest.

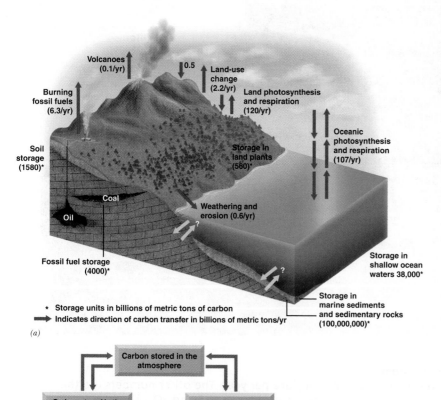

(a)

(b)

■ **FIGURE 3.12**
Carbon cycle.
(a) The generalized global carbon cycle. (b) Parts of the carbon cycle are simplified to illustrate the cyclic nature of the movement of carbon. Land-use change includes natural wildfires and human-caused fires from conversion of forestland to agriculture. [*Source:* Modified from G. Lambert, *La Recherche* 18: 782–783, 1987, with some data from R. Houghton, *Bulletin of the Ecological Society of America* 74 (4): 355–356, 1993, and R. Houghton *Tellus* 55B, (2): 378–390, 2003.]

Carbon leaves living organisms through respiration, a process that breaks down organic compounds to release gaseous carbon dioxide. For example, animals (including people) take in air, which has a relatively high concentration of oxygen. Oxygen is absorbed by blood in the lungs. Through respiration, carbon dioxide is released into the atmosphere. Figure 3.14 illustrates the role of photosynthesis and respiration, along with other processes, in the carbon cycle of a lake.

Carbon occurs in the ocean in several inorganic forms, including dissolved carbon dioxide. It also occurs in organic compounds of marine organisms and their products, such as seashells (calcium carbonate, $CaCO_3$). Carbon enters the ocean from the atmosphere by simple diffusion of carbon dioxide. The carbon dioxide then dissolves and is converted to carbonate and bicarbonate. Marine algae and photosynthetic bacteria obtain the carbon dioxide they use from the water in one of these three forms.

Carbon is transferred from the land to the ocean in rivers and streams as dissolved carbon, including organic compounds, and as organic particulates (fine particles of organic matter). Winds also transport small organic particulates from the land to the ocean. Globally, rivers and streams account for a relatively small fraction of the total transfer of carbon to the oceans. However, on a local and regional scale, input of carbon from rivers to near-shore areas, such as deltas and salt marshes, which are often highly biologically productive, is important.

Carbon may be stored in organic forms. Carbon enters the *biota* (a term meaning all life) through photosynthesis by green plants, algae, and photosynthetic bacteria, and is returned to the atmosphere or to the waters by the respiration of these organisms or by wildfire.

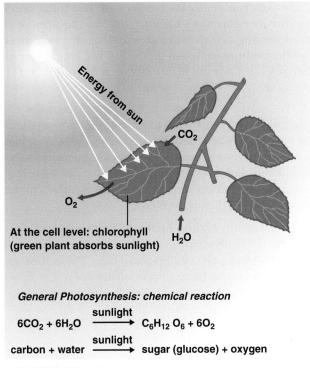

■ **FIGURE 3.13**
Photosynthesis.
An idealized diagram illustrating photosynthesis for a green plant (tree) and generalized reaction.

General Photosynthesis: chemical reaction

$$6CO_2 + 6H_2O \xrightarrow{\text{sunlight}} C_6H_{12}O_6 + 6O_2$$

$$\text{carbon} + \text{water} \xrightarrow{\text{sunlight}} \text{sugar (glucose)} + \text{oxygen}$$

Carbon dioxide enters biological cycles through photosynthesis, the process by which the cells of living organisms (such as plants) convert energy from sunlight into chemical energy through a series of chemical reactions. In the process, carbon dioxide and water are combined to form organic compounds, such as simple sugars and starch, with oxygen as a by-product (Figure 3.13).

■ **FIGURE 3.14**
Carbon cycle in a pond.
An idealized diagram showing the carbon cycle in a lake.

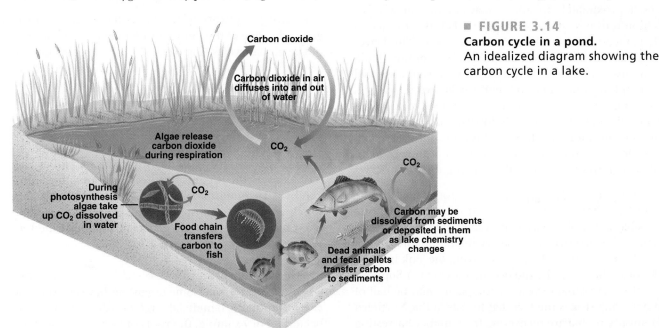

When an organism dies, most of its organic material decomposes to inorganic compounds, including carbon dioxide. But where there is not enough oxygen to make this conversion possible, or where temperatures are too low for decomposition, some carbon may be buried and stored in organic forms.

Over years, decades, and centuries, storage of carbon occurs in wetlands, including parts of floodplains, lake basins, bogs, swamps, forests, deep-sea sediments, and near-polar regions. Some carbon may be buried with sediments that, over thousands or even several million years, become sedimentary rocks. This carbon is transformed into fossil fuels, such as natural gas, oil, and coal. Nearly all of the carbon stored in the lithosphere exists as sedimentary rocks. Most of this is in the form of carbonates, such as limestone, much of which has a direct biological origin.

Life's role in the carbon cycle is a major factor in Earth's atmosphere. The cycling of carbon dioxide between land organisms and the atmosphere is a large flux, or flow. Each year, approximately 15% of all the carbon in the atmosphere is used for photosynthesis and released by respiration on land. Thus, as noted, life has a large effect on the chemistry of the atmosphere.

The Missing Carbon Sink

Because carbon forms two of the most important greenhouse gases—carbon dioxide and methane—much research has been devoted to understanding the carbon cycle. However, at a global level, some key questions remain unanswered. For example, monitoring of atmospheric carbon dioxide levels over the past several decades suggests that of the approximately 8.5 units released into the atmosphere each year by human activities, approximately 3.2 units remain there (one unit is a billion tons of carbon; see Figure 3.12). It is estimated that at least 2.4 units diffuse into the ocean. This leaves about 2.9 units unaccounted for.[8,9] Several hundred million tons of carbon are burned each year from fossil fuel and end up somewhere not entirely known to science. Inorganic processes do not account for this "missing carbon sink." Marine or land photosynthesis, or both, must provide the additional flux. At this time, however, scientists do not agree on which processes dominate or in what regions of the Earth this missing flux occurs.[10–12]

Terrestrial ecosystems, such as forests, could account for part of the mystery. It is believed that one possible sink, holding as much as half of the missing carbon, is the terrestrial ecosystems, including forests and, to a lesser extent, soils. (If this is true, the sink holding the other half must be the ocean ecosystems.) Supporting the theory that terrestrial ecosystems may be half of the missing sink is the fact that forests in the Northern Hemisphere that are recovering from timber harvesting of the past two centuries are growing fast today. Fast-growing forests remove carbon from the atmosphere at an increasing rate as biomass is added. However, fires and the melting of frozen ground in the boreal forests also release carbon, so it's not certain whether the forests are a net sink or a net source of carbon.[13]

The missing carbon sink has grown since the Industrial Revolution, due to changes in land use and the burning of fossil fuels. Uncertainties will be lessened in the future if we are more successful in measuring and monitoring land-use change (deforestation, burning, and clearing) and in estimating the flux of carbon in the ecosystems and the atmosphere.

The missing-carbon problem illustrates the complexity of biogeochemical cycles, especially ones in which the biota plays an important role. The carbon cycle will continue to be an important area of research because of its significance to global climate investigations, especially to global warming.[14,15]

The Nitrogen Cycle

Nitrogen is essential to life—it is needed to manufacture proteins and DNA. Free nitrogen (N_2 uncombined with any other element) makes up about 80% of Earth's atmosphere. However, many organisms cannot use this nitrogen directly. Some organisms, such as animals, require nitrogen in an organic compound. Others—including plants, algae, and bacteria—can take up nitrogen either as the nitrate ion (NO_3^-) or the ammonium ion (NH_4^+). Because nitrogen is a relatively unreactive element, few processes convert molecular nitrogen to one of these compounds. Lightning oxidizes nitrogen, producing nitric oxide. However, bacteria perform nearly all conversions of molecular nitrogen to biologically useful forms.

Nitrogen enters food chains through the **nitrogen cycle**, one of the most important and most complex of the global cycles (Figure 3.15). The process of converting inorganic, molecular nitrogen in the atmosphere to ammonia or nitrate is called *nitrogen fixation*. Once it is converted, nitrogen can be used on land by plants and in the oceans by algae. Bacteria, plants, and algae convert the inorganic nitrogen compounds into organic ones, and the nitrogen becomes available to ecological food chains. When organisms die, other bacteria convert the organic compounds containing nitrogen back to ammonia, nitrate, or molecular nitrogen, which enters the atmosphere. The process of releasing fixed nitrogen back to molecular nitrogen is called *denitrification* (see Figure 3.15).

Nearly all organisms depend on nitrogen-converting bacteria, and some organisms have evolved *symbiotic* (mutually beneficial) relationships with these bacteria. For example, the roots of the pea family have

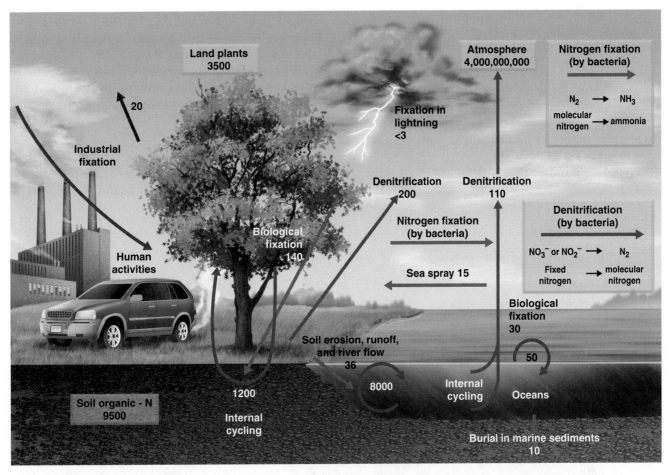

■ FIGURE 3.15

The global nitrogen cycle.
Numbers in boxes indicate amounts stored, and numbers with arrows indicate annual flux, in
10^{12} g N_2. Note that the industrial fixation of nitrogen is nearly equal to the global biological
fixation. [*Sources:* R. Söderlund and T. Rosswall. *The handbook of environmental chemistry,* Vol.
1, Pt. B, O. Hutzinger, ed. New York: Springer-Verlag, 1982; and W. H. Schlosinger.
Biogeochemistry: An analysis of global change. San Diego: Academic Press, 1997, p. 386.]

nodules that provide a habitat for the bacteria. The bacteria obtain food from the plants, and the plants obtain usable nitrogen. Such plants can grow in otherwise nitrogen-poor environments. When these plants die, they contribute nitrogen-rich organic matter to the soil, which improves the soil's fertility. Alder trees, too, have nitrogen-fixing bacteria as *symbionts* in their roots (symbionts are organisms in mutually beneficial relationships). These trees grow along streams, and their nitrogen-rich leaves fall into the water, increasing the supply of nitrogen in a form that is usable by freshwater organisms.

Nitrogen-fixing bacteria are also symbionts in the stomachs of some animals, particularly the cud-chewing animals. These animals—which include buffalo, cows, deer, moose, and giraffes—have a specialized four-chambered stomach. The bacteria provide as much as half of the total nitrogen needed by the animals; the rest is provided by protein in the green plants the animals eat.

Nitrogen has a gaseous phase but is not very reactive. In terms of its availability to living things, nitrogen falls somewhere between carbon and phosphorus. Like carbon, nitrogen has a gaseous phase and is a major component of Earth's atmosphere. Unlike carbon, however, it is not very reactive—its conversion depends heavily on biological activity. Thus, the nitrogen cycle is not only essential to life but also primarily driven by life.

Industrial processes are both helpful and harmful. In the early part of the 20th century, scientists discovered that industrial processes could convert molecular nitrogen into compounds usable by plants. Today, industrial fixation of nitrogen is a major source of commercial nitrogen fertilizer. The amount of industrial fixed nitrogen is about one-half of the amount fixed in the biosphere. Unfortunately, nitrogen in agricultural runoff is a potential source of water pollution.

In addition, many modern industrial combustion processes—including the burning of fossil fuels in gasoline and diesel engines—produce oxides of nitrogen, which are air pollutants that play a significant role in urban smog (see Chapter 14).

In sum, nitrogen compounds are both a bane and a boon for society and for the environment. Nitrogen is required for all life, and its compounds are used in many technological processes and in modern agriculture. But nitrogen is also a source of air and water pollution.

The Phosphorus Cycle

Phosphorus, one of the "big six" elements required in large quantities by all forms of life, is often a limiting nutrient for plant and algal growth—if it is not present in sufficient amounts, plants and algae will not thrive. However, if phosphorus is too abundant, it can cause environmental problems, as illustrated by the story of Lake Washington that opened this chapter.

No gaseous phase, so transfer is slow.   Unlike carbon and nitrogen, phosphorus does not have a gaseous phase on Earth (Figure 3.16). Thus, the phosphorus cy-

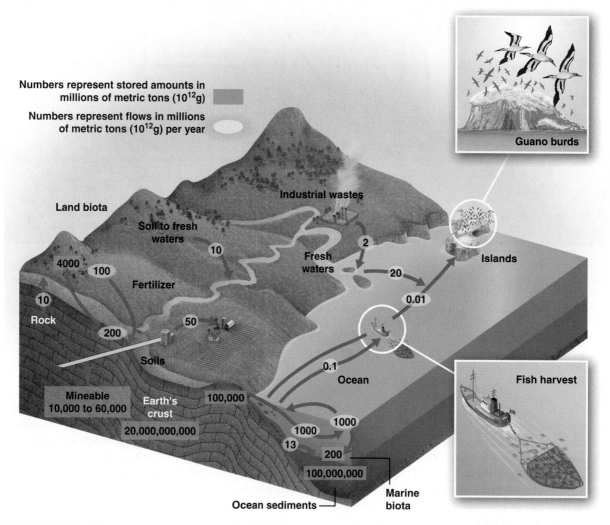

■ **FIGURE 3.16**
The global phosphorus cycle.
Phosphorus is recycled to soil and land biota by (1) geologic processes that uplift the land and erode rocks, (2) birds that produce guano, and (3) human beings. Although Earth's crust contains a lot of phosphorus, most of it is difficult to mine and thus expensive to produce. The amounts of phosphorus stored or in flux are estimates derived from various sources. [*Sources:* Based primarily on C. C. Delwiche and G. E. Likens. Biological response to fossil fuel combustion products in W. Stumm, ed., *Global chemical cycles and their alterations by man*, Berlin: Abakon Verlagsgesellschaft, 1977, pp. 73–88; and U. Pierrou. The global phosphorus cycle, in B. H. Svensson and R. Soderlund, eds., *Nitrogen, phosphorus and sulfur—Global cycles.* (Stockholm: Ecological Bulletin, 1976, pp. 75–88.]

cle is significantly different from the carbon and nitrogen cycles. The rate of transfer of phosphorus in Earth's system is slow compared with that of carbon or nitrogen. Phosphorus exists in the atmosphere only in small particles of dust. In addition, phosphorus tends to form compounds that do not easily dissolve in water. Consequently, phosphorus is not readily weathered chemically. It does occur commonly in an oxidized state as phosphate, which combines with calcium, potassium, magnesium, or iron to form minerals.

How phosphorus is taken up, and how it is lost. Phosphorus enters the biota when it is taken up as phosphate by plants, algae, and some bacteria. In a relatively stable ecosystem, much of the phosphorus that is taken up by vegetation is returned to the soil when the plants die. Nevertheless, some phosphorus is inevitably lost to ecosystems. Rivers carry it off to the oceans, either in a water-soluble form or as suspended particles.

The role of ocean-feeding birds. An important way in which phosphorus returns from the ocean to the land involves ocean-feeding birds, such as the Chilean pelican. These birds feed on small fish, especially anchovies, which in turn feed on tiny ocean plankton. Plankton thrive where nutrients are present, such as in areas of rising oceanic currents known as *upwellings*. Upwellings occur near continents where the prevailing winds blow offshore, pushing surface waters away from the land and allowing deeper waters to rise. These upwellings carry nutrients, including phosphorus, from the depths of the oceans to the surface.

The fish-eating birds nest on offshore islands, where they are protected from predators. Over time, their nesting sites become covered with their phosphorus-laden excrement, called *guano*. The birds nest by the thousands, and deposits of guano accumulate over centuries. In relatively dry climates, guano hardens into a rocklike mass. The guano results from a combination of biological and nonbiological processes: Without the plankton, fish, and birds, the phosphorus would have remained in the ocean; and without the upwelling, the phosphorus would not have been available. Guano deposits were once major sources of phosphorus for fertilizers. In the mid-1800s, nearly 10 million metric tons per year of guano deposits were shipped to London from islands near Peru.

Phosphate mines are important. Today, most phosphorus fertilizers come from mining of phosphate-rich sedimentary rocks containing fossils of marine animals. Fish and other marine organisms extract phosphate from sea water and it is incorporated in their bones and teeth. The richest phosphate mine in the world is Bone Valley, east of Tampa, Florida. About 10

million years ago, Bone Valley was the bottom of a shallow sea where marine organisms lived and died.[16] Through tectonic processes, Bone Valley was slowly uplifted, and in the 1880s and 1890s phosphate ore was discovered there. Today, Bone Valley provides more than one-third of the world's phosphate production and three-fourths of U.S. production.

Total U.S. reserves of phosphorus are estimated to be about 2 billion metric tons, enough to supply our needs for several decades. However, if the price of phosphorus rises as high-grade deposits are exhausted, phosphorus from lower-grade deposits can be mined at a profit. Florida is thought to have several billion metric tons of phosphorus that can be recovered with existing mining methods if the price is right.[16]

Mining's effects on the environment. Mining, of course, may have negative effects on the land and ecosystems. For example, in some phosphorus mines, huge pits and waste ponds have scarred the landscape, damaging biologic and hydrologic resources. Balancing the need for phosphorus with the adverse environmental impacts of mining is a major environmental issue. Figure 3.17 shows some land being reclaimed at an open-pit phosphate mine in Florida, as mandated by law.

William Felger/Grant Heilman Photography

■ **FIGURE 3.17**

Large open-pit phosphate mine in Florida (similar to Bone Valley), with piles of waste material. The land in the upper part of the photograph has been reclaimed and is being used for pasture.

Return to the Big Question:

?

Why are biogeochemical cycles essential to long-term life on Earth?

Biogeochemical cycles are essential to long-term, sustainable life on Earth for several reasons. First, living things require many chemical elements, and these must be available at the right time, in the right amounts, and in the right concentrations relative to each other. This is the essence of, and the importance of, biogeochemical cycles.

Second, sustained life on Earth is a function of ecosystems, and a primary ecosystem process is the cycling of chemicals necessary for life. On a larger scale, chemicals, nutrients, and trace elements necessary for life are made available from Earth through various parts of biogeochemical cycles. For example, soil and rock release nutrients to plants through weathering and biochemical processes; water infiltrates rock and soil to emerge as springs and streams necessary for life. In the ocean, single-cell algae release a sulfide compound that oxidizes in the atmosphere, producing condensation nuclei that are necessary to form clouds that transport water and solfur to the land.

Third, chemical reactions in biogeochemical cycles determine whether chemical elements and necessary compounds are available to living things. Photosynthesis involves the availability of carbon dioxide in the carbon cycle. Green plants use the carbon dioxide with sunlight and water to produce sugar. A by-product is oxygen, which is why we have free oxygen in the atmosphere. Without the carbon and water cycle, none of this would happen and life as we know it wouldn't be possible.

Summary

■ Biogeochemical cycles are the major way in which elements important to Earth processes and to life are moved through the atmosphere, hydrosphere, lithosphere, and biosphere.

■ Biogeochemical cycles can be described as a series of pathways, or fluxes, linking storage compartments.

■ In general, some chemical elements cycle quickly and are readily regenerated for biological activity. Elements whose biogeochemical cycles include a gaseous phase in the atmosphere tend to cycle more rapidly.

■ Our modern technology has begun to alter and transfer chemical elements in biogeochemical cycles at rates comparable to those of natural processes. Some of these activities are beneficial to society, but others pose dangers.

■ To be better prepared to manage our environment, we must recognize both positive and negative consequences of activities that transfer chemical elements, and we must deal with them appropriately.

■ Every living thing, plant or animal, requires a number of chemical elements, and these chemicals must be available at the appropriate time and in the appropriate form and amount.

■ Chemicals can be reused and recycled, but in any real ecosystem some elements are lost over time and must be replenished if life in the ecosystem is to persist.

■ Change and disturbance of natural ecosystems are the norm. A steady state, in which the net storage of chemicals in an ecosystem does not change over time, does not generally occur in nature.

■ There are many uncertainties in measuring either the amount of a chemical in storage or the rate of transfer between reservoirs. For example, the global carbon cycle includes a large sink that scientists have not yet been able to accurately define.

Key Terms

average residence time

biogeochemical cycle

carbon cycle

chemical reaction

flux

geologic cycle

hydrologic cycle

limiting factor

missing carbon sink

nitrogen cycle

phosphorus cycle

sulfur cycle

sink

source

Getting It Straight

1. Why is an understanding of biogeochemical cycles important in environmental science? Explain your answer using two examples.

2. What four main parts of the Earth system do all biogeochemical cycles involve?

3. Differentiate between *sink* and *flux* and give an example of each.

4. What are the "big six" elements that form the fundamental building blocks of life?

5. What factors determine whether a chemical element will cycle quickly or relatively slowly through a biogeochemical cycle?

6. What are the environmental effects of plate tectonics?

7. List the major storage compartments of water.

8. Name and describe the three different classifications of rock.

9. Identify the major processes of the carbon cycle and the environmental concerns associated with it.

10. What are the differences in the geochemical cycles for phosphorus and nitrogen, and why are the differences important in environmental science?

What Do You Think?

1. Sulfur forms several gases, including sulfur dioxide which is a major air pollutant and component of acid rain. Vehicle emissions are a source of sulfur dioxide. Do you think people who drive big trucks and SUV's which contribute more sulfur dioxides to the atmosphere should pay a usage tax to drive these vehicles? Explain your answer.

2. Although nitrogen and phosphorus are required for life, nitrogen and phosphorus in agricultural runoff is a potential source of water pollution. Do you think farmers who gross a certain amount of money or who raise a certain amount of animal or plant weight should be required to run their operations using a nutrient management plan that would address both nitrogen and phosphorus inputs? Explain your answer. Who would be responsible for holding farmers accountable under such a requirement?

3. Today, most phosphorus fertilizers come from mining of phosphate-rich sedimentary rocks containing fossils of marine animals, yet the process of mining has many negative environmental effects. How would you propose to balance the need for essential elements such as phosphorus against the adverse environmental impacts of mining?

Pulling It All Together

1. You are a member of a new Environmental Awareness Committee in your city. The committee is working on a public environmental awareness program and you have been given the task of developing a section of the program that deals with the environmental issues associated with biogeochemical cycles. How would you develop a plan to educate the public? What kinds of environmental concerns would you include in your awareness program? What practical suggestions could you offer the public to address these concerns?

2. Imagine you are a lawmaker who is going to introduce a bill that would require farmers in your area to utilize a nutrient management plan. How would you go about gathering information to support your claim that this requirement is warranted? How would you decide what to include in the plan? What would you propose to be the major components of the plan? Outline the elements of the plan including relevant information about the adverse effects of agricultural runoff to the environment. How would farmers be held accountable for their compliance in using such a plan?

3. Pretend you are suddenly in charge of the world. What are the most important actions you would take to manage each of the following major biogeochemical cycles: Carbon, Nitrogen, Phosphorus, Hydrologic? Consider the adverse effects of human activities on each of these biogeochemical cycles as you answer this question.

Further Reading

Berner, R. A., and E. K. Berner. 1996. *Global environment: Water, air, and geochemical cycles.* Upper Saddle River, NJ: Prentice-Hall.—A good discussion of environmental geochemical cycles, focusing on Earth's air and water systems.

Kasting, J. F., O. B. Toon, and J. B. Pollack. 1988. How climate evolved on the terrestrial planets. *Scientific American* 258(2):90–97.—This paper provides a good discussion of the carbonate–silicate cycle and why it is important in environmental science.

Lerman, A. 1990. Weathering and erosional controls of geologic cycles. *Chemical Geology* 84:13–14.—Natural transfer of elements from the continents to the oceans is largely accomplished by erosion of the land and transport of dissolved material in rivers.

Post, W. M., T. Peng, W. R. Emanual, et al. 1990. The global carbon cycle. *American Scientist* 78:310–326.—The authors describe the natural balance of carbon dioxide in the atmosphere and review why the global climate hangs in the balance.

Schlesinger, W. H. 1997. *Biogeochemistry: An analysis of global change*, 2nd ed. San Diego: Academic Press.—This book provides a comprehensive and up-to-date overview of the chemical reactions on land, in the oceans, and in the atmosphere of Earth.

Altrendo/Getty Images

Ecosystems

Big Question

**What Is Necessary
to Sustain Life on Earth?**

Learning Objectives

Life on Earth is sustained by ecosystems, which vary greatly but have certain features in common. After reading this chapter, you should understand . . .

- why the ecosystem is the basic system that supports life and enables it to persist;
- what food chains, food webs, and trophic levels are;
- how energy enters ecosystems and determines biological productivity;

- what a community-level effect is;
- what ecosystem management involves;
- how conservation and management of the environment might be improved through ecosystem management.

Case Study

The Acorn Connection

Lyme disease has become the most common tick-borne disease in the United States.[1, 2] How it spreads is a complicated ecological story, which begins with white-footed mice who live in hardwood forests of the eastern states, where one of their primary foods is acorns, rich in proteins and fats (Figure 4.1).

The mice carry the microorganisms responsible for Lyme disease in their bloodstreams. As the tick larvae feed on the blood of the mice, they pick up these microorganisms. Later in their life cycle, mature infected ticks attach to other animals, including deer and people. Deer deposit ticks on plants when they brush against them, and the ticks may then attach themselves to people who brush against the plants as they walk past. If an infected tick bites a person, the person may contract Lyme disease. It's all part of a forest food web, the web of who eats whom—in this case, who eats acorns, and who feeds on mice, deer, and people.

When acorns are abundant, the mice become abundant too. The amount of light and rain, the temperature patterns over the year, and the quality of the soil affect the production of acorns. The number of acorns produced varies from year to

(c)   O. Spielman/CNRI/Phototake   (e)   Runk/Schoenberger/
Grant Heilman Photography

(a)   Rick Seeney/iStockphoto

(b)   Bill Ivy/Ivy Images   (d)   David Meharey/iStockphoto   (f)   David Hughes/iStockphoto

■ **FIGURE 4.1**
The acorn connection.
The tick (c) that carries Lyme disease feeds on both the (b) white-footed mouse and the (d) white-tailed deer. Oak leaves (f) are an important food for the deer and for (e) gypsy moths, while (a) oak acorns are important food for the mouse. But the mouse also eats the moths. The more mice, the fewer gypsy moths, but the more ticks.

year, with "mast" years—years of high production (bumper crops)—occurring occasionally. In the years between bumper crops of acorns, mice populations decline. With the next bumper crop, many acorns survive to become oaks, the mouse population increases, and so does the number of ticks carrying Lyme disease.

The amount of forested land has increased in the eastern United States in the past century, so mice, deer, and ticks have become more abundant. Why has forest area increased? In colonial times, European settlers cleared the forests of the northeastern United States to make space for farming and settlements, to provide fuel, and to provide timber for commercial uses. As coal, oil, and gas replaced wood as a primary fuel, and as farming moved westward to the more fertile Great Plains, fields that had been cleared were abandoned. In many areas, the clearing of land peaked around 1900. Since then, forests have grown back.

The story gets even more complicated, because in addition to feeding on acorns (and other seeds), mice feed on insects, including larvae of the gypsy moth. Gypsy moth larvae feed on leaves of trees and are particularly fond of oak leaves. Studies suggest that in years when mouse populations are low—the years between bumper crops of acorns—gypsy moth populations can increase dramatically. During these periods, gypsy moth larvae can virtually denude an area, stripping the leaves from the trees. Oaks that have lost most or all of their leaves may not produce bumper crops of acorns.

Once the leaves are off the trees, more light reaches the ground, and seedlings of many plants that could not flourish in deep shade begin to grow. As a result, these other species of trees may gain a foothold in the forest and change its species composition (the number and kinds of species that exist in a particular forest). Of course, the next generation of gypsy moth larvae find little to eat, and the population of gypsy moths begins to decline again.

Abundant acorns draw deer into the woods, where they browse on small plants and tree seedlings. Ticks drop off the deer and lay eggs in the leaf litter. When the eggs hatch, the larvae attach to mice, and the cycle of Lyme disease continues. Deer do not eat ferns, however, so in areas where deer populations are dense, you will find many ferns but few wildflowers and tree seedlings.

As you can see, a person's chances of getting Lyme disease vary with weather, mast years, and changes in the abundances of oak trees, acorns, and white-footed mice. These also affect the abundance of deer, which help spread Lyme disease, and gypsy moths, which are another problem for people and for the forests.[1]

The acorn connection illustrates many of the basic characteristics of ecosystems and ecological communities. Everything is connected to everything else, and you can't change just one thing without affecting the others. The acorn connection shows us that life exists within an ecosystem—that is, within a set of local interacting species and their local environment. As we will learn in this chapter, an ecosystem is the basic unit of sustained life. *Life is sustained within and by ecosystems, not by individuals or single species.* But in the past, before ecosystems were understood, people took a "single-factor" approach to solving environmental problems. The acorn connection shows us that the single-factor approach does not always work; instead, we have to think about the entire system.

All of the living parts of the oak forest—called its "ecological community"—depend on the nonliving parts of the ecosystem for their survival: the water, soil, air, and the light that provides energy for photosynthesis. Members of the ecological community affect the nonliving parts of the ecosystem. When gypsy moths denude an area, for example, more sunlight can reach the forest floor. Relationships among the members of the ecological community are dynamic (constantly changing). Our chances of contracting Lyme disease depend on things beyond our control (such as changes in rainfall) and things that we do. To control Lyme disease, we need to understand how individual populations change over time, how they interact, and how they connect to their environment.

Managing something, whether a company or a forest, always involves trade-offs. In this case, if we manage the forest to protect people against Lyme disease, we increase the likelihood of gypsy moth damage. It seems as if you just can't win. But knowledge of ecosystems can help you win, or at least make your loss less harsh.

4.1 How Populations Change Over Time and Interact with Each Other

The acorn connection raises questions about how the size of a population changes over time, and what causes those changes. The answers to the questions are important to many environmental issues that we discuss throughout this book, including how to conserve endangered species, how to manage forests, fisheries, and wildlife, and how pollutants affect life. The first question we need to answer is: How and why does the abundance of a species change even without human influence? The answer will give us a baseline for determining our effects on species.

Competition, symbiosis, and predation/parasitism. One way that two species may interact with each other is by competing for resources. Competition is negative for both. A second way is to form a symbiotic relationship, in which each benefits the other. The third relationship, between predator and prey, or parasite and host, benefits one and harms the other.

Would nature remain in balance if we didn't interfere? Before modern science, the general belief was that all populations of all species were constant, creating a great balance of nature, with each species in its place, each following its role in the workings of the world, and each just abundant enough to do its job without doing serious harm to the other populations. Compare that idea with some of the longest observations of wildlife in an undisturbed ecosystem: the wolves and moose of Isle Royale National Park, Michigan (Figure 4.2). These observations are especially interesting because, globally, Isle Royale is one of the places least disturbed by human

influences, and because these are among the longest direct observations of two large mammals ever made, and also because the mammals being observed are predator and prey, a relationship whose interactions have long fascinated people.

More wolves, fewer moose. Fewer moose, fewer wolves, and more moose! Since the 1950s, scientists, first led by Durwood Allen of Purdue University, have studied Isle Royale's animal populations and plant-animal interactions. The wolf population has been counted since 1960, during the winter, when observers in light aircraft can easily see the animals against the snow and through the bare trees. The moose population appears to have increased from 1960 until the early 1970s, when it declined, only to increase again in the late 1980s, then decline slightly and increase once more in the mid-1990s. The wolf population appears to have grown or declined only slightly until the early 1970s, when it increased, though lagging behind the increase in the moose. When the wolf population was high in the late 1970s and early 1980s, the moose population declined. Then the wolf population declined, and the moose population increased.

The populations of wolves and moose change over time. Even without human interference, the populations are not constant. Do the two populations affect each other? One interpretation is that as the moose population increases, there is more food for the wolves, which then increase too. Once the wolf population has increased, there are too many wolves for the number of moose, which causes the moose population to decline. Perhaps the wolves are eating more moose than can be born, and thus are the cause, or at least one cause, of the decline of the moose. These are some simple inferences from the graphs, which are interesting to speculate about and have intrigued many ecologists for decades.

Do predators control the populations of their prey? This is one of the longest debates in ecology. In fact, the question long predates modern science. The ancient Greeks and Romans asked the same question, phrased a little differently: Why are vile and venomous creatures found on the Earth along with the benign and beneficial? The Greeks came up with an answer that persisted into modern times: that predators were put here to limit the population sizes of their prey. According to this, wolves and lions had a purpose—to keep the populations of deer, gazelles, sheep, etc., from becoming too abundant for their habitats, since these creatures were so productive that they could not control their own populations.

Ecologists agreed with this theory at first. About a hundred years ago, in the first decades of the 20th century, ecologists thought they had found the answer too,

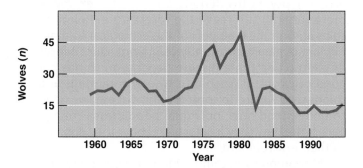

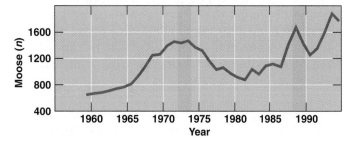

▪ **FIGURE 4.2**
The numbers of wolves and moose at Isle Royale National Park, Michigan, since 1960.
One of the longest scientific measurements of animal abundance. [*Source:* B. E. McLaren and R. O. Peterson. Wolves, moose, and tree rings at Isle Royale. *Science* 226:1555–1556, 1994.] Moose and wolves were not always on Isle Royale. Moose arrived there at about the turn of the 20th century, probably swimming from the mainland, and wolves arrived in the 1940s, probably running over the ice.

and it was the same answer the ancient Greeks had found: that predators had an ecological role, which was to control the populations of their prey. This was formalized in a mathematical theory named for its two originators, the Lotka-Volterra equations for predator-prey interactions. The prey supposedly grew exponentially in the absence of their predators, while the predators died off at a negative exponential rate—that is, a constant percentage of the population died in every time period (see Chapters 1 and 2) without any prey to eat. Together, the two regulated each other. Either both achieved a constant, unchanging abundance, or both oscillated regularly, becoming abundant and then less so and then abundant again, like clockwork, forever.

But is that what happened to the mule deer on the Kaibab Plateau? Whatever predators do or do not do was and is important to conservationists and wildlife managers. The issue became the focus of a widely known controversy in American conservation during the first decades of the 20th century due to a huge die-off of mule deer on the Kaibab Plateau, the edge of which forms part of the north rim of the Grand Canyon. According to an account made famous by the great American conservationist Aldo Leopold, the decline was the result of an earlier population explosion of the deer, during which these browsing animals destroyed the trees and shrubs that they depended on for food. Having destroyed much of their food, the deer starved and the population crashed, Leopold said, and he blamed the problem on "overcontrol" of the deer's major predator, the North American mountain lion. He believed that the mountain lion had kept the population of the deer in check, so that the two species had existed in a natural balance.[3]

Later ecologists reinvestigated this history, however, and found that other factors could have swelled the deer population. One was that sheep that had grazed on the plateau had mostly been removed, and without these competitors the deer population grew. Another factor was that some major fires led to regrowth of vegetation the deer preferred to eat.

So the answer to the question about the role of predators is, it depends. In some cases and some situations, big-game predators may reduce the abundance of their prey, and even affect the fluctuations in the size of the prey population. But not always.

The answer is clearer for parasites. It is well known that disease organisms can have huge effects on their prey. The Black Death is believed to have reduced the human population of Europe in the Middle Ages by one-third. Chestnut blight essentially eliminated the American chestnut from the forests of eastern North America.

In part the difference between parasites and predators is their longevity. Parasites that have a very short life compared to their hosts reproduce rapidly and die off rapidly, and thus can adjust quickly to changes in their hosts' abundance and "regulate" their hosts' population. But big-game predators that live about as long as their prey cannot increase fast enough to take advantage of a rapid population explosion in their prey, nor can their population easily decline when the prey becomes scarce.

In sum, whether predators control the population of their prey is still an active area of debate and research in ecology, with many implications for conservation and management of our living resources.

4.2 Professions and Places: The Ecological Niche and the Habitat

And what about the moose population without wolves? Would it be constant, perhaps, growing according to the logistic curve (discussed in Chapter 2) to a fixed carrying capacity, in balance with its vegetation food supply? This brings up questions about how a population of one species is affected by its habitat and is connected to that habitat. It takes us to the concept of the ecological niche.

What is a habitat, and what is a niche? Where a species lives is its **habitat**, but what it does for a living (its profession) is its **ecological niche**. Suppose you have a neighbor who is a bus driver. Where your neighbor lives and works is your town—that's his habitat. Driving a bus is his niche. If someone points to a picture and says "Look at the wolf," you think not only of a creature that inhabits the northern forests (its habitat) but also of a predator that feeds on large mammals (its niche).

Will a change in land use affect a species' niche? It's easy to damage a species' habitat so that the requirements of its niche are no longer available. Understanding the niche of a species helps us assess how a change in land use could affect it. A new highway that makes travel easier may eliminate your neighbor's bus route (an essential part of his habitat) and thereby eliminate his profession (his niche). Other things could also eliminate his niche. Suppose a new school were built so that all the children could walk to school. Then a bus driver would not be needed—his niche would no longer exist in your town. In the same way, cutting a forest may drive away prey and eliminate the niche of the wolf.

Measuring Niches

Can species share a niche? The distribution of two species of flatworm—a tiny worm that lives on the bottom of freshwater streams—illustrates some basic ideas

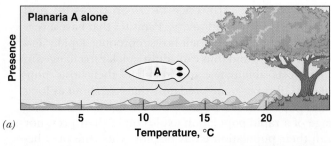

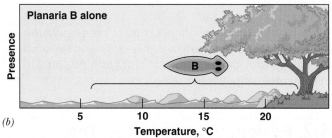

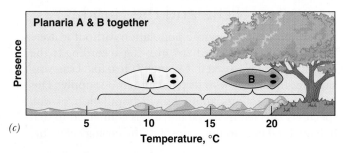

▪ **FIGURE 4.3**
The occurrence of freshwater flatworms in cold mountain streams in Great Britain.

about the ecological niche. A study of two species of these small worms in Great Britain found that some streams contained one species, some the other, and still others contained both[4,5] (Figure 4.3).

Temperature is an important factor. The streams are cold at their source in the mountains and get warmer and warmer as they flow down. Each species of flatworm lives within a specific range of water temperatures. When the two species live in the same stream, Species A lives in the colder upstream sections, while Species B lives in the warmer downstream areas (Figure 4.3c).

Competition between the two species affects where each can persist. If Species B isn't around, Species A will live within a greater range of water temperatures (Figure 4.3a). The temperature range in which Species A occurs when it has no competition from B is called its *fundamental temperature niche*. The temperature range it is confined to when it must share its habitat with Species B is called its *realized temperature niche*. Studying these flatworms shows us that species can divide up a habitat so that each uses resources from different parts of it.

Of course, temperature is only one aspect of the environment. Flatworms also have requirements in terms of the acidity of the water and other factors. We could draw graphs for each of these factors, showing the range within which Species A and Species B live.

Practical implication: Manage habitats and conserve niches. From the discussion above, you have learned something important about the conservation of species. If we want to conserve a species in its native habitat, we must make sure that all the requirements of its niche are present. Conservation of endangered species is more than a matter of putting many individuals of that species into an area. All the life requirements for that species must also be present—we have to conserve not only a population but also its habitat and its niche.

4.3 The Competitive Exclusion Principle

Complete competitors cannot coexist.[6] The flatworms show us that competing species have negative effects on each other's distribution. But how strong can that effect be? Scientists have learned from observations that two species that compete for the same life resources and have exactly the same requirements cannot coexist in exactly the same habitat—one will always go extinct. This is known as the **competitive exclusion principle**. From this we might expect that over a very long time, fewer and fewer species would survive. Taking this idea to its logical extreme, we could imagine Earth with very few species—perhaps one green plant on the land, one herbivore to eat it, one carnivore, and one decomposer. If we added four species for the ocean and four for freshwater, we would have only 12 species on our planet.

Being a little more realistic, we could take into account adaptations to major differences in climate and other environmental aspects. Perhaps we could specify 100 environmental categories: cold and dry, cold and wet, warm and dry, warm and wet, and so forth. Even so, we would expect that within each environmental category, competitive exclusion would result in the survival of only a few species. Allowing four species per major environmental category would result in only 400 species. Yet more than a million and a half species have been named, and scientists think many more millions may exist—so many that we do not have even a good estimate. How can they all coexist?

4.4 How Species Coexist

Experiments with flour beetles show how competitors can coexist. As their name suggests, flour beetles (*Tribolium*) live on wheat flour. They make good experimental subjects because they require only small containers of wheat flour to live and are easy to grow (in fact, too easy; if you don't store your flour at home properly, you will find these little beetles happily living in it).

The flour beetle experiments work like this.[7] A specified number of beetles of two species are placed in small containers of flour—each container has the same number of beetles of each species. The containers are then kept at various temperature and moisture levels—some are cool and wet, others warm and dry. Periodically, the beetles in each container are counted. This is very easy. The experimenter just puts the flour through a sieve that lets the flour pass through but not the beetles. Then he or she counts the beetles of each species and puts them back in their containers to eat, grow, and reproduce for another interval.

Eventually, one species always wins—some of the winning team continue to live in the container while the other team goes extinct. You may conclude that by now there should be only one species of *Tribolium* left. But which species survives depends on temperature and moisture. One species does better when it is cold and wet, the other when it is warm and dry (Figure 4.4). Curiously, when conditions are in between, sometimes one species wins and sometimes the other. But either way, one persists while the second becomes extinct. Both species can survive in a complex environment—one that has cold and wet habitats as well as warm and dry habitats. They cannot exist together in exactly the same location, but they can coexist by dividing up the habitat into different quantitative ranges, just as the flatworms do in streams.

4.5 Symbiosis

Our discussion up to this point may have given you the impression that species interact mainly through **competition**—by interfering with one another. But species are often necessary for each other's survival. The term **symbiosis**, derived from a Greek word meaning "living together," describes a relationship between two organisms that is beneficial to both and enhances each organism's chances of persisting. Each partner in symbiosis is called a *symbiont*.

Symbiosis is widespread and common; most animals and plants have symbiotic relationships with other species. We humans have

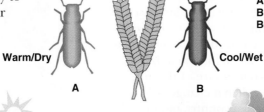

Two species of *Tribolium*

Warm/Dry A B Cool/Wet

A: Likes warm, dry conditions
B: Likes cool, wet conditions
Both: Like to eat wheat

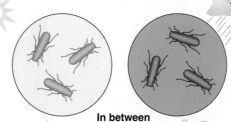

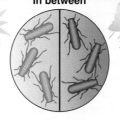

In a *uniform* environment, one will win out over the other. If the environment is warm and dry, A will win; if it is cool and wet, B will win.

In between

In a mixed environment, the beetles will use separate parts of the habitat.

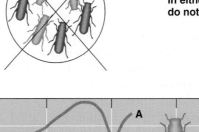

In either case, the beetles do not coexist.

(a)

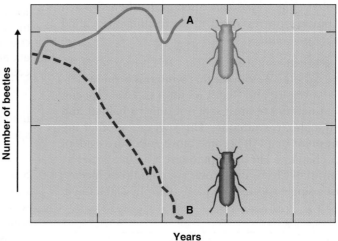

Number of beetles

Years

(b)

■ **FIGURE 4.4**

Competitive exclusion principle proven.
A classic experiment with flour beetles. (a) The general process illustrating competitive exclusion in these species; (b) results of a specific, typical experiment under warm, dry conditions. Two species of flour beetles are introduced into the same habitat. One always wins—the other goes extinct. But which one survives depends on the environmental conditions.

and are **symbionts**—microbiologists tell us that about 10% of a person's body weight is actually the weight of symbiotic microorganisms that live in the intestines. The bacteria help our digestion, and we in turn provide a habitat that supplies all their needs; thus both we and they benefit. Many of these are helpful to us but not absolutely necessary. We become aware of this intestinal community and its benefits when it changes—for example, when we travel to a foreign country and ingest new strains of bacteria. Then we suffer a well-known traveler's malady, gastrointestinal upset.

Some symbionts simply can't live without each other. An important kind of symbiotic interaction occurs between certain mammals and bacteria. An elk carries with it many companions. Like domestic cattle, the elk is a ruminant, with a four-chambered stomach (Figure 4.5) teeming with microbes (a billion per cubic centimeter). In this partially closed environment, certain species of bacteria digest woody tissue that the elk ingests but cannot digest by itself. Other bacteria in the elk's stomach take nitrogen from the air and convert it to organic compounds, essentially making amino acids. Some bacteria give off fatty acids that are also good food for the elk.

Many of these intestinal bacteria cannot live in the high-oxygen atmosphere outside, so they depend on the elk, because the inside of a ruminant's stomach is one of the few places on Earth's surface where the peculiar environment they require exists.[14] The bacteria and the elk are symbionts: Each provides what the other needs, and neither could survive without the other. They are therefore called *obligate symbionts*. Species that benefit each other but are not essential to each other are called *facultative symbionts*.

A practical implication: You can't have one without the others. We can see that symbiosis promotes biological diversity, and that if we want to save a species from extinction, we must save not only its habitat and niche but also its symbionts. This suggests another important point that will become more and more evident in later chapters: The attempt to save a single species almost invariably leads us to conserve a group of species, not just a single species or a particular physical habitat.

4.6 The Community Effect

Species can affect one another directly through food webs, through competition, and by helping one another (symbiosis). But species can also affect one another indirectly by what is called a *community-level effect*. The term refers to what happens when one species affects a third, a fourth, or many other species that, in turn, affect the second species. In addition, a species can affect the non-

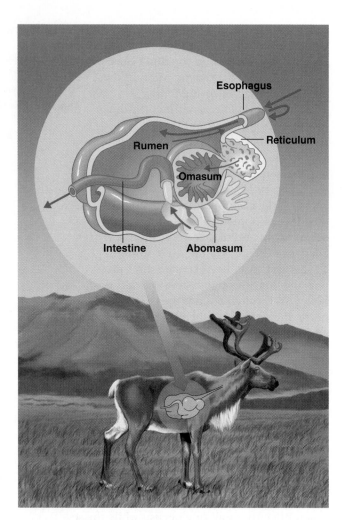

▪ **FIGURE 4.5**
An elk's stomach illustrates symbiosis.
In this stomach, bacteria digest woody tissue the elk could not otherwise digest. The result is food for the elk and food and a home for the bacteria, which could not survive in the local environment outside.

living environment, which then affects a group of species in the community.

Changes in that group affect another group. Sea otters of the Pacific Ocean offer a good example of interactions at the community level. In fact, the sea otters' community-level interactions are at the heart of some arguments in favor of conservation of this species.

Sea otters originally occurred throughout a large area of the Pacific Ocean coasts, from northern Japan northeastward along the Russian and Alaskan coasts, and southward along the coast of North America to Morro Hermoso in Baja California, to Mexico.[8] Sea otters feed on shellfish, including sea urchins and abalone (Figure 4.6). Because their fur is among the finest in the world, sea otters were hunted commercially. Hunters killed large numbers of them during the 18th and 19th

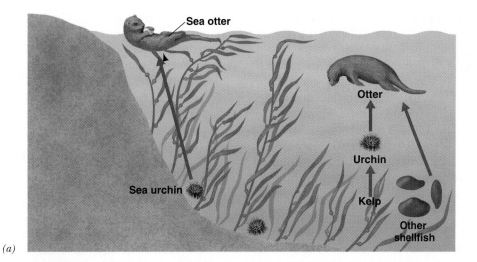

(a)

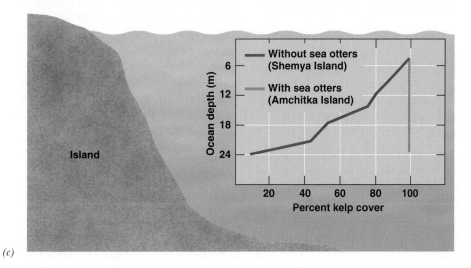

(b)

(c)

■ **FIGURE 4.6**

Sea otters and sea urchins.

(a) Sea otters affect kelp, but indirectly. Sea urchins feed on kelp, destroying kelp beds. (b) The otters eat the urchins, so when the otters are around, the kelp beds benefit. (c) In fact, with sea otters kelp covers the ocean at all depths, but without sea otters there is little kelp below 18 meters. Because kelp beds are home to many species, the sea otters indirectly benefit those species.

centuries. By the end of the 19th century, there were too few otters left for hunters to bother with, and the species was in danger of becoming extinct.

A small population survived and has increased since then, so that today there are approximately 111,000 sea otters. Approximately 2,000 sea otters live along the coast of California, a few hundred in Washington State and British Columbia, 100,000 along the Aleutian Islands of Alaska, and about 9,000 in Russian waters.[9] Legal protection of the sea otter by the U.S. government began in 1911 and continues under the U.S. Marine Mammal Protection Act of 1972 and the Endangered Species Act of 1973.

Otters are an example of "keystone species." The otters affect the community in many ways by feeding on sea urchins, their preferred food. The sea urchins feed on kelp—large brown algae that form underwater "forests" and provide important habitat for many species. Sea urchins graze along the bottoms of the kelp beds, feeding on the base of kelp, called *holdfasts,* which attach the kelp to the bottom. When holdfasts are eaten through, the kelp floats free and dies.

Where sea otters are abundant, as on Amchitka Island in the Aleutian Islands, kelp beds are abundant and there are few sea urchins (Figure 4.6b). At nearby Shemya Island, which lacks sea otters, sea urchins are abundant and there is little kelp (Figure 4.6c). Experimental removal of sea urchins has led to an increase in kelp.[8] Otters, then, benefit the kelp, but only indirectly. The otters don't care about the kelp, they just like to eat sea urchins. With fewer sea urchins, less kelp is destroyed. And with more kelp, there is more habitat for many other species. Thus, sea otters indirectly increase the diversity of species.

Such effects can occur through food chains and alter the distribution and abundance of individual species. A species such as the sea otter that has a large effect on its community or ecosystem is called a **keystone species**, or a key species. Its removal or a change in its role within the ecosystem changes the basic nature of the community.

The ecological community: a holistic view. Community-level effects demonstrate the reality behind the concept of an ecological community: They show us that certain processes can take place only when a set of species interact together. These effects also suggest that an ecological community is more than the sum of its parts—a perception called the *holistic view.*

How many otters are too many otters? The sea otter has been a focus of controversy and research. On the one hand, fishermen argue that the sea otter population has recovered—in fact recovered too well. They think there are too many sea otters today, and that they interfere with commercial fishing because they take large amounts of abalone. On the other hand, conservationists argue that sea otters play an important role at the community level, and in fact are necessary for the persistence of many oceanic species. They claim that actually there are still too few sea otters to perform this role adequately.

4.7 The Ecosystem: Sustaining Life on Earth

So far we have talked about species interacting in pairs, or, through the community effect, in larger groups, and we have discussed the dependence of species on characteristics of their environments. But there is a much more important connection between species and their environment: the ecosystem, which makes it possible for life to persist.

The oldest fossils are more than 3.5 billion years old, so life has persisted on Earth for an incredibly long time. We struggle to maintain threatened species for a decade or two and consider ourselves fortunate if we succeed. But for at least 4.5 billion years, life has been sustainable! What accounts for that?

Ecosystems are crucial to sustaining life. To understand how life persists on Earth, we have to understand ecosystems. We tend to think about life in terms of individuals, because it is individuals that are alive. But sustaining life on Earth requires more than individuals or even single populations or species. As we learned in Chapter 3, living things require 24 chemical elements, and these must cycle from the environment into organisms and back to the environment. Life also requires a flow of energy, as we will learn in this chapter. Although alive, an individual cannot by itself maintain all the necessary chemical cycling or energy flow. Those processes are maintained by a group of individuals of various species and their nonliving environment. We call that group and its local environment an **ecosystem.** Sustained life on Earth, then, is a characteristic of ecosystems, not of individual organisms or populations.[10]

4.8 Basic Characteristics of Ecosystems

Ecosystems have three fundamental characteristics: structure, processes, and change.

Structure. An ecosystem has two major parts: non-living and living. The nonliving part is the physical-chemical environment, including the local atmosphere, water, and mineral soil (on land). As we have already seen, the living part, called the **ecological community**, is the set of species interacting within the ecosystem.

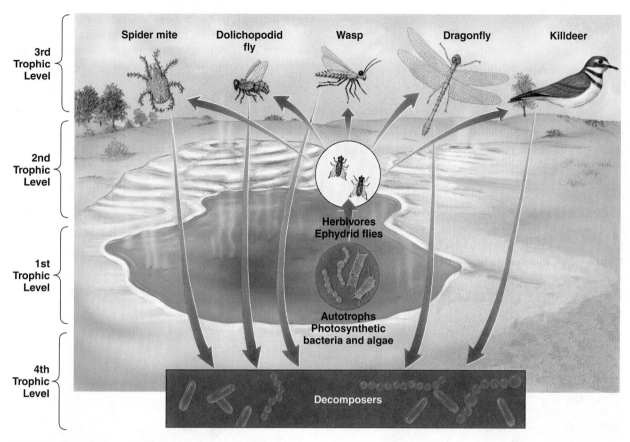

3rd Trophic Level — Spider mite, Dolichopodid fly, Wasp, Dragonfly, Killdeer

2nd Trophic Level

1st Trophic Level

Herbivores Ephydrid flies

Autotrophs Photosynthetic bacteria and algae

4th Trophic Level — Decomposers

■ **FIGURE 4.7**
Structure and function of an ecosystem.
Diagram of the food web and the trophic levels in a Yellowstone National Park hot spring.

Trees in a forest, grasses in a prairie, and kelp in an ocean create a biological structure that provides habitats for many species (Figure 4.6). The ecological community is a living part of an ecosystem, made up of individuals of a number of interacting species. The individuals interact by feeding on one another (*predation/parasitism*), by competing for resources (competition), and by helping one another (symbiosis). A diagram of who feeds on whom is called a **food web** (Figure 4.7). It shows us how chemicals cycle and energy flows within an ecological community. A food web is divided into **trophic levels**. A trophic level is all the organisms that are the same number of feeding levels away from the original source of energy.

Processes. Two basic kinds of processes must occur in an ecosystem: a cycling of chemical elements and a flow of energy. Related to this is the concept of *ecosystem function* which is the rates of chemical cycling and flow of energy.

Change. An ecosystem changes and develops through a process called *succession,* which is discussed in Chapter 6. How can an ecosystem be sustainable under such variable conditions?

To understand the idea of an ecosystem, it is helpful to consider one of the simplest ecosystems. Perhaps the simplest of all is a Yellowstone National Park hot spring that has among the fewest species (Figure 4.8).

■ **FIGURE 4.8**
One of Earth's simplest ecosystems: a hot spring at Yellowstone National Park. Peter Essiek/Getty Images

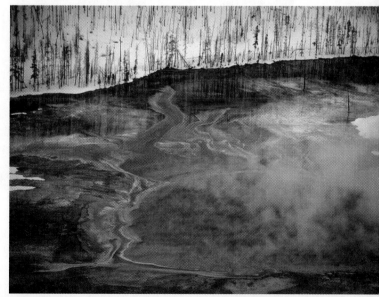

4.9 Food Webs

Food webs are even more complicated than they look. The introductory history of the acorn connection described one food web: acorns eaten by mice and deer, which in turn are food for a variety of carnivores, including people who hunt deer.

A diagram of a food web and its trophic levels seems simple and neat, as in the Yellowstone hot springs (Figure 4.8) or as usually shown for a forest ecosystem (Figure 4.9), but in reality food webs are complex, because most creatures feed on several trophic levels. For example, consider the food web of the harp seal (Figure 4.10). The harp seal is at the fifth trophic level.[11] It feeds on flatfish (fourth trophic level), which feed on sand lances (third level), which feed on euphausiids (second level), which feed on phytoplankton (first level). But the harp seal actually feeds at several trophic levels, from the

second through the fourth, so it feeds on predators of some of its own prey and thus is a competitor with some of its own food. (Note that a species that feeds on several trophic levels typically is classified as belonging to the trophic level above the highest level from which it feeds. Thus, we place the harp seal on the fifth level.)

4.10 Ecosystem Energy Flow

In a food web, energy and chemical elements are transferred up through trophic levels. All life requires energy, and the role of energy in life brings us to one of the most philosophical topics in ecology: life and the laws of thermodynamics.

Energy is the ability to do work, to move matter. Ecosystem energy flow is the movement of energy

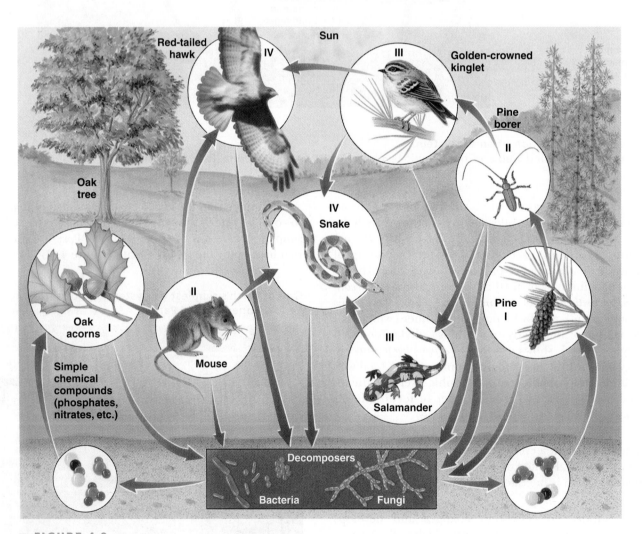

■ FIGURE 4.9
A food web in a forest ecosystem where mice eat acorns and there may be Lyme disease. The Roman numeral near the picture of an organism is its trophic level.

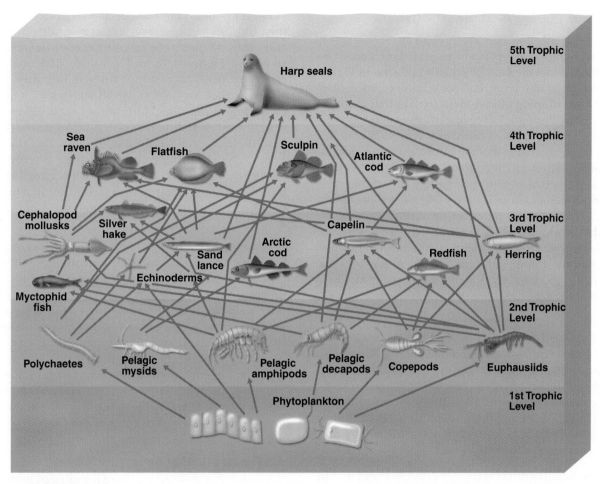

■ **FIGURE 4.10**
Food web of the harp seal.

through an ecosystem from the external environment, through a series of organisms, and back to the external environment (Figure 4.11). It is one of the fundamental processes common to all ecosystems. Energy enters an ecosystem when it is "fixed" by organisms—meaning that it is *put into and stored in organic compounds*. This fixation of energy is called **biological production**, which we will explain shortly.

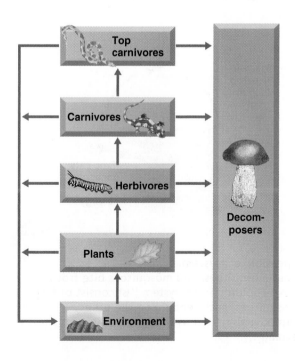

■ **FIGURE 4.11**
Energy pathways through an ecosystem. Usable energy flows from the external environment (the sun) to the plants, then to the herbivores, carnivores, and top carnivores. Death at each level transfers energy to decomposers. Energy lost as heat is returned to the external environment.

Life and the Laws of Thermodynamics

Energy can only flow one way through an ecosystem—it cannot be reused. If it could be reused, then it might be possible to have the kind of ecosystem shown in Figure 4.12, which would never require an input of energy but could keep running forever on its own. In that diagram, two facts are illustrated: Frogs eat insects, including mosquitoes, and mosquitoes bite frogs. Why then couldn't there be an ecosystem that was just frogs eating mosquitoes and mosquitoes eating frogs? This would be an ecological perpetual motion machine, which the laws of thermodynamics tell us is impossible, as we will explain here. (There are also other reasons that this ecosystem couldn't work, including that only female mosquitoes bite vertebrates to get proteins required for reproduction, and this biting, although a problem for us, is not the major food source for mosquitoes during their lifetimes.)

All life is governed by the laws of thermodynamics, which are fundamental physical laws about energy. There are three thermodynamic laws: (1) the conservation of energy, (2) the increase in entropy, and (3) what is usually called the "zeroth" law—the law of absolute zero temperature. We will discuss only the first two. These laws are important not only for the chapters in this book about ecosystems but also for the chapter on energy sources for human uses.

The law of conservation of energy states that in any physical or chemical change, energy is neither created nor destroyed but merely changed from one form to another. Here is a basic question that arises from the law of the conservation of energy: If the total amount of energy is always conserved—if it remains constant—then why can't we just recycle energy inside our bodies and why can't energy be recycled in ecosystems? Let us return to Figure 4.12 and consider that imaginary ecosystem consisting only of frogs, mosquitoes, water, air, and a rock for the frogs to sit on. Frogs eat insects, including mosquitoes. Mosquitoes suck blood from vertebrates, including frogs. In our imaginary ecosystem, the frogs get their energy from eating the mosquitoes, and the mosquitoes get their energy from biting the frogs (Figure 4.12). Such a closed system would be a biological perpetual-motion machine: It could continue indefinitely without an input of any new material or energy.

The law of entropy tells us that this is impossible. This second law of thermodynamics addresses how energy changes in form. It is a sad reality of our universe that *energy always changes from a more useful, more highly organized form to a less useful, disorganized form.* This means that energy cannot be completely recycled to its original state of organized, high-quality usefulness. Whenever useful work is done, heat is released to the environment, and the energy in that heat can never be completely recycled. The amount of usable energy gets less and less. For this reason, the mosquito-frog system will eventually stop working when not enough useful energy is left.

The net flow of energy through an ecosystem, then, is a one-way flow, from a source of usable energy to a place where heat can be released (Figure 4.11). *An ecosystem must lie between a source of usable energy and a sink for degraded energy (heat).* You can view the ecosystem as an intermediate system between the energy source and the energy sink. The energy source, ecosystem, and energy sink together form a thermodynamic system.

Producing New Organic Matter

Producing organic matter requires energy; organic matter stores energy. The total amount of organic matter in any ecosystem or area is called its *biomass*. Biomass increases through biological production (growth). Change in biomass over a given period is called *net production*. There are two kinds of biological production: primary and secondary.

Primary production. Some organisms make their own organic matter from a source of energy and inorganic compounds. These organisms are called *autotrophs* (meaning self-nourishing). The autotrophs include (1) green plants (plants containing chlorophyll), such as herbs, shrubs, and trees; (2) algae, which are usually found in water but occasionally grow on land; and (3) certain kinds of bacteria. The production carried out by autotrophs is called **primary**

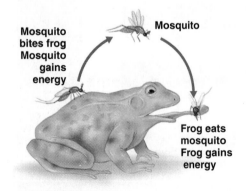

■ **FIGURE 4.12**

An impossible ecosystem.
Frogs eat mosquitoes, and mosquitoes bite frogs. Why then couldn't an ecosystem just consist of frogs and mosquitoes, each feeding on the other? The laws of thermodynamics tell us this is impossible, as the text explains.

production. Most autotrophs make sugar and oxygen from sunlight, carbon dioxide, and water in a process called *photosynthesis.*

Secondary production. Other kinds of life cannot make their own organic compounds from inorganic ones and must feed on other living things. These are called *heterotrophs*. All animals, including human beings, are heterotrophs, as are fungi, many kinds of bacteria, and many other forms of life. Production by heterotrophs is called **secondary production** because it depends on production by autotrophic organisms (Figure 4.11).

Living things use energy from organic matter through respiration. Once an organism has obtained new organic matter, it can use the energy in that organic matter to do things—to move, to make new kinds of compounds, to grow, to reproduce—or store it for future uses. In *respiration*, an organic compound combines with oxygen to release energy and produce carbon dioxide and water. The process is similar to the burning of fuels, like a fire in a fireplace, or gasoline burned in an automobile engine, but it takes place within cells at much lower temperatures with the help of organic chemicals called *enzymes*. Respiration is the use of biomass to release energy that can be used to do work. Complete respiration releases energy, carbon dioxide, and water into the environment. Incomplete respiration also releases a variety of organic compounds into the environment.

Gross and net production. The production of biomass and its use as a source of energy by autotrophs includes three steps:

1. An organism produces organic matter within its body.
2. It uses some of this new organic matter as a fuel in respiration.
3. It stores some of the newly produced organic matter for future use.

The first step, production of organic matter before use, is called *gross production*. The amount left over is called *net production.*

The difference between gross and net production is like the difference between a person's gross and net income. Your gross income is the amount you are paid. Your net income is what you have left after taxes and other fixed costs. Respiration is like the necessary expenses that are required in order for you to do your work.

Most primary production takes place through photosynthesis, which, as we have said, is the process by which sunlight, carbon dioxide, and water are combined to produce sugar and oxygen. Green plants, algae, and certain bacteria use photosynthesis.

Practical Implication I: Human Domination of Ecosystems

Many of Earth's ecosystems are dominated directly by human beings, and essentially no ecosystem in the oceans or on land is free of human influence. A recent study reached the following conclusions:[11]

▪ Human domination of ecosystems is not yet a global catastrophe, although serious environmental degradation has resulted.

▪ Earth's ecological and biological resources have been greatly modified by human use of the environment, and this modification and its impact are growing.

▪ An important human-induced alteration of Earth's ecosystems is land modification. Approximately 12% of the land surface of Earth is now occupied by agriculture (row crops such as corn, beans, or cotton) as well as urban-industrial uses. An additional 7% has been converted to pastureland. Although at first glance this may seem a small percentage of Earth's total land area, its impacts are large because much of Earth's land is not suitable for agriculture, pasture, or other urban uses.

Is there anything we can do to cause less damage? Having recognized that our activities can have significant global consequences for ecosystems, what can we do? First, we can reduce the rate at which we are altering Earth's ecosystems. This includes reducing the human population, finding ways of using fewer resources per person more efficiently, and better managing our waste. Second, we can try to better understand ecosystems and how they are linked to human-induced global change.

Practical Implication II: Ecosystem Management

Ecosystems can be natural or artificial or a combination of both. An artificial pond that is a part of a waste-treatment plant is an example of an artificial ecosystem. Ecosystems can also be managed, and management can include a large range of actions. Agriculture can be viewed as partial management of certain kinds of ecosystems (see Chapter 7), as can forests managed for timber production. Wildlife preserves are examples of partially managed ecosystems.

Sometimes, when we manage or domesticate individuals or populations, we separate them from their ecosystems. We also do this to ourselves (see Chapter 2). When we do this, we must replace the ecosystem functions of energy flow and chemical cycling with our own actions. This is what happens in a zoo, where we must provide food and remove the wastes for individuals separated from their natural environments.

The ecosystem concept is central to management of natural resources. When we try to conserve species or manage natural resources so that they are sustainable, we must focus on their ecosystem and make sure that it continues to function. If it doesn't, we must replace or supplement ecosystem functions ourselves.

Ecosystem management, however, involves more than just compensating for changes we make in ecosystems. It means managing and conserving life on Earth by considering chemical cycling, energy flow, **community-level interactions**, and the natural changes that take place within ecosystems.

Return to the Big Question

What is necessary to sustain life on Earth?

In this chapter we have discussed some of the basic features of ecosystems that make it possible for life to persist. We learned that certain kinds of interactions among species are necessary, in particular those across trophic levels—predation and parasitism that move energy and chemical elements up food webs. We learned that the ecosystem is the basic unit for the persistence of life, because life requires a cycling of chemical elements and a flow of energy, and this must take place among species and between species and their environment. The ecosystem is a set of species (an ecological community) and the local, nonliving environment.

One of the questions that will concern us most as we explore environmental issues about life on Earth is how a population's size changes over time. It is clear that populations do not remain constant, but change in abundance almost all the time. We want to be able to forecast changes in abundance, but to do that we need to know what factors cause populations to swell or shrink. We learned that populations respond to environmental change, and to changes in the abundance of other populations, especially changes in the abundance of species that one competes with or interacts with through predation/parasitism or symbiosis. We also learned that populations can change as a result of indirect interactions—community effects—the way sea otters affect kelp.

So the answer to the question of what is needed to sustain life on Earth is, in sum, an ecosystem with its chemical cycling and energy flow.

Summary

- Populations, once believed to be constant in abundance if people did not affect them, are now known to change continually.

- Populations of different species affect each other directly through competition, predation/parasitism, and symbiosis.

- Populations also affect each other indirectly through the community effect.

- An ecosystem is the simplest entity that can sustain life. At its most basic, an ecosystem consists of several species and a fluid medium (air, water, or both). The ecosystem must sustain two processes—the cycling of chemical elements and the flow of energy.

- Biological production is the production of new organic matter, which we measure as a change in biomass.

- In every ecosystem, energy flow provides a foundation for life.

- The living part of an ecosystem is the ecological community, a set of species connected by food webs and trophic levels. A food web or chain shows who feeds on whom. A trophic level consists of all the organisms that are the same number of feeding steps from the initial source of energy.

- Community-level effects result from indirect interactions among species, such as those that occur when sea otters reduce the abundance of sea urchins.

- Ecosystems are real and important, but it is often difficult to define the limits of a system or to pinpoint all the interactions that take place.

Key Terms

biological production
community-level interactions
competition
competitive exclusion principle
ecological community
ecological niche
ecosystem
energy flow
food web

habitat
keystone species
parasitism
predation
primary production
secondary production
symbiosis
symbionts
trophic level

Getting It Straight

1. What are some basic characteristics of ecosystems and ecological communities?
2. Define the terms *competition, symbiosis, predation/parasitism.*
3. What would be the impacts on nature if humans did not interfere?
4. What are the impacts of habitat destruction on ecological niches?
5. What is symbiosis and describe the benefits it has on organism sustainability.
6. Where do we fit in a food web? How many food webs are people involved in?

7. Which of the following are ecosystems? Which are ecological communities? Which are neither?
   a. Chicago
   b. a 1,000-acre farm in Illinois
   c. a sewage-treatment plant
   d. the Illinois River
   e. Lake Michigan
8. Is the wolf at Isle Royale National Park a keystone species?
9. What are the three fundamental characteristics of an ecosystem? Define and describe each.
10. Describe the two kinds of biological production.

What Do You Think?

1. Sea otters compete with fishermen for shellfish. Suppose a fisherman proposes a solution that he claims will work for both: Move all the sea otters to the Channel Islands National Park and its surrounding waters. Would this work? Explain your answer.
2. Describe your own ecological niche. Then describe the niche of a pet (if you have one) or of a specific domestic animal.
3. "Complete competitors cannot coexist," says the competitive exclusion principle.
   a. What does this mean?

   b. Do you think this applies to members of a football team? Explain.
4. You and a friend agree to treat the flour beetle experiments as a game, and you offer to bet your friend that you can predict which of two species will win. How would you modify the environment of the beetles so that you could be sure, or pretty sure, of winning?
5. What is "impossible" about the "impossible ecosystem" shown in 4.12? Explain.

Pulling It All Together

1. Refer to the introductory case study, the acorn connection. Suppose an expert on chemical pesticides says that the solution to Lyme disease is simple: Just use modern chemicals to kill the mites. Present arguments for and against this solution—why might it work and why might it not?

2. Develop a plan to bring back both deer and mountain lions to the Kaibab Plateau in a way that is sustainable for both species. Will this require active human management? Explain.

3. Three species of sea turtles—green, leatherback, and loggerhead—lay their eggs on the eastern beaches of South Florida. Use available research sources, including the Web, to learn more about the lives of these turtles. Then, explain how these three species are able to coexist in spite of the competitive exclusion principle.

4. Two professional baseball teams are considered competitors, but an ecologist points out that there is actually a lot of symbiosis going on between the teams. What could this ecologist mean? Explain your answer.

Further Reading

Molles, M. C. 2001. *Ecology: Concepts and applications*, 2nd ed. New York: McGraw-Hill.—A popular introductory college textbook.

Rickels, R. E., and G. Miller. 1999. *Ecology*, 4th ed. New York: W. H. Freeman.—Another introduction to ecology. Rickels' ecology texts have been among the most popular for the past 30 years.

Slobodkin, L. B. 2003. *A citizen's guide to ecology*. New York: Oxford University Press.—An introduction and overview of ecology by one of the pioneers of modern ecological research and thought.

Courtesy D. B. Botkin

5

Biological Diversity

Big Question ▶

Can We Save Endangered Species and Keep Biological Diversity High?

Learning Objectives

People have long wondered how the amazing diversity of living things on Earth came to be. This diversity has developed through biological evolution and is affected by interactions among species and by the environment. After reading this chapter, you should understand . . .

- what biological evolution means;
- how mutation, natural selection, migration, and genetic drift lead to the evolution of new species;
- why people value biological diversity;
- how people can affect biological diversity;

- what island biogeography is, and what it implies for the general geography of life, especially the geography of biological diversity;
- what we can do to reduce the rate of extinction.

Case Study

The Shrinking Mississippi Delta

Twice each year, America's longest bridge, the causeway over Lake Pontchartrain, a large body of water that surrounds New Orleans, Louisiana, becomes a favorite roosting place for hundreds of thousands of purple martins (Figure 5.1). This bird, a kind of swallow, stops at the lake and the surrounding wetlands to feed during migration to and from South America. In total, some 8 million purple martins pass through this area. What attracts them attracts many other species to this region: the highly productive wetlands that contain many kinds of habitats, including bald cypress swamps, home to alligators and great white egrets (Figure 5.2), treeless marshes, and open, flowing rivers and streams.[1]

Why is this area so rich in life of so many kinds? Part of the answer is that the Mississippi Delta (the area at the mouth of the river where the flowing water deposits silt, sand, gravel, and other material) contains the greatest coastal wetland in the lower 48 states, and the state of Louisiana has 40% of all the coastal wetlands in the lower 48 states. Frequent flooding and deposition of sediments by the Mississippi supply a renewable source of nutrient-rich soil. Another part of the answer is that subtle variations in elevation above and below the water table and variations in the soil promote high biological diversity. Land a meter above the water table could grow trees typical of dry climates; land frequently underwater provides habitat for specially adapted bog plants. And as you learned in Chapter 4, complex habitats allow many species to coexist.[2]

But like coastal wetlands throughout the United States, the Mississippi Delta wetlands have shrunk. The resulting loss of many kinds of habitats has reduced the abundance of many species and poses a threat to the diversity of life. The loss of wetlands stems partly from natural geological subsidence (lowering) of the land and a natural rise in the sea level, and partly from human actions, such as the building of levees and navigation channels on the Mississippi River. Since European settlement, 96% of the wetland forests of the lower valley of the Mississippi have been converted to agriculture or lost to U.S. Army Corps of Engineers shipping and flood-control projects. Shipping lanes alone have caused one-third of the wetland loss.[3] About 2 million hectares (5 million acres) are left, mostly in Louisiana, Mississippi, and Arkansas. But the wetlands are polluted by high levels of nitrogen from agricultural fertilizers flow-

▪ **FIGURE 5.1**
Purple martins roosting along America's longest bridge.

Associated Press

Photographs by D. B. Botkin, 2005

(a)

(b)

(c)

■ **FIGURE 5.2**

West Pearl River bald cypress swamp.
One of the many wetlands that form the delta of the
Mississippi River, this large forested swamp lies
alongside the Pearl River, which flows south into the
Gulf of Mexico. Not far is the Mississippi itself and
Lake Pontchartrain. (a) The bald cypress trees have
"knees"—so-called because they look like
aboveground bends in the roots—that provide
oxygen to the root tissue underwater. Among the
many species that live in this swamp are (b) alligators
and (c) great egrets.

ing off fields both nearby and throughout the Missouri–Mississippi River Valley.[2, 3]

The Mississippi River used to be the biggest land
builder in the world, carrying sediment from the combined Missouri–Mississippi River system and depositing
400 million tons in the delta each year—that's more
than enough to fill 33 million dump trucks. Today,
however, 13,000 kilometers of canals in Louisiana's
marshes transport the sediments through the wetlands and out into the Gulf of Mexico instead.

Various programs are attempting to improve the
situation. One, called "Coast 2050," has established a
prototype, or model, in Breton Sound, a body of water in the Gulf of Mexico just east of the Mississippi
Delta. This project has closed some canals and navigation channels so that sediments can once again be
deposited in the delta. Just three years after the project began, oysters in this area increased a thousand

times, marsh plants increased sevenfold, and there
are many more muskrats, alligators, and waterfowl.
However, the project is controversial. It is expensive—
the prototype at Breton Sound cost $26 million in
three years—and gulf-shrimp fishermen, lowland
farmers, shipping interests, and some towns along
the delta want to keep the levees and channels, and
oppose the restoration. And of course much of the
delta, including the Lake Pontchartrain Bridge, was
damaged by the hurricane of 2005, which therefore
threatened the purple martin's habitat.

*These delta wetlands restoration projects raise
the questions: Why does biological diversity matter?
Is it important to have many kinds of species? What
can we do to help endangered species persist?*

In this chapter we explore basic concepts of biological diversity. These provide a foundation for analyzing many issues covered here and later in this book.

5.1 What Is Biological Diversity?

Most simply, **biological diversity** is the wealth of species that live on Earth—a feature of our planet that has never ceased to amaze and impress people. Scientifically, biological diversity is commonly expressed as the number of species in an area, or the number of genetic types in an area. However, discussions about conserving biological diversity are complicated by the fact that people mean various things when they talk about it. They may mean conservation of a single rare species, of a variety of habitats, of the number of genetic varieties, of the number of species, or of the relative abundance of species. These concepts are interrelated, but each has a distinct meaning.

Newspapers and television programs frequently highlight the problem of disappearing species around the world and the need to conserve these species. But before we can intelligently discuss the issues involved in conserving the diversity of life, we must understand how this diversity came to be, and how ecological processes maintain it. Therefore, this chapter first discusses the principles of biological evolution. We then turn to biological diversity itself: its various meanings, how interactions among species increase or decrease diversity, and how the environment affects diversity.

5.2 Biological Evolution

An important question about biological diversity is, How did it all come about? This is a question that people have asked down through the ages. The diversity of life and the adaptations of living things to their environment seemed too amazing to have come about by chance. The only possible explanation seemed to be that this diversity was created by God (or gods). The great Roman philosopher and writer Cicero put it succinctly: "Who cannot wonder at this harmony of things, at this symphony of nature which seems to will the well-being of the world?" He concluded that "everything in the world is marvelously ordered by divine providence and wisdom for the safety and protection of us all."[4]

The evolution of new species. With the rise of modern science, other explanations presented themselves. In the 19th century, Charles Darwin proposed an explanation that became known as **biological evolution**. Biological evolution is the change in inherited characteristics of a population from generation to generation. Ultimately, it can result in new species—populations so different from the original species that the two populations can no longer reproduce with each other. Remember, we defined a species in Chapter 2 as all individuals capable of interbreeding. Along with self-reproduction, biological evolution is one of the features that distinguish living things from everything else in the universe.

The word *evolution* in the term *biological evolution* has a special meaning. Outside biology, *evolution* is used broadly to mean the history and development of something. Within biology, however, the term has a more specialized meaning. For example, geologists talk about the evolution of Earth, which simply means Earth's history and the geologic changes that have occurred over that history. Book reviewers talk about the evolution of the plot of a novel, meaning how the story unfolds. Biological evolution is a one-way process. Once a species is extinct, it is gone forever. You can run a machine, such as a mechanical grandfather clock, forward and backward, but when a new species evolves, it cannot evolve backward into its parents.

What causes evolution? According to the theory of biological evolution, new species arise as a result of (1) competition for resources and (2) the difference among individuals in their adaptations to environmental conditions. Because the environment continually changes, which individuals are best adapted changes too. As Darwin wrote, "Can it be doubted, from the struggle each individual has to obtain subsistence, that any minute variation in structure, habits, or instincts, adapting that individual better to the new [environmental] conditions, would tell upon its vigor and health? In the struggle it would have a better chance of surviving"[5]—as would its offspring, and theirs.

Sounds plausible, but how does this evolution occur? Four processes lead to evolution: *mutation, natural selection, migration,* and **genetic drift.**

Mutation

DNA: What it is and how it works. The chemical compound called **deoxyribonucleic acid (DNA)** carries inherited information from one generation of cells to the next. A DNA molecule carries information for many characteristics. This information is in smaller chemical units on the DNA strand. The chemical information for a single characteristic is called a **gene**, and the genetic makeup of an individual or group is called a **genotype**.

When a cell divides, the DNA is reproduced and each new cell gets a copy. Sometimes an error in reproduction changes the DNA and therefore changes inherited characteristics. Sometimes an external agent—including some pollutants, viruses, and radiation, such as X-rays and gamma rays—comes in contact with DNA and alters it.

When DNA is altered, it is said to have undergone *mutation.* In some cases, a cell or offspring with a mutation cannot survive. In other cases, the mutation simply adds variability to the inherited characteristics. But in still other cases, individuals with mutations are so different from their parents that they cannot reproduce with members of the original species, but only with others like themselves. These individuals are a new species.

Natural Selection

Change is not always for the better. Mutation can result in a new species whether or not that species is better adapted to the environment than its parent species is. And mutation can result in changes in individual offspring that may or may not be beneficial. Our knowledge of genetically linked diseases demonstrates that sometimes mutations lead to undesirable characteristics.

As a result of mutation there is variation within a species and some individuals may be better suited to the environment than others. Organisms whose biological characteristics make them better able to survive and reproduce in their environment leave more offspring than others. Their offspring are more "fit" for the environment and form a larger proportion of the next genera-tion. This process of increasing the proportion of better-adapted offspring is called **natural selection**. Which inherited characteristics lead to more offspring depends on the specific characteristics of an environment. As the environment changes over time, the "fit" characteristics will also change.

Migration

Geographic isolation can lead to divergent evolution. Sometimes two populations of the same species become geographically isolated from each other for a long time. During that time, the two populations may change so much genetically that they can no longer reproduce together even when they are brought back into contact. In this case, two new species have evolved from the original species. It can happen even if the genetic changes result in offspring that are not more fit but simply different enough to prevent interbreeding. This can lead to *divergent evolution*—that is, the two (or more) new species continue to evolve differently. Geographic isolation led to the divergent evolution of three large, flightless birds with similar niches on three different continents: the ostrich of Africa; the rhea of South America; and the emu of Australia (Figure 5.3).

Migration is an important evolutionary process over large areas and long times. For example, during intervals between recent ice ages and at the end of the

Ferrero/Labat/Auscape
International Pty. Ltd.

Toni Angermayer/Photo
Researchers

Gary Unwin/iStockphoto

(a)   *(b)*   *(c)*

▪ **FIGURE 5.3**
Divergent evolution.
These three large, flightless birds evolved from a common ancestor but are now found in widely separated regions: (a) the ostrich in Africa; (b) the rhea in South America; and (c) the emu in Australia.

last ice age, Alaska and Siberia were connected by a land bridge that permitted the migration of plants and animals. When the land bridge was closed off by a rising sea level, populations that had migrated to the New World were cut off, and new species evolved. The same thing happened much longer ago when marsupials reached Australia and were then cut off from other continents, leaving this group of animals as the sole mammalian inhabitants of that continent until much, much later.

Genetic Drift

Sometimes evolution results simply from chance. The term **genetic drift** refers to changes in the genetic makeup of a population due not to mutation, natural selection, or migration but simply to chance. Chance may determine which individuals in a population become isolated in a small group. They may not be better adapted to the environment—in fact, they may be more poorly adapted or neutrally adapted (neither better nor worse).

Genetic drift can be a problem for rare or endangered species, for two reasons: First, characteristics that are less well adapted to existing environmental conditions may dominate, making survival of the species less likely. Second, the small size of the population reduces genetic variability and thus reduces the ability of the population to adapt to future changes in the environment.

Biological Evolution in Action Today: Mosquitoes and the Malaria Parasite

Malaria is a serious disease caused by a parasite. Malaria poses a great threat to people in tropical and subtropical areas—2.4 billion people, over one-third of the world's population, living in more than 90 countries. In the United States, malaria used to be much more of a problem than it is now; Florida recently experienced a small but serious malaria outbreak. Worldwide, an estimated 300–400 million people are infected each year, and 1.1 million of these people die.[6] In Africa alone, more than 3,000 children per day die from this disease (Figure 5.4).

Once thought to be caused by filth or by bad air (the name *malaria* comes from the Latin for "bad air"), malaria is actually caused by parasitic microbes (four species of the protozoan *Plasmodium*) (Figure 5.5). These microbes affect and are carried by *Anopheles* mosquitoes, which then transfer the protozoa to people. One solution to the malaria problem, then, would be to eradicate *Anopheles* mosquitoes.

By the end of World War II, scientists had discovered that the pesticide DDT was extremely effective against these mosquitoes. They had also found chloroquine highly effective in killing *Plasmodium* parasites. (Chloroquine is an artificial derivative of quinine, a chemical from the bark of the quinine tree that was an early treatment for malaria.)

▪ **FIGURE 5.4**
Child with malaria. This mother and child from Indonesia are typical of many who contract malaria, a disease found in tropical and semitropical areas.

An anti-malaria campaign was initially successful. In 1957 the World Health Organization (WHO) began a $6 billion campaign to rid the world of malaria using a combination of DDT and chloroquine. At first, the strategy seemed successful. By the mid-1960s, malaria was nearly gone or had been eliminated from 80% of the target areas.

However, success was short-lived. The mosquitoes began to develop a resistance to DDT, and the protozoa became resistant to chloroquine. In many tropical areas, the incidence of malaria increased. For example, as a result of the WHO program, the number of cases in Sri Lanka had dropped from 1 million to only 17 by 1963, but by 1975 it increased to 600,000 cases. Resistance to DDT became widespread among the mosquitoes, and resistance to chloroquine was found in the protozoa in 80% of the 92 countries where malaria was a major killer.

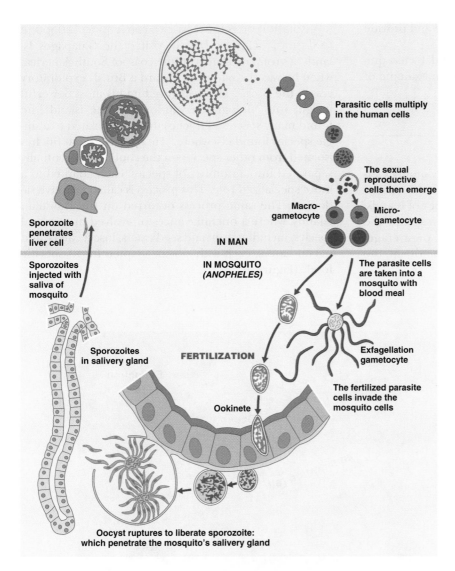

■ **FIGURE 5.5**
Life cycle of malaria.
The microrganism that causes malaria has a complex life cycle, which offers a number of paths to control the disease based on genetic information of both the *Plasmodium* parasite and the mosquito host. People and mosquitoes are hosts to the malaria-causing parasite, but at different stages in the life cycle. [*Source:* Malaria. Available at http://www-micro.msb.le.ac.uk/224/Malaria.html. Accessed March 7, 2005.]

The mosquitoes and protozoa developed resistance through natural selection. When they were exposed to DDT and chloroquine, the susceptible individuals died. The most resistant organisms survived and passed their resistant genes on to their offspring. Because the susceptible individuals died, they left few or no offspring, and any offspring they left were susceptible. Thus, a change in the environment—the human introduction of DDT and chloroquine—caused a particular genotype to become dominant in the populations. This demonstrates that natural selection functions and works.

It's either one knockout punch or a continual battle to stay ahead. We have learned that if we set out to eliminate a disease-causing species, we must attack it completely at the outset and destroy all the individuals before natural selection leads to resistance. Or we have to find ways to keep changing our attack on the species. But sometimes neither approach may work, partly because there is so much genetic variation in the target species.

Mutation often outstrips development of new drugs to prevent malaria. Since chloroquine is generally ineffective now, new drugs have been developed to prevent malaria. However, these second- and third-line drugs will eventually become unsuccessful, too, as a result of the same process of biological evolution by natural selection. This process is speeded up by the ability of *Plasmodia* to rapidly mutate. In South Africa, for example, the protozoa became resistant to one replacement, mefloquine, immediately after the drug became available as a treatment.

An alternative is to develop a vaccine. Biotechnology has made it possible to map the genetic structure of the malaria-causing *Plasmodium* protozoa. With this information, scientists expect to create a vaccine using a variety of the species that is benign in human beings but produces an immune reaction.[6] In addition, scientists are mapping the genetic structure of the carrier mosquito, with the hope that this genetic map could identify genes associated with insecticide resistance and provide clues to developing a new pesticide.

The development of resistance to DDT by mosquitoes and to chloroquine by *Plasmodia* is an example of biological evolution in action today.

5.3 Island Ecology

Jokes, stories, and movies about people becoming castaways on far-off islands and struggling to find food and shelter have a basis in facts about the ecology of islands. Islands have fewer species than continents, and the smaller the island, the fewer the species, on average. Also, the farther away an island is from a continent, the fewer species it will have (Figure 5.6).

Isolation on remote islands can lead to "adaptive radiation." Charles Darwin visited the Galapagos Islands, a group of islands off the coast of South America, when he was the naturalist aboard a British exploratory vessel, the *Beagle*. This visit gave him his most powerful insight into biological evolution. On the islands, he found many species of finches that were related to a single species found elsewhere.[8] He suggested that finches isolated from other species on the continents eventually separated into a number of species, each adapted to a more specialized role. This process is called **adaptive radiation**. The same process occurred on the Hawaiian Islands, where a finchlike ancestor evolved into several species, including fruit and seed eaters, insect eaters, and nectar eaters, each with a beak adapted for its specific food[9] (Figure 5.7).

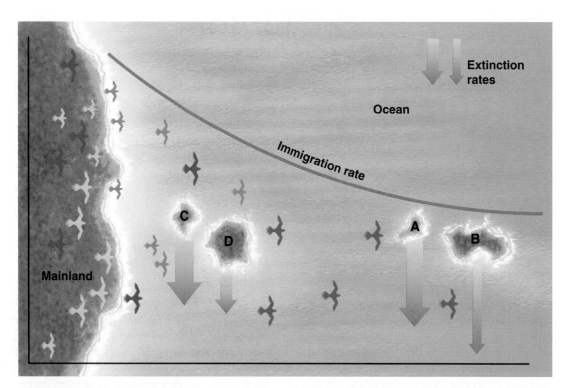

■ **FIGURE 5.6**

Island biogeography (idealized relation of an island's size, distance from the mainland, and number of species). The nearer an island is to the mainland, the more likely it is to be found by an individual, and thus the higher the rate of immigration. (Islands (C) and (D) will have more immigrants than islands (A) and (B)). The larger the island, the larger the population it can support and the greater the chance of persistence of a species—small islands have a higher rate of extinction. Island (B) can support more populations than (A), so that even through it is farther from the mainland it may have greater biological diversity. The same is true for islands (D) and (C). The average number of species therefore depends on the rate of immigration and the rate of extinction. Thus, a small island near the mainland may have a similar number of species as a large island far from the mainland. The thickness of the arrow represents the magnitude of the rate. [*Source:* Modified from R. H. MacArthur and E. O. Wilson. *The theory of island biogeography.* Princeton, NJ: Princeton University Press, 1967.]

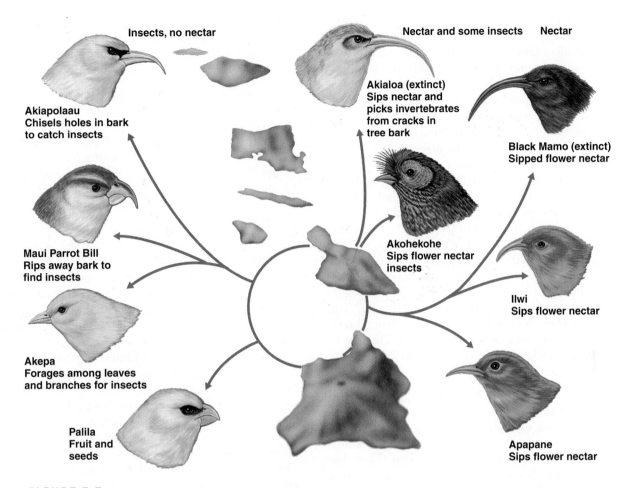

Insects, no nectar

Nectar and some insects   **Nectar**

Akiapolaau
Chisels holes in bark
to catch insects

Akialoa (extinct)
Sips nectar and
picks invertebrates
from cracks in
tree bark

Black Mamo (extinct)
Sipped flower nectar

Maui Parrot Bill
Rips away bark to
find insects

Akohekohe
Sips flower nectar
insects

Ilwi
Sips flower nectar

Akepa
Forages among leaves
and branches for insects

Palila
Fruit and
seeds

Apapane
Sips flower nectar

■ **FIGURE 5.7**
Adaptive radiation (evolutionary divergence) among honeycreepers in Hawaii.
Sixteen species of birds, each with a beak specialized for its food, evolved from a single ancestor.
Nine of the species are shown here (one is now extinct). The species evolved to fit ecological niches
that, on the North American continent, had previously been filled by other species not closely
related to the ancestor. [*Source:* From C. B. Cox, I. N. Healey, and P. D. Moore. *Biogeography.* New
York: Halsted, 1973.]

Small islands have fewer habitat types and greater
risk of extinction. Some habitats on a small island may
be too small to support a population large enough to have
a good chance of surviving for a long time. A small population might be easily wiped out by a storm, flood, or other
catastrophe or disturbance. Every species is at risk of extinction due to predators, disease (parasitism), competition, climate change, or habitat change. Generally, the
smaller the population, the greater the risk of extinction.
And the smaller the island, the smaller the population of
a particular species it can support.[9]

Also, the more distant the island, the less chance
of reaching it. The farther the island is from the mainland, the harder it will be for an individual to travel the
distance. In addition, a smaller island is a smaller "target," less likely to be found by individuals of any species.
Thus, fewer individuals migrate to small islands and distant islands than to large islands and nearby islands.

Lose a few, gain a few. Over a long time, an island
tends to maintain a rather constant number of species,
because the rate at which species are added is about the
same as the rate at which other species become extinct.
For any island, the number of species of a particular life-
form can be predicted from the island's size and its distance from the mainland.

On islands, smaller is better. Species evolve to a
smaller size on islands because islands often have a limited supply of food, fewer predators, and fewer species

competing for the same resources. As a result, chances of survival are best for individuals with smaller body sizes and thus smaller daily energy requirements. Island foxes on the Channel Islands of southern California are about the size of small house cats. Also on the islands are fossils of pygmy mammoths (extinct relatives of modern elephants) that evolved from the much larger mammoths who swam to the island when sea levels were lower and the islands closer to the mainland.

Another example is the recent discovery of fossils of dwarf humans believed to be a different species on the remote Indonesian island Flores, east of Bali. Named *Homo floresiensis* (modern humans are *Homo sapiens*), these dwarf humans were only about 1 meter (3 feet 4 inches) tall, considerably shorter than pygmies in Africa, who are closer to 1.5 meters tall. *Homo floresiensis* lived as recently as 20,000 years ago and may therefore have interacted with us. They may have become extinct along with pygmy elephants when a large volcanic eruption occurred.[10] The discovery of *Homo floresiensis* means that our evolutionary path is more complex than we had thought. There were other human species on Earth very recently!

Applying the concepts of islands to "ecological islands." The concepts of island ecology apply not just to a real island but also to an *ecological island*—a comparatively small habitat separated from a major habitat of the same kind. For example, a pond in the Michigan woods is an ecological island relative to the Great Lakes that border Michigan. A small stand of trees within a prairie is a forest island. The concept of ecological islands is important today for the conservation of biodiversity, especially because many human activities fragment habitats and make them much smaller than they used to be.

A city park is also an ecological island. Is a city park large enough to support a population of a particular species? To know whether it is, we can apply the concepts of island ecology.

5.4 Basic Concepts of Biological Diversity

Now that we have explored biological evolution, we can turn to *biological diversity*. To develop workable policies for conserving biological diversity, we must be clear about the meaning of the term.

Biological diversity involves the following concepts:

▪ *Genetic diversity:* the total number of genetic characteristics of a specific species, subspecies, or group of species, expressed as the total base-pair sequences in DNA, or the total number of genes, active or not, or the total number of active genes.

▪ *Habitat diversity:* the different kinds of habitats in a given unit area.

▪ **Species diversity,** which, in turn, has three aspects: *species richness* (the total number of species); *species evenness* (the relative abundance of species); and *species dominance* (the most abundant species).

To understand the differences among species richness, species evenness, and species dominance, imagine two ecological communities, each with 10 species and 100 individuals, as illustrated in Figure 5.8. In the first community (Figure 5.8a), 82 individuals belong to a single species, and the remaining 9 species are represented by 2 individuals each. In the second community (Figure 5.8b), all the species are equally abundant, so each has 10 individuals.

Which community is more diverse? At first, one might think the two communities have the same species diversity because they have the same number of species. However, if you walked through both communities, the second would appear more diverse. In the first community, most of the time you would see individuals of the dominant species (in the case shown in Figure 5.8a, elephants); you probably would not see many of the other species at all. In the second community, even a casual visitor would see many of the species in a short time. The first community would appear to have relatively little diversity unless you studied it carefully. You can test the probability of encountering a new species in these two communities by laying a ruler down in any direction on Figures 5.8a and 5.8b and counting the number of species that it touches.

This example shows that merely counting the number of species is not enough to describe biological diversity. Species diversity has to do not only with the actual number present but also with the relative chance of seeing species.

5.5 The Number of Species on Earth

Millions of species have come and gone during the several billion years that life has existed on Earth. How many exist today? No one knows the exact number because new species are discovered all the time, especially in little-explored areas such as tropical rain forests. For example, since 1992, five new mammals have been discovered in Laos: (1) the spindle-horned oryx (which is not only a new species but also represents a previously unknown genus); (2) the small black muntjak; (3) the giant muntjak (the muntjak, also known as "barking deer," is a small deer; the giant muntjak is so called because it has large antlers); (4) the striped hare (whose nearest relative lives in Sumatra); and (5) a new species of civet cat. Some 1.5 million species have been named, but biologists estimate that the total number is probably considerably higher, from 3 million to as many as 10 million.

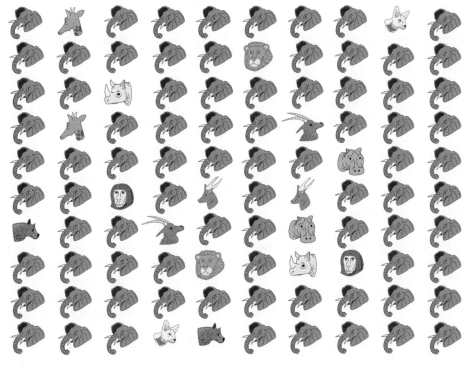

(a)

(b)

■ **FIGURE 5.8**
The difference between species evenness and species richness.
Figures (a) and (b) have the same richness but different evenness. Lay a ruler across
each diagram and count the number of species the edge crosses. Do this several
times, then determine which figure has greater evenness. See text for an
explanation of the results.

Insects and plants make up most of the known species—there are approximately 500,000 insect species and 230,000 plant species. Many of the insects are tropical beetles in rain forests. In contrast, our kind of life, mammals, are much less diverse, with only about 4,500 species. And although bacteria are crucial to sustaining life because they carry out many essential chemical reactions, there are fewer than 10,000 species of bacteria.[11]

5.6 Why Are There Many Species in Some Places and Not in Others?

Why so many species, and why such great biodiversity in some places? If the competitive exclusion principle is true (see Chapter 4)—that complete competitors cannot coexist—how can so many species coexist? The answer, as we learned in Chapter 4, is that species coexist by dividing up the environment into specific ranges of such things as temperature. Another way to put this is that species with similar requirements coexist by having different **ecological niches**—perhaps doing the same "job" but under different environmental conditions.

But here's a puzzle that has long intrigued people: The tropics generally have much greater biological diversity than other areas, and biodiversity generally declines with latitude. This is generally true for oceans as well as for land. And there are many differences in biological diversity from place to place. For example, Yellowstone Lake, a large freshwater lake in Yellowstone National Park, has many more species than one of the hot springs in the same park. What accounts for the difference? Here are a few of the explanations.

In general, habitat complexity increases biodiversity. If species coexist by dividing up the environment, then the more opportunities there are to do so, the more species can be "packed" into a given area. A large farm where acre after acre has been plowed the same way, treated with the same fertilizers, and planted with the same crops offers fewer kinds of habitats than a prairie where there are subtle differences in elevation, small streams that flow across the land, and soils that differ slightly depending on drainage and elevation.

Using this knowledge, ecologists have tried to characterize relationships between the environment and the kinds of species that live in an area. Some of the possible interrelationships are illustrated in Figure 5.7.

Complex topography also increases biodiversity. Where there are hills and valleys, there are slopes that face north, south, east, and west, and each of these provides different environments. South-facing slopes are hotter and drier than north-facing slopes, for example. A hilly landscape can offer more habitats than a flat one.

Disturbances, such as wildfires, can increase biodiversity by increasing the number and kinds of habitats. However, very intense fires have the opposite effect, decreasing both variation and biodiversity.

Life's diversity further increases diversity. If there are many kinds of trees in a forest, then there are more kinds of habitats and more niches for animals, so the diversity of plants can increase the diversity of animals.

Biodiversity changes over time. As a forest ecosystem develops from bare ground, there are changes in the dominant species of plants, and these provide habitats for different animals. An old forest—with dead trees, logs on the ground, old stumps, and subtle variations in the soil as trees age and die—provides many kinds of habitats.

People affect biodiversity. In general, urbanization (see the opening case study about New Orleans), industrialization, and agriculture decrease diversity, reducing the number of habitats and simplifying habitats. In addition, we favor specific species and manipulate populations for our own purposes, as when a person plants a lawn or when a farmer plants a single crop over a large area.

Urbanization can decrease biodiversity several ways. Cities have typically been located at good sites for travel, such as along rivers or near oceans, where biological diversity is high. But cities can contribute in important ways to the conservation of biological diversity, as in small "kitchen gardens" in cities in developing nations, where plants can provide food for birds.

Change in relative abundance of species can occur over an area or distance. This is an **ecological gradient**. Such a change in species of vegetation can be seen with changes in elevation in mountainous areas like those at the Grand Canyon and the nearby San Francisco Mountains of Arizona (Figure 5.9).Although we can see these patterns most easily in vegetation, they occur for all organisms. See, for example, the pattern of distribution of African mammals on Mount Kilimanjaro (Figure 5.10).

5.7 What Can We Do to Save Endangered Species?

How many species are threatened with extinction? The International Union for the Conservation of Nature (IUCN) maintains a list of threatened and endangered animals in a publication known as *The Red Book*. The IUCN *Red Book of Threatened Species* reports that about 1% of all species are threatened, and that 23% of mammalian species, 12% of birds, 4% of reptiles, 31% of amphibians, and 3% of fish (primarily freshwater fish) are at risk of extinction. The IUCN *Red List of Threatened Plants* estimates that more than 11,000 of the plants—3% of those known—are threatened with extinction.

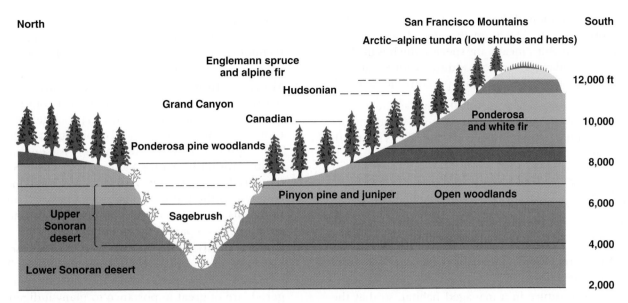

■ **FIGURE 5.9**

Change in the relative abundance of a species over an area or a distance is referred to as *an ecological gradient*. Such a change can be seen with changes in elevation in mountainous areas. The altitudinal zones of vegetation in the Grand Canyon of Arizona and the nearby San Francisco Mountains are shown. [*Source:* From C. B. Hunt. *Natural regions of the United States and Canada.* San Francisco: W. H. Freeman, 1974. © 1974 by W. H. Freeman.]

What does it mean to call a species "endangered" or "threatened"? The terms can have a strictly biological meaning, or they can have a legal meaning. A scientific definition is phrased in terms of the likelihood that a species will go extinct within a certain time. For a legal definition, we can turn to the U.S. Endangered Species Act of 1973. The Act says, "The term **endangered species** means any species which is in danger of extinction throughout all or a significant portion of its range other than a species of the Class Insecta determined by the Secretary to constitute a pest whose protection under the provisions of this Act would present an overwhelming and overriding risk to man."[23] In other words, legally, if certain endangered insect species are pests, we'll be happy to be rid of them. (It is interesting that insect pests can be excluded from protection by this legal definition, but

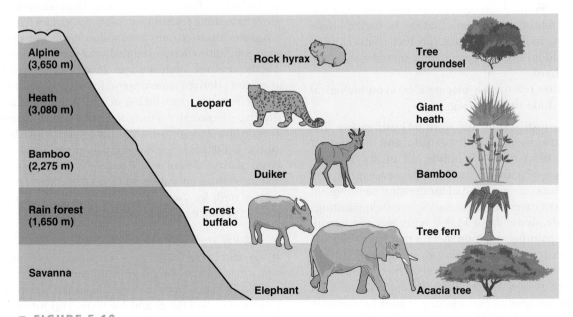

■ **FIGURE 5.10**

Changes in the distribution of animals with elevation on a typical mountain in Kenya.
[*Source:* From C. B. Cox, I. N. Healey, and P. D. Moore. *Biogeography.* New York: Halsted, 1973.]

there is no mention of disease-causing bacteria or other microorganisms.) "The term **threatened species,**" according to the Act, "means any species which is likely to become an endangered species within the foreseeable future throughout all or a significant portion of its range."[23]

Can our knowledge about biodiversity help save endangered species? Or can we only stand on the sidelines and watch biodiversity decline? Let's start with the basics.

5.8 Why Save Endangered Species?

When we say we want to save a species, what is it we really want to save? There are four possible answers:

1. A wild creature in a wild habitat, as a symbol to us of wilderness.
2. A wild creature in a managed habitat, so that the species can feed and reproduce with little interference and so that we can see it in a naturalistic habitat. (The recovery of the Aleutian goose fits this goal.)
3. A population in a zoo, so that the genetic characteristics are maintained in live individuals.
4. Genetic material only—frozen cells containing DNA from a species for future scientific research.

Which goals we choose involve not only science but also values. There are two basic kinds of reasons for conserving the diversity of life: Either this diversity is necessary for the workings of the world—for the persistence of life on Earth—or for one reason or another we want that diversity. The first kind is called the *ecological justification*. For example, all life depends on certain bacteria that convert nitrogen in the atmosphere to ammonia and nitrate, compounds that can be used to make proteins essential to life. If these species became extinct, all life would go extinct. Fortunately, these bacteria are extremely hardy.

Our other reasons for placing a value on biological diversity include the following.

▪ **Aesthetic justification and spiritual justification:** that we find the diversity of life beautiful and that contact with the diversity of life uplifts our spirits. This appreciation of nature is ancient. Whatever other reasons Pleistocene people had for creating paintings on the walls of caves in France and Spain, their paintings of wildlife, done about 14,000 years ago, are beautiful. The paintings include species that have since become extinct, such as mastodons. Poetry, novels, plays, paintings, and sculpture often celebrate the beauty of nature. Appreciation of nature's beauty is a very human quality and a strong reason for conserving endangered species.

▪ **Recreational justification:** that we enjoy activities that bring us in contact with the diversity of life.

▪ **Utilitarian justification:** that many species are useful to us, providing food, medicine, shelter, materials for technology, such as wood for boats, and the potential for new sources of these.

▪ **Moral justification:** the belief that all forms of life have a right to exist, and that therefore, in our role as global stewards, we are obligated to promote the continued existence of species and of biological diversity. This right to exist was stated in the UN General Assembly World Charter for Nature, 1982. The U.S. Endangered Species Act also includes statements concerning the rights of organisms to exist. Thus, a moral justification for the conservation of endangered species is part of the intent of the law.

▪ **Cultural justification:** that the loss of some species may cause the loss of one of the human cultures. This is because certain species, some threatened or endangered, are of great importance to many indigenous peoples, who rely on these species for food, shelter, tools, fuel, materials for clothing, and medicine.

Moral justification has deep roots within human culture, religion, and society. Those who focus on cost–benefit analyses tend to downplay moral justification, but as more and more citizens of the world assert the validity of moral justification, more actions are taken to defend a moral position, even if those actions have negative economic effects.

Medicines are one example of the utilitarian value of species diversity. Of 275 species found in less than a half acre of a tropical Peruvian forest, 72 yielded products with direct economic value. Many important chemical compounds come from wild organisms. Digitalis, an important drug in treating certain heart ailments, comes from purple foxglove; aspirin comes from willow bark; and a powerful cancer-fighting chemical named taxol was discovered in the Pacific yew tree (genus name *Taxus*). Other well-known medicines derived from tropical forests include anticancer drugs from rosy periwinkles, steroids from Mexican yams, antihypertensive drugs from serpent wood, and antibiotics from tropical fungi. Some 25% of prescriptions dispensed in the United States today contain ingredients extracted from plants.[12] Scientists are testing marine organisms for use in pharmaceutical drugs. Coral reefs offer a promising area of study for such compounds because many coral-reef species produce toxins to defend themselves.

Many species help to control pollution. Plants, fungi, and bacteria remove toxic substances from air, water, and soils. Carbon dioxide and sulfur dioxide are removed by vegetation, carbon monoxide is reduced and oxidized by soil fungi and bacteria, and nitric oxide is incorporated into the biological nitrogen cycle. Because different species do different things, diversity of species provides the best range of pollution control.

Tourism provides yet another utilitarian justification. Ecotourists value nature, including its endangered species, for aesthetic or spiritual reasons. As a result, ecotourism, with tours led by wildlife experts, is a growing source of income for many developing countries.

5.9 How a Species Becomes Endangered and Extinct

Our next step in finding out what science can do to help endangered species is to understand how species become endangered. Extinction is the rule of nature—in a finite world subject to frequent chance events, the eventual fate of every species is extinction. But the chance of extinction can be so low and eventual extinction so far off—perhaps billions of years away—that the possibility may not be important to us. Let's refine our terms a little more, distinguishing between local and global extinction.

Where have all the flowers gone? *Local extinction* is when a species disappears from a part of its range but can still be found elsewhere. *Global extinction* is when a species can no longer be found anywhere.

Rates of extinctions have varied greatly over geologic time. From 580 million years ago until the beginning of the Industrial Revolution, about one species per year, on average, became extinct. Over much of the history of life on Earth, the rate of evolution of new species equaled or slightly exceeded the rate of extinction. The average longevity of a species has been about 10 million years. However, the fossil record suggests that there have been several periods of mass extinction and other periods of rapid evolution of new species (Figure 5.11). Interspersed with the episodes of mass extinctions, there seem to have been periods of hundreds of thousands of years with comparatively low rates of extinction.

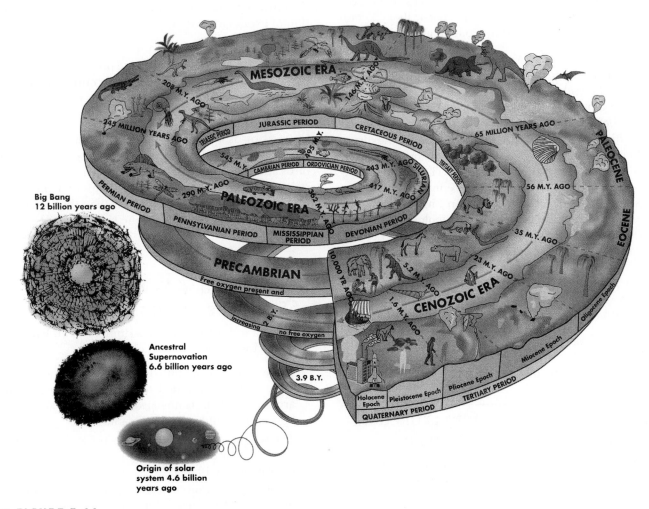

■ FIGURE 5.11
The history of life on Earth shows an overall increase in the number of species and six episodes of mass extinctions.

5.10 Causes of Mass Extinction

Six major mass extinctions occurred during the past 550 million years. The earliest, about 450 million years ago, may have been related to global cooling. About 100 families and their associated species became extinct. In the second, about 90 million years later, 70% of all marine invertebrate species went extinct, probably from climate change. In the third, 245 million years ago, about 95% of marine species died. Although there is growing evidence that Earth may have been hit by a large asteroid, scientists don't believe that this mass extinction came about quickly from a single catastrophe but instead that it may have taken about 7 million years. Global cooling, followed by rapid global warming with large variations in climate, may have been responsible.

The end of the dinosaurs. About 65 million years ago, another tremendous mass extinction occurred. Most dinosaurs became extinct. This time, the event was sudden, and evidence suggests that it was caused by the impact of a giant asteroid. Another mass extinction occurred near the end of the Eocene epoch, about 35 million years ago, also probably from an asteroid impact.

More-recent mass extinctions. An intriguing example of extinction occurred about 20,000–10,000 years ago, at the end of the last great continental glaciation. At that time, mass extinctions of large birds and mammals occurred: 33 genera of large mammals—those weighing 50 kg (110 lb) or more—became extinct, whereas only 13 genera had become extinct in the preceding 1 or 2 million years (Figure 5.11). Smaller mammals were not so affected, nor were marine mammals. As early as 1876, Alfred Wallace, an English biological geographer, noted that "we live in a zoologically impoverished world, from which all of the hugest, and fiercest, and strangest forms have recently disappeared."[13] Some have suggested that these sudden extinctions coincided with the arrival, on different continents at different times, of Stone Age people and therefore may have been caused by hunting.[14]

The rate of extinctions has increased greatly since the Industrial Revolution, as a result of widespread human activities.

5.11 How People Cause Extinctions and Affect Biological Diversity

People have become an important cause of extinction. Among the ways we cause extinction are the following:

▪ hunting or harvesting—for commercial purposes, for sport, or to control a species that is considered a pest;

▪ disrupting, fragmenting, or eliminating habitats;

▪ introducing exotic species, including new parasites, predators, or competitors of a native species;

▪ polluting.

The earliest people probably caused extinctions through hunting. People still hunt, especially for valuable animal products, such as elephant ivory and rhinoceros horns.

When people learned to use fire and then to farm, they began to change habitats over large areas. The development of agriculture and the rise of civilization led to rapid deforestation, to habitat changes, and to fewer and simpler habitats.

As people explored new areas, they introduced exotic species. These became major causes of extinction. The Polynesian people sailed widely throughout the Pacific and settled many islands, bringing with them crops, pigs, and other domestic animals. After Columbus's voyage to the New World, Magellan's circumnavigation of the globe, and the resulting spread of European civilization and technology, human introduction of exotics increased.

The introduction of thousands of novel chemicals into the environment made pollution an increasing cause of extinction in the 20th century. Pollution control has proved to be a successful way to help species.

People have caused about 75% of the extinctions of birds and mammals since 1600. Hunting is estimated to have caused 42% of the extinctions of birds and 33% of the extinctions of mammals. The current extinction rate among most groups of mammals is estimated to be 1,000 times greater than the extinction rate at the end of the Pleistocene epoch (about 10,000 years ago).

5.12 The Good News: The Status of Some Species Has Improved

As a result of human actions, a number of endangered species have recovered. For example:

▪ The elephant seal, hunted almost to extinction in the 19th century, dwindled to about a dozen animals around 1900 but is now protected by law and numbers in the hundreds of thousands.

▪ The sea otter, also hunted for its fur and reduced in the 19th century to several hundred, now numbers approximately 10,000.

▪ Many species of birds were endangered because the insecticide DDT caused thinning of eggshells and failure of reproduction. With the elimination of use of DDT in the United States, many bird species recovered, including the bald eagle, brown pelican, white pelican, osprey, and peregrine falcon.

■ The blue whale was thought to have been reduced to about 400 when whaling was still actively pursued by a number of nations. Today, 400 blue whales are sighted annually in the Santa Barbara Channel along the California coast, and that is just a fraction of the total population.

■ The gray whale, which was hunted to near-extinction, has recovered and is abundant along the California coast and in its annual migration to Alaska.

Since the U.S. Endangered Species Act became law in 1973, 43 species have recovered sufficiently to be either reclassified from "endangered" to "threatened" or removed completely from the list. In addition, the U.S. Fish and Wildlife Service—which, along with the National Marine Fisheries Services, administers the Endangered Species Act—lists 33 species that have the potential for reclassification to an improved status.

5.13 Can a Species Be Too Abundant? If So, What Should We Do?

Sometimes we succeed too well in increasing the numbers of a species. All marine mammals are protected in the United States by the Federal Marine Mammal Protection Act of 1972, which has led to improvement in the status of many marine mammals. Sometimes, however, we end up with a bit too much of a good thing. Case in point:

Sea lions have become so abundant as to be local problems. For example, in San Francisco Harbor and in Santa Barbara Harbor, sea lions haul out and sun themselves on boats and pollute the water with their excrement near shore. In one case, so many sea lions hauled out on a sailboat in Santa Barbara Harbor that they sank the boat, and some of the animals were trapped and drowned.

Mountain lions, too, have become locally overabundant. In the 1990s, California voters passed an initiative that protected the endangered mountain lion but contained no provisions for management of mountain lions if they became abundant, except in cases where they threatened human life and property. Few people thought the mountain lion could ever recover enough to become a problem, but in several cases in recent years mountain lions have attacked and even killed people. These attacks become more frequent as the mountain lion population grows and as the human population grows and people build houses in what was mountain lion habitat.

5.14 The Kirtland's Warbler and Environmental Change

Environmental change is necessary for some species. Many endangered species are adapted to natural environmental change and require it. When people eliminate that change, a species can become threatened with extinction. This happened with the Kirtland's warbler, which nests in jack-pine forests in Michigan. In 1951 the Kirtland's warbler became the first songbird in the United States to be subject to a complete census, and about 400 nesting males were found. Concern about the species grew in the 1960s and increased when only 201 nesting males were found in the third census, in 1971.[15]

Kirtland's warblers are known to nest only in young jack-pine woodlands. Jack pine is a "fire species," which persists only where there are periodic forest fires. Cones of the jack pine open only after they have been heated by fire. The trees are intolerant of shade, able to grow only when their leaves can reach into full sunlight; so even if seeds were to germinate under mature trees, the seedlings could not grow in the shade and would die. Jack pine produces an abundance of dead branches, which may be an evolutionary adaptation to promote fires, essential to the survival of the species.

Kirtland's warblers require fairly frequent change. Forest fires occurred approximately every 20–30 years in jack-pine woods before settlers arrived.[21] At the time of the first European settlement of North America, jack pine may have covered a large area in what is now Michigan. Even as recently as the 1950s, the pine was estimated to cover nearly 500,000 acres in that state. Commercial loggers considered the small, poorly formed jack pines a trash species and left them alone, but large fires often followed logging operations when quantities of "slash"—branches, twigs, and other economically worthless parts of trees—were left in the woods. Elsewhere, fires were set to clear jack-pine areas and promote the growth of blueberries. Some experts think that the population of Kirtland's warblers peaked in the late 19th century as a result of these fires. After 1927, fire suppression became the practice, and people were encouraged to replace jack pine with economically more useful species. One result was that the warblers' nesting areas shrank. Realizing this, managers have introduced controlled burning in the warblers' habitat.

5.15 Ecological Islands and Endangered Species

The history of the Kirtland's warbler illustrates that a species may inhabit isolated jack-pine stands of the right age range for the bird. The stands are an example of an

ecological island (a term introduced earlier), an area that is biologically isolated, so a species living there cannot mix (or only rarely mixes) with any other population of the same species. Mountaintops and isolated ponds are ecological islands. Real geographic islands are also ecological islands. Insights gained from studies of the biogeography of islands have important implications for the conservation of endangered species and for the design of parks and preserves for biological conservation.

Almost every park is an ecological island for some species. A small city park between buildings may be an island for trees and squirrels. At the other extreme, even a large national park is an ecological island. For example, the Masai Mara Game Reserve in the Serengeti Plain, which stretches from Tanzania to Kenya in East Africa, and other great wildlife parks of eastern and southern Africa are becoming islands of natural landscape surrounded by human settlements. Lions and other great cats exist in these parks as isolated populations, no longer able to roam completely freely and to mix over large areas. Other examples are islands of uncut forests left by logging operations, and oceanic islands, where intense fishing has isolated parts of fish populations.

How large must an ecological island be to ensure survival of a species? The size varies with the species but can be estimated. Some islands that seem large to us are too small for species we wish to preserve. For example, a preserve was set aside in India in an attempt to reintroduce the Indian lion into an area where it had been eliminated by hunting and by changes in land use. In 1957 a male and two females were introduced into a 95-km^2 (36-mi^2) preserve in the Chakia forest known as the Chandraprabha Sanctuary. The introduction was carried out carefully and the population was counted annually. There were four lions in 1958, five in 1960, seven in 1962, and eleven in 1965. After that, they disappeared and were never seen again.

Why did they go? Although 36 mi^2 seems large to us, male Indian lions have territories of 50 mi^2. Within that territory, females and young also live. A population that could persist for a long time would need a number of such territories, so an adequate preserve would require 640–1,300 km^2 (247–500 mi^2). Various other reasons were suggested for the disappearance of the lions, including poisoning and shooting by villagers. But regardless of the immediate cause, the lions needed a much larger area for long-term survival.

Return to the Big Question:

Can we save endangered species and keep biological diversity high?

Yes, if we are vigilant. As we have seen, the major ways that people have caused extinctions are by overhunting; habitat disruption and habitat elimination; and the introduction of exotic species and pollutants. Therefore, the first steps we can take in trying to save endangered species are to stop such practices:

- Hunt and fish, but not to the point where wildlife populations become endangered. Let scientific knowledge help determine what should be a safe catch (a topic we take up in a later chapter).
- Restore damaged habitats and reconstruct habitats that have been widely eliminated.
- Remove exotic species where they are causing trouble for other species, and prevent new introductions.
- Control pollutants.

We can also take a more aggressive approach to monitoring agricultural and forestry practices and to the management of our natural resources in general, as discussed in later chapters.

Summary

- Biological evolution—the change in inherited characteristics of a population from generation to generation—is responsible for the development of the many species of life on Earth. Four processes that lead to evolution are mutation, natural selection, migration, and genetic drift.

- Biological diversity involves three concepts: genetic diversity (the total number of genetic characteristics), habitat diversity (the diversity of habitats in a given unit area), and species diversity. Species diversity, in turn, involves three ideas: species richness (the total number of species), species evenness (the relative abundance of species), and species dominance (the most abundant species).

- Species engage in three basic kinds of interactions: competition, symbiosis, and predation–parasitism. Each type of interaction affects evolution, the persistence of species, and the overall diversity of life. It is important to understand that organisms have evolved together so that predator, parasite, prey, competitor, and symbiont have adjusted to one another. Human interventions frequently upset these adjustments.

- The number of species in a given habitat is affected by many factors, including latitude, elevation, topography, the severity of the environment, and the diversity of the habitat. Predation and moderate disturbances, such as fire, can actually increase the diversity of species. The number of species also varies over time. Of course, people affect diversity as well.

- The good news is that many species once endangered have been successfully restored to an abundance that suggests they are unlikely to become extinct. Success depends on restoring the habitat to conditions required by a species. The conservation and management of wildlife presents great challenges but also offers great rewards that hold deep meaning for people.

- The study of island life has led to a view of island ecology that includes several important concepts. One is that islands have fewer species than mainlands because of their smaller size and distance from the mainland. Another is that the smaller an island and the farther it is from the mainland, the fewer species the island will contain.

- Ecological islands—habitats separated from the main part of a biome—show the same diversity as physical islands. The smaller the ecological island and the farther it is from its "mainland," the fewer species it can support.

Key Terms

adaptive radiation

biological diversity

biological evolution

competitive exclusion principle

deoxyribonucleic acid (DNA)

divergent evolution

ecological gradient

ecological island

ecological niche

endangered species

extinction

gene

genetic diversity

genetic drift

habitat

habitat diversity

island biogeography

landscape perspective

logistic carrying capacity

mass extinction

mutation

natural selection

species diversity

threatened species

Getting It Straight

1. Related to the theory of biological evolution, what are the causes of new species development?

2. What is divergent evolution?

3. Does evolution of species result simply by chance? Explain.

4. What is an ecological island? Give an example.

5. What three concepts contribute to biological diversity?

6. What is an approximate total number of the following species on Earth?
   a. Insects   b. Mammals   c. Bacteria

7. Define endangered and threatened species and identify one species that falls into each category.

8. Why should we save endangered species?

9. Why do introduced species often become pests?

10. Identify three utilitarian values of species diversity.

11. How do species become endangered or extinct?

What Do You Think?

1. In the past several thousand years, more species of birds and mammals have gone extinct on islands than on continents. Referring to the theory of island biogeography, explain why this has happened.

2. You are going to conduct a survey of national parks. What relationship would you expect to find between the number of species of trees and the size of the parks?

3. A city park manager has run out of money to buy new plants. How can the park's labor force alone be used to increase the diversity of (a) trees and (b) birds in the parks?

4. There are more than 600 species of trees in Costa Rica, most of them in the tropical rain forests. Explain how so many species with similar needs manage to coexist.

5. What is the importance of biological diversity to ecological sustainability? What does the case study at the beginning of this chapter convey about biological diversity?

Pulling It All Together

1. Suppose another hurricane hits New Orleans, similar to the one of 2005. Where in that city and its surroundings might biodiversity increase? Where might it decrease? Explain your answers in terms of changes in habitats and the possibility of the introduction of exotic species.

2. Apply Darwin's theory of evolution to the following two groups: motorcycle riders who do not wear helmets and are unmarried and without children, motorcycle riders who do wear helmets and are unmarried and without children, and motorcycle riders who do not wear helmets and are married with children. If Darwin's theory applies, how might the use of helmets by motorcyclists change in the future if these three groups continue and remain distinct from one another?

3. Develop a plan to control malaria, taking into account the fact that both the malaria parasite and mosquitoes evolve quite rapidly in response to new chemicals. Make use of ecological principles from this chapter and from previous chapters as well.

4. Develop a plan to conserve mountain lions in California so that they do not threaten people.

Further Reading

Leveque, C., and J. Mounolou. 2003. *Biodiversity*. New York: John Wiley.

Charlesworth, B., and C. Charlesworth. 2003. *Evolution: A very short introduction*. Oxford: Oxford University Press.

Darwin, C. A. 1859. *The origin of species by means of natural selection, or the preservation of proved races in the struggle for life*. London: Murray. Reprinted variously.—A book that marked a revolution in the study and understanding of biotic existence.

Dawkins, R. 1996. *Climbing Mount Improbable*. New York: Viking.—A discussion of some implications of modern discoveries in genetics and evolution.

Margulis, L., and D. Sagan. 1995. *What is life?* New York: Simon & Schuster.—A beautifully illustrated and well-written introduction to the major forms of life on Earth and the effect of life's diversity on the global environment.

Wilson, E. O., ed. 1992. *The diversity of life*. New York: Norton.—A book outlining the story of evolution of life on Earth, how species became diverse, and the scope of the current threat to that diversity.

Will and Deni McIntyre/Stone/Getty Images

Restoration Ecology

Big Question

Can We Restore Damaged Ecosystems?

?

Learning Objectives

If we have damaged Earth's ecosystems, then we need to figure out how to
fix them, how to restore them. In this chapter we consider . . .

- how an ecosystem restores itself through ecological succession after a disturbance;

- what it means to "restore" ecosystems—since they are always changing, restore to what?

- the role that disturbances play in the persistence of ecosystems;

- what kinds of goals are possible for ecological restoration.

Case Study

Restoring a Ponderosa Forest

At Fort Valley Experimental Forest, about seven miles northwest of Flagstaff, Arizona, a ponderosa pine forest was in trouble—in danger of suffering a seriously damaging wildfire. The reason the forest was in trouble was ironic: It had been *too protected from fire.* Ponderosa forests burn naturally, and the species that live in them have evolved with and adapted to fire over millions of years. Natural fires occur frequently, so when they start, there is less to burn. As a result, natural fires are light, clearing out many young trees but not burning through the bark of most of the large, thick-barked, seed-producing trees. Nor are the fires intense enough to destroy much of the organic matter in the soil. Ponderosa seeds germinate after a fire. The young trees do well in bright light, not in the shade of other trees, and therefore require the openings that result from wildfires.

A century ago the woodlands were open, with scattered trees, mostly large and mature, among which grasses grew. These characteristics of the forest were the result of frequent but light fires. However, throughout most of the 20th century, people thought fire was bad for forests, so they generally tried to prevent them, and fight them if they started. The ponderosa pine forest at Fort Valley had been protected from fire for over a century, and during that time the forest had changed—it had become a thicket of small and large stems. In 1876 the forest averaged 57 trees per hectare (23 trees per acre). In 1992, after a century of fire suppression, density had increased to almost 3,000 trees per hectare (1,200 trees per acre). This high density posed a severe wildfire danger. Not only were fires more likely, but they were more likely to be intense and damaging.[1]

A restoration project was launched at Fort Valley with the goal of returning the forest to the way it was around 1876. But this could not be done simply by reintroducing fire into the forest, which was now not only dense but also had a large accumulation of organic matter in the soil and on the surface. The kind of fire that would occur under these conditions could be highly destructive, hot enough to kill even the mature, seed-bearing trees and burn through the soil, destroying its organic matter. Such a fire could either permanently eliminate ponderosa pines from the burned area or so damage the trees and the soil that recovery would take a very long time.

Before fire could be reintroduced, the forest had to be returned to a lower fuel condition—that is, a lot of the material that could fuel a really big wildfire had to be removed. Under the direction of Professor Wallace Covington of Northern Arizona University, in Flagstaff, trees were carefully removed to re-create conditions typical in the 19th century. The researchers thinned the forest, removing, one by one, 2,200 trees per hectare (900 trees per acre), leaving clumps of trees around large, grassy openings. They also raked organic matter from the forest floor. This was a hard work.

When these modified forested areas were burned, the fire was of low intensity, with flames averaging only about six inches high. Figure 6.1a shows part of the Coconino National Forest in Arizona prior to thinning. The same area is shown in Figure 6.1b after thinning and controlled burning. The fires did not kill the mature trees and left some of the younger trees to replace the older ones. Since then, grass has returned to the openings. Now the forest can once again follow its historical pattern of frequent light fires. The forest has been restored, but the case study of Fort Valley illustrates that ecological restoration can be complex and can require great care and considerable effort.[1]

Gus A. Pearson

Martos Hoffmann

(a) *(b)*

■ **FIGURE 6.1**

Photographs of the restoration project at Fort Valley, Arizona.
(a) The forest protected from fire had become a dense thicket, (b) a similar forest after thinning to restore the forest to a more natural condition.

6.1 Restoration Ecology

Wherever people have lived, they have changed their environment. In North America, this was true of the original people to arrive on the continent, just as it has been since European settlement. While some of the changes that people make are beneficial and desirable, others are considered degradation. For lands and waters that have been degraded, the question is . . .

Can we restore them? **Restoration ecology** is a new field. Its goal is to return damaged, degraded, or destroyed ecosystems to some set of conditions considered functional, sustainable, and "natural." Whether restoration attempts can always be successful is still an open question.

And restore to what? Restoration generally means putting something back the way it was—like taking an old painting or piece of furniture that has been damaged over the years and making it look the way it used to. But we did not invent ecosystems, and as we learned in Chapter 5, ecosystems are not constant—over time, they keep changing. So the question is, Which set of conditions do we want to re-create? For example, is there an original state to which the ecosystem must return?

Is there really a "balance of nature"? Until recently, most people and most scientists would have said yes, there is an original state of nature, a single, perfect, best state, to which nature would always return if it were left undisturbed by people. An ecosystem that reached this state was said to be *pristine, primordial,* or at a *climax state.* Until the second half of the 20th century, the predominant belief in Western civilization was that any natural area—a forest, a prairie, an intertidal zone—left undisturbed by people achieved a single condition that would persist indefinitely. If it was disturbed but then was left alone, it would return exactly to that same permanent state. In this permanent state of nature, all species would exist in a "great chain of being" with a place for each creature—a habitat and a niche—and each creature in its appropriate niche and habitat. Such an ecosystem was believed to have maximum organic matter, maximum storage of chemical elements, and maximum biological diversity. This idea that nature is unchanging and constant is called **the balance of nature**.

If all this were true, the answer to the question "Restore to what?" would be simple: Restore to the original, natural, perfectly balanced state. The way to do this would be simple, too: Get out of the way and let nature take its course. In North America, this often means "Put nature back the way it was before Columbus got here."

But think about Fort Valley's ponderosa forests, always changing—burning, regrowing, burning again. Since change has been a part of natural ecological sys-

tems as long as life has existed on Earth, species have adapted to change, and many, like ponderosa pine, require specific kinds of change in order to survive.

Nature is always changing, and we must make sure it continues to do so. Ecologists know now that nature is not constant and that even if left alone, forests, prairies—all ecosystems—change. Putting nature back together appears to require that we keep it changing, that we become active agents of change. Many people find this a difficult concept and insist that people should stay out of nature, that if we interfere, even when we try to do good, we only cause problems.

At the extreme are those who argue that all human impacts on nature are "unnatural" and therefore undesirable and that the only goal of restoration should be to put nature back the way it was before people arrived. The anthropologist Paul S. Martin takes this position. He proposes that the only truly "natural" time was before any significant human influence occurred. Going back before Columbus, he argues, isn't nearly good enough. He believes we should restore nature in North America to conditions of 10,000 B.C.—before farming and before hunting and gathering in North America. He even suggests introducing the African elephant into North America to replace the mastodon, whose extinction, he argues, was the result of hunting by Indians and therefore unnatural and bad. Ironically, although he argues that people have interfered too much, introducing African elephants, as he proposes, would be a novel interference.[2, 3]

This raises the interesting question: Are we part of nature or not? Are people natural? Are human actions natural? Are we part of nature, or outside it? This is a troubling question that has disturbed people for a very long time. Knowledge of how nature restores itself after disturbances can help us answer this question.

6.2 How Nature Restores Itself

Disturbed ecosystems recover naturally through **ecological succession**. There are two kinds of succession: primary and secondary. **Primary succession** is the initial establishment and development of an ecosystem where one did not exist (Figure 6.2). **Secondary succession** is reestablishment of an ecosystem following disturbances.

Succession is one of the most important ecological processes, and the patterns of succession teach us a lot about ecosystem management. We see examples of succession all around us. When a house lot is abandoned in a city, weeds begin to grow. After a few years, shrubs and trees can be found; succession is taking place. A farmer weeding a crop and a homeowner weeding a lawn are both fighting against the natural processes of secondary succession.

Masha Nordbye/Bruce Coleman, Inc.

©Grant Heilman Photography

(a)

(b)

■ **FIGURE 6.2**
Primary succession.
(a) Forests developing on new lava flows in Hawaii and (b) at the edge of a retreating glacier in Alaska.

Patterns in Succession

Succession occurs in most kinds of ecosystems and follows general patterns. We can see these patterns in three of the first places that ecologists studied succession: (1) on dry sand dunes along the shores of the Great Lakes in North America, (2) in a northern freshwater bog, and (3) in an abandoned farm field.

Dune Succession

Dunes form, are destroyed, and form again. Sand dunes are continually being formed along sandy shores, and then breached and destroyed by storms. Soon after a dune forms, dune grass invades. This grass has special adaptations to the shifting sands of the unstable dune. Just under the surface, it puts out runners with sharp ends (if you step on one, it will hurt), which allow the grass to grow through the sand. The dune grass rapidly forms a complex network of underground runners, crisscrossing like a coarsely woven mat. Above the ground, the green stems carry out photosynthesis, and the grasses grow (Figure 6.3a).

Once the dune grass is established, its runners stabilize the sand, and seeds of other plants are less apt to be buried too deep to germinate or so close to the surface that they blow away. The seeds germinate and grow, and an ecological community of many species begins to develop (Figure 6.3b). At first the plants tend to be small, grow well in bright light, and withstand harshness of the environment—high temperatures in the summer, low temperatures in the winter, and in-

tense storms. These plants are thus well adapted to the environment of early succession.

Slowly, larger plants—along the Great Lakes of North America, eastern red cedar and eastern white pine—are able to grow on the dunes. Eventually, a forest develops, which may include such species as beech and maple. This kind of forest can persist for many years, but at some point a severe storm breaches even these heavily vegetated dunes, and the process begins again (Figure 6.3c).

Bog Succession

A bog is a body of water with surface inlets—usually small streams—but no surface outlet. As a result, the waters of a bog are quiet, flowing slowly. Many bogs that exist today began as lakes when water filled depressions that glaciers left in the land during the Pleistocene ice age.

Bog succession begins with plants that live on top of the water. In a northern bog, such as Livingston Bog in Michigan (Figure 6.4a), succession begins when a sedge (a grasslike herb) puts out floating runners that allow the plant to live and grow on the surface of the water (Figure 6.4b). These runners form a mat on the water's surface, a complex network similar to the one formed by dune grass. The stems of the sedge grow on the runners and carry out photosynthesis.

Winds blow particles onto the mat until a kind of soil builds up. Seeds of other plants land on the mat and germinate there atop the water. The floating mat

Jonnie Miles/Photographer's Choice/Getty Images

Daniel B. Botkin

(a)

Photodisc Blue/Getty Images, Inc.

(b)

(c)

▪ **FIGURE 6.3**
Succession on sand dunes.
(a) Soon after a dune forms, dune grass invades and creates a foundation for other plants. (b) Mature forests develop eventually on the dunes. (c) Eventually a storm breaches the dune, clearing the forest. A new dune forms, and the process of succession starts again. Here an old tree trunk, part of a previous succession, is exposed as the dune sand is moved away from it.

becomes thicker, and small shrubs and trees, adapted to wet environments, grow. In the North, these include species of the blueberry family. The pattern of development of a bog is similar to what occurs on dunes; only the species and their niches are different.

The bog also fills in from the bottom as streams carry fine particles of clay into it (Figure 6.4c, d). At the shoreward end, the floating mat and the bottom sediments meet, forming a solid surface. But farther out

there is a "quaking bog." You can walk on the quaking bog mat; and if you jump up and down, all the plants around you bounce and shake, because the mat is really floating.

From bog to forest. Eventually, as the bog fills in from the top and the bottom, trees that can withstand wetter conditions—such as northern cedar, black spruce, and balsam fir—grow. Over time, what began as an open body of water becomes a wetland forest (Figure 6.4d).

■ **FIGURE 6.4**

Succession in bogs.

(a) Livingston Bog in northern Michigan, one of the locations of a classic study of ecological succession. The entire process of succession is visible in this picture, from open water in the center to a floating sedge mat, the earliest successional vegetation, to the forest in the background. The patterns of bog succession are shown in (b), (c) and (d).

Daniel B. Botkin

(a)

Sedge puts out floating runners
Open water

(b)

Sedge forms a floating mat that supports other plants

Sedge mat almost closed

(c)

Sediments

Original soil

(d)

Photos by Daniel B. Botkin

(a)   *(b)*

▪ **FIGURE 6.5**
Old farm field to forest.
Abandoned farmland returns to forest through ecological succession which takes several hundred years. (a) Land that was a farm field and abandoned about ten years before this picture was taken is returning to forest. Small trees adapted to open areas with much sunlight are common, including the column-shaped red cedar, along with grasses and other flowering plants typical of early stages in forest succession in this region. (b) The interior of the nearby Hutchenson Memorial Forest, the last remaining uncut forest in New Jersey, where widely scattered large oaks are common, most trees are tall, shade is deep. This is forest succession in the eastern deciduous forest of North America, but the patterns of succession are general.

Old-Field Succession

Forest to farmland to forest again. In the eastern United States, a great amount of land was cleared and farmed in the 18th and 19th centuries. Today, much of this land has been abandoned for farming and allowed to grow back to forest. The first plants to enter the abandoned farmlands are small, short-lived annuals or perennials, adapted to the harsh and highly variable conditions of a clearing—a wide range of temperatures and precipitation (Figure 6.5a). As these plants become established, other, larger plants

enter. Eventually, large trees grow, such as oaks and hickories (Figure 6.5b), forming a dense forest. The pattern is probably beginning to sound familiar to you, because it is similar to the pattern of succession on dunes and in bogs—only the species and their niches are different.

General Patterns of Succession

Whether in dunes, bogs, or fields, the patterns are similar. Even though the environments are different, you can see common elements in these three examples of ecological succession.

1. An initial stage with kinds of vegetation that are specially adapted to the unstable conditions. These plants are typically low-growing, with adaptations that help to stabilize the physical environment.

2. A second stage with plants still of small stature but rapidly growing and with seeds that spread quickly.

3. A third stage in which larger plants, including trees, enter and begin to dominate the site.

4. A fourth stage in which mature forest develops.

Although we list four stages, the first two are usually combined so that there are typically three stages: early, middle, and late.

Early, middle, and late successional stages. These general patterns of succession can be found in most ecosystems. We describe them here in terms of vegetation, but similarly adapted animals and other life-forms are associated with each stage. Later in this chapter we will discuss other general properties of the process of succession.

Early-successional species are characteristic of the early stages. Often called "pioneers," they have evolved in conditions similar to those of early succession and have developed adaptations to those environmental conditions.

Late-successional species are dominant in the late stages of succession. These species, which have evolved in environments similar to those of late successional stages, are adapted to those conditions and thus tend to be slower-growing and longer-lived. Late-successional plants, for example, grow well in shade and have seeds that, although not as widely dispersing, can survive for a rather long time.

Knowing when species occur during succession is helpful to us. For example, a typical sequence of commercially valuable trees in the eastern deciduous (leaf-shedding, not evergreen) forests of North America, from New England south down the Appalachian Mountains, is cherry trees followed by white ash followed by sugar maple and beech. If we want to grow cherry trees for their wood and fruit, then we would want to *clear-cut* the forest (cut all the trees) comparatively often, say every 30 or 50 years. If the forest is in Pennsylvania, where the best white ash grows for making baseball bats, we might want the successional procedure to be longer, say 50–60 years. And if we want to harvest sugar maple for its wood but also save enough large maple trees for making sugar, we might not cut any maple trees for a longer period, perhaps 70 or 100 years or even more (the exact number of years in each case depends on local conditions, such as average temperatures and soil types).

6.3 During Succession, Does One Species Prepare the Way for Another?

Sometimes yes . . . In dune and bog succession, the first plant species—dune grass and floating sedge—prepare the way for other species to grow. This is called **facilitation**, because the early-successional species facilitate (help) the establishment of later successional species (Figure 6.6).

. . . and sometimes no. Some early-successional species, unlike the facilitators, actually interfere with the entrance of other species. For example, in the old fields in the eastern United States, a prairie grass called little bluestem often gets established and forms a mat so dense that seeds of other plants that fall onto it cannot reach the ground and therefore do not germinate. This process is called, naturally enough, **interference**.

Interference does not last forever. Eventually, some breaks occur in the grass mat—perhaps a patch of grass dies from a disease or is eaten away by an herbivore, or perhaps water erodes a patch away, or fire burns a small clearing. Whichever way they occur, openings in the grass mat allow seeds of trees such as red cedar to reach the ground. Red cedar is adapted to early succession in two ways: (1) Its seeds are spread rapidly and widely by birds who feed on them, and (2) this species can grow well in the unshaded sun and otherwise harsh conditions of early succession. Once it gets started, red cedar soon grows taller than the grasses, shading them so much that they either cannot grow or else grow poorly (Figure 6.5a). As a result, more ground opens up, and the grasses are eventually replaced by shrubs and trees and by woodland perennials that can grow in the shade.

But some plants can interfere longer than others. In parts of tropical and subtropical Asia and Southeast Asia, such as the Philippines, bamboo and another grass, *Imperata*, form dense mats that interfere with other plants. Like little bluestem in the United States, these grasses form stands so dense that seeds of other, later-successional species cannot reach the ground, germinate, or get enough light, water, and nutrients to survive. Once established, *Imperata* and bamboo seem able to persist for a long time. *Imperata* either replaces itself or is replaced by bamboo, which then replaces itself. Once again, when and if breaks occur in the cover of these grasses, other species can germinate and grow, and a forest eventually develops. But replacement can take much longer than in eastern North America.

Life-History Differences

In some cases, the life histories of species affect the time of succession. Differences in their life histories allow some to arrive first and grow quickly, while

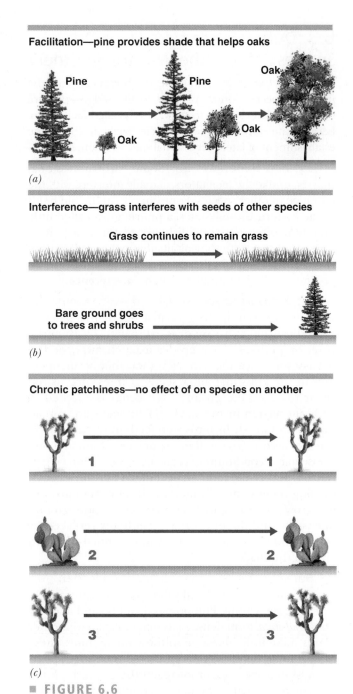

Facilitation—pine provides shade that helps oaks

(a)

Interference—grass interferes with seeds of other species

Grass continues to remain grass

Bare ground goes to trees and shrubs

(b)

Chronic patchiness—no effect of on species on another

(c)

■ **FIGURE 6.6**

How species might affect each other during succession. Two of the three patterns of interaction among species in ecological succession. (a) Facilitation. As Henry David Thoreau observed in Massachusetts more than 100 years ago, pines provide shade and act as "nurse trees" for oaks. Pines do well in openings. If there were no pines, few or no oaks would survive. Thus the pines facilitate the entrance of oaks. (b) Interference. Some grasses that grow in open areas form dense mats that prevent seeds of trees from reaching the soil and germinating. The grasses interfere with the addition of trees. (c) Chronic patchiness. Earlier-entering species neither help nor interfere with other species; instead, as in a desert, the physical environment dominates.

others arrive later and grow more slowly. For example, seeds of early-successional species are typically light and are easily blown about by the wind, or are adapted to be carried by animals (as with red cedar mentioned earlier). Some seeds have prickles that cling to an animal's fur or to a person's clothes. As a result, they reach clearings sooner and are able to germinate sooner than seeds of late-successional species, whose seeds tend to be larger. In many forested areas of eastern North America, birds eat the fruit of cherries and red cedar, and their droppings contain the seeds, which are spread widely. Sugar maples can grow in open areas, but their seeds take longer to travel and the seedlings can tolerate shade. Beech trees produce large nuts that store a lot of food for a newly germinated seedling. This helps the seedling establish itself in the deep shade of a forest until it is able to feed itself through photosynthesis. But these seeds are heavy and are moved relatively short distances.

Chronic Patchiness

Succession does not always take place. In some cases, species do not interact and succession, as it has been described, does not take place. The result is called **chronic patchiness**, and it may occur in some deserts where individual plants grow apart from each other and do not interact very much or at all (Figure 6.6c). In the warm deserts of California, Arizona, and Mexico, the major shrub species grow in patches, which often consist of mature individuals and few seedlings. These patches tend to persist for long periods. Similarly, in highly polluted environments, a sequence of species replacement may not occur.

What kinds of changes occur during succession depend on the complex interplay between life and its environment. Life tends to build up, or aggrade, whereas nonbiological processes in the environment tend to erode and degrade. In harsh environments, where energy or chemical elements required for life are limited and disturbances are frequent, the physical, degrading environment dominates, and succession does not occur.

Other Changes During Succession

Biomass, biological diversity, and chemical cycling change. In early stages of succession, biomass and biological diversity increase. But then the net production decreases, approaching zero in late successional stages. Chemical cycling also changes. On land, the storage of chemical elements (including nitrogen, phosphorus, potassium, and calcium, essential for plant growth and function) generally increases during succession. There are two reasons for this.

First, organic matter stores chemical elements; as long as there is an increase in organic matter within the ecosystem, there will be an increase in the storage of chemical elements. This is true for live and dead organic matter. Additionally, many plants have root nodules containing bacteria that can assimilate atmospheric nitrogen, which is then used by the plant in a process known as **nitrogen fixation**.

Second, live and dead organic matter retards erosion by wind and water. The amount of chemical elements stored in a soil depends not only on the total volume of soil but also on its storage capacity for each element. Chemical storage capacity of soils varies with the average size of the soil particles. Soils composed mainly of large, coarse particles, like sand, have a smaller total surface area and can store a smaller quantity of chemical elements. Clay, which is made up of the smallest particles, stores the greatest quantity of chemical elements.

Chemical elements in living tissue are more easily available to other living things. Soils contain greater quantities of chemical elements than do live organisms. However, much of what is stored in soil may be relatively unavailable, or may only become available slowly because the elements are tied up in complex compounds that decay slowly. In contrast, the elements stored in living tissues are readily available to other organisms through food chains.

Left undisturbed for a long time, an ecosystem slowly loses its stored elements. The increase in chemical elements that occurs in the early and middle stages of succession does not continue indefinitely. If an ecosystem persists for a very long time with no disturbance, it will gradually lose its stored chemical elements. The ecosystem will slowly run down and become *depauperate*—literally, impoverished—and thus less able to support rapid growth, high biomass density, and high biological diversity (Figure 6.7).

Disturbance changes the chemical cycling in an ecosystem. When an ecosystem is disturbed by fire, storms, or human activities, changes occur in its chemical cycling.

For example, when a forest is burned, complex organic compounds, such as wood, are converted to smaller inorganic compounds, including carbon dioxide, nitrogen oxides, and sulfur oxides. Some of the inorganic compounds from the wood are lost to the ecosystem during the fire as vapors that escape into the atmosphere and are distributed widely, or as particles of ash that are blown away. However, some of the ash falls directly onto the soil. These compounds are highly soluble in water and readily available for vegetation uptake. Therefore, immediately after a fire, the availability of chemical elements increases.

Plants that survive a fire take up the newly available elements rapidly, especially if the fire is followed by a moderate amount of rain (enough for good vegetation growth but not enough to cause excessive erosion). The pulse (increase) in inorganic nutrients can increase the growth of vegetation and thereby increase the amount of stored chemical elements in the

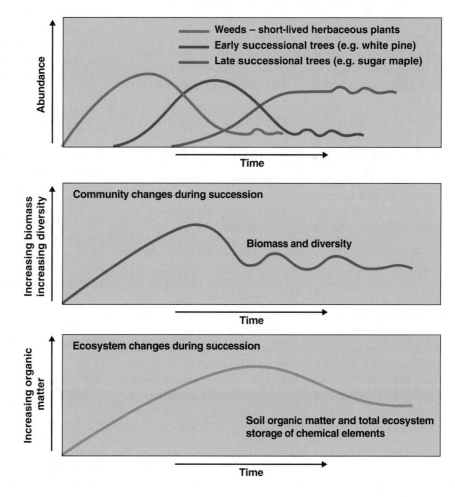

■ **FIGURE 6.7**
Ecosystem changes with succession.
Graphs showing changes in biomass and diversity with succession.

vegetation. This in turn boosts the supply of nutritious food for herbivores, and as a result their populations may increase as well. As you can see, a pulse in chemical elements in the soil can have effects throughout the food chain.

Other disturbances on the land produce effects similar to those of fire. For example, severe storms, such as hurricanes and tornadoes, knock down and kill vegetation. The vegetation decays, increasing the concentration of chemical elements in the soil, which are then available for vegetation growth. Storms also have another effect in forests: When trees are uprooted, chemical elements that were near the bottom of the root zone are brought to the surface, where they are more readily available.

Restoration requires knowledge of changes in chemical availability. We know that nutrients must be available within the rooting depth of the vegetation, and that the soil must have enough organic matter to hold on to nutrients. Restoration will be more difficult where the soil has lost its organic matter and has been leached. A leached soil has lost nutrients as water drained through it, especially acidic water, dissolving and carrying away chemical elements. Heavily leached soils, such as those subjected to acid rain and acid mine drainage, pose special challenges to those attempting to restore the land.

6.4 Can Nature Ever Be Constant?

We think not. Although the discussion of succession so far may suggest that it does lead to a final, constant state, several examples suggest that it can never happen—that nature can never be completely constant.

Australia's ancient dunes are unusual. In Australia, some sand dunes have existed for 100,000 years, which is uncommon because dunes are usually subjected to intense storms and are eventually blown open by the wind. The Australian dunes form a sequence, with the youngest dunes nearest the shore and the oldest dunes farthest inland. (This is because the sand blows inland, forming new dunes nearest the ocean.) You can trace the history of a typical area by walking inland from the coast, from one dune to another, a walk that is a journey back through time, taking you to older and older stages in ecological succession.

At first, the vegetation seems to follow the classical model of succession: Walking inland from the shore, from dune to dune, one sees at first that the plants become larger and denser. Near the shore are a few scattered hardy plants, small and with shallow roots. These hold the sand in place, allowing seeds of other plants to sprout. Farther inland grow woodland plants that are larger and more diverse.

But at the oldest dunes, the pattern deviates from the classical one: When you reach the dunes farthest inland—the oldest dunes—the vegetation becomes smaller and less diverse as woodlands give way to a shrubland.

Why does this happen? Near the shore, chemical elements needed for life can build up in the sandy soil from salty ocean spray and from organic matter added by the plants as they die. At the same time, rain leaches chemical elements down through the soil, until eventually most of the chemicals settle in a layer below the reach of most kinds of plants. The soil within reach becomes infertile, and a scrubland of relatively few species survives. Only a severe disturbance will turn the sand over and bring the chemical elements necessary for life back to the surface.

The same is true for rain forests on the west coast of New Zealand's South Island. This is one of the rainiest parts of the world, so chemical elements in the soil leach downward fast. Once again you can take a walk to follow ecological succession. In this case you will take just a short walk toward the ocean from the edge of a glacier toward the shore. Farthest inland, near the glaciers, where plants are just beginning to become established, the normal pattern of ecological succession occurs: first lichens and mosses, then low flowering plants, then trees, then temperate rain forest with a dense growth of many species. But when you reach the land that has been ice-free the longest, nearest the shore, the rain forest disappears and is replaced by scrubby grass and shrubs (Figure 6.8). In this case, as on the Australian dunes, chemical elements necessary for the plants have leached down below the reach of the trees, and only a few plants adapted to very low-nutrient soils can grow. Another glaciation that would turn over the soil could bring these elements to the surface.

A third example is the regrowth of forests along the Alaskan coast as the glaciers melted back in the 20th century. Again, the succession of plants at first seems to fit the classical pattern. Alders are among the first trees to become established; they have nitrogen-fixing symbiotic bacteria in their roots, and together the alders and their bacteria enrich the soil. In the enriched soil grow other trees, including spruce, which eventually grows taller than the alders, shading that species and preventing young alders from growing. With fewer and fewer alders and their bacteria, which add nitrogen to the soil, the soil gradually becomes less fertile. When spruce trees die, beds of sphagnum moss develop. They make the soil acid and soak up water, making the area uninhabitable for trees. In the end, what was forest be-

(a)

(b)

■ **FIGURE 6.8**
New Zealand temperate rain forest.
(a) At its "mature" stage and (b) the soil of an older stage, where trees have been replaced by low shrubs.

comes bog, not at all like the mature forest of spruce that was supposed to be the final stage.

In all three cases, environmental change could restart the process. Another ice age would start the Alaskan forests again. Even a slight cooling period—perhaps no greater than the little ice age that occurred from the mid-16th to the mid-19th century—could rebuild the glaciers enough to cover Glacier Bay and renew forest succession. Similarly, the New Zealand rain forests will not occur again without another ice age. A smaller, shorter change in the environment, perhaps a period of severe storms, might be enough to restart the succession of dunes along the Australian coast. But in all three cases, an environmental change is necessary for the recurrence of what used to be thought of as the climax (final) stage.

We can help nature along by performing some of its functions ourselves. If we want to keep some of these stages frozen in their current conditions, we can change the soils by turning them over or adding fertilizers. That is, we can substitute our own energy, time, and resources to replace nature's natural processes. We can do this in some cases for a long time and in all cases for a short time, but we cannot freeze all of nature indefinitely in a single state.

6.5 Examples of Restoration

How should we approach restoration? As you now see, an ecosystem passes naturally through many different states, and all of them are "natural." So if change itself—including certain kinds of wildfire—is natural, then what can it mean to "restore" nature? And how can a restoration plan that involves such things as wildfires be carried out without causing undue damage to human life and property?

Steps in Ecological Restoration: Planning

Before attempting an ecological restoration, you must do the following things:[3]

■ Understand why restoration is needed.
■ Describe ecologically the area to be restored.

▪ List the goals of the restoration. Sometimes objectives and goals are as simple as removing invasive exotic species and planting native species. Other times restoration has complex goals and objectives, as, for example, restoring a field to a freshwater wetland.

▪ Develop performance standards and plans for monitoring the project.

▪ Develop strategies to ensure long-term (sustainable) maintenance and protection of the restored ecosystem. That is restore ecosystem process and function.

A control plot is desirable. The restoration project will be more scientific if we maintain an untreated control plot. Comparing the ecosystem we are restoring with the untreated ecosystem will help us determine how far we have progressed toward the goals of the transformation.

Prairie Restoration

Prairies once occupied more land in the United States than any other kind of ecosystem. Today, only a few small remnants of prairie remain.

Prairie restoration is of two kinds. In a few places, there is still original prairie that has never been plowed. Here, the soil structure is intact, and restoration is simpler: The primary task is to restore the prairie animals and plants that have been removed, and to re-create the proper habitats for them. One of the best known of these areas is the Konza Prairie near Manhattan, Kansas. In other places, where the land has been plowed, restoration is more complicated. Nevertheless, restoration of prairies has gained considerable attention in recent decades, and in many midwestern states farmland is being restored to prairie.

The Allwine Prairie, within the city limits of Omaha, Nebraska, is an example. It has been undergoing restoration from farm to prairie for many years. In 1970, 65 hectares (158 acres) were seeded with a variety of prairie grass, and 5 of the 65 were set aside to test the effects of fire in maintaining the prairie vegetation. Experiments indicated that summer fires increased the diversity of the vegetation.

Unplowed strips of prairie are useful sources of seeds. Although most prairie land was converted to agriculture, long, narrow strips of unplowed native prairie remain on the rights-of-way along roads and railroads. In Iowa, for example, prairie once covered more than 80% of the state—11 million hectares (28 million acres). More than 99.9% of the prairie land has been converted to other uses, primarily agriculture, but the strips of surviving prairie along roadsides add up to 242,000 hectares (600,000 acres)—more than in all of Iowa's county, state, and federal parks. These roadside and railway stretches of prairie provide some of the last surviving habitats for native prairie plants, and restoration of prairies elsewhere in Iowa is making use of these habitats as seed sources.[4]

Restoration of the Florida Everglades

The Florida Everglades is home to thousands of species. More than 11,000 species of plants, several hundred bird species, and many species of fish live in the Everglades, and it is the last remaining habitat for 70 threatened and endangered species, including the Florida manatee, the Florida panther, and the American crocodile (Figure 6.9).

A century of draining the Everglades has shrunk the area by half. People have been draining the Everglades wetlands since the beginning of the 20th century for agriculture and urban development, so that today only half of the original wetlands remain. A complex system of canals and levies controls much of the flow of water for a variety of purposes, including flood control, water supply, and land drainage.[5]

The world's largest wetlands restoration project. Today the Everglades is the focus of the world's largest environmental project on a wetlands, a 30-year endeavor that will cost more than $10 billion. Restoration goals include:

▪ restoration of a more natural flow of water into and within the Everglades;

▪ enhancement and recovery of native and endangered species;

▪ improvement of water quality, especially control of nutrients from agricultural and urban areas;

▪ restoration of habitat for all wildlife that uses the Everglades.

The restoration plan is an aggressive one that involves federal, state, and American Indian tribal partners, as well as numerous other groups interested in the Everglades. The project has already reduced pollution of water flowing into the Everglades from agricultural lands by about one-half, and thousands of acres have been treated to remove invasive exotic species. The goal is to improve and conserve habitat for a variety of endangered species, including the key deer and the red-cockaded woodpecker.

Related to the Everglades project is restoration of the Kissimmee River, one of the largest and most expensive river restoration projects in the United States: The river was channelized by the U.S. Army Corps of Engineers to provide ship passage through Florida. However, although channelizing greatly altered the river and its adjacent ecosystems, shipping never developed, and now several hundred million

Rob Rayworth/Alamy Images

(a)

Susan Greenwood/Getty Images, Inc.

(b)

Fritz Poelking/Age Fotostock America, Inc.

(c)

■ **FIGURE 6.9**
The Everglades.
Inside the Everglades, the National Park Service operates tours in (a) open boats past mangrove islands, (b) alligators, and (c) many species of birds.

dollars must be spent to put the river back as it was before. The task will include restoring the meandering flow of the river channel and replacing the soil layers in the order in which they had lain on the bottom of the river before channelization.

The Everglades restoration is a long-term project and involves many years of scientific research yet to be completed. The program is complicated by the fact that over 5 million people live in South Florida, and the area has a rapidly growing economy. As a result, many urban issues related to water quality and land use need to be addressed. The really big issue is the water. The planners will need to carefully consider restoration that delivers water in the proper

amount, quantity, and place to support ecosystems in the Everglades.

Restoration of California's Channel Islands and Their Strange Island Foxes

Exotic species are the main problem. The Channel Islands lie offshore of southern California, stretching in a line from the Mexican border to west of Santa Barbara. The islands were settled by California Indians and were used after European settlement for ranches and, in a few cases, for recreation. In the 1800s, ranches on the islands had pigs, sheep, and horses—all introduced, not native—as well as a variety of introduced plants, including fennel, a European herb that displaced native island plants.

The goals are to remove exotic species and protect island species. Several of the native species are in danger of extinction—especially the odd little island fox found on eight islands (Figure 6.10). This fox evolved over the past 20,000 years to become a separate species from its ancestors, the California gray fox (see the discussion of geographic isolation in Chapter 5).

Isolated on the islands, the fox evolved into a smaller animal, so that today they are about the size of a house cat.[6] Native Americans arrived about 12,000 years ago. Indian burial sites include foxes buried with human beings, which suggests that the animals were pets. Another odd thing happened to the island fox: Isolated without natural enemies, the island fox began to live a long time, so long that some became blind in old age, perhaps from cataracts or accident. The blind foxes were able to find their way around and could be seen feeding on beaches and other areas despite their handicap.

In the 1990s, the fox populations declined suddenly on several islands. On San Miguel Island, there were approximately 400 foxes in 1994 but only about 15 just 5 years later. Similar declines occurred on Santa Rosa and Santa Cruz islands. At first it was thought that some disease must be spreading rapidly through the fox population—and in fact there was an outbreak of canine distemper on Catalina Island.[7] On other islands, however, the explanation was not so clear. Ecologists eventually solved the mystery by discovering that foxes were being killed and eaten by golden eagles, which had only recently arrived on the islands after the islands' bald eagles disappeared. The bald eagles eat primarily fish and did not bother the island fox, but golden eagles feed on land animals.

The use of DDT led to the bald eagles' disappearance. In the 1970s we saw increasing concentrations of that pesticide in fish that the bald eagles ate. This interfered with reproduction by making their eggshells soft, and the bald eagles became endangered. Then golden eagles arrived and colonized the islands in the 1990s, apparently attracted by abundant food that they could hunt without competition now that the bald eagles were gone. The golden eagles found young feral pigs much to their liking and evidently also found the island foxes easy targets. Remains of island foxes have been found in eagle nests, and it is now generally agreed that the golden eagles are responsible for the decline in populations of the fox. In fact, of 21 fox carcasses studied on Santa Cruz Island in the 1990s, 19 were apparently victims of golden eagles.[8]

A program to restore the islands' foxes. Three of the Channel Islands are now part of the Channel Islands National Park, and the park staff has developed a management and restoration program that has five steps.[9]

- Capture remaining island foxes and place them in protected areas.
- Begin a captive breeding program to rebuild populations of the fox.
- Capture and transfer golden eagles to the mainland, far from the islands and in suitable habitat so that they will not return.

Courtesy Channel Island National Park/NPS

Courtesy Channel Island National Park/NPS

(a)

(b)

▪ **FIGURE 6.10**
(a) Channel Islands fox and (b) its habitat.
The Channel Islands are off the coast of Southern California near Santa Barbara.

■ Reintroduce bald eagles into the island ecosystem, in the hope that they will establish territories that will prevent the return of golden eagles.

■ Remove feral pigs, which are food for golden eagles and attracted them to the islands.

If all the steps necessary to save the island fox are successful, then the island fox will again take its place as one of the interesting species on the islands.

Restoring Land Damaged by Lead Mines in England

Approximately 55,000 hectares (136,000 acres) of land in Great Britain have been heavily damaged by years of mining—indeed, some of the mines have been used since medieval times. Now, people are trying to remove toxic pollutants from the mines and mine tailings, to restore these damaged lands to useful biological production, and to restore the beauty of the landscape.[10]

First efforts failed. One area damaged by a long history of mining lies within the British Peak District National Park, where lead has been mined since the Middle Ages and waste tailings are as much as 5 meters (more than 16 feet) deep. The first attempts to restore this area used a modern agricultural approach: a heavy application of fertilizers and the planting of fast-growing agricultural grasses. These grasses grow quickly on the good soil of a level farm field, and it was hoped that, with fertilizer, they would do the same in this situation. But after a short period of growth, the grasses died. On the poor soil, leached of its nutrients and lacking organic matter, erosion continued, and the fertilizers that had been added were soon leached away by water runoff. As a result, the areas were soon barren again.

Knowledge of ecological succession helped. When the agricultural approach failed, an ecological approach was tried, using knowledge about ecological succession. Instead of planting fast-growing but vulnerable agricultural grasses, ecologists planted slow-growing native grasses that were better adapted to soils poor in minerals and to the harsh conditions in cleared areas. In choosing these plants, the ecologists relied on their observations of what plants were the first to appear in areas of Great Britain that had undergone succession naturally.[10] The result of the ecological approach has been successful restoration of the damaged lands.

You can find heavily damaged landscapes in many places. Similar restoration is being done in the United States to reclaim lands damaged by strip mining. Today, in such cases, restoration sometimes begins during the mining process, rather than afterward. Similar methods could be used to restore areas once occupied by buildings in cities.

Return to the Big Question

Can we restore damaged ecosystems?

The answer is yes. In fact, ecological restoration has become a major activity and a subdiscipline of the science of ecology. Some of the best-known successes in ecological restoration are the restoration of prairies, wetlands, and certain forested areas. Much remains to be done, however, especially in learning how to restore ecosystems that require certain kinds of changes, such as wildfires.

Summary

■ Restoration of damaged ecosystems is a major new emphasis in environmental sciences and is developing into a new field. Restoration ecology is the science behind the practice of ecological restoration. Restoration involves a combination of human action and natural processes of ecological succession.

■ Disturbance, change, and variation in the environment are natural, and ecological systems and species have evolved in response to these changes.

■ When ecosystems are disturbed, they undergo a process of recovery known as ecological succession, the establishment and development of an ecosystem.

Knowledge of succession is important in the restoration of damaged lands.

▨ During succession there is usually a clear, repeatable pattern of changes in species. Early-successional species are adapted to the first stages, when the environment is harsh and variable but necessary resources may be abundant. In late stages of succession, biological effects have modified the environment and reduced some of the variability but also have tied up some resources. Typically, early-successional species are fast-growing, whereas late-successional species are slow-growing and long-lived.

▨ Biomass, production, diversity, and chemical cycling change during succession. Biomass and diversity peak in midsuccession, increasing at first to a maximum, then declining and varying over time.

▨ Changes in the kinds of species found during succession can be due to facilitation, interference, or simply life-history differences. In facilitation, one species prepares the way for others. In interference, an early-successional species prevents the entrance of later-successional ones. Life-history characteristics of late-successional species sometimes slow their entrance into an area.

Key Terms

balance of nature
chronic patchiness
early-successional species
ecological restoration
ecological succession
facilitation
interference

late-successional species
life history difference
nitrogen fixation
primary succession
restoration ecology
secondary succession
successional stages

Getting It Straight

1. Redwood trees reproduce successfully only after disturbances (including fire and floods), yet individual redwood trees may live more than 1,000 years. Is redwood an early- or late-successional species?

2. Why could it be said that succession does not take place in a desert shrubland (an area where there is very little rain and the only plants are certain drought-adapted shrubs)?

3. What is the difference between primary and secondary succession?

4. Describe the general patterns of succession as they relate to the formation of dune grass in a dune ecosystem.

5. What is the goal of restoration ecology?

6. Discuss facilitation as it relates to successional species. What is the difference between early- and late-successional species?

7. What are the steps in ecological restoration planning? Can nature be restored using this planning process?

8. What is ecological restoration and how can it be linked to improving island fox populations in California's Channel Islands?

What Do You Think?

1. What does the case study about saving the island fox suggest about the importance of environmental unity?

2. Farming has been described as managing land to keep it in an early stage of succession. What does this mean, and how is it achieved?

3. Oil has leaked for many years from the underground gasoline tanks of a gas station. Some of the oil has oozed to the surface. As a result, the gas station has been abandoned and revegetation has begun to occur. What effects would you expect this oil to have on the process of succession?

Pulling It All Together

1. Develop a plan to restore an abandoned lot in your town to natural vegetation for use as a park. The following materials are available: bales of hay; artificial fertilizer; and seed of annual flowers, grasses, shrubs, and trees.

2. List at least five ways in which humans modify natural ecosystems. Discuss the impact these activities have on the principles of sustainability. Are you willing to modify your lifestyle in order to ensure a sustainable lifestyle for your future? If so, what are you willing to change and how will you get others to follow your actions?

3. What dangers are involved when new species are introduced into an ecosystem? Who would survive, who might die, and would there be a need to create a new habitat to support the new species?

4. Who is responsible for restoring ecosystem sustainability? Do you have an ethical responsibility to ecosystem function to live within the limits of the natural environment? How might you convince others to avoid anthropogenic activities (human-caused) that contribute to ecosystem destruction and habitat demise?

5. Examine the balance between nature and humans that exists in your own community. Based on restoration ecology thinking, describe how your community might return damaged ecosystems back to their original condition or avoid anthropogenic activities (human-caused) that cause ecosystem destruction.

Further Reading

Berger, J. J. 1990. *Environmental restoration: Science and strategies for restoring the Earth.* Washington, DC: Island Press.—An informed and lively overview of the beginning of the restoration movement. It includes scientific and technical papers given at the first national conference on restoration, held in 1933.

Botkin, D. B. 1992. *Discordant harmonies: A new ecology for the 21st century.* New York: Oxford University Press.

Botkin, D. B. 2001. *No man's garden: Thoreau and a new vision for civilization and nature.* Washington, DC: Island Press.

Cairns, J., Jr., ed. 1995. *Rehabilitating damaged ecosystems,* 2nd ed. Boca Raton, FL: Lewis Publishers.—Discussions of natural and human-assisted restoration of various ecosystem types after either natural or human-caused disturbance.

Foster, D. R., and J. F. O'Keefe. 2000. *New England forests through time: Insights from the Harvard Forest diora-* *mas.* Cambridge, MA: Harvard University Press.—A beautifully illustrated short book that discusses secondary succession of forests in New England, using as the centerpiece a famous set of miniatures at the Harvard Forest.

Petts, G., and P. Calow, eds. 1996. *River restoration.* London: Blackwell Science.—An overview of international efforts to restore rivers, designed for undergraduates as well as other audiences. The book begins with a general introduction to rivers and includes chapters on the control of weeds, the conservation and restoration of fish, and the relationships between disturbance and recovery.

Stevens, W. K. 1995. *Miracle under the oaks, the revival of nature in America.* New York: Pocket Books.—The story of several citizen action groups' efforts to restore damaged ecosystems.

Lugris/Alamy Images

7

Forests and Wildlife

Big Question

Can We Have Forests and Wildlife and Use Them Too?

Learning Objectives

This chapter deals with the conservation and management of two kinds of commercially valuable living resources: forests and wildlife. These are combined because historically both forests and wildlife have been part of our "hunter-gatherer" activities. Mostly we have harvested these as they grow in the wild, but there is more and more emphasis on the "farming" of forests (in plantations) and wildlife. (For example, today some American bison are raised on farms, and in Africa there is what is called "game ranching," where native wildlife are maintained in naturalistic habitats and harvested from these.)

We will first discuss forestry, then wildlife management. The question that confronts us about both is how to use them sustainably. There are two basic kinds of ecological sustainability: sustainability of the harvest of a resource that grows within an ecosystem; and sustainability of the entire ecosystem, with its many species, habitats, and environmental conditions. For forests, this translates into sustainability of the harvest of timber and sustainability of the forest as an ecosystem. And for wildlife, this translates into sustainability of a wildlife population and sustainability of its ecosystem. When you finish this chapter, you should understand . . .

- why people value forests;
- how forests and wildlife have been managed;
- the difference between preservation and conservation of nature;
- the difference between a sustainable product and a sustainable ecosystem;
- what the conservation of forests and the conservation of wildlife have in common;

- what an old-growth forest is, and what a second-growth forest is;
- how understanding ecological succession can help us manage and conserve forests;
- what is meant by maximum sustainable yield of a wildlife population, and why the concept is outdated;
- the ecosystem context for wildlife management.

Case Study

Trying to Save a Small Owl from Extinction

Every year, each of us uses more than 700 pounds of paper—the weight of about 175 of the textbooks you are reading. We are living in the computer age, an age that was supposed to lead to the paperless office and a great reduction in the world's overall use of paper. But the opposite has happened: Whereas in 1900, U.S. paper companies produced 14,000 tons of paper a day, today they produce 250,000 tons a day. Together, you and the rest of Americans get 250 million magazines a year, 2 billion books, and 24 billion newspapers. Paper is crucial to our way of life, to our civilization.

Most paper is made from wood, and wood products have always been vitally important for people and civilization.[1] But while we must cut down trees to obtain most of these products, we also value forests that are standing. So the big question is: Can we continue to have these products and still have our forests? This raises another question: What is it that we value so much about forests that are intact?

The story of a small owl illustrates what people in America value about forests. In the early 1990s the spotted owl (Figure 7.1), which many people had never heard of, was listed as threatened under the U.S. Endangered Species Act. Experts claimed that the spotted owl nested and lived only in old-growth forests (forests that have never been cut and thus consist mainly of large, ancient trees) of the Pacific Northwest. The experts said that widespread clear-cutting of these forests—the usual logging practice in the Pacific Northwest at that time—was going to cause extinction of these owls.

The federal government funded a study involving 600 scientists, who produced a Northwest Forest Plan that required conservation of existing old-growth forest. This plan caused one of the greatest economic changes of any American conservation policy, putting a stop to logging on 2.4 million acres in the state of Washington alone, and ending an estimated 30,000 jobs in that state. Towns like Forks, Washington, on the Olympic Peninsula, which had been a center for the timber industry, found themselves with many people out of work and forced to seek other kinds of jobs.

As you can imagine, the whole issue caused fierce arguments. Opponents claimed that the owl could live just as well in second-growth forest (forests that have grown back after logging, thus having young and relatively small trees and lacking many other features of an old-growth forest). They said the owl and the loggers could continue together, and that in any case the owl was not worth the cost. They claimed that the real goal of the so-called supporters of the spotted owl was to stop the logging and preserve old-growth forests, and they cried, "Fowl!"

Ironically, although much old-growth forest has been conserved for the owl, today this species is suffering from other threats. The barred owl, a competitor, is moving into the spotted owl's habitat, and as we learned in our earlier discussion of the competitive exclusion principle (see Chapter 4), the two species cannot coexist. Not only is the barred owl driving out the spotted owl, but in addition a new disease called "sudden oak disease" is killing many oaks and may wipe out the owl's preferred tree, the tanoak. As well, the recently introduced West Nile virus is attacking the spotted owl.

As inquiring students who probably have not spent a lot of time in forests, nor worked in them,

Michael Townsend/Photographer's Choice/Getty Images

■ FIGURE 7.1
Spotted owl.

you may be wondering, first of all, why clear-cutting was considered so bad? Was old-growth forest actually necessary for the owl? And even more basic: What's all the fuss about? Do we really need all that forestland, and who cares about some little owls that most of us will never see? Was all this worth it? This chapter will help you answer these questions.

Section I: Forestry

7.1 Keeping Our Living Resources Alive

Forests have always been important to people, both as a source of products valuable to civilizations—wood for construction and fuel—and as spiritual places where people could go for inner renewal or religious rituals. The conservation and management of forests and commercially valuable wildlife have a lot in common, and learning about one will help us understand about the other.

For both, the traditional goal has been the maximum sustainable yield—the largest harvest that a forest, or wildlife population (or other resource), is able to replace. And until now, the management of both forests and wildlife has been based on traditional concepts—in particular, the belief in the balance of nature (see Chapter 6). According to this belief, the only changes that take place in these resources are the result of human actions; without our interference, wildlife and forests and all living resources would achieve the largest biomass and greatest diversity, and remain constant. This belief in the balance of nature still underlies much of the approach to conservation and management of our living resources.

As we do throughout this book, we approach the subject through science. To decide what is useful and what is worth doing, our first task is to understand certain principles of forest management, especially the scientific basis for forestry.

7.2 Modern Conflicts over Forestland and Forest Resources

Conservation of forests has become a popular cause the world over, especially conservation of our remaining old-growth forests and most especially the old-growth forests of the giant trees in the Pacific Northwest of America—the forests of redwoods, sequoia, Douglas fir, western hemlock, western white pine, and such—and of many tropical rain forests. At the heart of the matter is a debate about whether forests left undisturbed achieve a single, permanent condition, the balance of nature that we have run

into before, referred to here as virgin forests or old-growth forests. (We should point out that although the term *old-growth forest* has gained popularity in several well-publicized disputes about forests, it is not a scientific term and does not yet have an agreed-on, precise meaning.)

What's "natural" and what isn't? There is a long tradition of believing that only this kind of forest is "natural." But as we learned in our discussion of restoration ecology in Chapter 6, ecosystems are always undergoing change, and all the stages in forest succession are therefore natural. Today the question has to be rephrased as, What percentage of the land is naturally in each stage of succession? A related question is, Does this "naturalness" really matter?

Growing trees has become a profession called *silviculture* (combining the Latin word *silvus* for "forest," and *culture*, as in "agriculture"). People have long practiced silviculture, much as they have grown crops, but not until the late 19th and early 20th centuries did forestry develop into a science-based activity and into what we today consider a profession.

Civilizations have literally been built on wood. Forests have always been vital resources for civilization, as major providers of construction materials and fuel. Forest products of all kinds are worth about $150 billion a year as exports around the world, let alone the value of these products used within each nation.[2] Nearly two-thirds of all wood produced in the world is used for firewood. Although today firewood provides only 2% of total commercial energy in developed countries, it provides 15% of the energy in developing countries and is the major source of energy for most countries of Sub-Saharan Africa, Central America, and continental Southeast Asia.[3] As the human population grows, the use of firewood increases. With such worldwide demands for timber, management is essential.

Forests also have had religious, spiritual, and aesthetic importance—nonmaterial, nonutilitarian value—to people throughout history, as illustrated by the customs in many societies and by what people have written over the ages. From the aesthetic point of view, clear-cuts are particularly offensive (Figure 7.2), looking like scars on the landscape.

At the heart of the conflict are the two different kinds of values, utilitarian and nonutilitarian, both of which make forests matter to a lot of people. The question is, are forests there for us to harvest, or are they there to raise our spirits, and because they have a right to exist independent of us (the same reasons we learned about in Chapter 5 for the conservation of endangered species)? The long-standing debate between these two points of view is perhaps best illustrated by the conflict between two famous American conservationists, John Muir

Photodisc/Getty Images

■ FIGURE 7.2
Clear-cut forestland in the Pacific Northwest.
One reason people oppose clear-cutting is that the effect is ugly; these large clear-cuts look like scars on the landscape. But is clear-cutting always bad for forest ecosystems or their species?

Library of Congress

Library of Congress

(a)

(b)

■ FIGURE 7.3
(a) John Muir (standing on a mountain to the right of President Teddy Roosevelt) and (b) Gifford Pinchot.

(1838–1914) and Gifford Pinchot (1865–1946), pictured in Figure 7.3.

Muir and Pinchot personify the two viewpoints. Pinchot was the first head of the U.S. Forest Service and founder of the Yale School of Forestry (America's first professional forestry school), and is generally considered the father of modern professional forestry. Pinchot, along with President Theodore Roosevelt, coined the term *conservation,* by which they meant the preservation and wise use of all natural resources.[4] Pinchot believed in making good use of forests, while Muir believed in the preservation of nature. The two met in 1893 to survey problems in the forests of America's West.

Good friends at first, but not for long. Pinchot had been quoted in a newspaper as saying that sheep grazing in western forests did no harm. Muir, in contrast, called sheep "hoofed locusts," and when he found out what Pinchot had said, told him he wanted nothing more to do with him.[5] Muir went on to write eloquently about the preservation of nature undisturbed. "Everybody needs beauty as well as bread, places to play in and pray in, where nature may heal and give strength to body and soul alike,"[6] he wrote. "Keep close to Nature's heart . . . and break clear away, once in awhile, and climb a mountain or spend a week in the woods. Wash your spirits clean."[7] Meanwhile, Pinchot wrote, "Conservation is the fore-sighted utilization, preservation and/or renewal of forests, waters, lands and minerals, for the greatest good of the greatest number for the longest time," and, "The earth and its resources belong of right to the people."[8]

For Muir, concern for nature overrode concern for the common man. Thus, Muir campaigned to preserve forests without human use while Pinchot worked for what we would call today "sustainable use and harvest." Today we distinguish these points of view by calling Muir's the *preservation* of nature and calling Pinchot's use of forests for the good of people the *conservation* of nature.

The dam at Hetch Hetchy led to one of their greatest arguments—and also especially illustrates our modern conflicts. The dam was to be built in Hetch Hetchy Valley, a glacial valley in California through which the Tuolumne River flowed and which had become part of Yosemite National Park in 1890 (Figure 7.4). Pinchot approved of the dam when he was head of the U.S. Forest Service, saying that it "would not injure the National Park or detract from its beauties or natural grandeur."[9] Muir, meanwhile, said that Hetch Hetchy was "a grand landscape garden, one of nature's rarest and most precious mountain temples," and that a dam could only be built by "devotees of ravaging commercialism," who "have a perfect contempt for Nature,"[10] a description that Pinchot would hardly have applied to himself.

■ **FIGURE 7.4**
Hetch Hetchy Valley, 1987, by Galen Rowell.
Hetch Hetchy Valley with its dam and Hetch Hetchy reservoir.

The dam was built, but the debate goes on. Although the dam has been a source of water for San Francisco since 1934, the debate over Hetch Hetchy still rages, illustrating the depth of feelings and the intensity of conflict today over the material and nonmaterial values of forests and forested landscapes (and their wildlife).

Forests also provide indirect benefits. In addition to the two kinds of values that Muir and Pinchot argued over—direct benefits—today we realize that forests benefit people and the environment indirectly, through what we call *public-service functions*, such as retarding soil erosion (forests are maintained around reservoirs for that purpose). Forests also are habitats for endangered species and other wildlife. They are important for recreation, hiking, hunting, and bird and wildlife viewing, and may affect the climate both regionally and globally.

To summarize, old-growth forests are at the heart of modern forestry controversies, which focus on the following questions:

- Should a forest be used only as a resource to provide materials for people and civilization?
- Should a forest be used only to conserve natural ecosystems and biological diversity, including specific endangered species?
- Can a forest be managed for timber harvest and also provide recreation and landscape beauty and meet the spiritual needs of people? And does this matter anyway?
- Can we achieve sustainable forests?
- What role do forests play in our global environment, such as their effects on climate?
- What is "natural" in a forest?
- How much old growth do forests need?

7.3 A Modern Forester's View of a Forest

Timber is no longer the sole concern. In the early days of the 20th century, the goal of silviculture was generally to maximize the yield in the harvest of a single resource. The ecosystem was a minor concern, as were those species of trees and wildlife that had no commercial value. Today, most foresters take a much broader view, considering the *sustainability* of the timber harvest and of the ecosystem, and considering the entire range of goals for managing forests.

So why would a forester want to clear-cut a forest? To better understand this section, you may want to review the ideas about ecological succession in Chapter 6 on restoration ecology. In brief, two major reasons for clear-cutting are: (1) it's cheaper—once you get all the equipment and people in place, it's cheaper to cut all the trees than to only cut some; and (2) if the species you want to harvest grows in early-successional stands, you have to open up the forest enough for that species to regenerate and grow. Many commercially valuable trees are early-successional. Even Douglas fir, the dominant species of the spotted owl's habitat, requires an opening to regenerate. So do redwood trees, sequoia, white pine (one of the most important timber trees of eastern North American forests), and most species of pine, as well as cherry and other fruit trees whose hard wood is valuable.

Why wouldn't you clear-cut? For the following reasons:

- If the species you want to harvest is late-successional, you do not want to clear the forest.
- Clear-cutting can be very hard on the ecosystem, increasing erosion, for example.
- Even if the species you want to harvest is early-successional, you may believe that natural clearing disturbances, such as fires and storms, are frequent enough anyway, and that additional clearing as part of logging may reduce the old-growth habitat too much and threaten species that depend on old-growth forest conditions.

The Famous Hubbard Brook Experiment

What happens to an ecosystem when you clear-cut? We know a lot about what happens to a forest when you clear-cut it, because several major experimental clear-cuts have been done by the U.S. Forest Service and university scientists, starting in the second half of the 20th century.

Courtesy of US Forest Service, Northern Research Station

■ **FIGURE 7.5**
The Hubbard Brook Ecosystem study.
An entire watershed was clear-cut in northern New Hampshire, and the response of the cut ecosystem was compared to an uncut watershed on the same mountain.

The best known of these experiments is at Hubbard Brook Experimental Forest in northern New Hampshire. There, in 1965, all the trees and shrubs were cut in a small *watershed*—about 16 hectares, or 40 acres. (A watershed is an area of land in which every drop of rain that falls flows out through the same stream. So a watershed makes a natural water unit and therefore a natural ecological unit.) Herbicides were broadcast on the watershed to kill regrowth for four years (Figure 7.5).

One dramatic result was a much greater concentration of nitrate in the stream flowing out of the watershed. This happened because of a sequence of events. Clear-cutting left lots of dead vegetation and opened the ground to sunlight. The sunlight warmed the ground and hastened the decomposition of the dead vegetation, causing the proteins in the dead plants to release nitrogen in the form of nitrate. Nitrate reached a level considered unsafe for drinking water!

Stream water runoff increased 30% for the first three years after clearing. This happened because trees and shrubs take water up from the soil and evaporate it through their leaves. Without the forest trees, evaporation of water in the clear-cut region decreased. (Interestingly, you can conclude from this that if you were to manage the forest only for water supply, and your only goal therefore was to increase water runoff, without concern about soil erosion or about the nitrate concentration in the water, then you might opt to keep the trees from regrowing. But this would be a shortsighted approach.)

What happened as the forest grew back? Four years after the clear-cut, the stream flow decreased as the forest regrew. And 12 years after the clear-cut, the stream flow actually dropped below the pre-experiment level.[11]

7.4 Clear-Cutting That Really Did Not Work: The Sad Story of Michigan's "Stump Barrens"

Between 1840 and 1910, 19 million acres of white pine forests were clear-cut in Michigan, the Paul Bunyan folktale country. The area was so huge that the foresters who started the cutting believed they would never run out of timber. They were sure that by the time the last acre was cut, the first would have regrown and be ready to cut again. Unfortunately, they underestimated the power of their tools and the size of the markets for the timber, much of which was used to build the houses of the eastern United States and the Midwest. Look at one of those wood frame houses of the eastern and midwestern United States and you are probably seeing what was once a tall white pine of Michigan. Maybe you live in one or grew up in one.

Poor logging practices made a bad situation even worse. Not only were 19 million acres clear-cut, but little

Courtesy Dan Botkin

▪ **FIGURE 7.6**
A Stump barren in Michigan, an area that was clear-cut and probably burned a century or more ago. The soil was badly damaged and has never regrown to forest. Bracken ferns, grasses, small flowering plants, and shrubs, typically of the blueberry family, are all that grow. The stumps of white pine are still visible, cut perhaps more than a century ago.

care was taken to protect the soils and the forest ecosystem. Careless practices by foresters led to frequent forest fires, and these were fueled by parts of trees that had been left lying around on the ground because nobody wanted them—branches, leaves, limbs, twigs, and so forth. As a result, the fires were much more intense than natural fires are (fires started by lightning, for example). They were so intense that they burned through the organic soil, destroying it and killing the few remaining seed-bearing trees. The result: large areas that even to this day have never regrown, and are called "stump barrens" (Figure 7.6). You can see these when you travel through Michigan's Upper Peninsula. They are testimony to poor logging practices and have helped to give clear-cutting a bad name.

Clear-cutting can be useful, but not on such a large scale. It's just not necessary to clear-cut as extensively as was done in Michigan. However, where the ground is level or slightly sloped, where rainfall is moderate, and where the desirable species are early-successional ones (see Chapter 6) that require open areas for growth, clear-cutting on an appropriate scale is a useful way to regenerate desirable species. The key here is that clear-cutting is neither all good nor all bad for timber production and for forest ecosystems. The use of clear-cutting must be evaluated on a case-by-case basis, taking into account the size of the cuts, the environment, and the available species of trees.

7.5 Are There Other Ways to Harvest Trees?

We have several alternatives to clear-cutting: shelterwood cutting, seed-tree cutting, selective cutting, and strip-cutting.

Shelterwood cutting is the practice of cutting dead and less desirable trees first, and later cutting mature trees. As a result, there are always young trees left in the forest.

Seed-tree cutting removes all but a few seed trees (mature trees with good genetic characteristics and high seed production) to promote regeneration of the forest.

In *selective cutting*, individual trees are marked and cut. Sometimes smaller, poorly formed trees are selectively removed; this practice is called *thinning*. At other times, trees of specific species and sizes are removed. For example, some forestry companies in Costa Rica cut only some of the largest mahogany trees, leaving other, less valuable, trees to help maintain the ecosystem, and permitting some of the large mahogany trees to continue to provide seeds for future generations.

In *strip-cutting*, narrow rows of forest are cut, leaving wooded corridors. Strip-cutting offers several advantages. The uncut strips protect regenerating trees from wind and direct sunlight, and these remaining trees provide seeds. In addition, a strip-cut looks better than a clear-cut and serves better as a wildlife habitat and as a place for recreation because of its wooded buffer zones and corridors of forest.

Traditionally, foresters have managed trees at a local level in *stands,* an informal term that foresters use to refer to a group of trees. If you go for a walk in the woods with a professional forester, he or she will likely point to a group of trees and say, "See that stand?" Stands can be small (half a hectare) to medium-size (several hundred hectares). Foresters classify stands on the basis of tree composition. The two major kinds of commercial stands are *even-aged stands*, where all live trees began growth from seeds and roots germinating the same year (that is, after a clear-cut or other clearing event, such as a storm or fire), and *uneven-aged stands*, which have at least three distinct age classes. In even-aged stands, trees are approximately the same height but differ in girth and vigor. Another important management term is *rotation time*, which is the time between cuts of a stand.

Somdutt Prasad/iStockphoto

■ **FIGURE 7.7**
Annual growth rings visible on a cut tree.

Other tree groupings. Foresters and forest ecologists group the trees in a forest into the *dominants* (tallest, most common, and most vigorous); *codominants* (fairly common, sharing the *canopy*—the top part of the forest); *intermediate* (forming a layer of growth below the dominants); and *suppressed* (growing in the *understory*—the area of trees and shrubs between the canopy and the groundcover).

Forests offer one advantage over other ecosystems. Although forests are complex and difficult to manage, one advantage they have versus many other ecosystems is that trees provide easily obtained information that can be a great help to us. For example, the age and growth rate of trees can be measured from tree rings, which most trees produce at the rate of one a year (Figure 7.7).

7.6 International Aspects of Forestry

Forests were cut even in ancient times, by the first civilizations. We find evidence of this in the Near East, Greece, the Roman Empire before the modern era, and northward in Europe as civilization advanced. Fossil records suggest that prehistoric farmers in Denmark did so much clearing of forests that early-successional weeds occupied large areas. In medieval times, Great Britain's forests were cut, and many forested areas were eliminated. With colonization of the New World, much of North America was cleared.

Robert Frerck/Odyssey Productions

NASA/Science Source/Photo Researchers

▪ **FIGURE 7.8**
The Indus River in northern India carries a heavy load of sediment, as shown by the sediments deposited in and along the flowing water and by the color of the water itself. This scene, near the headwaters, shows that erosion takes place at the higher reaches of the river.

▪ **FIGURE 7.9**
A satellite image showing clearings in the tropical rain forests in the Amazon in Brazil.
Rivers are black. The bright red is the leaves of the living rain forest. Straight lines of others colors, mostly light blue to gray, are deforested clearings extending from the roads. Much of the clearing is for agriculture. The distance across this image is about 100 kilometers (63 mi).

Cutting forests in one country affects other countries. For example, deforestation is estimated to have increased erosion and caused the loss of 562 million hectares (1.4 billion acres) of soil worldwide, and the estimated annual loss is more than 7 million hectares (more than 17 million acres).[12] Nepal, one of the most mountainous countries in the world, lost more than half its forest cover between 1950 and 1980. Such cutting destabilizes soil, increasing the frequency of landslides, the amount of runoff, and the sediment load in streams. Many Nepalese streams feed rivers that flow into India (Figure 7.8). Recent heavy flooding in India's Ganges Valley has caused $1 billion a year in property damage and may have resulted from the loss of large forested watersheds in other countries.[13]

Flooding in India could become a permanent problem. The loss of forest cover in Nepal continues at a rate of about 100,000 hectares (247,000 acres) per year. Reforestation efforts replace less than 15,000 hectares (37,050 acres) per year. If present trends continue, little forestland will remain in Nepal, which will make India's flooding problems permanently worse.[14] This in turn will have international economic consequences.

Today, deforestation occurs largely in the developing world.[15] Many of these forests are in the tropics, mountain regions, or high latitudes, places that were difficult to exploit before the arrival of modern transportation and machines. The problem is especially severe in the tropics because of high human popula-

tion growth, which poses many threats for forests. For example, deforestation due to fires rose sharply in Indonesia in 1997.

Satellites are a new and useful tool to monitor deforestation. The good news is that satellite images provide a new way to detect deforestation (Figure 7.9), and that, based on a United Nations analysis of 300 satellite images, the rate of deforestation in tropical rain forests appears to be slowing.[15]

Areas where tropical deforestation is a major concern include the Amazon basin, western and central Africa, and Southeast Asia and the Pacific regions, including Borneo, the Philippines, and Malaysia. Between 1960 and 1990, Asia cleared 30% of its tropical forests, Africa 18%, Latin America 18%, and the world as a whole 20%.[16]

7.7 Plantation Forestry

Why not just plant trees for harvesting and leave "natural" forests alone? In Western Europe, most of the forests you see are plantations, with trees of a sin-

■ **FIGURE 7.10**
Even the poplars behind this person on horseback riding along the *Canal de Midi* in France are part of a plantation.

gle species planted in neat rows, just like any crop (Figure 7.10). Even in the Alps, whose scenery is famous for its grandeur and "natural" beauty, most of the forests are plantations.

But plantations have caused another big argument. Those opposed to them complain that since plantations are not natural, not part of the original balance of nature, they must be bad. Critics also claim that plantations are bad land use. Typically in modern North American forest plantations, the land is fertilized, sometimes by helicopter, and modern machines do a speedy job of harvesting— some remove the entire tree, root and all. Indeed, plantation forestry is a lot like modern-day farming. Intensive management like this is common in Europe and parts of the northwestern United States.

On average, plantations are ten times as productive as other forests. It has been estimated that to meet the demand for international trade in timber products, about one-quarter of all the world's "natural" forests would have to be subjected to harvest. In contrast, the same amount of timber could be harvested from only about 4% of the kind of high-production trees raised on plantations. If so, then widespread use of forest plantations could offer a sizable environmental advantage, leaving more than 90% of the world's forestland for conservation of biodiversity and the many woodland activities that people enjoy. In short, Gifford Pinchot's goals might be met by just 4% of the world's forests, leaving the rest for John Muir's goals.[17]

7.8 Are the World's Forests Shrinking, Growing, or Neither?

Some experts argue that there is a worldwide net increase in forests because large areas in the temperate zone, such as the eastern and midwestern United States, were cleared in the 19th and early 20th centuries and are now regenerating. Most experts disagree. Because few forests are successfully managed to achieve sustainability, it seems likely that the world's forests are undergoing a net decline, perhaps rapidly.

The fact is that we lack enough information to accurately evaluate the situation. Because forests cover large areas, often in remote places that are seldom visited or studied, it has been difficult to assess the total amount of forest area. According to our best information, forests covered an estimated one-quarter of Earth's land area in 1950 but only one-fifth in 1980.[18]

But this has begun to change. Modern technologies, including remote sensing and geographic information systems, now make it possible to assess the world's total forest area and track future changes (Figure 7.11). Only recently have programs begun to obtain accurate estimates of the distribution and abundance of forests, and these suggest that past methods overestimated forest area by 100–400%.[18]

7.9 Indirect Deforestation

Chopping trees down isn't the only way to shrink forests. So far we have talked only about direct deforestation— cutting trees. A more subtle cause of the loss of forests is indirect deforestation—the death of trees from pollution or disease. Acid rain and other pollutants may be killing trees in and near industrial areas, or even near industrial countries. In Germany, there is talk of *Waldsterben* ("forest death"). The German government estimates that one-third of the country's forests have suffered damage: the death of standing trees, yellowing of needles, or poorly formed shoots.

Why trees are dying appears to involve a number of factors, including acid rain, ozone, and other air pollutants that tend to weaken trees and make them more susceptible to disease. This problem extends throughout Central Europe and has been especially acute in Poland, the Czech Republic, and Slovakia. In the New England area of the United States, similar, curious damage has affected red spruce.

Global warming could cause widespread damage. If global warming occurs as projected by global climate models, indirect forest damage could occur over

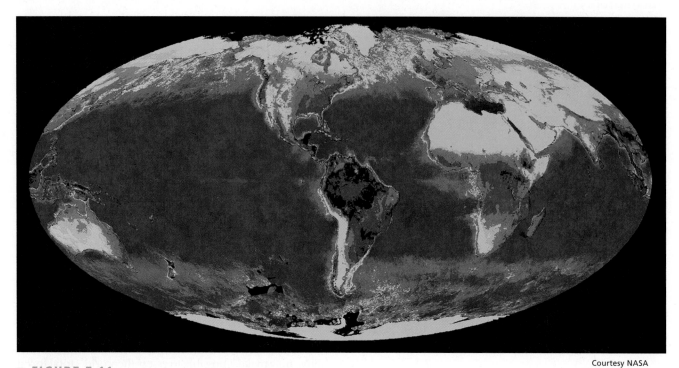

▪ **FIGURE 7.11**
The world's forests based on NASA satellite data.

vast regions, with major die-offs in many areas and major shifts in the areas of potential growth for each species of tree.[16] Global warming could alter the combination of temperature and rainfall required for various tree species in many areas, so that some species may no longer be able to grow in their current locations. Such changes would be much more widespread than damage from other causes, affecting the habitats of many endangered species.

Section II: Wildlife Management

7.10 Traditional Wildlife Management

Like forests, wildlife has had special value for people, not only for the meat and hides it provides for our food and clothing, but also because we enjoy knowing that we share the Earth with other living creatures, enjoy observing them, and in some cases have made them cultural and even sacred symbols.

Modern conflicts about wildlife are similar to those about forests. On one side are those who believe that wildlife is there for people to harvest and that the greatest good for the greatest number of people will be the largest harvest that is sustainable. On the other side are those who believe that wildlife is important because living creatures are beautiful, interesting, an important feature of the landscape, raise our spirits and have religious value to us, and, moreover, that they have an intrinsic right to exist, just as we do.

How has wildlife been faring in this modern world? One thing is clear: Many species of wildlife have declined greatly in abundance, some have become endangered, and some have gone extinct. The first European settlers in North America found an almost unbelievable abundance of wildlife here. Even Manhattan Island, which is now the middle of New York City and today seems home primarily to squirrels, rats, pigeons, and a few other species of birds, once teemed with wildlife. A Dutch settler on that island in the 17th century wrote that "birds fill also the woods so that men can scarcely go through them for the whistling, the noise and the chattering. Whoever is not lazy can catch them with little difficulty." On Manhattan Island and in the surrounding countryside were "incredibly numerous" deer "as fat as any Holland cow can be," as well as "many wolves, bear, mink, otter, muskrats, raccoon, and beaver." In 1693 one Dutchman wrote that 80,000 beaver were killed "in this quarter of the country" every year.[19]

FIGURE 7.12
Buffalo in the 19th century.
Herd of Bison on the Upper Missouri plate 40 from Volume 2 of "Travels in the Interior of North America". Painting done in 1838 by Karl Bodmer.

In sum, wildlife hasn't been faring very well. The general attitude of the early European settlers was that this abundance would last forever and was there for the taking—the same thing people believed about forests in Michigan, as discussed earlier. The result was a period of rapid exploitation—a kind of mining—of the living resources. This continued well into the 19th century, as illustrated by the fate of the buffalo (also called bison) in the second half of that century (Figure 7.12).

Bison on the Range and Then Mostly Off the Range

How many bison were there to begin with? Well, in the fall of 1868, one observer said a train traveled 120 miles between Ellsworth and Sheridan, Wyoming, through a continuous, browsing herd of bison, packed so thick that the engineer had to stop several times, mostly because the animals would "scarcely get off the tracks for the whistle and the belching smoke."[20] That spring, a train had been delayed for eight hours while a single herd passed "in one steady, unending stream." We can't be exactly sure how many bison were in that herd, but we can use observations like these to make an educated estimate. At the highest extreme, we can assume that the train bisected a circular herd with a diameter of 120 miles. Such a herd would cover 11,310 square miles, or more than 7 million acres. Even if we suppose that people exaggerated the density of the buffalo, and there were only ten per acre—a moderate density for a grazing herd—this single herd would have numbered 70 million animals!

Some might say that this estimate is probably too high, because the herd would more likely have formed a broad, meandering, migrating line, rather than a circle. The impression remains the same—there were huge numbers of buffalo in the American West even as late as 1868, numbering in the tens of millions and probably 50 million or more.

How could that many buffalo actually disappear? Ominously, that same year, the Kansas Pacific Railroad advertised a "Grand Railway Excursion and Buffalo Hunt."[20] Some say that many hunters believed the buffalo could never be brought to extinction because there were so many—just as was believed for eastern white pine in Michigan at the same time.

Environmentalists existed long before the 20th century. We tend to think of environmentalism as a 20th-century issue, but in fact, after the Civil War, there were angry protests in every legislature over the slaughter of the buffalo. In response to these concerns, in 1871 the U.S. Biological Survey sent biologist George Grinnell, who later became head of what is today the U.S. Fish and Wildlife Service, to survey the herds along the Platte River. He estimated that only 500,000 buffalo remained there, and said that at the rate the buffalo were then being killed, the animals would not last long. As late as the spring of 1883, a herd of about 75,000 buffalo crossed the Yellowstone River near Miles City, Montana, but fewer than 5,000 reached the Canadian border—the rest were slaughtered.[20]

From millions down to just 50. By the end of that year—only 15 years after the Kansas Pacific train was delayed for eight hours by a huge herd of buffalo—only a thousand or so buffalo could be found: 256 in captivity and about 835 roaming the plains. A short time later, there were only 50 buffalo wild on the plains. Even Buffalo Bill, named and famed for being one of the great buffalo hunters, began to collect the remaining bison and grow them on his ranches. That was the beginning of wildlife conservation in America.

Should we harvest wildlife for our food and clothing? Or do all living creatures have an intrinsic right to exist? Should we be content to leave them alone to enjoy their lives and be enjoyed by us, to raise our spirits?

Aldo Leopold struggled with these two points of view. Leopold came to be called the father of wildlife management and was one of the great conservationists of 20th-century America. He was educated at the Yale School of Forestry, and in this sense was an intellectual descendent of Gifford Pinchot. In his early work for the

U.S. government, he was in charge of getting rid of mountain lions and wolves—predators that were considered pests at the time. Then he met and was educated by Charles Elton, one of the most important British ecologists of the 20th century.

After meeting Elton, Leopold's point of view changed completely. Instead of seeing mountain lions and wolves as vermin that killed deer that people otherwise would have hunted, he wrote that "when wolves are removed from mountains the deer multiply . . .I have seen every edible bush and seedling browsed . . . to death . . . In the end, the bones of the hoped-for deer herd, dead of its own too-much, bleach with the bones of dead sage, or molder under the high-lined junipers."[21] In other words, deer died *en masse* unless controlled by predators. Predators played an essential role in nature, as did all creatures, and it was our obligation to sustain them all. Leopold developed "the Land Ethic," which states that all Earth's resources—animals, plants, even bedrock—have the right to exist in a natural state, and that our role should shift from conqueror to citizen and protector.[21]

Whichever goal was chosen for wildlife conservation and management, thinkers from Grinnell to Leopold were confronted with the same challenge: How could modern science help? From the ecological ideas of Charles Elton and other animal ecologists, a beginning was made.

The use of science to manage animal populations began with the logistic growth curve. The first approach followed along with the early ideas about human population growth that we discussed in Chapter 2—specifically, that wildlife populations, as well as the human population, must grow according to the S-shaped logistic curve. This assumption, you will recall, came from a belief that there is and must be a balance of nature, in this case for wildlife. Consider what it would mean for the spotted owl, discussed earlier: Old-growth forests would remain in place forever, with a constant number, size, and species of trees, a constant number of standing dead trees, and an unchanging abundance of small mammals on which the owls fed.

The logistic growth curve says how much you can harvest. It gives a simple answer to what is the maximum sustainable harvest (usually called the maximum sustainable yield, or MSY) of a wildlife population. According to the mathematics behind this curve, the maximum population growth occurs when the population is at exactly one-half of its carrying capacity (it defines the carrying capacity as the maximum abundance that occurs in the logistic growth—the constant number of individuals to which a population grows). Wildlife policies based on this idea were straightforward: Find out the carrying capacity of a population and keep it at exactly one-half that number by allowing a harvest of exactly the number of animals that will bring it back down to this level each year.

How well did this approach work? Early in the 20th century, wildlife managers found an ideal situation to test this approach: on two islands in the Aleutians off the coast of Alaska, where northern caribou (also called reindeer) had been introduced. Here's what happened when this approach to maximum sustainable yield was tried.

Pribilof Island Reindeer

The Pribilof Islands lie in the cold Bering Sea between Alaska and Siberia. When the United States purchased Alaska from Russia in 1867, the Pribilof Islands and their inhabitants came under U.S. jurisdiction. In 1911 the U.S. government introduced small groups of reindeer on two of the islands: 4 bucks and 21 does on St. Paul, the largest of the islands, covering 12,000 hectares; and 3 bucks and 12 does on St. George, the second-largest island.[28] The reindeer were meant to provide a much-needed source of food for the islands' inhabitants, a group of Aleuts who had been settled there in 1787 by Gerasim Pribilof, the Russian explorer for whom the islands are named. The Aleuts had survived primarily on what the sea offered, but people concerned with management of the islands felt that some additional source of protein was necessary.

The islands seemed perfect for the reindeer—lots of plants and no predators. The islands had abundant vegetation and no wolves or other predators large enough to affect reindeer. In 1922, G. D. Hanna, a wildlife expert, wrote in an article in *Scientific Monthly*, "It would seem that here is the place to maintain model reindeer herds and to determine many of the needed facts for the propagation of these animals on a large scale. At no other place are conditions so favorable."[22]

Even so, something went very wrong with the Pribilof Islands reindeer. At first, the introductions seemed a success. In the spring of 1912, 17 fawns were born on St. Paul and 11 on St. George (Figure 7.13). But, ironically, in the year that Hanna wrote his enthusiastic report, the population on St. George had reached a peak of 222 individuals, and from there on the reindeer population declined, never to rebound. By the 1940s, the reindeer herd on St. George numbered 40–60, but in the 1950s the herd became extinct. On St. Paul, the herd reached a peak of 2,000 animals in 1938, when there was 1 deer for every 6 hectares (14 acres) of the island and 1 for every 5 hectares (12 acres) of range-

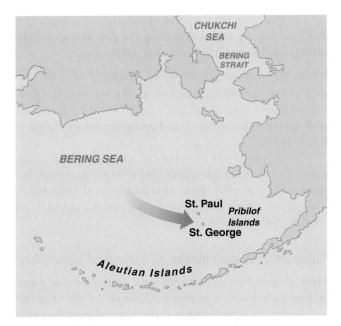

(a)

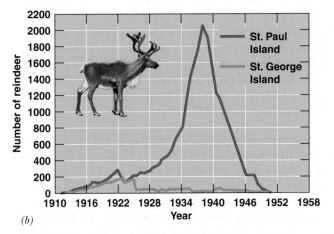

(b)

■ **FIGURE 7.13**
Pribilof Islands and their reindeer.
(a) Location of the Pribilof Islands. (b) The fate of the introduced reindeer.

land. Soon after, the reindeer herd rapidly declined. The St. Paul reindeer population numbered only 8 in 1950 and 2 in 1951.

What happened? Scientific managers had monitored the herds throughout the period, attempting to manage them so they would provide a sustained food supply for the inhabitants. Where did this management go wrong? Much later in the 20th century, another ecologist, Victor Sheffer, reviewed the history of the Pribilof Islands reindeer herds and shed some light on what might have caused the decline.[22]

There was plenty of their favorite food in the summer. Grasses and small flowering plants and shrubs found in the interiors of the islands are the main foods for the reindeer population during spring, summer, and fall. These plants remained abundant during those seasons throughout the entire period of the rise and fall of the reindeer population, so the decline was not caused by a lack of the grasses and herbs that provided the bulk of their diet.

But the situation was different in winter. Winter is the time of greatest stress for the reindeer. Seeking food, they paw through the snow to feed on lichens called "reindeer moss." Because lichens are very slow-growing, they were rapidly depleted by the reindeer. A particularly cold winter in 1940 worsened matters—island records indicate that a crust of ice remained on the snow for weeks. Although reindeer can paw through as much as a meter of soft snow, they had difficulty digging through this crust. In early spring, 150 dead reindeer, primarily females, were found on St. Paul Island.

Why did more females than males die? In contrast to a hypothetical logistic population, the reindeer did not adjust instantaneously to changes in their food supply. Also contrary to the assumptions of the logistic, not all the individuals were identical: More females died because they were carrying calves and required additional nutrition.

It didn't happen overnight. Reindeer live a relatively long time, and they starve to death slowly—slowly enough that a large population can have a great effect on future food supplies. The Pribilof Islands reindeer herds grew rapidly when all their food was in great abundance, but the population ultimately outstripped the capacity of the reindeer moss to sustain the reindeer over a long time. The decline of the reindeer took a number of years. During those years, the supply of slow-growing lichens grew smaller and smaller, so even though the reindeer population also grew smaller and smaller, the supply of winter food continued to be less than they needed.

What can we learn from the story of the Pribilof Islands reindeer? (1) There was a lag effect in the population's response to changes in its habitat (for several years the reindeer population continued to be larger than the number that could be supported by the lichens in winter); (2) the death rate was higher for certain parts of the population (after a particularly hard winter in 1940, deaths occurred mainly among females); and (3) the population was controlled by a brief but crucial aspect of its life—an *environmental bottleneck* (limited food available in late winter).

The logistic equation does not take these into account. It does not take into account such factors as time lags or differences among individuals in a population and kinds of food. Yet, the Pribilof Island reindeer are typical of most wildlife. Wildlife populations usually have complicated life histories and are made up of individuals in many different stages of their lives.

Even so, the logistic curve has been important. It has been used in managing populations and in the study of population dynamics (how populations function). For many years the logistic curve has been the basis for managing many biological resources, especially marine fisheries, endangered species such as the great whales (see Chapter 13 about ocean life), and game populations such as the large grazing mammals of the great African savannas.

Although real populations do not follow it, the logistic curve is appealing to those of us who want to believe in a balance of nature—a nature that sustains itself in a constant and desirable condition and returns to this condition after being disturbed (as discussed in Chapter 6). Because the logistic population is constant and stable, balanced and in harmony with its environment, many have argued that the logistic curve provides a basis for us to believe in a harmonious, balanced nature. Knowing that actual populations don't follow this curve, we must search for another basis for our view of what undisturbed populations in nature are really like.

7.11 Improved Approaches to Wildlife Management

The U.S. Council on Environmental Quality (an office within the executive branch of the federal government), the World Wildlife Fund United States, the Ecological Society of America, the Smithsonian Institution, and the International Union for the Conservation of Nature (IUCN) have proposed four principles of wildlife conservation:

1. A safety factor in terms of population size, to allow for limitations of knowledge and imperfections in procedures. We should not harvest a population to the point where it has been reduced to the lowest limit of what we think is probably safe. Instead, we should assume that our estimate of that lower limit is crude, and leave more animals than that minimum would suggest.

2. Concern for the entire community of organisms and all the renewable resources, so that policies developed for one species are not wasteful of other resources.

3. Maintenance of the ecosystem of which the wildlife are a part, minimizing the risk of irreversible change and long-term harm as a result of use.

4. Continual monitoring, analysis, and assessment. We should continue to apply science to the pursuit of knowledge about the wildlife we are concerned with and their ecosystems, and make the results available to the public.

These principles broaden the scope of wildlife management from a narrow focus on a single species to inclusion of the ecological community and ecosystem. This is a starting point for an improved approach to wildlife management.

Time Series and Historical Range of Variation

Here's a problem we run into once we realize that there is no simple balance of nature: How do we decide what is a sustainable population if the natural population is always changing?

One answer is to consider a range of population levels natural. But then we have to find out what that range is, which we can do if we have an estimate of population over a number of years. This set of estimates is called a *time series* and could provide us with a measure of the *historical range of variation*—how the population or species varied in number, from most to fewest, over some past time period.

Such records exist for only a few species. One is the American whooping crane (Figure 7.14), America's tallest bird, standing about 1.6 meters (5 feet) tall. Because this species became so rare and because it migrated as a single flock, people began counting the total population in the late 1930s. At that time, they saw only 14 whooping cranes. Not only was the total number counted, but also the number born that year. The difference between these two numbers tells us the number dying each year as well. So we know the historical range of variation in the whooping crane population.

We can use this historical range to estimate the probability of extinction. The first estimate of the probability that the whooping crane would become extinct, made in the early 1970s and based on the bird's historical range of variation, was a surprise. Even though there were only a few whooping cranes, the probability of extinction was less than one in a billion. How could this number be so low? The explanation is that when you use the historical range of variation as a basis for predicting future trends, you are assuming that variation in the future will be caused only by the same things that caused variation during the historical period. If the whooping cranes suffered one catastrophe that had not occurred during the

Ken Lucas/Visual Unlimited

(a)

■ **FIGURE 7.14**

Whooping cranes.
The whooping crane (a) is one of many species that
appear always to have been rare. Rarity does not
necessarily lead to extinction, but a rare species,
especially one whose population has undergone a
rapid and large decline, needs careful attention and
assessment as to whether it is threatened or
endangered. (b) Migration route and (c) change in
population from 1940.

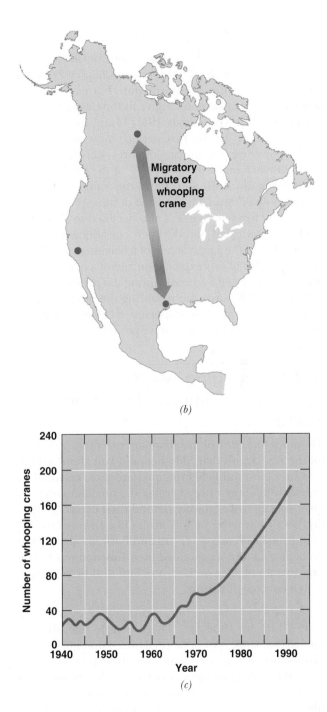

(b)

(c)

period studied—for example, a long drought on the wintering grounds—this could cause a population decline
not observed in the past.

Even so, this method provides valuable information.
As the method predicted, whooping cranes have continued to increase, and a record number of 189 reached
the wintering grounds at Aransas National Wildlife
Refuge in Texas in December 2003.[23]

Breeding programs have further boosted the
number of whooping cranes. With help from a variety
of governmental and nongovernmental organizations,
eastern populations of whooping cranes have been established through breeding programs. The numbers reached
77 nonmigrating birds in Florida and 36 birds migrating
between Florida and Wisconsin, so the total number of
wild whooping cranes in the world is now more than 300.
In addition, the U.S. Fish and Wildlife Service breeds
whooping cranes in Patuxent, Maryland, and has a total of
49, and the International Crane Foundation does the
same in Wisconsin, where 29 cranes live, so counting both
wild and captive cranes, more than 420 are alive.[23]

Age Structure as Useful Information

Monitoring age structure is another key to successful wildlife management (see Chapter 2). It can provide many kinds of information. For example, the age structures of salmon caught on the Columbia River during two different periods, 1941–43 and 1961–63, were quite different. In the first period, most of the catch (60%) consisted of four-year-olds; three-year-olds and five-year-olds each made up about 15% of the catch. Twenty years later, in 1961 and 1962, the catch was much younger—half the catch consisted of three-year-olds; the number of five-year-olds had also declined to about 8%, and the total catch had declined considerably as well. Such a shift in the age structure of a harvested population is an early sign of overexploitation and of a need to lower the allowable catch. As proof, during the period 1941–43, 1.9 million fish were caught. During the second period, 1961–63, the total catch dropped to 849,000, just 49% of the number caught in the earlier period.

7.12 Managing Two or More Species at a Time: Do Predators Matter?

Do predators control the abundance of their prey? One of the revelations that Aldo Leopold had about wildlife was his newfound belief that predators are essential for the survival of their prey, because the predators keep the prey species from growing beyond their carrying capacity. We all learned from folk stories like *Peter and the Wolf* that predators such as wolves and lions are dangerous, to be feared. But we assumed their function was to control the abundance of their prey and thus help to maintain a balance of nature.

It appears that predators probably play a smaller role than we thought. There is little scientific evidence that large vertebrate predators—mammals, birds, fish—have as much control over their prey as was assumed in the classic balance-of-nature idea. Some of our best information about the history of wildlife populations comes from harvests. For example, trappers in northern Canada provide us with the longest continual records of animal populations, beginning around 1770.

The Hudson's Bay Company (today no longer involved in fur trading) entered the fur trade in Canada around 1830. They and their predecessors kept records for 220 years. This history of trapping shows that the number of Canadian lynx pelts obtained varied widely, from fewer than 1,000 in the 1790s to more than 80,000 in 1885, but the variations were fairly regular, with peaks about every decade. Other fur-bearing animals showed the same kind of population cycles, and fell into two groups: herbivores (such as the arctic hare, which is a prey of the lynx), whose abundance cycled approximately every four years; and carnivores (such as the lynx), whose abundance cycled about every ten years (Figure 7.15). These data suggest that predators do not neatly control the abundance of their prey, at least not now in large areas of wilderness.[33]

Predators can have large effects in some cases. In most cases, predators do not seem to control the abundance of their prey in the exact sense consistent with the balance-of-nature idea. Instead, in some cases predators and parasites seem to reduce the abundance of their prey but don't keep the numbers constant.[24]

For example, mosquito fish can greatly reduce mosquito abundance. Mosquito fish—predators that feed on mosquito larvae and on other small animals—are introduced into rice fields in many parts of the world to

National Geographic Society

▪ **FIGURE 7.15**

Photograph of a Canadian lynx. Do predators control the abundance of their prey?
Did lynx control the abundance of the arctic hare, one of their prey species, during several centuries of observations? See text for the answer.

control mosquitoes. In the United States, these fish are used in the large rice fields of the great Central Valley of California, in southern Nevada, in Louisiana and Alabama, and elsewhere. Mosquitoes lay eggs in freshwater, and the larval stages live in the water. In an experimental study at the Rice Experiment Station, in Crowley, Louisiana, mosquito fish reduced the mosquito larvae population by 96%.

In this second example, an *absence* of predators proves the point. In 1839 the Asian water buffalo was introduced as a beast of burden into northern Australia, an area with no native grazers on floodplain vegetation and no major predators of large mammals. The water buffalo arrived with only a few of its natural parasites. We know this because only 11 species of parasitic worms occur in the Australian water buffalo, whereas 77 species are found in Asia. Transported to a habitat that lacked major disease organisms and predators, but offered a suitable climate and plentiful food, the water buffalo population exploded, increasing beyond the capacity of its food resources. The result was several large die-offs, with many animals dying of slow starvation. In their native habitat in Asia, death would have come more quickly to the buffalo, as weakened individuals would have been easy targets for predators and parasites.

In this case, not just the water buffalo but also their habitat suffered. A slow death from starvation is much worse for the habitat, because slowly starving animals continue to eat all possible vegetation. As a result, in some areas of Australia during the dry season, only bare soil was left.[25, 26] The water buffalo population persists in Australia today but is less stable than the populations in its native Asia.

Return to the Big Question

Can we have our forests and wildlife and use them too?

We can—if we understand how they work. This means that we have to apply a scientific approach to forestry and wildlife management, and recognize that forests and the wildlife that live in them are always changing. We can—but we may achieve this best by seeking the two goals in different places at the same time. We can—but only if we seek long-term historical records. This means that long-term scientific monitoring is necessary.

Summary

About Forestry . . .

■ **Debate continues about the role of forests.** Forestry has become a profession with a scientific base, but the issues that Gifford Pinchot and John Muir argued over more than a century ago are still debated: Are forests there primarily to provide lumber and other resources for people, or are they there for their own sake and to raise our spirits, and thus should be left alone?

■ **Change of certain kinds is natural and necessary.** In the 20th century many believed that the only "natural" forest was old-growth. Today we know that forests, like all ecosystems, are dynamic and always changing. Some kinds and rates and intensity of change in a forested landscape are natural and work well for us and for the creatures that live in forests. Other kinds can be damaging and, where possible, should be avoided.

■ **The goal for harvesting timber is sustainable yield in a changing environment.** We know a great deal about how trees grow, how forests develop and change over time, and how various kinds of logging affect forests. But we still need a better understanding of how changes in forest ecosystems affect all the forms of life in a forest—such as the spotted owl, whose fate remains in doubt. The problem is much more complicated than simply growing trees and harvesting them without concern for the forest ecosystem.

About Wildlife Management . . .

■ **People have greatly affected wildlife around the world,** as illustrated by the exploitation of buffalo and other wildlife during European settlement of North America.

■ **People value wildlife for much the same reasons as they value forests:** for their products of economic value, and for the joy, beauty, and recreation they add to our lives, raising our spirits. The conflict about wildlife management, like that about forests, has to do with whether both kinds of uses, utilitarian and nonutilitarian, can be served.

▪ **As with forestry, a scientific basis for wildlife management developed in the 20th century.** At first, this scientific basis centered around the idea of a balance of nature, represented by the use of the logistic growth curve for wildlife populations, and by the belief that predators controlled their prey.

▪ **By the end of the century we saw that balance-of-nature assumptions did not work.** We realized that wildlife populations were continually changing. New approaches to wildlife conservation and management take into consideration (1) historical range of abundance; (2) the probability of extinction based on historical range of abundance; (3) age-structure information; and (4) better use of harvests as sources of information. These, along with an understanding of the ecosystem and landscape in which populations live, are improving our ability to conserve wildlife.

Getting It Straight

1. From an ecosystem point of view, are predators ever necessary? Explain.
2. Forests provide what two products of value to civilization?
3. What are the indirect benefits of a forestry biome?
4. What is clear-cutting? How does it impact forestry sustainability?
5. What are the four alternatives to clear-cutting forests? Identify and define each.
6. What is the one advantage over all other ecosystems that forests offer?
7. What is the impact of cutting forests in one country or another country?
8. Where does most global deforestation occur?
9. Are the world's forests shrinking or growing? Explain.
10. What does traditional wildlife management consist of?
11. What are some of the current approaches being used to help manage wildlife in the United States?

What Do You Think?

1. Who do you agree with more? John Muir or Gifford Pinchot? Why?
2. Attempts to manage forests and wildlife are complicated by disagreement as to what it means to say that an area or an ecosystem is "natural." Based on what you have learned in this chapter and others in this text, develop a new definition of *natural* that could be used to manage forests and wildlife.
3. Some people suggest that we stop using forests at all. Imagine a civilization without wood. Describe how houses would be built, records kept, and music and art created.
4. Suppose you are in charge of water resources for the state of California. Would you argue for the removal of Hetch Hetchy Dam? Explain your reasons.
5. White-tailed deer have become a problem in many parts of the United States where people live in suburbs and even in cities, for example, in Helena, Montana. Focus on a specific town or city and describe how you would manage the deer so that both they and people's gardens are sustained.
6. How would you use, if at all, predators to control the population of Asian water buffalo that were introduced into Australia, as described in this chapter?

Pulling It All Together

1. Based on what you have learned in this chapter, create an ecosystem management plan to conserve the spotted owl, as discussed in the opening case study. In answering this question, first define what ecosystem management means, then develop your plan with this definition in mind. Refer specifically to how you would deal with new problems caused by the arrival of the barred owl, the new "sudden oak disease," and West Nile virus attacking the spotted owl.

2. This chapter states that a traditional goal in managing both forests and wildlife is a maximum sustainable yield. Suggest an alternative goal that takes into account larger issues than a single species. Apply this goal to management of a specific species that lives near you and is considered either threatened or endangered, or is of commercial value and is harvested at present.

3. The United States Department of Defense is now attempting to practice ecosystem management on its military bases. Consider as an example a tank-training area in a coastal forest. Devise a management plan for sustainable forestry that also takes into account tank training that will knock over trees.

Further Reading

Botkin, D. B. 2001. *No man's garden: Thoreau and a new vision for civilization and nature.* Washington, DC: Island Press.—A work that discusses deep ecology and its implications for biological conservation, as well as reasons for the conservation of nature, both scientific and beyond science.

Caughley, G., and A. R. E. Sinclair. 1994. *Wildlife ecology and management.* London: Blackwell Scientific.—A valuable textbook based on new ideas of wildlife management.

Leopold, A. 2000. *A Sand County almanac.* New York: Oxford University Press.

Miller, C. 2004. *Gifford Pinchot and the making of modern environmentalism.* Washington, DC: Shearwater Books.

Muir, John. 1997. *Nature writings.* W. Cronon (ed.). New York: Library of America.

Perlin, J. 1989. *A forest journey: The role of wood in the development of civilization.* New York: Norton.—A fascinating presentation of the story of wood, forests, and people, covering a period of 5,000 years and spanning five continents.

Mangel, M., L. M. Talbot, G. K. Meffe, M. T. Agardy, D. L. Alverson, J. Barlow, and D. B. Botkin et al. 1996, Principles for the Conservation of Wild Living Resources, *Ecol. Applications* 6(2): 338–362.

Digital Vision/Getty Images

8

Environmental Health, Pollution, and Toxicology

Big Question

Why Are Even Very Small Amounts of Some Pollutants a Major Concern?

Learning Objectives

Serious health problems and diseases may arise from toxic elements in water, air, soil, and even the rocks on which we build our homes. After reading this chapter, you should understand . . .

■ how the terms *toxic, pollution, contamination, carcinogen, synergism,* and *biomagnification* are used in environmental health;

■ the classification and characteristics of major groups of pollutants in environmental toxicology;

■ why there is controversy and concern about synthetic organic compounds such as dioxin;

■ whether we should be concerned about exposure to human-produced electromagnetic fields;

■ the dose-response concept, and how it relates to ecological gradients, and tolerance;

■ how the process of biomagnification works and why it is important in toxicology;

■ why the threshold effects of environmental toxins are important;

■ the process of risk assessment in toxicology, and why such processes are often difficult and controversial.

Case Study:

Demasculinization and Feminization of Frogs in the Environment

The story of wild leopard frogs (Figure 8.1) from a variety of areas in the midwestern United States sounds something like a science-fiction horror story. In affected areas, up to 92% of male frogs exhibit gonadal abnormalities, including retarded development and hermaphroditism (they have both male and female reproductive organs). Other frogs have underdeveloped vocal sacs. Since vocal sacs are used to attract female frogs, these frogs are less likely to mate.

What is apparently causing some of the changes in the male frogs is exposure to atrazine, the most widely used herbicide in the United States today. The chemical is used to kill weeds, primarily in agricultural areas. The region of the United States with the highest known frequency of sex reversal in frogs is along the North Platte River in Wyoming. Although it is not near any large agricultural area and the use of atrazine there is not particularly significant, hermaphrodite frogs are common in this region because the North Platte River flows from areas in Colorado where atrazine is commonly used.

The amount of atrazine released into the environment in the United States is estimated at more than 7 million kilograms (15 million pounds) per year. Atrazine degrades (breaks down) in the environment,

Stephen Dalton/Photo Researchers

■ **FIGURE 8.1**
Wild leopard frogs in America have been affected by human-made chemicals in the environment.

but often not before the chemical is applied again. Because it is applied every year, the waters of the Mississippi River basin, which drains about 40% of the lower United States, discharges approximately 0.5 million kilograms (1.2 million pounds) of atrazine per year into the Gulf of Mexico. Atrazine attaches easily to dust particles and has been found in rain, fog, and snow. As a result, it has contaminated groundwater and surface water even in regions where it isn't used.

The Environmental Protection Agency (EPA) states that up to 3 parts per billion (ppb) of atrazine in drinking water is acceptable for people, but at this concentration it definitely affects frogs that swim in the water. Other studies around the world have confirmed this. For example, in Switzerland, where atrazine is banned, the chemical generally occurs only in very low concentrations of about 1 ppb, but even that is sufficient to change some male frogs into females. In fact, atrazine can apparently cause sex change in frogs when its concentration in the water is as low as one-thirteenth of the level set by the EPA for drinking water.

Of particular interest and importance is the process that causes the changes in the leopard frogs. We begin the discussion with the *endocrine system,* which is composed of glands that internally secrete hormones directly into the bloodstream. The blood carries endocrine hormones to various parts of the body, where the hormones regulate and control growth and sexual development. Testosterone and estrogen are examples of hormones. Testosterone in male frogs is partly responsible for the development of male characteristics. However, scientists believe that atrazine switches on a gene that turns testosterone into estrogen, a female sex hormone. It's the hormones, not the genes, that actually regulate the development and structure of reproductive organs.

Frogs are particularly vulnerable during their early development, as they metamorphose from tadpoles into adult frogs. This change occurs in the spring, when atrazine often reaches a maximum level in surface water. It appears that even a single exposure to the chemical may affect the frog's development. Thus, the herbicide is known as a *hormone disrupter.*

Substances that interact with the hormone systems of an organism, whether or not they are linked to disease or abnormalities, are known as **hormonally active agents** (HAAs). These HAAs have the ability to trick the organism's body (in this case, a frog's body) into believing that the chemicals have a role to play in the body's development. You are probably familiar with computer viruses, which fool the computer into accepting them as part of its working system. Similar to computer viruses, HAAs interact with an organism and its mechanisms for regulating growth and development, thus disrupting normal growth.

What happens when HAAs—in particular, hormone disrupters (such as pesticides and herbicides)—are introduced into the system is shown in Figure 8.2. Natural hormones produced by the body send chemical messages to receptors on the outsides and insides of cells and transmit instructions to the cells' DNA, directing development and growth. We now know that chemicals, such as some pesticides and herbicides, can also bind to the receptors and either mimic or obstruct the role of the natural hormones. Thus, hormonal disrupters may also be known as HAAs.[1-4]

The story of wild leopard frogs in America dramatizes the importance of carefully evaluating the role of human-made chemicals in the environment. Frogs

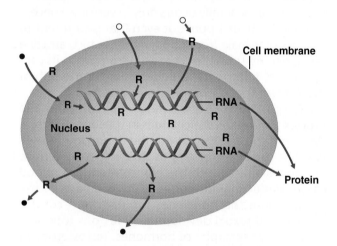

R	Receptor molecule
$\bigvee\!\!\bigwedge$	DNA
●→	Path of hormonally active agent = for example, pesticide DDT or PBC
○→	Path of natural hormone
RNA	Transmission of genetic information

▪ **FIGURE 8.2**
How HAAs work.
Idealized diagram of hormonally active agents (HAAs) binding to receptors on the surface of and inside a cell. When HAAs, along with natural hormones, transmit information to the cell's DNA, the HAAs may obstruct the role of the natural hormones, which produce proteins that in turn regulate the growth and development of an organism.

and other amphibians are declining globally, and a great deal of research has focused on finding out why. Studies of past or impending extinctions of organisms often center on global processes such as climate change, but the story of leopard frogs leads us down another path, one associated with human use of the natural environment. It also raises a number of more disturbing questions: Does exposure to chemicals bring about change only in certain plants and animals, or are these changes a forerunner of what we might expect in the future on a much broader scale? Indeed, are we participating in an unplanned experiment on how human-made chemicals, such as herbicides and pesticides, might transform the bodies of many living beings, even people? Perhaps we will look back on this moment of understanding as a new beginning in meaningful studies that will answer some of these important questions.

In this chapter we will explore selected chemicals, and processes related to their harmful (toxic) effects. Our purpose is to emphasize principles of toxicology, not to discuss all classes of pollutants. For example, nuclear radiation is not discussed here (it can be found in our discussion of energy in Chapter 10 and air pollution in Chapter 14).

8.1 Some Basics

Can the environment make you sick? Disease is often due to a poor adjustment between the individual and the environment.[5] However, disease seldom has a one-cause/one-effect relationship with the environment. Rather, the incidence of a disease depends on several factors, including physical environment, biological environment, and lifestyle. Linkages between these factors are often related to other factors, such as local customs and the level of industrialization. More-primitive societies that live directly off the local environment are usually plagued by environmental health problems that are different from those that afflict our urban society. Industrial societies have nearly eliminated such diseases as cholera, dysentery, and typhoid.

Are we the only problem? Many people believe that soil, water, and air in a so-called natural state are clean and good and become bad only if people contaminate and pollute them.[6] This is by no means the entire story; many natural events, including dust storms, floods, and volcanic processes, can introduce materials harmful to humans and to other living things into the soil, water, and air. For example, on the Big Island, Hawaii, volcanic gases, including sulfur dioxide, emitted into the air produce a type of smog known as "vog" (volcanic gases and fog). The vog is acidic and is a potential hazard to people, plants, and other living things.

Terminology

What do we mean when we speak of pollution, contamination, toxins, toxicology, and carcinogens? The term **pollution** refers to an unwanted change in the environment caused by introducing harmful materials or by producing harmful conditions (heat, cold, sound). A polluted environment is impure, dirty, or otherwise unclean. **Contamination** has a meaning similar to that of pollution and implies making something unfit for a particular use through the introduction of undesirable materials—for example, the contamination of water by hazardous waste. A *toxin* is a substance that is poisonous (toxic) to people and other living things. **Toxicology** is the science that studies chemicals that are or could be toxic, and *toxicologists* are scientists in this field. A **carcinogen** is a toxin that increases the risk of cancer. Carcinogens are among the most feared toxins in our society and therefore the ones subject to the greatest regulation.

Synergism is an important concept in considering pollution problems. **Synergism** is the interaction of different substances, resulting in a combined effect that is greater than the effects of the separate substances. For example, both sulfur dioxide and coal dust particulates are air pollutants. Either one by itself may cause health problems, but when they combine, as when sulfur dioxide (SO_2) adheres to the coal dust, the dust with SO_2 is inhaled deeper than sulfur dioxide alone, causing greater damage to lungs. Another aspect of synergistic effects is that the body may be more sensitive to a toxin if it is subjected to other toxins at the same time.

Prof. Ed Keller

■ **FIGURE 8.3**
Coastal pollution.
This Southern California urban stream flows into the Pacific Ocean at a coastal park. The stream water often carries high counts of fecal coliform bacteria. As a result, the stream is a point source of pollution for the beach, which is sometimes closed to swimming after runoff events.

How do pollutants get into the environment? Pollutants are commonly introduced into the environment by way of *point sources*, such as smokestacks, pipes discharging into waterways (Figures 8.3 and 8.4) or accidental spills. *Area sources* (also called *non-point sources*) are more spread

Bill Brooks/Masterlife

(a)

©Mike Grandmaison Photography

(b)

■ **FIGURE 8.4**
Point source pollution.
Lake St. Charles, Sudbury, Ontario. (a) Note high stacks (smelters) in the background emitting sulfur dioxide and toxic metals in the 1960s. Lack of vegetation in the foreground resulted from air pollution by acid and heavy metals. (b) Recent photo showing regrowth and restoration after emissions were reduced by as much as 85% in recent years.

out over the land and include urban runoff and *mobile sources,* such as automobile exhaust (Figure 8.5). Area sources are difficult to isolate and correct because the problem is often widely dispersed over a region, as in agricultural runoff that contains pesticides.

How We Measure the Amount of Pollution

It depends on what you're measuring. The amount of treated wastewater entering Santa Monica Bay in the Los Angeles area is a big number reported in millions of gallons per day. Emission of nitrogen and sulfur oxides into the air is also a big number reported in millions of tons per year. Small amounts of pollutants or toxins in the environment, such as pesticides, are reported in units as parts per million (ppm) or parts per billion (ppb). When dealing with water pollution, units of concentration for a pollutant may be milligrams per liter (mg/l) or micrograms per liter (μg/l). A milligram is one-thousandth of a gram, and a microgram is one-millionth of a gram. Units such as ppm, ppb, or μg/l reflect very small concentrations. For example, if you were to use 3 g (one-tenth of an ounce) of salt to season popcorn in order to have salt at a concentration of 1 ppm by weight of the popcorn, you would have to pop approximately 3 metric tons of kernels!

8.2 Categories of Pollutants

We list some categories of pollutants below: infectious agents; toxic heavy metals; organic compounds; hormonally active agents; nuclear radiation; thermal pollution; particulates; electromagnetic fields; and noise pollution. Other pollutants are discussed in other parts of the book.

Infectious Agents

Airplane travel has made this a bigger problem than ever. Infectious diseases, spread by the interactions between individuals and food, water, air, or soil, are some of the oldest health problems that humans face. Today, infectious diseases have the potential to pose rapid local to global threats by spreading in a matter of hours via airplane travelers. Terrorist activity may also spread diseases. Inhalation anthrax, caused by a bacterium, sent in a powdered form in envelopes through the mail in 2001, killed several people. New diseases are emerging, and previous ones may emerge again. Although we have cured many diseases, we have no reliable vaccines for others, such as HIV, bird flu, hantavirus, and dengue fever.

Toxic Heavy Metals

These are all around us. The major **heavy metals** (metals with relatively high atomic weight) that pose health hazards to people and ecosystems include mercury, lead, cadmium, nickel, gold, platinum, silver, bismuth, arsenic, selenium, vanadium, chromium, and thallium. Each of these elements may be found in soil or water that has not been contaminated by people. However, each of these metals has uses in our modern industrial society, and each is also a by-product of the mining, refining, and use of other elements. Heavy metals often have direct physiological toxic effects. Some are stored or incorporated in living tissue, sometimes permanently. Heavy metals tend to accumulate over time in fatty body tissue. As a result, a little arsenic each day may eventually result in a fatal dose (the plot of more than one murder mystery).

Mercury, thallium, and lead are very toxic to people. They have long been mined and used, and their toxic properties are well known. Mercury, for example, is the "Mad Hatter" element. At one time, it was used in making felt hats stiff, and because mercury damages the brain, hatters were known to act peculiarly in Victorian England. Thus, the Mad Hatter in Lewis Carroll's *Alice in Wonderland* had real antecedents in history.

Toxic Pathways

One pathway is biomagnification. Chemical elements released from rocks or human processes can become concentrated in humans through many

Taxi/Getty Images

▪ **FIGURE 8.5**
Mobile sources.
Cars, trucks, and buses are all mobile sources of air pollution.

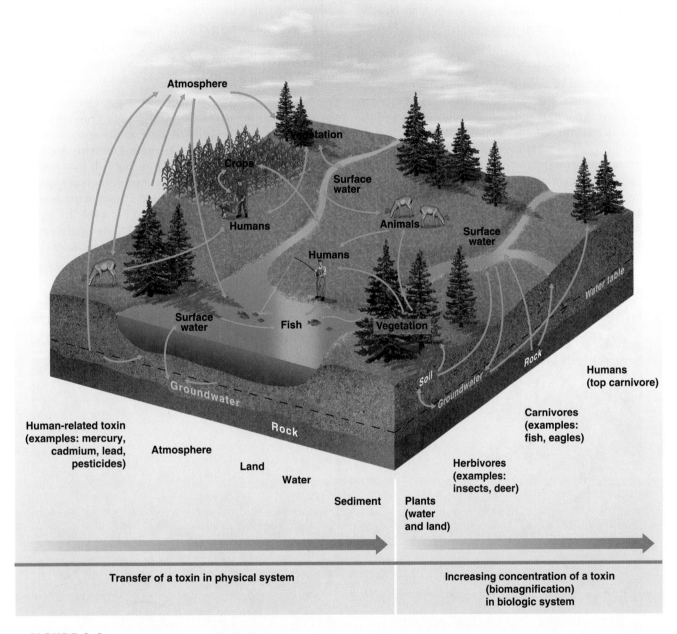

Atmosphere

Vegetation

Crops

Surface
water

Humans

Animals

Surface
water

Humans

Surface
water

Fish

Vegetation

Water table

Groundwater

Soil

Groundwater

Rock

Humans
(top carnivore)

Rock

Carnivores
(examples:
fish, eagles)

Human-related toxin
(examples: mercury,
cadmium, lead,
pesticides)

Atmosphere

Land

Water

Sediment

Herbivores
(examples:
insects, deer)

Plants
(water
and land)

Transfer of a toxin in physical system

**Increasing concentration of a toxin
(biomagnification)
in biologic system**

■ **FIGURE 8.6**

Pathways for toxic materials.
Potential complex pathways for toxic materials through the living and nonliving environment. Note the many arrows into humans and other animals, sometimes in increasing concentrations, as they move through the food chain.

pathways (Figure 8.6). These pathways may involve what is known as **biomagnification**—the accumulation or increase in the concentration of a substance in living tissue as it moves through a food web (also known as *bioaccumulation*). For example, cadmium, which influences the risk of heart disease, may enter the environment via ash from burning

coal. The cadmium in coal exists in very low concentrations. After coal is burned in a power plant, the ash is collected in a solid form and disposed of in a landfill. The landfill is covered with soil and revegetated. The low concentration of cadmium in the ash and soil is taken into the plants as they grow. But the concentration of cadmium in the plants is three to five times greater than the concentration in the ash. As the cadmium moves through the food chain, it becomes more and more concentrated. By the time it is incorporated into the tissue of people and other carnivores, the concentration is approximately 50–60 times the original concentration in the coal.

Mercury in aquatic ecosystems offers another example of biomagnification. Mercury is a potentially serious pollutant of aquatic ecosystems, such as ponds, lakes, rivers, and the ocean. Natural sources of mercury in the environment include volcanic eruptions and the erosion of natural mercury deposits. However, we are most concerned with human input of mercury into the environment through burning coal in power plants, incinerating waste, and processing metals, such as gold. Although we are unable to measure it precisely, it is estimated that human activities have doubled or tripled the amount of mercury in the atmosphere, and that it is increasing at about 1.5% per year.[7]

Bacterial activity leads to methylation. A major source of mercury in many aquatic ecosystems is precipitation from the atmosphere—rain and snow. Most of what is deposited is inorganic mercury, but once this mercury is in surface water, a process known as *methylation* may occur. Methylation changes inorganic mercury into methyl mercury through bacterial activity. Methyl mercury is much more toxic than inorganic mercury, and it is eliminated more slowly from animals' systems.

The higher up on the food chain, the higher the mercury concentrations. As the methyl mercury works its way through food chains, biomagnification occurs, so that higher concentrations of methyl mercury are found farther up the food chain. Thus, big fish that eat little fish contain higher concentrations of mercury than do smaller fish and the aquatic insects that the fish feed on.

Selected aspects of the mercury cycle in aquatic ecosystems are shown in Figure 8.7. The figure emphasizes the input side of the cycle, from deposition of in-

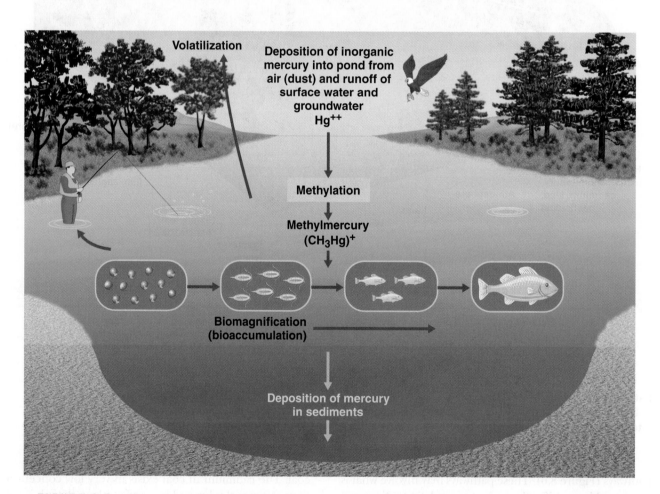

■ **FIGURE 8.7**
Pathways for mercury.
Idealized diagram showing selected pathways for the movement of mercury into and through an aquatic ecosystem. [*Source:* Modified from G. L. Waldbott. *Health effects of environmental pollutants,* 2nd ed. St. Louis: C. V. Mosby, 1978.]

organic mercury through formation of methyl mercury, biomagnification, and sedimentation of mercury at the bottom of a pond. On the output side of the cycle, the mercury that enters fish may be taken up by animals that eat the fish; and sediment may release mercury by a variety of processes, including resuspension in the water, where eventually the mercury enters the food chain or is released into the atmosphere through conversion of liquid mercury into a vapor.

Biomagnification also occurs in the ocean. Large fish, such as tuna and swordfish, have elevated mercury concentrations, which is why today we are advised to limit our consumption of these fish. Indeed, pregnant women are advised not to eat them at all.

Mercury and Minamata, Japan

Several serious incidents of methyl mercury poisoning occurred during the 20th century. For instance, in Iran, after a methyl-mercury fungicide was used to treat wheat seeds, many people suffered from mercury poisoning.[7] The most notorious example, however, occurred in the Japanese coastal town of Minamata, on the island of Kyushu, when a strange illness began to affect animals and people in the middle of the 20th century.

It was first recognized in birds that lost their co-ordination and fell to the ground or flew into buildings, and in cats that went mad, running in circles and foaming at the mouth.[8] The affliction, called by local fishermen the "disease of the dancing cats," subsequently affected people, particularly families of fishermen. The first symptoms were subtle: fatigue, irritability, headaches, numbness in arms and legs, and difficulty in swallowing. More severe symptoms involved the sensory organs: vision was blurred, and the visual field was restricted. Afflicted people became hard of hearing and lost muscular coordination. Some complained of a metallic taste in their mouths. Their gums became inflamed, and they suffered from diarrhea. Eventually, about 800 people were diagnosed with the disease, 43 people died, and 111 were severely disabled. In addition, 19 babies were born with congenital defects. Those affected lived in a small area, and much of the protein in their diet came from fish from Minamata Bay.

A vinyl chloride factory on the bay used mercury in an inorganic form in its production processes. The mercury was released in waste that was discharged into the bay. Mercury forms few organic compounds, and it was believed that the mercury, although poisonous, would not get into food chains. But the inorganic mercury released by the factory underwent methylation—bacteria in the bay converted it into methyl mercury, an organic compound that turned out to be much more harmful. Unlike inorganic mercury, methyl mercury readily passes through cell membranes. Red blood cells transport it throughout the body, and it enters and damages brain cells.[9] Fish absorb methyl mercury from water 100 times faster than they absorb inorganic mercury. (This was not known before the epidemic in Japan.) Once absorbed, methyl mercury is retained two to five times longer than inorganic mercury.

Harmful effects of methyl mercury depend on a variety of factors, including the amount and route of intake, the duration of exposure, and the species affected. The effects of the mercury are delayed from three weeks to two months from the time of ingestion. If mercury intake ceases, some symptoms may gradually disappear, but others are difficult to reverse.[9]

Four lessons from Minamata. The mercury episode at Minamata taught us that four major factors must be considered in evaluating and treating toxic environmental pollutants.

- *Individuals vary in their response to the same dose, or amount, of a pollutant.* Not everyone in Minamata responded in the same way, even among those most heavily exposed. Because we cannot predict exactly how any individual will respond, we need to find a way to state the expected response of a particular percentage of individuals in a population.
- *Pollutants may have a threshold*—that is, until they reach a certain level, their effects may not be observable. In Minamata, symptoms appeared in individuals who had concentrations of 500 ppb of mercury in their bodies. No measurable symptoms appeared in individuals with significantly lower concentrations.
- *Some effects are reversible.* Some people recovered when the mercury-filled seafood was eliminated from their diet.
- *The chemical form of a pollutant, its activity, and its potential to cause health problems may be changed markedly by ecological and biological processes.* In the case of mercury, its chemical form and concentration changed as the mercury moved through the food webs.

Lead and the Urban Environment

Lead is one of the most common toxic metals in our inner-city environments. It is found in all parts of the urban environment. Due to our past use of leaded gasoline, urban air and soil contain lead. Other sources of lead include old plumbing pipes and some paint. It is also found in biological systems, including people (Figure 8.8), even though there is no apparent biological need for lead.

Lead affects nearly every system of the body. Lead poisoning probably causes widespread stillbirths, deformities, and brain damage. Symptoms of lead poisoning

■ **FIGURE 8.8**
The lead in urban soils (from the years before we stopped using lead in gasoline) is still concentrated where children are likely to play. Lead-based paint in older buildings, such as these in New York, also remains a hazard to young children, who sometimes eat flakes of paint.

may include anemia, mental retardation, palsy, coma, seizures, apathy, poor coordination, loss of recently acquired skills, and bizarre behavior. Lead is particularly a problem for young children, who apparently are more susceptible than adults to lead poisoning. Some children suffering from exposure to lead may become aggressive and difficult to manage.[10–12]

Lead may have contributed to the decline of the Roman Empire. Over 2,000 years ago, the Romans produced and used tremendous amounts of lead over the course of several hundred years. They used lead for a wide variety of things. For example, lead was used in pots in which grapes were crushed and processed into a syrup for making wine, in cups and goblets from which wine was drunk, in plates to eat from, and as a base for cosmetics

and medicines. In the homes of Romans wealthy enough to have running water, lead was used to make the pipes that carried the water. Today, some people believe that lead poisoning among the upper class in Rome was partly responsible for Rome's decline. Studies of the lead content of bones of ancient Romans tend to support this.[13]

Greenland's glaciers still contain lead from the days of ancient Rome. Glaciers develop a new layer of ice every year. The older layers are buried by younger layers, which helps us to identify the age of each layer. Researchers drill into glaciers and take samples of the layers. The samples look like long solid rods of glacial ice and are called *cores*. Measurements of the concentration of lead in cores show that lead concentrations in glacial ice dating from the Roman period (from about 500 B.C. to A.D. 300) are about four times higher than before and after this period. This suggests that the mining and smelting of lead in the Roman Empire added small particles of lead to the atmosphere that eventually settled out in the glaciers of Greenland.[14]

Even small amounts may affect children. Today, in some populations, over 20% of children have blood concentrations of lead that are higher than the levels we believe are safe. It is sufficiently concentrated in the blood and bones of children living in inner cities to cause health problems, and a new and significant hypothesis is that in children, lead concentrations lower than the levels known to cause physical problems may cause behavior problems, such as antisocial, delinquent behavior. This is a testable hypothesis (see Chapter 1 for a discussion of hypotheses). If the hypothesis is correct, then some of our urban crime may be traced to environmental pollution![10]

A recent study in children aged 7 to 11 years measured the amount of lead in bones and compared it with data concerning behavior over a 4-year period. The study took into account such factors as maternal intelligence, socioeconomic status, and quality of child rearing. In the end, the conclusion was that an above-average concentration of lead in children's bones was associated with an increased risk of attention-deficit disorder, aggressive behavior, and delinquency.[10]

Organic Compounds

What are they? **Organic compounds** are carbon compounds that are produced naturally by living organisms or synthetically by human industrial processes. It is difficult to generalize about the environmental and health effects of artificially produced organic compounds because there are so many of them, they have so many uses, and they can produce so many different kinds of effects.

Synthetic organic compounds are used in industrial processes, pest control, pharmaceuticals, and food additives. We have produced over 20 million synthetic chemicals, and new ones are appearing at a rate of about 1 million per year! Most are not produced commercially, but up to 100,000 chemicals are now being used, or have been used in the past. Once used and dispersed in the environment, they may produce a hazard for decades or even for centuries.

Persistent organic pollutants. Some synthetic compounds are called **persistent organic pollutants,** or **POPs**. Many were first produced decades ago, when their harm to the environment was not known, and they are now banned or restricted (Table 8.1). POPs have several properties that define them:[15]

■ They have a carbon-based molecular structure, often containing highly reactive chlorine.

■ Most are manufactured by people—that is, they are synthetic chemicals.

■ They are persistent in the environment—that is, they do not easily break down.

■ They are polluting and toxic.

■ They are soluble in fat and therefore likely to accumulate in living tissue.

■ They occur in forms that allow them to be transported by wind, water, and sediments for long distances.

One example is polychlorinated biphenyls (PCBs), which are heat-stable oils originally used to insulate electric transformers.[15]

■ A factory in Alabama manufactured PCBs in the 1940s, shipping them to a General Electric factory in Massachusetts. They were put in insulators and mounted on poles in thousands of locations.

■ The transformers deteriorated over time. Some were damaged by lightning, and others were damaged or destroyed during demolition. The PCBs leaked into the soil or were carried by surface runoff into streams and rivers. Others combined with dust and were transported by wind around the world.

□ TABLE 8.1 SELECTED COMMON PERSISTENT ORGANIC POLLUTANTS (POPs)

Chemical	Example of Use
Aldrin[a]	Insecticide
Atrazine	Herbicide
DDT[a]	Insecticide
Dieldrin[a]	Insecticide
Endrin[b]	Insecticide
PCBs[a]	Liquid insulators in electric transformers
Dioxins	By-product of herbicide production

Source: Data in part from Anne Platt McGinn, "Phasing Out Persistent Organic Pollutants," in Lester R. Brown et al., *State of the World 2000* (New York: Norton, 2000).

[a] *Banned in the U.S. and many other countries.*

[b] *Restricted or banned in many countries.*

■ The dust containing PCBs was deposited in ponds, lakes, or rivers, where it entered the food chain. First it entered algae along with nutrients it combined with. Insects ate the algae, which were eaten by shrimp and fish. In each stage up the food web, the concentration of PCBs increased.

■ Fish were caught by fishermen and eaten. The PCBs were then passed on to people, where they are concentrated in fatty tissue and mother's milk.

Another example is dioxin, a persistent organic pollutant that may be one of the most toxic of the human-made chemicals in the environment. Dioxin is a colorless crystal made up of oxygen, hydrogen, carbon, and chlorine. It is classified as an organic compound because it contains carbon. Many types of dioxin (and dioxin-like compounds) are known. The history of the scientific study of dioxin and its regulation illustrate once again the interplay of science and values. Although science isn't entirely certain about the toxicity of dioxin to humans and ecosystems, society has made a number of value judgments involving regulation of dioxin. This has led to continuing controversy.

Dioxin is a by-product, not usually manufactured intentionally. It results from chemical reactions, including the combustion of compounds that contain chlorine in the production of herbicides.[16] These compounds are discharged into the air through such processes as incineration of municipal waste (the major source), incineration of medical waste, burning of gasoline and diesel fuels in vehicles, burning of wood as a fuel, and refining of metals such as copper.

From town to ghost town to no town. The dioxin problem became well known in 1983 when Times Beach, Missouri, a town on the Meramec River just west of St. Louis, with a population of 2,400, was evacuated and purchased for $36 million by the government. The evacuation and purchase occurred after the discovery that oil sprayed on the town's roads to control dust contained dioxin and the entire area had been contaminated. Times Beach was labeled a dioxin ghost town (Figure 8.9). The buildings were bulldozed, and all that was left was a grassy and woody area enclosed by a barbed-wire-topped chain-link fence. Some scientists, including the person who ordered the evacuation, view the event as an overreaction by the government to a perceived dioxin hazard. Today the land has been cleaned up, planted with trees, and is a state park and bird refuge.

We know that dioxin is harmful to many living things. Studies suggest that some fish, birds, and other animals are sensitive to even small amounts of dioxin. As a result, it can cause widespread environmental damage to wildlife, including birth defects and death.

O. Franken/Corbis Sygma

▪ **FIGURE 8.9**
Dioxin-contaminated soil.
Soil samples from Times Beach, Missouri, thought to
be contaminated by dioxin.

However, how dioxin affects people is less certain.
Although dioxin is known to be extremely toxic to mammals, its actions in the human body are not well known.
What is known is that sufficient exposure to dioxin (usually from meat or milk containing the chemical) produces
a skin condition (a form of acne) that may be accompanied by weight loss, liver disorders, and nerve damage.[17]
The concentration at which it poses a hazard to human
health is still uncertain. Studies suggest that workers exposed to high concentrations of dioxin for longer than a
year have an increased risk of dying of cancer.[18]

The good news is that dioxin emissions have decreased significantly. However, we are only beginning to
understand the many ways dioxin gets into the air, water,
and land, the linkages and rates of transfer from airborne
transport to deposition in water, soil, and the biosphere. In
too many cases, estimates of the amounts emitted are based
more on expert opinion than on high-quality data or even
limited data.[19, 20] As a result of scientific uncertainty, the
controversy concerning dioxin is sure to continue.

Hormonally Active Agents (HAAs)

HAAs are also POPs. The opening case study discussed the feminization of frogs resulting from exposure to
the herbicide atrazine. There is increasing scientific evidence that certain chemicals in the environment, known
as **hormonally active agent (HAAs),** can cause developmental and reproductive abnormalities in animals, including humans. HAAs include a wide variety of chemicals,
such as some herbicides, pesticides, and phthalates (compounds found in many chlorine-based plastics). The evidence comes from studies of wildlife in the field and from
laboratory studies of human diseases, such as breast,
prostate, and ovarian cancer, as well as abnormal testicular
development and thyroid-related abnormalities.[2]

Studies link HAAs to reproductive abnormalities
among wildlife. Alligator populations in Florida exposed to pesticides such as DDT exhibit genital abnormalities and low egg production. Pesticides have also
been linked to reproductive problems among several
species of birds, including gulls, cormorants, brown pelicans, falcons, and eagles. Studies are also ongoing on
Florida panthers, which apparently have abnormal ratios
of sex hormones that may be affecting their reproductive capability. In summary, the major disorders that
have been studied in wildlife have centered on abnormalities including thinning of eggshells of birds, decline
in populations of various animals and birds, reduced viability of offspring, and changes in sexual behavior.[1]

Do HAAs play a role in human diseases? Research on linkages between HAAs and breast cancer has
been exploring relationships between environmental estrogens and cancer. Finally, there is concern that exposure to phthalates in chlorinated-based plastics is also
causing problems. The consumption of phthalates in
the United States is considerable, with the highest exposure in women of childbearing age. The products being tested as the source of contamination include
perfumes and other cosmetics, such as nail polishes and
hairsprays.[1]

Thermal Pollution

Releasing heat into water or air can cause harm.
Thermal pollution, also called heat pollution, occurs
when heat released into water or air produces undesirable effects. Heat pollution can be a sudden, one-time
event or be a long-term, chronic problem. Sudden heat
releases may result from natural events, such as brush
fires, forest fires, and volcanic eruptions, or from human-induced events, such as agricultural burning. The
major sources of chronic heat pollution are electric
power plants that produce electricity in steam generators, releasing heat into the environment.

There are several solutions to chronic thermal discharge into bodies of water. The heat can be released
into the air by cooling towers (Figure 8.10), or the heated
water can be temporarily stored in artificial lagoons until
it cools to a normal temperature. There have been some
attempts to use the heated water to grow organisms of
commercial value that require warmer water temperatures. Waste heat from a power plant can also be used for
a variety of purposes, such as warming buildings.

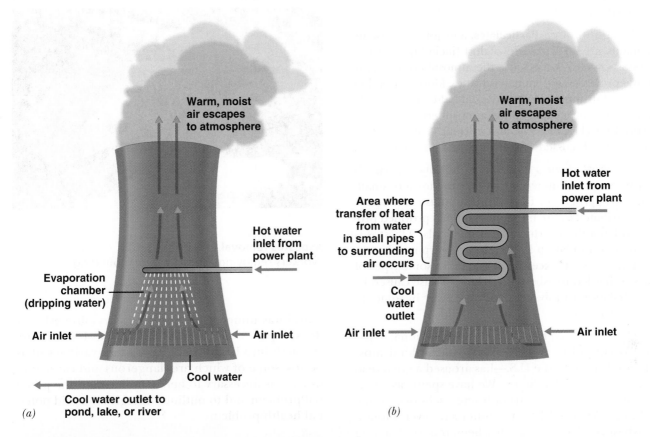

■ **FIGURE 8.10**

Two types of cooling towers.
(a) Wet cooling tower: Air circulates through the tower; hot water drips down and evaporates, cooling the water. (b) Dry cooling tower: Heat from the water is transferred directly to the air, which rises and escapes the tower. (c) Cooling towers emitting steam at Didcot power plant, in Oxfordshire, England. Red and white lines are vehicle lights resulting from long exposure time (photograph taken at dusk).

Particulates

What are they? **Particulates,** a major group of air pollutants, are small particles of dust (including soot and asbestos fibers) released into the atmosphere by many natural processes and human activities. Modern farming and the burning of oil and coal add considerable amounts of particulates to the atmosphere, as do dust storms, fires (Figure 8.11), and volcanic eruptions. Particulates are discussed with air pollution in Chapter 14.

Asbestos—how dangerous is it really? *Asbestos* is a term for several minerals that take the form of small, elongated particles or fibers. Industrial use of asbestos has contributed to fire prevention and has provided protection from the overheating of materials. Asbestos is also used as insulation for a variety of purposes. Unfortunately, however, excessive contact with some types of asbestos has led to asbestosis (a lung disease caused by the inhalation of asbestos) and to cancer in some industrial workers.[21]

More research needed. Nonoccupational exposure to chrysolite (white) asbestos—the kind most commonly used in the U.S.—has aroused a great deal of fear in the United States. We have spent vast sums of money to remove it from homes, schools, public buildings (Figure 8.12), and other sites even though no asbestos-related disease has been reported among those exposed to chrysolite in nonoccupational circumstances. It is now thought that much of the re-

John Chiasson/Liason/Getty Images

▪ **FIGURE 8.12**
Asbestos removal.
Workers removing asbestos from a building.

moval was unnecessary and that chrysolite asbestos doesn't pose a significant health hazard. Additional research into health risks from other varieties of asbestos, some of which are dangerous and cause lung disease, is necessary to better understand the potential problem and to outline strategies to avoid potential health problems.

Electromagnetic Fields

Electromagnetic fields (EMFs) are part of every-day urban life. Electric motors, electric transmission lines for utilities, and electrical appliances—such as toasters, electric blankets, and computers—all produce **electromagnetic fields**. The question is, do these fields pose a health risk?

Investigators initially did not believe that magnetic fields were harmful, because fields drop off quickly with distance from the source, and the strength of the fields that most people come into contact with are relatively weak. For example, the magnetic fields generated by power transmission lines or by a computer terminal are normally only about 1% of Earth's magnetic field. Directly below power lines, the electric field induced in the body is about what the body naturally produces within cells.

Early studies did indicate risk, but later studies did not. Several early studies concluded that children exposed to EMFs from power lines are at increased risk of contracting leukemia, lymphomas, and nervous-system cancers.[22] Investigators concluded that children so exposed are about one-and-a-half to three times more likely to develop cancer than children with very low exposure. But the results were questioned on the basis of

Michael Yasmashita

▪ **FIGURE 8.11**
Particulates.
Fires in Indonesia in 1997 caused serious air-pollution problems. People here are purchasing surgical masks in an attempt to breathe cleaner air.

the research design—problems in sampling, tracking children, and estimating exposure to EMFs. A later study analyzed over 1,000 children, approximately half of whom suffered from acute leukemia. It was necessary to estimate exposure to magnetic fields generated by nearby power lines in the children's present and former homes. Results of that study, which is the largest such investigation to date, concluded that there is no association between childhood leukemia and measured exposure to magnetic fields.[22, 23]

In summary, despite many studies, the jury is still out. There seems to be some indication that magnetic fields may cause diseases—including brain cancer and leukemia for electric-utility workers—but so far the risks appear relatively small and difficult to quantify.

Noise Pollution

Noise pollution is unwanted sound. Sound is a form of energy that travels as waves. We hear sound because our ears respond to sound waves through vibrations of the eardrum. How loud something sounds depends on the intensity of the energy carried by the sound waves and is measured in units of decibels (dB). Table 8.2 gives examples of sound levels. The threshold for human hearing is close to 0 dB; the average sound level in the interior of a home is about 45 dB; the sound of an automobile, about 70 dB; the sound of a jet aircraft taking off, about 120 dB; and a rock concert can reach 110 dB (Figure 8.13). A tenfold increase in the strength of a particular sound adds 10 dB units on

Kevin Winter/Getty Images

■ **FIGURE 8.13**
Loud music.
A rock concert can reach a very loud sound level. Exposure longer than half an hour may damage hearing.

the scale. An increase of 100 times adds 20 units.[8] The decibel scale is logarithmic: It increases exponentially as a power of 10. For example, 50 dB is 10 times louder than 40 dB and 100 times louder than 30 dB. Very loud noises (more than 140 dB) cause pain, and high levels can cause permanent hearing loss.

But loudness isn't the only problem. Environmental effects of noise depend also on the sound's pitch, frequency, and pattern, as well as the time of day (or night) and length of time we must listen to it.

□ TABLE 8.2 EXAMPLES OF SOUND LEVELS

Sound Source	Intensity of Sound (dB)	Human Perception
Threshold of hearing	~0	Threshold of human hearing
Rustling of leaf	10	Very quiet
Faint whisper	20	Very quiet
Average home	45	Quiet
Light traffic (30 m away)	55	Quiet
Normal conversation	65	Quiet
Chain saw (15 m away)	80	Moderately loud
Walkman at maximum volume	100	Very loud
Rock music concert (close)	110	Very loud
Thunderclap (close)	120	Uncomfortably loud
Jet aircraft takeoff at 100 m	125	Uncomfortably loud
Takeoff of fighter jet (close)	140	Threshold of pain
Rocket engine (close)	180	Traumatic injury

Note: cronic exposure to very loud sound levels can cause hearing problems such as "ringing" in the ears.

Voluntary Exposure

People sometimes expose themselves to harmful pollutants on purpose. Voluntary exposure to toxins and potentially harmful chemicals is sometimes referred to as exposure to personal pollutants. The most common of these are tobacco, alcohol, and other so-called recreational drugs. Use and abuse of these substances have led to a variety of human ills, including death and chronic disease, criminal activity such as reckless driving and manslaughter, loss of careers, street crime, and the straining of human relations at all levels.

8.3 General Effects of Pollutants

Almost every part of the human body is affected by one pollutant or another, as shown in Figure 8.14a. For example, lead and mercury (remember the Mad Hatter) affect the brain; arsenic, the skin; carbon monoxide, the heart; and fluoride, the bones. Wildlife is affected as well. How some of the major pollutants affect an animal is shown in Figure 8.14b. The effects that pollutants have on wildlife populations are listed in Table 8.3.

The lists of potential toxins and affected body sites for humans and other animals in Figure 8.14 may be

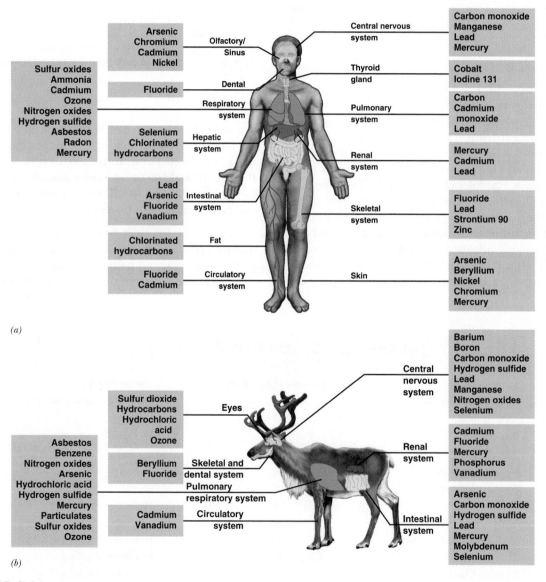

■ **FIGURE 8.14**
Pollution effects on people and wildlife.
(a) Effects of some major pollutants in human beings. (b) Known sites of effects of some major pollutants in wildlife. [*Source:* (a) G. L. Waldbott. *Health effects of environmental pollutants,* 2nd ed. St. Louis: Mosby, 1978. ©1978 by C. V. Mosby. (b) J. R. Newman. Effects of air emissions on wildlife resources. U.S. Fish and Wildlife Services Program, National Power Plant Team, FWS/OBS-80/40. Washington, DC: U.S. Fish and Wildlife Service, 1980.]

□ TABLE 8.3 EFFECTS OF POLLUTANTS ON WILDLIFE

Effect on Population	Examples of Pollutants
Changes in abundance	Arsenic, asbestos, cadmium, fluoride, hydrogen sulfide, nitrogen oxides, particulates, sulfur oxides, vanadium, POPs[a]
Changes in distribution	Fluoride, particulates, sulfur oxides, POPs
Changes in birth rates	Arsenic, lead, POPs
Changes in death rates	Arsenic, asbestos, beryllium, boron, cadmium, fluoride, hydrogen sulfide, lead, particulates, selenium, sulfur oxides, POPs
Changes in growth rates	Boron, fluoride, hydrochloric acid, lead, nitrogen oxides, sulfur oxides, POPs

[a] *Pesticides, PCBs, hormonally active agents, dioxin, and DDT are examples (see Table 8.1).*
Source: J. R. Newman, *Effects of Air Emissions on Wildlife*, U.S. Fish and Wildlife Service, 1980. Biological Services Program, National Power Plant Team, FWS/OBS-80/40, U.S. Fish and Wildlife Service, Washington, D.C.

somewhat misleading. For example, chlorinated hydrocarbons, such as dioxin, are stored in the fat cells of animals, but they cause damage not only to fat cells but to the entire organism through disease, damaged skin, and birth defects. Similarly, a toxin that affects the brain, such as mercury, causes a wide variety of problems and symptoms, as shown by what happened in Minamata, Japan. Nevertheless, Figure 8.14 is helpful in pointing out the general adverse effects of excess exposure to chemicals.

Dose and Response

"Everything is poisonous, yet nothing is poisonous," said the physician and alchemist Paracelsus five centuries ago. By this he meant that large amounts of any substance can be dangerous, while in an extremely small amount the same substance can be relatively harmless. Even water, if you drink too much of it too fast, can kill you by diluting the salt in your blood, upsetting the electrolyte balance that controls your heart and brain. This recently happened on a radio show where contestants competed to see who could drink the most water. Every chemical element has a spectrum of possible effects on a particular organism. For example, living things require selenium in small amounts, but in high concentrations selenium may be toxic or increase the probability of cancer in cattle and wildlife. Copper, chromium, and manganese are other chemical elements required by animals in small amounts but toxic in higher amounts.

But not everyone responds the same way to the same dose. People realized many years ago that how a chemical affected an individual depended on the dose. This concept is called **dose response**. However, individuals differ in their response to chemicals, and it is difficult to predict what dose will cause a particular response in a particular individual. For this reason, it is practical to use dose-response curves that predict instead the percentage of a population that will generally respond to a specific dose of a chemical.[24] Generalized dose-response curves for two toxins (A and B) are shown in Figure 8.15. TD-50 is the dose that produces a toxic response in 50% of the population.

Threshold Effects

Is there always a threshold? When we discussed mercury toxicity, we said that a *threshold* is a level below which no effect occurs and above which effects begin to

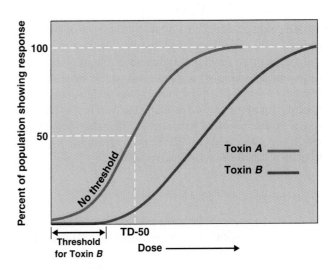

■ **FIGURE 8.15**

Dose response with and without a threshold effect.
In this hypothetical toxic dose-response curve, toxin *A* has no threshold—even the smallest amount has some measurable effect on the population. The TD-50 for toxin *A* is the dose required to produce a response in 50% of the population. Toxin *B* has a threshold (flat, lower part of curve) where response is constant as dose increases. After the threshold dose is exceeded, the response increases.

occur. If we determine that there is a threshold dose of a chemical, that means we consider any concentration of that chemical in the environment below the threshold safe (see toxin B in Figure 8.15). If there is no threshold dose, then even the smallest amount of the chemical has some negative toxic effect (see toxin A in Figure 8.15).

Whether or not there is a threshold for environmental toxins is an important issue. For example, an original goal of the U.S. Federal Clean Water Act was to reduce to zero the discharge of pollutants into water. This goal implied that there is no threshold—that no level of pollution should be legal. However, it is not realistic to think we can achieve zero discharge of water pollutants or a zero concentration of carcinogenic chemicals (chemicals believed to cause or promote cancer).

And what about combinations of toxins? A problem in evaluating thresholds for toxic pollutants is that we don't know very much about whether or how thresholds may change if an organism is exposed to more than one toxin at the same time or to a combination of toxins and other chemicals, some of which may actually be beneficial. Our own exposures to chemicals in the environment are complex, and we are only beginning to understand and study the possible interactions and consequences of multiple exposures.

Ecological Gradients

Dose response differs among species. For example, the kinds of vegetation that can live nearest to a toxic source are often small plants that have relatively short lifetimes and are adapted to harsh and highly variable environments—such plants include grasses, sedges, and weedy species usually regarded as pests. Farther from the toxic source, trees may be able to survive. The changes in vegetation you observe as you move farther from the source of toxicity is called the *ecological gradient*.

Tolerance

This, too, varies among populations. The ability to withstand exposure to a pollutant or other harmful condition is referred to as *tolerance*. Some populations can develop a tolerance for some pollutants, but no populations can tolerate all pollutants.

Acute and Chronic Effects

Pollutants can have immediate and/or long-term effects. An *acute effect* is one that occurs soon after exposure, usually to large amounts of a pollutant. A *chronic effect* takes place over a long period, often as a result of prolonged exposure to low levels of a pollutant. For ex-

ample, a person exposed all at once to a high dose of poison may be killed soon after exposure (an acute effect). However, that same total dose, received slowly in small amounts over an entire lifetime, may instead lead to disease later or affect the person's DNA and offspring (a chronic effect).

8.4 Risk Assessment and Risk Management

Risk assessment is the process of determining potential health effects on people exposed to environmental pollutants and potentially toxic materials. Such an assessment generally includes four steps:[25]

1. *Identification of the hazard.* Identification consists of testing the substance to determine whether exposure is likely to cause health problems.

2. *Dose-response assessment.* This involves figuring out whether and how the dose of a chemical (therapeutic drug, pollutant, or toxin) affects people's health.

3. *Exposure assessment.* In this step we evaluate the intensity, duration, and frequency of human exposure to a particular chemical pollutant or toxin. The hazard to society depends on how much of the population is exposed to the chemical.

4. *Risk characterization.* Using what we learned in the first three steps, we attempt to determine the percentage of the population at risk and the probability of an individual suffering ill effects.

Risk management requires us to make scientific judgments and decide what actions we should take to help minimize health problems related to exposure to pollutants and toxins. Risk management takes into account our risk assessment plus technical, legal, political, social, and economic issues.

Risk assessment and risk management can lead to arguments. Scientific opinions about toxicity of a substance are often open to debate, and so are opinions about what actions to take. The appropriate action may be to apply the *Precautionary Principle*, discussed in Chapter 1—that is, take cost-effective measures to protect ourselves even when we are not entirely certain about the risk. For example, although we cannot be absolutely certain how hormonally active substances such as the weed killer atrazine may affect us, that should not keep us from taking cost-effective steps to protect ourselves and the environment from the pesticide. The Precautionary Principle is emerging as a powerful ideology that is shifting the burden of proof from those who claim a substance is dangerous to those who manufacture, distribute, and use it. In short, it's not up to you to prove it's harmful, it's up to them to prove it's not—*before* they use it.

Return to the Big Question

Why are even very small amounts of some pollutants a major concern?

First of all, because once released into the environment, toxins can travel far and wide, and, as we saw in our earlier discussion of leopard frogs, even very small amounts can turn up far from their original site and produce wholly unexpected and undesirable effects.

Second, some toxins can remain in the environment for hundreds, even thousands of years.

Third, there is evidence that even amounts we have considered safe may have unexpected results in some members of a population—in children, for example—which can cause problems for the society as a whole.

Fourth, very small amounts of some toxins that are stored in the body become larger and larger amounts as they accumulate over time.

Summary

- Pollution makes air, water, and soil impure and unclean. Contamination by undesirable substances makes something unfit for a particular use. Toxic materials are poisonous to people and other living things; toxicology is the study of toxic materials.

- An important concept in studying pollution problems is synergism, whereby the combined effect of different substances is greater than the sum of their individual effects.

- We commonly express the concentration of pollutants in parts per million (ppm) and parts per billion (ppb). Air pollutants are commonly measured in units such as micrograms of pollutant per cubic meter of air ($\mu g/m^3$).

- Categories of environmental pollutants include toxic chemical elements (particularly heavy metals), organic compounds, persistent organic pollutants, hormonally active agents, radiation, heat, particulates, electromagnetic fields, and noise.

- Organic compounds of carbon are produced by living organisms or synthetically by humans. Synthetic organic compounds may have physiological, genetic, or ecological effects when introduced into the environment. The potential hazards of organic compounds vary: Some are more readily degraded in the environment than others; some are more likely to undergo biomagnification; and some are extremely toxic even at very low concentrations. Those that are a serious concern include persistent organic pollutants, such as pesticides, dioxin, PCBs, and hormonally active agents.

- How a chemical or toxic substance affects an individual depends on the dose and on the individual's tolerance. Effects may be acute or chronic.

- Risk assessment involves hazard identification, assessment of dose response, assessment of exposure, and risk characterization.

Key Terms

biomagnification

carcinogen

contamination

dose response

electromagnetic fields (EMFs)

heavy metals

hormonally active agents (HAAs)

noise pollution

organic compounds

particulates

persistent organic pollutants (POPs)

pollution

risk assessment

synergism

thermal pollution

toxic

toxicology

Getting It Straight

1. What kinds of life-forms would most likely survive in a highly polluted world? What would be their general ecological characteristics?

2. What is biomagnification, and why is it important in toxicology?

3. Describe the difference between acute and chronic effects of pollutants.

4. What are the roles of human-made chemicals in our environment?

5. What are the impacts of "Vog" on the environment?

6. Define the following terms. Identify something in your community that can be categorized by each of the terms.
   a. Pollution   **b.** Contamination   **c.** Toxin

7. What is the difference between point source and non-point source pollutants?

8. What category of pollutant do each of these pollutants fall into?
   a. Anthrax   **b.** Arsenic   **c.** Selenium   **d.** Lead
   e. PCBs   **f.** Asbestos

9. What is biomagnification? How are food webs impacted by this principle?

10. What chemical exposure was the cause of the deaths, disabilities, and congenital defects in Minamata Bay?

11. What chemical was the cause of the evacuation, destruction, and abandonment of Times Beach, Missouri?

12. What activities cause release of particulates into the environment?

What Do You Think?

1. Do you think the hypothesis that some crime is caused in part by environmental pollution is valid? Why? How might the hypothesis be further tested?

2. Some environmentalists argue that there is no such thing as a threshold for pollution effects. What do they mean? How would you determine whether it was true for a specific chemical and a specific species?

3. Why is it difficult to establish standards for acceptable levels of pollution? In giving your answer, consider physical, climatological, biological, social, and ethical reasons.

4. Design an experiment to test whether tomatoes or cucumbers are more sensitive to lead pollution.

5. Do you think the Precautionary Principle is necessary to protect the environment? Why? Why not?

Pulling It All Together

1. You are lost in Transylvania while trying to locate Dracula's castle. Your only clue is that the soil around the castle has an unusually high concentration of the heavy metal arsenic. You wander in a dense fog, able to see only the ground a few meters in front of you. What changes in vegetation warn you that you are nearing the castle?

2. A new highway is built through a pine forest. Driving along the highway, you notice that the pines nearest the road have turned brown and are dying. You stop at a rest area and walk into the woods. One hundred meters away from the highway, the trees look fine. Could you make a crude dose-response curve from direct observations of the pine forest?

What else would be necessary to devise a dose-response curve from direct observation of the forest? What else would be necessary to devise a dose-response curve that could be used in planning the route of another highway?

3. How might you apply the case study on wild leopard frogs to concerns humans face with regard to environmental issues? What environmental toxicology risks can affect humans without them even recognizing the initial concerns?

Further Reading

Amdur, M., J. Doull, and C. D. Klaasen, eds. 1991. *Casarett & Doull's toxicology: The basic science of poisons,* 4th ed. Tarrytown, NY: Pergamon.—A comprehensive and advanced work on toxicology.

Carson, R. 1962. *Silent spring.* Boston: Houghton Mifflin.—A classic book on problems associated with toxins in the environment.

Schiefer, H. B., D. G. Irvine, and S. C. Buzik. 1997. *Understanding toxicology: Chemicals, their benefits and risks.* Boca Raton, FL: CRC Press.—A concise introduction to toxicology as it pertains to everyday life, including information about pesticides, industrial chemicals, hazardous waste, and air pollution.

Travis, C. C., and H. A. Hattemer-Frey. 1991. Human exposure to dioxin. *The Science of the Total Environment* 104:97–127.—An extensive technical review of dioxin accumulation and exposure.

iStockphoto

9

Agriculture and Environment

Big Question

Can We Feed the World Without Destroying the Environment?

Learning Objectives

Agriculture is one of humanity's and civilization's greatest triumphs, but it is also the source of some of our greatest environmental problems. Agriculture has an ancient lineage, going back thousands of years, and it has always changed the local environment. Today it changes the environment both locally and globally in many ways. After reading this chapter, you should understand that . . .

- farming erodes the soil, but we can do a lot to slow this process;

- although we need to use fertilizers because farming uses up the nutrients in soil, we can find ways to use less of these;

- some lands are best used for grazing, but overgrazing can damage land;

- widespread farming on marginal land—land that is already suffering from a harsh climate and/or poor soil—can create deserts;

- a farm is a great habitat, especially for insects that love to eat plants (both crops and weeds) that grow well in the open fields;

- newer agricultural methods—such as integrated pest management, no-till agriculture, mixed cropping, and other methods of soil conservation—can provide major environmental benefits;

- genetic modification of crops could improve food production and benefit the environment—but perhaps also create new environmental problems.

Case Study

Clean-Water Farms

Steve Burr, a farmer near Salina, Kansas, has 300 acres of crops and 400 acres of grassland on which he raises cattle. In 1994 Steve became concerned about the amount of erosion on his cropland and the effects of fertilizers, pesticides, and livestock wastes on water quality. He decided to make two major changes in the management of his farm. He converted some of his cropland to pasture grass, which not only reduced erosion but also enabled him to cut down on his use of fertilizer and pesticides, thus reducing chemical pollution in the runoff from his land. At the same time, he divided his pastureland into sections, called *paddocks,* and rotated his animals through them (Figure 9.1). Rotating through the paddocks meant they had a constant supply of fresh forage, and their waste was evenly distributed rather than piling up in feedlots, where they had posed a major disposal problem. Steve found that his animals had fewer parasites. Moreover, the paddock rotation system also prevented overgrazing, soil fertility improved, and he spent less money on dry feed for his cows, so he was able to increase the size of his herd and generate more income. This method of raising livestock is known as *intensive rotational grazing* or *management-intensive grazing.*

Steve Burr is not the only farmer trying out new farm-management practices to reduce pollution, improve water quality, and also improve his balance sheet. The Burr farm is one of 36 participants in the Kansas Rural Center's Clean Water Farms Project, which began in 1995. The goal of the project is to farm in a manner beneficial both to the environment and to the economics of farming.[1] More and more farmers in many other states are using intensive rotational grazing systems, and U.S. farmers are not the only ones adopting this approach to raising livestock. In fact, the first system of rotational grazing was developed in France, and a version of it was pioneered in South Africa. New Zealand ranchers are known for their expertise in intensive rotational grazing, and the world-famous Argentine beef comes from cattle raised by similar methods.

According to Missouri farmer David Shafer, consumers, too, reap benefits from this system of raising beef—they get higher-quality products at lower prices. On a small scale, farmers like David Shafer and Steve Burr are re-creating migration patterns similar to those of bison and elk. Shafer hopes that over time rotational grazing will restore the prairies that once existed in the central United States.

Intensive rotational grazing is just one of many sustainable approaches to farming. Others include crop rotation, use of cover crops (which we will explain later) to reduce fertilizer needs and erosion, composting of livestock wastes, no-till farming, integrated pest and weed management (also explained later), and redesigning livestock waste management and watering systems.

This case study shows that practices that benefit the environment can be profitable as well.

Courtesy Steve Burr

■ **FIGURE 9.1**
Steve Burr on his farm.
Steve Burr rotates his cattle on this grassland to prevent overgrazing on his farmland.

9.1 How Agriculture Changes the Environment

Of all human activities, agriculture has proved to be perhaps the most sustainable. The fact is, people have farmed the Nile Valley, the fertile crescent of the Middle East, rice fields in China, and elsewhere for thousands of years. Few, if any, other human activities have been maintained in the same place for such a long time.

How much do we actually consume each year? Every year the average American consumes more than 410 pounds of vegetables, 200 pounds of grains, 275 pounds of fruit, almost 170 pounds of red meat and chicken, more than 200 pounds of milk and milk products, 33 pounds of eggs, and 16 pounds of fish and shellfish—more than half a ton of food a year! That's a lot of food, and the environmental effects of producing and processing it are important to us.[2]

Farmers feed the more than 6 billion people in the world, a tremendous feat. This is why a surprisingly large percentage of the world's land area is in agriculture—approximately 11% of the total land area of the world, excluding Antarctica. All told, agriculture occupies an area about the size of South and North America combined—enough to make agriculture a human-induced biome.

Interestingly, most of the world's food is provided by only 14 plant species. In approximate order of importance, these are wheat, rice, maize, potatoes, sweet potatoes (which aren't actually potatoes), manioc, sugarcane, sugar beet, common beans, soybeans, barley, sorghum, coconuts, and bananas (Figure 9.2). The first six provide more than 80% of the total calories consumed by human beings either directly or indirectly[3] (Figure 9.3). Only about 3,000 of Earth's half-million plant species have been used as agricultural crops, and

Kevin Morris/Stone/Getty Images

JC Carton/Bruce Coleman, Inc

Thomas Horland/Grant Heilman Photography

(a)

(b)

(c)

■ **FIGURE 9.2**
Some of the world's major crops:
(a) wheat, (b) rice, and (c) soybeans.

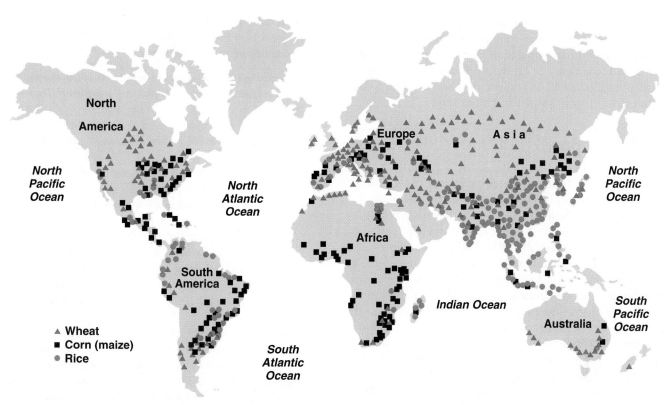

Where the world's major crops grow.

only 150 species have been cultivated on a large scale. In the United States, 200 species are grown as crops.

Now for the bad news. In the process of feeding the world, farming degrades soil; fertilizers and pesticides affect soil, water, and downstream ecosystems; irrigation of farmland can lead to *salinization* (the buildup of salts in the soil to the point that crops can no longer grow) and to the accumulation of toxic metals, and ultimately to a loss of biodiversity.

And here's another big problem: In the future, if the human population doubles as expected, agriculture will need to double its production just to keep feeding people the amount they're accustomed to eating today—and for some people, the food supply even now isn't adequate. Since the end of World War II, rarely has a year passed without a famine somewhere in the world[4] (Figure 9.4). Food emergencies affected 34 countries at the end of the 20th century. Examples include famines in Brazil (1979–1984), Ethiopia (1984–1985), Somalia (1991–1993), and the 1998 crisis in Sudan (Figure 9.4). Varying weather patterns in Africa, Latin America, and Asia, along with inadequate international trade in food, contributed to these emergencies.[5] Africa remains the continent with the most acute food shortages, due to adverse weather and civil strife.[6]

Where would we produce all that additional food? If we can't increase the amount of food produced

by the land area that is being used for agriculture today, we would need to use an additional area as big as the entire New World. Where will we find it? Which biomes should we alter greatly? Land best suited for agriculture

Photograph of children suffering from kwashiorkor, one kind of malnutrition.
Their extended stomachs are one symptom of this ailment.

has already been put to this use, and much of it is already under pressure for conversion to cities, towns, and suburbs. As the human population increases, cities and suburbs will continue to grow, and more of the best agricultural land will be converted to housing, shopping centers, schools, hospitals, and other uses.

9.2 Dust Bowls and Our Eroding Soils

Soils are keys to sustainable farming, but farming damages soils. When farmers clear land of its natural vegetation, such as forest or grassland, the soil begins to lose its fertility. Some of this is due to erosion. In the United States, about 1 million hectares (2.5 million acres)—an area larger than Rhode Island—are lost each year to urbanization and soil erosion. About one-third of the country's topsoil has been lost—80 million hectares (198 million acres) either totally ruined by soil erosion or no longer very productive. Since the end of World War II, farming has seriously damaged more than 1 billion hectares (2.47 billion acres, or about 10.5% of the world's best soil), an area equal in size to China and India combined. Overgrazing, deforestation, and destructive crop practices have so seriously damaged about 9 million hectares (22 million acres) that recovery will be difficult. Restoration of the rest will require serious actions.[7]

Soil erosion became a national issue in the U.S. in the 1930s, when a major drought left loose, dry soil over large areas. The wind created dust storms that blew the soil away, burying automobiles and houses, destroying many farms, and impoverishing many people. One result was a large migration of farmers from Oklahoma and other western and midwestern states to California. The human tragedies of the Dust Bowl were made famous by the "Dust Bowl Ballads," songs written by Woody Guthrie, including "The Great Dust Storm," whose verses describe what it was like to be in one of the dust storms: [8]

Our relatives were huddled into their oil boom shacks,
And the children they was cryin' as it whistled through the cracks.
And the family it was crowded into their little room,
They thought the world had ended, and they thought it was their doom.
The storm took place at sundown, it lasted through the night,
When we looked out next morning, we saw a terrible sight.
We saw outside our window where wheat fields they had grown
Was now a rippling ocean of dust the wind had blown.
It covered up our fences, it covered up our barns,
It covered up our tractors in this wild and dusty storm.
We loaded our jalopies and piled our families in,
We rattled down that highway to never come back again.

What caused the Dust Bowl? The land that became the Dust Bowl had been part of America's great prairie, where grasses rooted deep, creating a heavily organic soil a meter or more down. The deep-rooted grasses protected the soil from water and wind. When the plow turned over those roots, the soil was exposed directly to sun and wind, which dried it out and blew it away. It was a great tragedy of that time and a lesson people thought would be remembered forever. But soil continues to erode. The introduction of heavy earth-moving machinery after World War II added to the problem by compacting the soil, damaging the soil structure so important for crop production (Figure 9.5). This erosive effect of agriculture is one of the reasons that the farming practices discussed in the opening case study have become popular.

Corbis Betmann

(a)

U.S. Dept. of Agriculture Photography Ctr

(b)

▪ **FIGURE 9.5**
Poor agricultural practices and a major drought created the Dust Bowl, which lasted about ten years in the 1930s. Heavily plowed land lacking vegetation blew away easily in the dry winds, (a) creating dust storms and (b) burying houses and trucks.

Things are getting better. Although soil loss in the United States continues, measurements suggest that it has slowed considerably thanks to improvements in plowing and the use of no-till agriculture (discussed later in this chapter). On average, soil erosion in the United States has declined from 17 metric tons per hectare per year to about 13 tons per hectare per year.[9]

One example is the drainage area of Coon Creek, Wisconsin, an area of 360 square kilometers, which has been heavily farmed. This stream's watershed was the subject of a detailed study in the 1930s by the United States Soil Conservation Service, and was studied again in the 1970s and 1990s. Measurements at these three times indicate that the amount of soil lost to erosion in the 1990s was only 6% of the amount lost in the 1930s.[10, 11]

The Plow Puzzle. Here is a curious puzzle about agriculture and the plow. There are big differences be-tween the soils of an unplowed forest and soils of previously forested land that has been plowed and used for crops for several thousand years—in Italy, for example, iron plows were pulled by oxen many centuries ago. These differences were observed and written about by one of the originators of the modern study of the environment, George Perkins Marsh. Born in Vermont in the 19th century, Marsh became the American ambassador to Italy and Egypt. While in Italy, he was so struck by the differences between the soils of the forests of his native Vermont and the soils that had been farmed for thousands of years on the Italian peninsula that he made this a major theme in his landmark book *Man and Nature*. The farmland he observed in Italy had once been forests. But while the soil in Vermont was rich in organic matter and had definite layers, the soil of Italian farmland had little organic matter and lacked definite layers (Figure 9.6).

■ **FIGURE 9.6**
Soils.
In forests and prairie, soils develop a definite structure with layers affected by animals and plants. Plowing destroys this soil structure, leading to undesirable effects on plant growth. Here is a diagram of a mature forest soil.

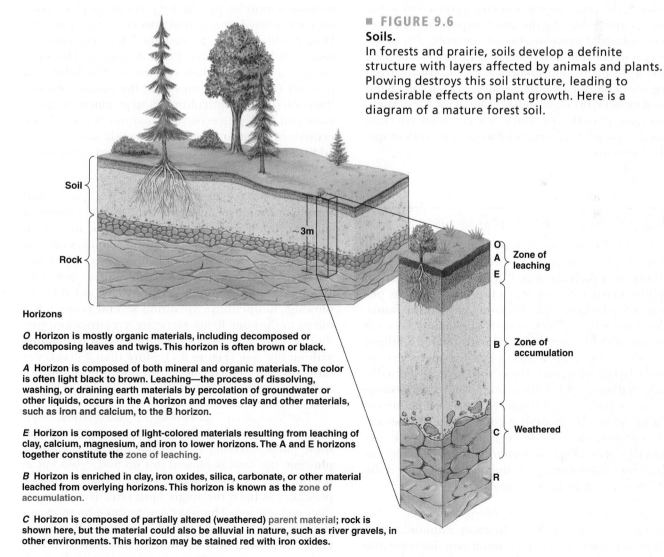

Horizons

O Horizon is mostly organic materials, including decomposed or decomposing leaves and twigs. This horizon is often brown or black.

A Horizon is composed of both mineral and organic materials. The color is often light black to brown. Leaching—the process of dissolving, washing, or draining earth materials by percolation of groundwater or other liquids, occurs in the A horizon and moves clay and other materials, such as iron and calcium, to the B horizon.

E Horizon is composed of light-colored materials resulting from leaching of clay, calcium, magnesium, and iron to lower horizons. The A and E horizons together constitute the zone of leaching.

B Horizon is enriched in clay, iron oxides, silica, carbonate, or other material leached from overlying horizons. This horizon is known as the zone of accumulation.

C Horizon is composed of partially altered (weathered) parent material; rock is shown here, but the material could also be alluvial in nature, such as river gravels, in other environments. This horizon may be stained red with iron oxides.

R Unweathered (unaltered) parent material.

One would expect that farming in such soil would eventually become unsustainable, but much of the farmland in Italy and France has been used continuously since pre-Roman times and is still highly productive. How can this be? And what has been the long-term effect of such agriculture on the environment?

Farmers kept farming by putting nutrients back into the soil. Traditionally, farmers combated the decline in soil fertility by using organic fertilizers, such as animal manure. These have the advantage of improving both chemical and physical characteristics of soil. But organic fertilizers can have drawbacks, especially under intense agriculture on poor soils. In such situations, they do not provide enough of the chemical elements needed to replace what is lost.

Chemical fertilizers were an important development. Industrially produced fertilizers, commonly called "chemical" or "artificial" fertilizers, were a major factor in the great increases in crop production in the 20th century. Among the most important advances were industrial processes to convert molecular nitrogen gas in the atmosphere into nitrate that can be used directly by plants. Phosphorus is mined, usually from organic deposits (sometimes fossilized), such as guano (bird excrement) on islands that birds use for nesting. Nitrogen, phosphorus, and other elements are combined in proportions beneficial to specific crops in specific locations.

9.3 Where Eroded Soil Goes

Soil eroded from one location has to go somewhere else. A lot of it travels down streams and rivers and is deposited at their mouths. U.S. rivers carry about 3.6 billion metric tons per year (4 billion U.S. tons per year) of sediment, 75% of it from agricultural lands. That's more than 25,000 pounds of sediment for each person in the United States. Of this total, 2.7 billion metric tons per year (3 billion U.S. tons per year) are deposited in reservoirs, rivers, and lakes. Eventually, these sediments fill in these bodies of water, destroying some fisheries. In tropical waters, sediments entering the ocean can destroy coral reefs near a shore. The sediment deposits on the reefs block out the sunlight that photosynthetic reef organisms need, and can also cause other damage to the reefs, especially if the sediments contain toxic chemicals.

Soil eroded from farms carries chemicals that affect the environment. Nitrates, ammonia, and other fertilizers carried by sediments increase the growth of algae in water downstream (a process called *eutrophication*) just as they boost the growth of crops, but people generally don't want algae in their water. The water develops a thick, greenish-brown mat, unpleasant for recreation and for drinking. In addition, because the dead algae are decomposed by bacteria that remove oxygen from the water, fish can no longer live in that water. Sediments also can carry toxic chemical pesticides. Efforts to limit soil erosion have reduced the amount of agricultural sedimentation since the 1930s. Even so, taking into account the costs of dredging and the decline in the useful life of reservoirs, sediment damage costs the United States about $500 million a year.

9.4 Making Soils Sustainable

It's not enough for crops to be sustainable—the ecosystem must be, too. At this point in our discussion, we have arrived at a partial answer to the question: How could farming be sustained for thousands of years, while the soil has been degraded? However, there is a difference between the sustainability of a product (in this case crops) and the sustainability of the ecosystem. In agriculture, crop production can be sustained while the ecosystem may not be. And if the ecosystem is not sustained, then people must provide additional input of energy and chemical elements to replace what is lost.

Several ways to make soils sustainable. Soil forms continuously. In ideal farming, the amount of soil lost would never be greater than the amount of new soil produced. Production of new soil is slow—on good lands, the formation of a layer of soil 1 millimeter deep (thinner than a piece of paper) may take 10–40 years. Sustainability of soils can be aided by fall plowing; multiculture (planting several crops intermixed in the same field), terracing, crop rotation, contour plowing, and no-till agriculture—that is, planting without plowing (Figure 9.7). More than 250 acres of farmland are treated one way or another to improve soil conservation (Figure 9.8)

Contour plowing. Plowing creates furrows, and if the furrows go downhill, then the water pours down these paths, carrying a lot of soil with it. In **contour plowing**, the land is plowed not up and down but as horizontally as possible across the slopes. Contour plowing has been the single most effective way to reduce soil erosion. This was demonstrated by an experiment on sloping land planted in potatoes. Part of the land was plowed in rows running downhill, and part

Science Vu/Visuals Unlimited

(a)

Jeri Gleiter/Peter Arnold Inc.

(b)

■ **FIGURE 9.7**

Growing crops but avoiding or limiting plowing.
(a) Contour strip crops in the midwestern United States; (b) no-till soybean crop planted in wheat stubble on a Kansas farm.

was contour-plowed (Figure 9.7). The up-and-down section lost 32 metric tons of soil per hectare (14.4 tons per acre). The contour-plowed section lost only 0.22 metric ton per hectare (0.1 ton per acre), as shown in Figure 9.7a. It would take almost 150 years for the contour-plowed land to erode as much as the traditionally plowed land eroded in a single year!

In addition to greatly reducing soil erosion, contour plowing uses less fuel and time because the plowing machines move faster and require less energy than when they must travel uphill. Even so, today contour plowing is used on only a fraction of the land in the United States. For example, of Minnesota's 4 million hectares (10 million acres) of cropland, only 530,000 hectares (1.3 million acres) are contour-plowed.

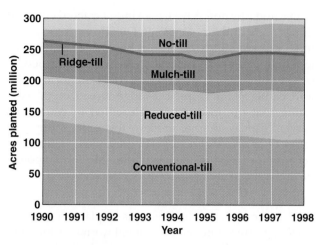

■ **FIGURE 9.8**

Conservation tillage.
This graph shows the acres planted using different ways to conserve soil (or not). About half the acreage in the United States is now planted using some kind of soil conservation method. (*Source:* USDA. Available at http://www.ers.usda.gov/Briefing/AgChemicals/soilmangement.htm#conservationtillage. Accessed May 21, 2005.)

No-till agriculture. An even more efficient way to slow erosion is to avoid plowing altogether. No-till agriculture (also called conservation tillage) involves not plowing the land, using herbicides and integrated pest management (discussed later in this chapter) to keep down weeds, and allowing some weeds to grow. Stems and roots that are not part of the commercial crop are left in the fields and allowed to decay in place (Figure 9.7b). In contrast to standard modern approaches, the goal in no-till agriculture is to suppress and control weeds but not to eliminate them if doing so would harm the soil. Worldwide, no-till agriculture is increasing. Paraguay leads the world with 55% of its farmland in no-till. The United States, with 17.5% in no-till, lags behind many other nations. Argentina has 45%, Brazil 39%, and Canada reached 30% in 2001, up from 24% a decade earlier. Of course, like so many things we do, no-till involves trade-offs—for example, it requires greater use of pesticides. But decreased erosion means that a small percentage of these pesticides will be transported off the agricultural fields, and the pesticides will have a longer time in which to decompose in place.

9.5 Farm Pests

All that food is bound to attract party crashers. As long as agriculture has existed, it has had pests—insects, mammals, and birds that eat crops and seeds; weeds competing with crops; diseases attacking the crops. And why not? A farm is a kind of cafeteria of good things to eat and live on, planted neatly one species at a time, so easy to find. Insect pests have been such a problem that in a medieval French village the townspeople went to the judge and asked him to try the locusts who were eating all the crops. The judge agreed, on the condition that the insects be represented in court by a lawyer.[12] Today, despite modern technology, the total losses from all pests are huge. In the United States, pests account for an estimated loss of one-third of the potential harvest and about one-tenth of the harvested crop.[13]

Some of the major agricultural pests are insects that feed mainly on the live parts of plants, especially leaves and stems; nematodes (small worms), which live mainly in the soil and feed on roots and other plant tissues; bacterial and viral diseases; and vertebrates (mainly rodents and birds) that feed on grain or fruit.

Weeds are the worst pests. We tend to think that the major agricultural pests are insects, but in fact weeds are the major problem. Farming produces special environmental and ecological conditions that tend to promote weeds. Remember that the process of farming is an attempt to (1) hold back the natural processes of ecological succession, (2) prevent migrating organisms from entering an area, and (3) prevent natural interactions (including competition, predation, and parasitism) between populations of different species.

Weeds thrive in early-successional croplands. Because a farm is maintained in a very early stage of ecological succession and is enriched by fertilizers and water, it is a good place not only for crops but also for other early-successional plants. These noncrop and therefore undesirable plants are what we call "weeds." A weed is just a plant in a place we do not want it to be. Recall that early-successional plants tend to be fast-growing and have seeds that are easily blown by the wind or spread by animals. These plants spread and grow rapidly in the inviting habitat of open, early-successional croplands (see Chapter 6).

There are about 30,000 species of weeds, and in any year a typical farm field is infested with anywhere from 10 to 50 of them. Weeds compete with crops for all resources: light, water, nutrients, and just plain space to grow. The more weeds, the less crop. Some weeds can have a devastating effect on crops. For example, the production of soybeans is reduced by 60% if a weed called cocklebur grows three individuals per square meter (one individual per square foot).[14] In the United States, weeds cause agricultural losses of more than $16 billion a year, and U.S. farmers spend an additional $3.6 billion for chemical weed control, amounting to 60% of all pesticide sales.

9.6 How Much Pesticide Do We Release into the Environment? And Where Does It Go?

We use a lot. About 500 million kilograms (1.1 billion pounds) of pesticides of more than 600 kinds were used in the United States and $40 billion worth were purchased worldwide in 2005.[15]

About 60% of pesticides found in U.S. waters are herbicides (weed killers). Eventually, the toxic compounds decompose, but for some chemicals, this can take a very long time. How long do they last in the environment, both where they were first used and downstream and downwind? What is the concentration of these in our waters? We don't know. Surprisingly little is known about past and present concentrations of pesticides in the major rivers of America. For example, there is no well-established program to monitor changes in the concentration of pesticides in the Missouri River, one of the longest rivers in the world and a river that drains one-sixth of the United States, much of it from the major agricultural states.

We need a wide-scale program to monitor pesticides in our water. The United States Geological Survey established a network for monitoring 60 sample watersheds throughout the nation. These are medium-size watersheds, not the entire flow from the nation's major rivers. One such watershed is that of the Platte River, a major tributary of the Missouri River. The most common herbicides used for growing corn, sorghum, and soybeans along the Platte River are alachlor, atrazine, cyanazine, and metolachlor, all artificial organic compounds. Along this river during heavy spring runoff, concentrations of some herbicides may be reaching or exceeding levels considered safe according to established public-health standards. But this research is just beginning, and it is too early to reach conclusions as to whether present concentrations are causing harm in public water supplies or to wildlife, fish, algae in freshwaters, or vegetation. A wider and better program to monitor pesticides in water and soil is important to provide a sound scientific basis for dealing with pesticides.

9.7 The Search for a Magic Bullet

Before the Industrial Revolution, farmers could do little to prevent pests except remove them when they appeared or use farming methods that tended to decrease their density. One of those methods, "slash-and-burn agriculture," is a traditional practice in which people cut away some but not all vegetation in small patches in a forest (Figure 9.9).

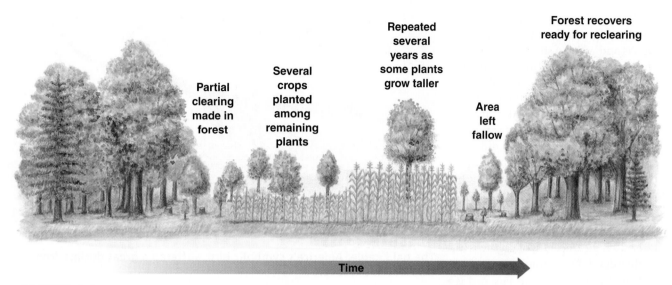

Partial clearing made in forest

Several crops planted among remaining plants

Repeated several years as some plants grow taller

Area left fallow

Forest recovers ready for reclearing

Time

■ **FIGURE 9.9**
Slash-and-burn agriculture.
Over time, secondary succession will take place on land partially cleared by slash-and-burn.

Slash-and-burn agriculture works pretty well. Some of the plants left are those that produce fruits and other edible products or produce chemicals that pests prefer to avoid. Several crops are planted together among existing vegetation, creating a more complex environment. Crops are harvested for a few years, then the land is allowed to grow back to a forest. The natural process of secondary succession—the redevelopment of the ecosystem as discussed in Chapter 5—is allowed to occur. This farming practice tends to work against pests and has other benefits. It helps to conserve chemical elements in the ecosystem.

Slash-and-burn agriculture has many other names. It is sometimes called cultivation with forest or bush fallow. In Latin America, it is called *milpa* agriculture; in Great Britain, *swidden* agriculture; in western Africa, *fang* agriculture. Whatever its name, people using this method harvest a mixture of crops, including root, stem, and fruit crops. For example, in western Africa, yams (a root crop) are harvested along with maize; in Southeast Asia, root crops are grown with rice and millet or with rice and maize.

Toxic chemical elements as pesticides. With the beginning of modern science-based agriculture, people began to search for chemicals that would reduce the abundances of pests. More important, they searched for a "magic bullet"—a chemical (referred to as a narrow-spectrum pesticide) that would have a single target, just one pest, and not affect anything else. But this proved elusive. We have already seen that living things have many chemical reactions in common, so a chemical that is toxic to one species is likely to be

toxic to another. The story of the scientific search for pesticides is the search for a better and better magic bullet. The earliest pesticides were inorganic compounds that were widely toxic (broad-spectrum). One of the earliest was arsenic, a chemical element that is toxic to all life, including people. It was certainly effective in killing pests, but it killed beneficial organisms as well and was very dangerous to use.

Nicotine and other natural organic compounds. A second stage in the development of pesticides, begun in the 1930s, used petroleum-based sprays and natural plant chemicals. Many plants produce chemicals as a defense against disease and herbivores, and these chemicals are effective pesticides. Nicotine, from the tobacco plant, is the primary agent in some insecticides that are still widely used today. However, although natural plant pesticides are comparatively safe, they are not as effective as desired.

Artificial organics. The third stage in the development of pesticides was the development of artificial organic compounds. Some, like DDT, are broad-spectrum but are more effective than natural plant chemicals. These chemicals have been important to agriculture, but they have had unexpected environmental effects. In sum, the magic bullet remains to be found.

DDT

At first it looked like the magic bullet. The real revolution in chemical pesticides began with the end of World War II and the discovery of DDT and other chlorinated hydrocarbons. When DDT was first developed in the 1940s, it seemed to be the long-sought magic bullet, deadly only to insects and with no short-term effects on people.[16] At the time, scientists believed that a chemical could not be readily transported from its original site of application unless it was water-soluble. DDT was not very soluble in water and therefore did not appear to pose an environmental hazard. DDT was used very widely until three things were discovered.

▪ It has long-term effects on desirable species. Most spectacularly, it decreased the thickness of eggshells as they developed within birds.

▪ It is stored in oils and fats and is transferred up food chains as one animal eats another. As it is passed up food chains, it becomes more and more concentrated, so that the higher an organism is on a food chain, the greater its concentration of DDT. This process is known as *food-chain concentration* or *biomagnification*. (We discuss it in greater detail in Chapter 15.)

▪ The storage of DDT in fats and oils allows the chemical to be transferred biologically even though it is not very soluble in water.

DDT's effect on birds' eggs caused it to be banned. DDT and the products of its chemical breakdown made eggshells so thin and soft that they broke easily, making reproduction less successful. The problem was especially severe in birds that are high on the food chain—predators that feed on other predators, such as the bald eagle, the osprey, and the pelican, which mainly eat fish that eat other fish. As a result, DDT was banned in most developed nations—it was banned in the United States in 1971.

Since then, these bird populations have recovered dramatically. The brown pelican of the Florida and California coasts had become rare and endangered, and their reproduction had been restricted to offshore islands where DDT had not been used. But since the banning of DDT, they have become common again. The bald eagle, too, became abundant again. They can be seen even flying in Potomac River inlets near Reagan National Airport, in Washington, D.C., or nesting in large groups along the coast of Alaska and very many other places in America (Figure 9.10).

DDT has some benefits and is still produced in the United States for use in the developing and less developed nations, especially as a control for mosquitoes that spread malaria. It has been primarily responsible

▪ **FIGURE 9.10**
The bald eagle, America's symbolic bird, suffered a great decline from DDT but has recovered.

for eliminating malaria and yellow fever as major diseases, reducing the incidence of malaria in the United States from an average of 250,000 cases a year prior to the spraying program to fewer than 10 per year in 1950. Even for these uses, DDT's effectiveness has declined over the years because many species of insects have developed a resistance to it. Nevertheless, DDT continues to be used because it is cheap and sufficiently effective and because people have become accustomed to using it. About 35,000 metric tons of DDT are produced annually in at least five countries, and it is legally imported and used in dozens of countries, including Mexico.

Although its use is banned in the U.S., it finds it way back here. While people in developed nations believe they are safe from the effects of DDT, in fact this chemical is transported back to industrial nations in several ways. One way it returns is in agricultural products from nations that still use DDT. Another way is that migrating birds that spend part of the year in malarial regions are still subject to DDT. For this reason, although it is banned in the developed nations, DDT remains an important pest-control issue worldwide. (The developing nations' use of pesticides that are banned in other nations is an issue not only for DDT but also for other chemicals.)

With DDT banned in developed nations, other chemicals came into use, chemicals that were less persistent in the environment. Among the next generation of insecticides were organophosphates (chemicals that contain phosphorus). These are more specific (narrow-spectrum) and decay rapidly in the soil, so they do not persist in the environment as long as DDT. But they are toxic to people—they affect the nervous system—and must be handled very carefully by those who apply them.

Chemical pesticides have revolutionized agriculture, but have major drawbacks. In addition to the negative environmental effects of chemicals such as DDT, there is the problem of secondary pest outbreaks. These occur after extended use of a pesticide and can come about in two ways: (1) Reducing one target species reduces competition with a second species, which then flourishes and becomes a pest, or (2) the pest develops resistance to the pesticides through evolution and natural selection, which favor those who have a greater immunity to the chemical.

Resistance has developed to many pesticides. For example, Dasanit (fensulfothion), an organophosphate first introduced in 1970 to control maggots that attack onions in Michigan, was originally successful but is now so ineffective that it is no longer used for that crop. Two commonly used insecticides today are aldrin and dieldrin. These have been widely used to control termites as well as pests on corn, potatoes, and fruits. Dieldrin is about 50 times as toxic to people as DDT.

9.8 Ecological Approaches to Pest Control

With the failure to find a magic bullet, research to control agriculture pests shifted in the second half of the 20th century to a fourth stage, **biological control**, and a fifth stage, **integrated pest management**. Both of these make use of ecological knowledge, and they have proved effective and comparatively benign environmentally.

Biological control uses predators and parasites to control pests. One of the most effective biological control agents is *Bacillus thuringiensis* (BT). BT causes a disease that affects caterpillars and the larvae of other insect pests. BT has been one of the most important ways to control epidemics of gypsy moths, an introduced moth whose larvae periodically strip most of the leaves from large areas of forests in the eastern United States. BT has proved safe and effective—it attacks only specific insects and is harmless to people and other mammals; and because it is a natural biological "product," its presence and its decay are nonpolluting. You can buy spores of BT at your local garden store and use this method for your home garden.

Another group of biological control agents are small wasps that kill caterpillars. The wasps are parasites—they lay their eggs on the caterpillars, and when the eggs hatch, the larval wasps feed on the caterpillars, killing them. These wasps tend to have very specific relationships—one species of wasp will be a parasite of one species of pest—so they are both effective and narrow-spectrum.

And let's not forget ladybugs. Ladybugs are predators of many pests. You can buy these, too, at many garden stores and release them in your garden.

Another technique to control insects involves the use of sex pheromones, chemicals released by most species of adult insects (usually the female) to attract members of the opposite sex. In some species, pheromones have been shown to be effective up to 4.3 kilometers (2.7 miles) away. These chemicals have been identified, synthesized (produced artificially), and used as bait in insect traps, in insect surveys, or simply to confuse the mating patterns of the insects involved.

Integrated Pest Management

While biological control works well, it has not solved all problems. As a result, a fifth stage developed, known as integrated pest management (IPM). IPM uses a combination of methods, including biological control, certain chemical pesticides, careful scouting of fields to check for infestations, and some methods of planting crops, such as paying careful attention to the timing of planting and the timing of pesticide use[17] (Figure 9.11). IPM is an ecosystem approach to pest management, because it makes use of the characteristics of ecological communities and ecosystems.

A key idea is that the goal can be control rather than complete elimination. This is justified for several reasons. It costs more and more to eliminate a greater

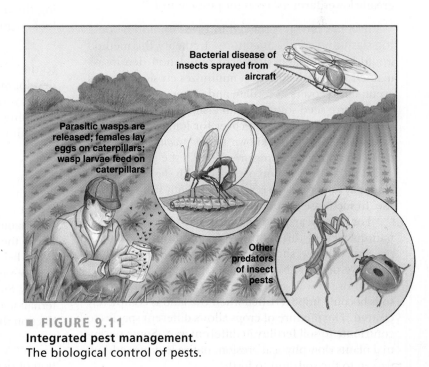

■ FIGURE 9.11
Integrated pest management.
The biological control of pests.

and greater percentage of a pest, but the size of the harvest does not increase at the same rate as the costs, so the farmer ends up spending more and more for less and less. It makes economic sense to eliminate only enough pests to provide a net benefit. In addition, allowing a small and controlled portion of a pest population to remain does less damage to ecosystems, soils, water, and air.

IPM moves away from monoculture (the planting of just one crop). This is because a complex habitat slows the spread of parasites—it becomes a maze that they are forced to find their way through slowly in search of their prey, unlike a field planted in regular rows of nothing but what the pest likes best.

No-till or low-till agriculture is another feature of IPM, because this helps the natural enemies of some pests to build up in the soil. One example of IPM is control of the oriental fruit moth by introducing a wasp, *Macrocentrus ancylivorus*, that lives on the moth caterpillars.[18] Interestingly, in peach fields the wasp was more effective when strawberry fields were nearby. Why? Because it made for a more complex habitat. The strawberry fields provided an alternative habitat for the wasp, especially important for overwintering.[19]

IPM does use artificial organic pesticides, but sparingly and specifically, because they are used along with the other techniques. A study by the U.S. Office of Technology Assessment concluded that IPM could reduce the use of pesticides by as much as 75% and reduce pre-harvest pest-caused losses by 50%. This would also greatly lower farmers' costs for pest control.[20]

IPM has led to renewed interest in, and respect for, slash-and-burn agriculture. In theory, this method would be sustainable if human population density remained low. Slash-and-burn minimizes erosion, the soil eventually recovers its fertility, and uncut vegetation provides future seed sources. With a smaller population to feed, sites can be left to recover for longer periods between plantings. This is known as a *long rotation period*. Under high population pressure, which is the case in many places today, the rotation period is much shorter, and the land may not be able to recover sufficiently from previous use. In such cases, production is not sustainable.

For many years, agricultural experts from developed nations viewed this method of agriculture as a poor process with low, short-term productivity, used only by primitive peoples. Now we understand that this kind of agriculture is well suited to high-rainfall lands, where soils become impoverished when the land is completely cleared. The mixture of crops allows different species to contribute to soil fertility in different ways. Some perennial plants slow physical erosion; native legumes add nitrogen to the soil; and so forth.

U.S. agricultural practices today use a combination of approaches, but in most cases they are more restricted than an IPM strategy. Biological control methods are used to a comparatively small extent. They are the primary tactics for controlling vertebrate pests (mice, voles, and birds) that feed on lettuce, tomatoes, and strawberries in California, but they are not major techniques for grains, cotton, potatoes, apples, or melons. Chemicals are the principal control methods for insect pests. For weeds, the principal controls are methods of land culture. The use of genetically resistant stock (discussed below) is important for disease control in wheat, corn, cotton, and some vegetable crops, such as lettuce and tomatoes.

9.9 Hybrids and Genetic Modification: Creating Better Crops

Genetic modification is a different approach to pest management. The goal is to make changes in crops so they grow better and are better able to resist diseases, pests, and competitors.

A scientific approach began with an experiment in 1926. A high-school student who grew up on a farm began a scientific experiment comparing different strains of corn. At the time, most people thought corn that looked good—a nicely shaped ear of corn with straight rows of kernels—was the most productive. The young man, Henry Wallace (who later became U.S. secretary of agriculture), got curious about this belief and divided a farm field into sections, growing good-looking corn in one part and planting the other sections with varieties of corn that didn't look so good but which local farmers said grew well.

The experiment showed that looks had nothing to do with productivity, and this began the scientific development of hybrid corn. In 1926, Wallace and eight others founded the Pioneer Seed Company, which is still a leading developer of hybrid crops. Modern scientific agriculture was on its way.

Hybridization was so successful that it led to the Green Revolution. This was focused in South America with the establishment of the International Maize and Wheat Improvement Center in Mexico, and in Asia with the establishment of the International Rice Research Institute in the Philippines. These led to new strains with higher yields, better resistance to disease, better ability to grow under poor environmental conditions, and higher nutritional quality. Superstrains of rice that were developed at the International Rice Research Institute have changed food production throughout the region, from China to India.

More is not necessarily better. Although hybridization of rice vastly increased rice production per acre, the

Eric Cabanis/Getty Images, Inc.

■ **FIGURE 9.12**
Farming with genetically modified crop remains controversial.
It can greatly increase yields but many people are afraid that it will have a variety of negative consequences, as shown in this photograph of people in France demonstrating against these crops.

new strains required more fertilizers and as much as four to seven times as much water. And in some cases they produced a rice that was not considered good to eat.

But hybridization improved as it went along. By the second half of the 20th century, new hybrids of each of the major crops were developed each year, based on climate predictions for the year and then-current crop diseases. Today one of the main reasons that farmers the world over can feed all the people is hybridization.

Biotech Comes to the Farm

What is biotech? Biotech is genetic engineering, and it stems from the discovery that DNA is the universal carrier of genetic information. Now that the chemistry of inheritance is understood, scientists have been able to develop ways to transfer specific genetic characteristics from one individual to another, from one population to another, and from one species to another. This has enabled us to genetically modify crops, which holds major implications for agriculture and the promise of increased agricultural production, but also has given rise to new environmental controversies.

Acreage planted with genetically modified crops (GMCs) has grown rapidly since the first plantings in 1996. In 2005 the billionth acre was planted (adding up all the acres planted in GMCs since 1996) (Figure 9.13).[21] One-quarter of all the world's cropland is now in GMCs, and two-thirds of all GMCs are grown in the United States.[22] Argentina, Canada, Brazil, China, and

South Africa are the other large users of GMCs; together with the United States, these six nations account for 99% of all the GMCs produced. Europe and India are

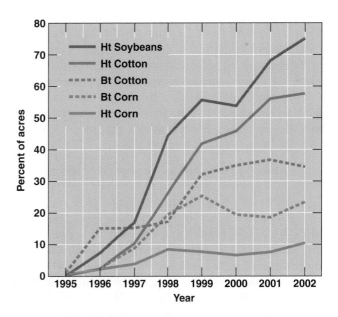

■ **FIGURE 9.13**
Adoption of biotech crops has increased greatly, in the United States, with more than 70% of soybean acreage in genetically modified forms.
(*Source:* USDA. Available at http://www.ers.usda.gov/Briefing/AgChemicals/pestmangement.htm#pesticide. Accessed May 21, 2005.)

absent from this list because they formally oppose the use of GMCs due to ecological, economic, and cultural concerns (Figure 9.12).[23]

It is hard to avoid food and nonfood agricultural products from GMCs. Among the major GMCs are corn, cotton, soybeans, canola, squash, and papaya.[24] Three-quarters of the soybeans, almost a third of the corn, and more than half of the plants that produce canola oil grown in the United States are genetically modified. In addition, at present it is impossible to separate GMC products from non-GMC products when they reach the retail market.

Genetic engineering in agriculture involves several different practices: (1) faster and more efficient ways to develop new hybrids; (2) the introduction of the "terminator gene" (discussed below); and (3) the transfer of genetic properties from widely divergent kinds of life. The first of these, hybridization, is the least novel. Crop hybridization became common practice in the 20th century. Biotechnology adds a new way to create hybrids, but it is not an entirely new process. In contrast, the second and third practices were never previously possible; they arise from our understanding of biological evolution and its mechanisms.

Bioteching New Hybrids

The development of hybrids within a species is a natural phenomenon, part of biological evolution (see Chapter 5), and the development of hybrids of major crops, especially of small grains, was a major factor in the great increase in productivity of 20th-century agriculture. So, strictly from an environmental perspective, genetic engineering to develop hybrids within a species is likely to be as benign as the development of agricultural hybrids has been with conventional methods.

However, there are concerns about superhybrids and superweeds. Some people worry that the great efficiency of genetic-modification methods may produce "superhybrids" that are so productive they can grow where they are not wanted and become pests. "Superhybrids" might also require much more fertilizer, pesticide, and water, which could increase pollution and require more irrigation. In addition, some of the new hybrid characteristics could be transferred by interbreeding with closely related weeds. This could inadvertently create a "superweed" whose growth, persistence, and resistance to pesticides would make it difficult to control.

Some also worry about potential effects on marginal lands. New hybrids might be developed that could grow on poor land. Growing such crops on those lands could increase erosion and sedimentation and cause biological diversity to decline in certain biomes.

On the other hand, it could work the other way around. Genetic engineering could lead to hybrids that require less fertilizer, pesticide, and water. For example, at present, only legumes (peas and their relatives) have symbiotic relationships with bacteria and fungi that allow them to fix nitrogen. Scientists are trying to transfer this capability to other crops. If they succeed, more kinds of crops will enrich the soil with nitrogen and require much less nitrogen fertilizer.

Why are we trying to genetically modify crops? Following are some of the major things that scientists hope to achieve through genetic engineering of crops.

- Improve crop resistance to diseases and pests.
- Increase crop growth rates.
- Make crops more nutritious.
- Create crops with greater environmental tolerance—tolerance for heat, cold, drought, toxic chemical elements, and so forth, so that crops can grow in a greater range of habitats.
- Create crops that produce their own pesticides.
- Create new crops that can "fix" nitrogen (convert atmospheric gaseous nitrogen to a form that can be used by green plants), just as legumes (peas and their relatives) do now. The bacteria live on substances produced by the legumes; in turn, the bacteria fix nitrogen.

Why is nitrogen-fixing important? Because legumes have a symbiotic relationship with nitrogen-fixing bacteria, farmers often rotate legumes with other crops to enrich the soil with nitrogen. If we can develop new strains of corn and other crops that can form a symbiotic nitrogen-fixing relationship with new strains of bacteria, this would increase the production of these crops and reduce the need for fertilizers.

We currently have three categories of genetically modified crops, according to which of three methods we use: (1) introduction of the "terminator gene," (2) faster and more efficient development of new hybrids, and (3) transfer of genetic properties from widely differing kinds of life. Each of these poses different potential environmental problems.

Keep in mind a general rule of environmental actions: *If the changes we make are similar to natural changes, then the effects on the environment are likely to be benign.* This is because species have had a long time to evolve and adapt to these changes. In contrast, changes that are novel—that do not occur in nature—are more likely to have undesirable environmental effects, both direct and indirect. We can apply this rule to the three categories of genetically engineered crops.

The Terminator Gene

Plants with this gene cannot reproduce. One of the main reasons European nations and India resist the use of GMCs is the introduction of the **terminator gene** into crop seeds to make the seeds sterile. This is done for both environmental and economic reasons. In theory, it prevents a GMC from spreading. It also protects the market for the corporation that developed the seeds: Farmers cannot avoid buying new seeds each year by using seeds from last year's hybrid crops.

This poses social and political problems. Traditionally, farmers save some of their harvested seeds to plant the next year. But corporations do not make a profit on reused seeds and consider their GMCs proprietary, patented products. Farmers in less-developed nations and governments of nations that lack genetic-engineering capabilities are concerned that the terminator gene will allow the United States and a few of its major corporations to control the world food supply. Many people believe that farmers in poor nations must be able to grow next year's crops from their own seeds because they cannot afford to buy new seeds every year. This is not directly an environmental problem, but it can become an environmental problem indirectly by affecting total world food production, which then affects the human population and how land is used in areas that have been in agriculture.

Transfer of Genes from One Form of Life to Another

This does not occur in nature, so could be troublesome. That's why most of the concern about genetic modification of crops centers on this method. One of the most controversial uses of this method is the modification of potatoes and corn so that they will produce a chemical that is toxic to caterpillars that eat them. As discussed earlier, the bacterium *Bacillus thuringiensis* (BT) is a successful pesticide, causing a disease in many caterpillars. With the development of biotechnology, agricultural scientists found out what the toxic chemical was, and which gene caused its production within the bacteria. This gene was then transferred from one kind of life, bacteria, to another kind of life, potatoes and corn, so that the biologically engineered plants would produce their own pesticide.

Good news . . . and bad news. The good news is that with the toxin produced and embedded in the crop, it is no longer necessary to spray a pesticide. The bad news is that the genetically engineered potatoes and corn produced the toxic BT substance in every cell—not just in the leaves that the caterpillars eat, but also in the potatoes and corn sold as food, and in the flowers and pollen. Pollen from fields of these GMCs spread widely, and people began to worry that the toxic pollen could be ingested by monarch butterflies as they migrated through areas where the crops were grown. Monarch butterflies are among the most popular insects in the United States, with organizations and Web sites devoted to their famous long migrations. Our enjoyment of this species leads to special concerns about its status. In experiments, the butterflies did ingest the toxic pollen. They did not ingest it purposely—this pollen is not their food; they feed on milkweed, but they ingested pollen that fell on the milkweed, and they died.

Unexpected effects are typical of ecosystems and ecological interactions and are a very real cause for concern when considering the transfer of genes from one kind of life to another. This is a novel effect and therefore one that should be carefully monitored.

Gene transfer can make a crop more nutritious. Scientists have found the genes in daffodils that enable that species to make beta-carotene, an important vitamin food supplement. They transferred that genetic information to a strain of rice so that the rice produces beta-carotene. This rice seems especially valuable for the poor of the world who depend on rice as a primary food. This genetic transfer would not occur by itself. It actually required the introduction of four specific genes and would likely be impossible without genetic-engineering techniques, because genes were transferred between plants that would not exchange genes in nature.

The jury is still out. Although the genetically modified rice appears to have beneficial effects, the government of India will not allow it to be grown in that country. There is worldwide concern about the political, social, and environmental effects of genetic modification of crops, and, again, the rule of natural change suggests that we should monitor such actions carefully. This is a story in process, one that will change rapidly in the next few years. You can check on these fast-moving events on the textbook's Web site.

9.10 Grazing on Rangelands: An Environmental Benefit or Problem?

Farm animals affect the environment, in part because there are so many of them. Worldwide, people keep 14 billion chickens, 1.3 billion cattle, more than 1 billion sheep, more than a billion ducks, almost a billion pigs, 700 million goats, more than 160 million water buffalo, and about 18 million camels.[25]

Almost half of Earth's land area is used as rangeland, and much of that is arid, easily damaged by grazing, especially during drought. Indeed, much of the world's rangeland is in poor condition from **overgrazing**. Land near streams fares the worst. In the United States, where

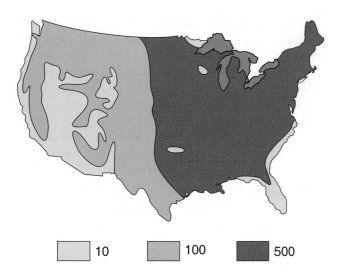

10   100   500

▪ **FIGURE 9.14**
Carrying capacity of pastureland and rangeland in the United States shown as the average number of cows per square kilometer. (*Source:* U.S. Department of Agriculture http://www.csrees.usda.gov/nea/nre/in_focus/range-lands_if_health.html.)

more than 99% of rangeland is west of the Mississippi River, rangeland conditions have improved since the 1930s (Figure 9.14), especially in upland areas. However, land near streams and the streams themselves continue to be heavily affected by grazing. Grazing cattle trample stream banks and release their waste into stream water. Maintaining a high-quality stream environment requires that cattle be fenced behind a buffer zone.

Retracing the steps of Lewis and Clark reveals the damage done since then. A celebration of the 200th anniversary of the Lewis and Clark expedition began in 2004 and has led to renewed interest in their journey, with many people retracing the steps of these explorers. A large number of these steps were along the Missouri River, and the upper Missouri River in Montana runs through grazing land. So many cattle come down to the river to drink that they have damaged the land along the river, and the river itself runs heavy with manure. These effects extend to an area near a federally designated wild and scenic portion of the upper Missouri River, and Lewis and Clark buffs traveling on the Missouri have complained about it. In recent years, fencing along the upper Missouri River has increased, with small openings to allow cattle to drink, but otherwise restricting what they can do to the shoreline[26] (Figure 9.15).

Traditional and Industrialized Use of Grazing Lands and Rangelands

Industry use of feedlots is particularly damaging to the environment. Traditional herding prac-

▪ **FIGURE 9.15**
Cattle grazing along the Missouri River, the route of the 1804–1806 Lewis and Clark expedition west. The cattle have polluted the river and destroyed the land nearby.

tices and industrialized production of domestic animals have different effects on the environment. In modern industrialized agriculture, cattle are first raised on open range and then transported to feedlots, where they are fattened for market. Feedlots have become widely known in recent years as sources of local pollution. The penned cattle are often crowded and are fed grain or forage that is transported to the feedlot. Manure builds up in large mounds. When it rains, the manure pollutes local streams. Feedlots are popular with meat producers because they are an economical way to rapidly produce good-quality meat. However, large feedlots require intense use of resources and have negative environmental effects.

Traditional herding chiefly affects the environment through overgrazing. Goats are especially damaging to vegetation, but all domestic herbivores can destroy rangeland. How greatly domestic herbivores affect the land depends on the amount of rainfall and how fertile the soil is.

At low to moderate densities, the animals may actually be helpful. They aid the growth of aboveground vegetation by fertilizing the soil with their manure and by clipping off the tips of plants, just as pruning and cutting back stimulates plant growth. But at high densities, the vegetation is eaten faster than it can grow. This results in the loss of some species and greatly reduced growth of others.

The Geography of Agricultural Animals

Fast-growing populations of domestic animals bring special problems. People have distributed cattle, sheep, goats, horses, and other domestic animals around the world and then helped their numbers grow at a rate that has changed the landscape. People began doing this long before modern industrial agriculture. For example, Polynesian settlers brought pigs and other domesticated animals to Hawaii and other Pacific islands. Since the age of exploration by Western civilization, starting in the 15th century, Europeans have introduced their domestic animals into Australia, New Zealand, and the Americas. Horses, cows, sheep, and goats were brought to North America after the 16th century. The spread of domestic herbivores around the world is one of the major ways we have changed the environment through agriculture. The spread of cattle brought new animal diseases and new weeds, which arrived on the animals' hooves and in their manure, and introducing domestic animals into new habitats has many other environmental effects.

Two important environmental effects are that (1) native vegetation, not adapted to the introduced grazers, may be greatly reduced and threatened with ex-

tinction; and (2) introduced animals may compete with native herbivores, reducing their numbers to a point at which they, too, may be threatened with extinction.

Another problem is the clearing of tropical forests for agriculture. A recent important issue in cattle production is the opening up of tropical forest areas and their conversion to rangeland—for example, in the Brazilian Amazon basin. In a typical situation, the forest is cleared by burning, and then crops are planted for about four years. After that time, the soil has lost so much fertility that crops can no longer be grown economically. Ranchers then purchase the land, already cleared, and run cattle that are bred to survive in the hot, humid conditions. After another four years or so, the land can no longer support even grazing and is abandoned. Land depleted to this degree can no longer support many uses, including forest growth.[27, 28] Clearly, this is an unsustainable approach to agriculture and therefore undesirable.

More people and higher living standards mean greater demand for meat. As the human population grows and income and expectations rise, the demand for meat increases. As a result, we can expect greater demand for rangeland and pastureland in the next decades. A major challenge for agriculture will be to develop ways to make the production of domestic animals sustainable.

How Many Grazing Animals Can the Land Support?

The carrying capacity of grazing lands, like the carrying capacity of our planet, discussed in Chapter 2 on human population growth, is the maximum number of a species per unit area that can persist without decreasing the ability of that population or its ecosystem to maintain that density in the future. The carrying capacity of land for cattle varies with rainfall, topography (flat, hilly, other natural or human-made features), soil type, and soil fertility. When the number of individuals exceeds the carrying capacity, the land becomes overgrazed.

What happens when land is overgrazed by cattle? Overgrazing slows the growth of the vegetation, reduces the diversity of plant species, leads to dominance by plant species that cattle aren't fond of, hastens the loss of soil by erosion as the plant cover is reduced, and subjects the land to further damage from the cattle's trampling on it (Figure 9.16). The damaged land can no longer support the same density of cattle. Areas with moderate to high rainfall evenly distributed throughout the year can support cattle at high densities. In arid and semiarid regions, the carrying capacity drops greatly.

R. de la Harpe/Biological Photo Service/PO

▪ FIGURE 9.16
Soil erosion in northern Natal caused by overgrazing and other land-use practices.

9.11 Organic Farming

Organic farming combines a variety of methods with a minimum of artificial chemicals, without genetically modified crops, and with the least damage to soils and ecosystems. An organic farm produces food without artificial pesticides or bioengineering, and farm animals are grown without antibiotics and hormones. The result has implications for human health as well as for the environment. The U.S. Department of Agriculture certifies a farm as organic if its farming practices meet such requirements. The area in organic farms, although still small, is growing (Figure 9.17).

Organic farming involves trade-offs. A 21-year study comparing conventional and organic farming in Central Europe showed that fertilizer use was reduced 34% and energy consumption reduced by more than half on organic farms. However, crop yields were 20% lower on the organic farms because pests were less well controlled and because fertilizers were used less intensely and less efficiently. Although total income from

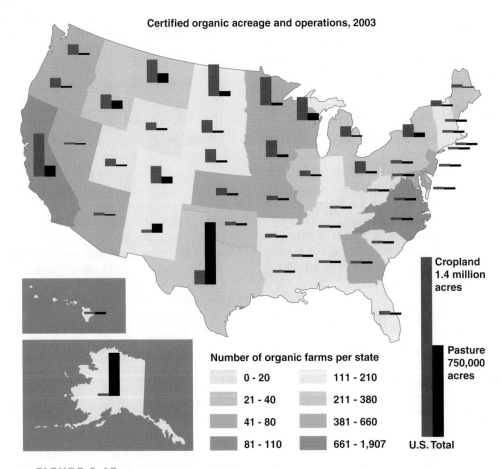

▪ FIGURE 9.17
Number of organic farms in the United States.
(*Source:* USDA. Available at http://www.ers.usda.gov/Briefing/AgChemicals/
sustainability.htm#agriculturalsustainability. Accessed May 21, 2005.)

crops may be lower for an organic farm, the net income to the farmer may be equal to or greater than that of a conventional farm because of reduced costs, similar to what we saw in this chapter's opening case study. But because the land is less productive, feeding the world with organic farming requires greater acreage.

So which is actually better for the global environment? Is it better to use less pesticide and artificial fertilizer, less industrial plowing and other land-moving equipment, but convert more land to agriculture? Or is it better to use those industrial methods but farm less land? The answer is not clear today.

9.12 Deserts: What Are They and What Causes Them?

Deserts occur naturally where there is too little water for most plants. Because the plants that do grow are too sparse and unproductive to create a soil rich in organic matter, desert soils are mainly inorganic, coarse, and typically sandy. It doesn't rain often, but when it does, it is often heavy and causes severe erosion. The warmer the climate, the more rainfall is needed to keep the land from becoming a desert, or to change desert to nondesert, such as grassland. Even in cooler climates and at higher altitudes, deserts may form if there is too little rain or snow to support more than sparse plant life. The crucial factor is the amount of water in the soil available for plants to use. Factors that destroy the ability of a soil to store water can create a desert.

Earth has five natural warm desert regions, all of which lie mainly between latitudes 15° and 30° north and south of the equator. They include the deserts of the southwestern United States and Mexico; Pacific Coast deserts of Chile and southern Ecuador; the Kalahari Desert of southern Africa; the Australian deserts that cover most of that continent; and the greatest desert region of all—the desert that extends from the Atlantic coast of North Africa (the Sahara) eastward to deserts of Arabia, Iran, Russia, Pakistan, India, and China. Only Europe lacks a major warm desert; it lies north of the desert latitudinal band.

These natural deserts are caused by global climate dynamics. On average, it works like this: The air at the equator is warmed by the sun. The warm air rises (warm air always rises) and moves northward, cooling as it moves. As it cools, it becomes heavy and descends. Its water condenses and falls as rain before reaching latitudes 15° to 30°.

Desertification is an environmental problem caused partly by people. Based on climate, about one-third of Earth's land area should be desert. However, it is estimated that much more—43%—of the land is desert. This additional desert area is believed to be a result of human activities.[29, 30] **Desertification** is the deterioration of land in arid and semiarid areas due to climate changes and human activities.[31] Desertification is a serious global problem. It affects one-sixth of the world's population (about 1 billion people), 25% of the world's total land area, and 70% of all dry lands (3.6 billion hectares). Land degradation caused by people has altered 73% (3.3 billion hectares) of drier rangelands and the soil fertility and structure of 47% of dryland areas that have marginal (meaning barely sufficient) rainfall for crops. Land degradation also affects 30% of dryland areas that are densely populated and have agricultural potential.

A large part of desertification occurs in the poorest countries. These regions include Asia, Africa, and South America. Worldwide, 6 million hectares (14 million acres) of land per year are lost to desertification, with an estimated economic loss of $40 billion per year. The recovery of these lands could cost as much as $10 billion per year.[32]

Bad farming practices are the leading human causes of desertification. Some kinds of landscapes are vulnerable to becoming deserts—they are marginal lands that could easily be turned into deserts by poor farming practices. Such practices include overgrazing, too much farming, the failure to use contour plowing, the conversion of rangelands to croplands in areas where rainfall is not sufficient to support crops over the long term, and poor forestry practices, including cutting all the trees in an area that is barely able to support tree growth (Figure 9.18).

How it happens. In northern China, areas that were once grasslands were overgrazed, and then some of

Nicholas DeVore/Getty Images, Inc.

■ **FIGURE 9.18**

Camels and other animals trample the soil in the semiarid Sahel of Africa as they move to water holes, such as this one in Chad, and as a result promote desertification.

these rangelands were converted to croplands. Both practices caused the land to become desert. Some 65,000 square kilometers (25,100 square miles, an area larger than Denmark) became desert between 1949 and 1980, and an additional 160,000 square kilometers (61,760 square miles) are in danger of becoming desert. As a result of desertification, the frequency of sandstorms increased from about three days per year in the early 1950s to an average of 17 days per year in the next decade and to more than 25 days per year by the early 1980s.[33]

Poisoning the soil can create desertlike areas anywhere. The soil may be poisoned by persistent pesticides, by improper disposal of toxic chemicals from industrial processes, by airborne pollutant acidification, excessive manuring in feedlots, and oil or chemical spills. All of these can force us to abandon certain land or reduce its agricultural use. Worldwide, toxic chemicals account for about 12% of all soil degradation, affecting agriculture worldwide.

Ironically, irrigation in arid areas can also lead to desertification. When irrigation water evaporates, a residue of salts is left behind. Although these salts may have been in very low concentrations in the irrigation water, over time they can build up in the soil to the point at which they become toxic. This effect can sometimes be reversed by greatly increasing irrigation, so that the larger volume of water redissolves the salts and carries them with it as it percolates down into the water table.[34]

Preventing Desertification

The first step is spotting early symptoms. The major symptoms of desertification are the following:

- Lowering of the water table (wells have to be dug deeper and deeper).
- Increase in the salt content of the soil.
- Reduced surface water (streams and ponds dry up).
- Increased soil erosion (the dry soil begins to be blown and washed away by wind and rains).
- Loss of native vegetation (not adapted to desert conditions, native vegetation can no longer survive).

Preventing desertification begins with monitoring these factors. Monitoring aquifers and soils is important in marginal agricultural lands. When we spot undesirable changes, we can try to control the activities that are causing these changes. Proper methods of soil conservation, forest management, and irrigation can help prevent the spread of deserts.

Good soil conservation also includes the use of windbreaks, narrow lines of trees that help slow the wind, preventing wind erosion of the soil. A landscape with trees is a landscape with a good chance of avoiding desertification. Practices that lead to deforestation in marginal areas should be avoided. Reforestation, including the planting of windbreaks, should be encouraged.

Fortunately, if we make a desert, we can unmake it—that is, we can restore the land using restoration methods discussed in Chapter 6. For example, in China's Turpan Depression area, one of the world's marginal lands for agriculture, land that became desert from poor practices has been restored to vineyards by helping to replenish the *aquifer*—the underground area where water is naturally stored.

9.13 Does Farming Change the Biosphere?

Viewing Earth from space gave rise to the idea of global ecology. People have long recognized local and regional impacts of agriculture, but it is a recent idea that farming might affect Earth's entire life-support system. This possibility came to people's attention in the 20th century, first with events like the American Dust Bowl, discussed earlier in this chapter, which led some to speculate that such disasters could become worldwide. The idea gained supporters in the late 20th century when satellites and astronauts gave us views of Earth from space and the idea of a global ecology began to develop.

How might farming change the biosphere?

First, agriculture changes land cover, resulting in changes in the reflection of light by the land surface, the evaporation of water, the roughness of the surface, and the rate at which chemical compounds (such as carbon dioxide) are produced and removed by living things. Each of these changes can have regional and global effects on climate.

Second, modern agriculture increases carbon dioxide (CO_2) in two ways. As a major user of fossil fuels to run farm machinery, modern agriculture increases the amount of CO_2 in the atmosphere, adding to the buildup of greenhouse gases (discussed in detail in Chapter 14). In addition, clearing land for agriculture speeds the decomposition of organic matter in the soil, transferring the carbon stored in the organic matter into CO_2, which also increases the CO_2 concentration in the atmosphere.

Agricultural use of fire and fertilizers can also affect climate. Using fire to clear land for agriculture (a method used especially in tropical countries) can affect the climate because fires add small particulates to the atmosphere. Another global effect of agriculture results from the artificial production of nitrogen compounds for use in fertilizer, which may be leading to significant changes in global biogeochemical cycles (see Chapter 3).

Finally, agriculture affects species diversity. The loss of competing ecosystems (because of agricultural land use) reduces biological diversity and increases the number of endangered species.

Return to the Big Question

Can we feed the world without destroying the environment?

The jury is still out. On the positive side, we have learned many agricultural practices that tend to make farming more sustainable, and in the second half of the 20th century erosion from farming decreased considerably. This is especially crucial as the human population grows—and you will recall that it is expected to double! On the negative side, we are not certain whether artificial fertilizers and pesticides, as they are generally used today, will turn out to provide a net benefit on a global scale.

If we cannot make the land more productive, we will need more land for farming. Indeed, if the population doubles, we will need to put an additional area equal to the entire New World into farmland. And if that production per acre declines, we will need to put even more land into agriculture.

Where will we find it? Which biomes should we alter greatly? Farmers do an amazing job of feeding most of the world's 6 billion people, and farming is perhaps the most sustainable of all human activities, at least in terms of the production of economically valuable products, if not in terms of the sustainability of the original ecosystems.

The jury is especially undecided about bioengineering. It does greatly increase production, but it could also produce some dangerous side effects.

And will local benefits from organic farming lead to global benefits? Are we better off with more ecologically sound farming on more land area, or with highly artificial methods that use less acreage?

The answer to this chapter's Big Question is one of the most important for the people of the world, as well as for biodiversity. So seeking the answer should be one of our top priorities.

Summary

- The Industrial Revolution and the rise of agricultural sciences have led to a revolution in agriculture, with many benefits and some serious drawbacks. These drawbacks have included an increase in soil loss, erosion, and resulting downstream sedimentation, as well as the pollution of soil and water with pesticides, fertilizers, and heavy metals that are concentrated as a result of irrigation.

- Modern fertilizers have greatly increased the yield per unit area. Modern chemistry has also led to the development of a wide variety of pesticides that have reduced, but not eliminated, the loss of crops to weeds, diseases, and herbivores.

- Most 20th-century agriculture has relied on machinery and the use of abundant energy, paying relatively little attention to the loss of soils, the limits of groundwater, and the negative effects of chemical pesticides.

- Overgrazing has caused severe damage to lands. It is important to properly manage livestock, such as using appropriate lands for grazing and keeping livestock at a sustainable density.

- Desertification is a serious problem that can be caused by poor farming practices and by the conversion of marginal grazing lands to croplands. We can help to prevent further desertification by improving farming practices, planting trees as windbreaks, and monitoring land for symptoms of desertification.

- Two revolutions are occurring in agriculture, one ecological and the other genetic. In the ecological approach, pest control will be dominated by integrated pest management, and agriculture will be approached in terms of ecosystems and biomes, taking into account the complexity of these systems. This approach will emphasize soil conservation through no-till agriculture and contour plowing, along with water conservation.

- Although it is a new technology, genetic modification of crops is now widespread and is already the subject of controversy, because while it offers benefits, it may pose some environmental dangers. The dangers lie in using genetic modification without considering the ecosystem, landscape, biome, and global implications.

Key Terms

biological control

contour plowing

desertification

dust bowl

integrated pest management

no-till agriculture

overgrazing

terminator gene

Getting It Straight

1. What are the environmental effects of producing and processing food?

2. What is the total world population that needs to be feed by world farmers?

3. What plant species accounts for most of the world's food production?

4. Does our current world food production have capabilities of feeding the growing world population? Explain.

5. What are some environmental concerns of farming?

6. What is the primary method used by farmers to combat soil fertility issues?

7. List five ways to make soil sustainable for future farm production.

8. Where does most used pesticide end up in our environment?

9. What is DDT? Identify its benefits and concerns with use.

10. What is integrated pest management and how can it help improve pest control in farming?

11. What trade-offs are involved in organic farming?

12. What are the major concerns with desertification?

What Do You Think?

1. If most of the farmers in the 1930s in the Midwest had practiced the kind of agriculture that Steve Burr practices today (as discussed in the introductory case study), would the Dust Bowl have been prevented? Present arguments for and against this outcome.

2. Should genetically modified crops be considered acceptable for "organic" farming?

3. Some parks in Great Britain have traditionally used grazing sheep to keep the lawns mowed. Based on what you have learned in this chapter, compare the environmental effects of this kind of lawn maintenance with American-style power mowing. Plan a scientific experiment to compare these two kinds of lawn control and discuss why

the results would or would not lead you to recommend sheep as lawn mowers in U.S. parks.

4. This chapter points out that agriculture has been practiced in the Nile Valley for thousands of years. The major 20th-century environmental change to this river was the building of the Aswan High Dam. Read up on this dam and discuss how it might affect agriculture in the Nile Valley. Is it likely to (a) increase production of crops? (b) Increase the sustainability of agriculture there?

5. Could changes in agricultural practices be an important way to combat global warming? Use standard literature sources and consider the global effects of agriculture as discussed in this chapter.

Pulling It All Together

1. Design an integrated pest-management scheme for a small vegetable garden in a city lot behind a house. How would this scheme differ from integrated pest management (IPM) used on a large

farm? What aspects of IPM could not be used? How might we use the artificial structures of a city to benefit IPM? (Include in your discussion scientific experiments to test your proposal.)

2. In the summer of 2006, the Environmental Protection Agency found that a genetically engineered grass was growing in the wild. The grass was engineered to be resistant to 35 weed-killing herbicides, so the herbicides could be used on lawns and golf courses without harming the grass. Design a scientific experiment (refer to Chapter 2 on the scientific method) to test what might be the outcome of the spread of this grass in (a) wild prairies; (b) suburban lawns.

3. Consider modern agricultural practices in California, where large areas that were once desert are now among the world's most agriculturally productive. Using standard reference sources, consider whether any other methods used in California might help slow desertification in Africa. Pick a specific African nation for your explanation.

4. Imagine a future NASA mission to create a human colony on Mars. What kind of agriculture would you recommend, including the crops to be grown and the kind of production methods? In particular, consider that the area available would be extremely limited because it would have to be indoors so that an Earth-like atmosphere could be maintained. Consider also the trade-offs between high production and sustainability of that agriculture.

Further Reading

Terrence, J., George Toy, George R. Foster, and Kenneth G. Renard. 2002. *Soil erosion: Processes, prediction, measurement, and control.* New York: Wiley.

Grainger, A. 1982. *Desertification: How people make deserts, how people can stop and why they don't,* 2nd ed. London: Earthscan Books.—A book that provides examples and discussion of the connections between desertification and human activity, particularly in nonindustrialized countries.

Matthews, A. 1992. *Where the buffalo roam.* Weidenfeld, NY: Grove.

Rissler, J., and M. Mellon. 1996. *The ecological risks of engineered crops.* Cambridge, MA: MIT Press.

Young, M. D. 1991. *Towards a sustainable agricultural development.* New York: Wiley.—A broad survey of the use of agricultural chemicals, intensive animal production, soil erosion, land-use patterns, and the impact on agriculture of pollution from other sources.

Alberto Estevez/epa/Corbis Images

10

Energy and Environment

Big Question

Can We Assure a Sustainable Supply of Energy?

Learning Objectives

The shift from fossil fuels—coal, oil, and natural gas—to other energy sources, such as wind and solar power, will be one of the milestones of the 21st century. For all practical purposes, with increasing use of automobiles worldwide, we could run out of usable petroleum in this century. Even if we do not, we will run into supply problems when we pass the point when half of Earth's available oil is used up. That time, you may be surprised to learn, will likely come within the next ten years or so. To avoid serious economic and social problems due to oil shortages, we need to find alternatives before we reach that point. The need is great, and the time to act is now. After reading this chapter, you should understand . . .

■ when we will have to begin shifting from fossil fuels to alternative energy;

■ what are the most likely replacements for fossil fuels;

■ the major problems with nuclear energy, and

why some people think this form of energy has advantages;

■ how we can best store and transfer energy;

■ the environmental impacts of generating and using energy.

Case Study

Winds of Change in Iowa

The shift from fossil fuels to other energy sources is already happening in some places, and in many instances the environmental statement *"Think globally, act locally"* is the byword.

The Spirit Lake Community School District in northwestern Iowa made a decision in 1991 to act locally to reduce its dependence on fossil fuels. The goals were to (1) help reduce air pollution and other environmental impacts of fossil-fuel power plants; (2) help reduce our nation's dependence on foreign oil and achieve energy independence; (3) provide a source of income at the local level; and (4) provide a hands-on learning experience for students.

To achieve these goals, the school district turned to wind power, one of the fastest-growing energy sources in the world today. The district has two wind turbines. One of them, near the elementary school (Figure 10.1), has been on-line since 2001 and is producing a profit from selling excess power. When the second turbine comes on-line (scheduled for 2007), the electricity needs of the district's schools and offices—including the middle school, high school, and football stadium—will be supplied by the wind that blows over the schools' land. The wind turbine at the elementary school is already giving students a valuable learning experience about the role of energy in society. Students learn that their windmill, very visible from the playground, saves oil and coal, helps reduce air pollution, and brings income to the school for maintenance and improvements.

The Spirit Lake School District's energy transformation is one small step for Iowa and a larger symbolic step in the global transformation that is coming in this century.

Courtesy Spirit Lake Community School

■ **FIGURE 10.1**

Wind power at Spirit Lake elementary and middle schools in Iowa.
The wind turbine behind the playground is a symbol of the transition from oil and gas to alternative energy.

10.1 World Energy Supply and Use

We, the people of the world, use a lot of energy from five major kinds of sources: oil, natural gas, and coal (the three forms of fossil fuels), nuclear, and "renewables," which we will also refer to as "alternative energy sources"—solar, wind, etc. (Figure 10.2).

North America, mostly the U.S., uses about one-fifth of the world's energy (Figure 10.3). However, with the rapid economic and industrial development of China and India, whose citizens are just as fascinated by automobiles as the rest of the world is, the percentage of energy used by industrialized Asia will grow at the expense of North America.

The amount of energy the world uses is usually expressed in BTUs (British thermal units), originally a measure of the heat energy released from a fuel. In 2003 the people of the world used 425 quadrillion BTUs.[1] For simplicity, energy experts refer to a quadrillion BTUs as

a "quad," but let's put this in more familiar terms. It is the amount of energy that would be used by more than 190 million automobiles with 100-horsepower engines running nonstop for the entire year at maximum power (that is, trying to accelerate and run at top speed without

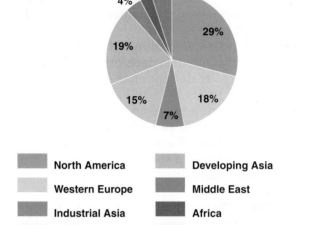

Energy use 1995 by region

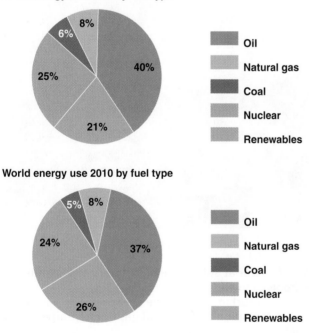

World energy use 1995 by fuel type

World energy use 2010 by fuel type

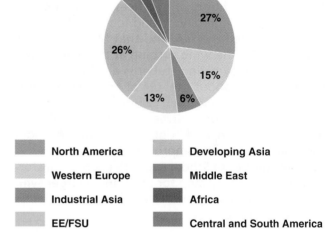

Energy use 2010 by region

■ **FIGURE 10.2**
World energy use by fuel type.
(a) 1995; (b) projected, based on current trends, for 2010. The forecasts are conservative in that they are extrapolations and do not take into account potential changes discussed in this chapter regarding what are called "renewables" and what we will call "alternative" energy sources: wind, solar, etc. (*Source: International Energy Outlook* 1998, National Energy Information Center, EI-30 Energy Information Administration Forrestal Building, Room 1F-048 U. S. Department of Energy.)

■ **FIGURE 10.3**
World energy use by major region.
(a) 1995; (b) projected, based on current trends, for 2010. Current world energy use is 425 quads a year (see text for an explanation of this quantity). (*Source: International Energy Outlook* 1998, National Energy Information Center, EI-30 Energy Information Administration Forrestal Building, Room 1F-048 U.S. Department of Energy.) (Note that EE/FSU stands for Eastern Europe/Former Soviet Union.)

even stopping to refuel).[2] If these cars ran only eight hours a day, seven days a week, and only at a more realistic average of half power, then the world's total annual energy use would be equal to the energy used by more than 1.1 billion of those cars. It is somewhat comforting to note, however, that this is still a small fraction of the total energy that our planet receives annually from the sun.

Energy use by the world's people is increasing rapidly. Standard estimates, based on past rates of increase, suggest that over the next 15 years energy use may rise 2.3% per year. This does not take into account the recent, very rapid increase in energy use by China and India, with a third of the world's human population. Even at 2.3%, however, energy use would double by 2036, and we can expect it to double before that. What does this imply for the energy supply and the energy available to each of us in the near future? The simple answer is that our societies, governments, and peoples are not ready for this rapid increase, especially with the end in sight for fossil fuels. One purpose of this chapter is to suggest what we can do to get ready.

10.2 Energy and Work

The concept of energy is closely tied to the concept of work. We tend to take energy pretty much for granted. We are used to seeking more energy in batteries for our electronic devices, used to filling up our gasoline tanks so our cars will have the energy to keep running. If you are sitting quietly and studying this book, you would say you are working. If your car stalls and you use all your strength trying to push it, you feel that you are working hard even if the car doesn't budge. But to a physicist, and in the scientific meaning of this book, you are not working if the car doesn't move. To a physicist, *energy is the ability to move matter.*

How do we measure energy? First, we have to define a *force*, because energy is a force applied over a distance. The basic unit of force is the *newton (N)*, which is the force necessary to accelerate a mass of 1 kilogram by 1 meter per second per second (m/s^2). This means that the speed of that kilogram increases by 1 meter per second for every second it travels. The fundamental energy unit is the *joule:* 1 joule is the force of 1 newton applied over a distance of 1 meter. Power is energy per unit time. In the metric system, power is a joule per second, called a *watt.*

When we need to refer to larger power units, we can use multipliers, such as *kilo-* (thousand), *mega-* (million), and *giga-* (billion). For example, the rate of production of electrical energy in a modern nuclear power plant is 1,000 megawatts (MW) or 1 gigawatt (GW). Sometimes it is useful to use a hybrid energy unit, such as the watt-hour, Wh (remember, energy is power multiplied by time). Electrical energy is usually expressed and sold in kilowatt-hours (kWh, or 1,000 Wh). This unit of energy is 1,000 W applied for 1 hour (3,600 seconds), the energy equivalent of 3,600,000 J (3.6 MJ).

To put these terms in perspective, an electric clock typically uses 2 watts, an incandescent bulb usually about 100 watts, a washing machine 700 watts, a hair dryer about 1,000 watts, and a modern high-definition TV between 100 and 250 watts. Today's television sets use about 4% of residential electricity, and this is expected to rise to 10% over the next few years.[3]

In general, we use energy for four kinds of activities: for lighting; for heating/cooling (changing the temperature); for doing physical work (moving objects around); and for transferring and converting information (this includes the music you listen to on your iPod and using your computer).

TVs, computers, iPods, cars, air conditioning—we love using energy! Abundant energy lights up the world and allows each of us not only to do more, but also to do things never possible before, from listening to digital music to running a personal computer to relaxing in an air-conditioned movie theater. Modern civilization, especially in the developed nations, uses more energy, and more energy per person, than people did at any other time in history.

The higher the standard of living, the greater the use of energy. There is a direct relationship between a country's standard of living (as measured by gross national product) and energy consumption per capita. People in industrialized countries are a small percentage of the world's population but consume a disproportionate share of the total energy produced in the world. North America, with only 5% of the world's population, uses about 20% of the total energy consumed in the world (Figure 10.3).

Ways of obtaining and using energy have different environmental effects. A lot of our energy use involves burning fuel to heat a gas or liquid. Fuels have stored (potential) energy, and heating them releases it to do other work. There are various ways of heating a gas or liquid, and different ways have different effects on the environment. In addition, there are four stages in making energy usable to us—discovery, extraction, transport and/or storage, and use—and each of these can affect the environment. Our goal is to seek a combination of methods that minimizes undesirable environmental effects.

10.3 Types of Fuels

Fuels are divided into two types: conventional and alternative. We will talk about conventional fuels and their environmental effects first, then discuss alternatives.

Conventional fuels are fossil fuels: oil, natural gas, and coal. We call these **fossil fuels** because they were formed in the Earth from dead plants, algae, and bacteria that did not decompose completely, and over millions of years were converted into coal, petroleum, and natural gas. We call them **conventional** because we are accustomed to using them. Actually, though, oil was not widely used until the gasoline engine was invented in the late 19th century. The modern use of petroleum began in 1859, when an oil well was drilled in Titusville, Pennsylvania. The kerosene obtained from that well replaced whale oil as the major lamp fuel (thus contributing to the end of the era of Yankee whaling). Today, approximately 90% of the energy consumed in the United States and 80% worldwide comes from burning petroleum, natural gas, and coal. However, fossil fuels are forming in the Earth today at a slow rate relative to human needs; as a result they are considered a fossil **nonrenewable resource**.

Alternative energy sources are wind, solar, water, wood, nuclear, and geothermal. Some **alternative sources** are **renewable**—wind, solar energy (energy from the sun), water, and wood are rapidly replaced. Before the Industrial Revolution, the primary fuels for heating things were wood and other organic matter (such as dung in areas where big domestic animals grazed). The primary source of energy to do work in the scientific sense—that is, moving matter—was animals and people. Horses, camels, and even elephants transported people and heavy loads, horses and oxen pulled plows and turned mills. Wind and water also have a long history as sources of energy for this kind of work, especially pumping water from one place to another, grinding grain, and, in the Middle Ages in Europe, doing more and more sophisticated industrial tasks, such as ironwork and sawing wood.

Wood was the main source of heat in the U.S. until the end of the 19th century, really quite recently, and horsepower continued into the early 20th century until electrical motors and gasoline and diesel engines replaced animals. The first urban light railways, the equivalent of trolleys, were pulled by horses in New York City until the late 19th century, when Thomas Edison and his competitors made the first great electrical inventions, including electric generators and motors.

Fossil fuels were more efficient and at first seemed more environmentally friendly. In cities such as New York, horse-drawn carriages and trolleys were quickly replaced by electrical trolleys and fossil-fuel engines in cars and trucks because these are much more efficient and also seemed cleaner. The occasional horse-drawn carriage is a picturesque sight for tourists in today's cities, but when the streets were crowded with this form of transportation, animal manure was a major urban pollutant that had to be dealt with continually. Moreover, horses are inefficient sources of energy, because even when they are not pulling a load, their metabolism still keeps going, and they still have to eat. In New York City, for every horse pulling a railcar, two others were resting and eating. In contrast, fossil-fuel engines and electric motors use no energy when they are shut off.

We turned to water power and steam with the rise of the industrial age. At first, water power consisted of no more than a big wooden wheel moved by water flowing on one side of a stream. But in the 19th century, water power underwent two major advances. First, the picturesque wooden waterwheel was replaced by a metal turbine wheel that extracted energy from the water much more efficiently. Second, with the invention of electric generators, those same turbines, scaled up to huge sizes, spun the generators and produced electricity. In the United States, the water turbines at Niagara Falls started this energy revolution.

The success of water turbines led to the next major step: big dams and reservoirs. People came to realize that even rivers that did not have natural waterfalls like Niagara could become major sources of electricity if we built dams across them. The building of huge dams and their reservoirs was one of the great modern social and environmental changes of the 20th century. In the United States, it took place in full force in the 1930s and 1940s, led by two federal programs of President Franklin Roosevelt's administration. The Bonneville Power Administration built big dams on the Columbia River System (the Columbia and Snake rivers), and the Tennessee Valley Authority did the same in the Southeast. Private power companies and state governments participated also in constructing such facilities.

They seemed environmentally friendly at first. The huge reservoirs created by the big dams were used not only for power generation but also for water supply and recreation. All went well until the second half of the 20th century, when the negative effects of dams on migrating fish became major environmental issues. We discuss these issues in other chapters of this book.

Steam remains an important source of electricity. Steam engines, invented in the 18th century by James Watt of England, first directly powered many industries and then provided another way to spin the turbines of electrical generators. Today steam-driven turbines are a major source of electrical energy. In terms of the environment, however, the question is, what kind of fuel should we use to create that steam? The primary choices are fossil fuels (generating steam is a major present use of coal, but also a use of oil and natural gas), nuclear energy, and to a much lesser extent geothermal energy.

Conventional Energy Sources: The Environmental Impacts of Extracting Them, Delivering Them, and Using Them

10.4 Petroleum Products: Oil and Natural Gas

Oil

Oil is very abundant, but known supplies are dwindling. Next to water, oil is the most abundant fluid in Earth's upper crust, but most of the proven oil reserves are in a few fields. We are so used to fossil fuels and the devices they power that it is hard to imagine a world without them, but at the rate we are using them and are expected to use them, the reserves will last only a few more decades.[4, 5, 6]

What's the difference between resources and reserves? When geologists discuss sources of fossil fuels (or any useful mineral), they distinguish **resources** from **reserves**. A mineral *resource* is the entire amount on Earth—sometimes called the *total resource*. A *reserve* is what we can get at now economically, the portion of the resource that we can extract now at a profit. This is sometimes called the *proven reserve*.

The main question is, when will we reach peak production? This is more important than how long oil will last, because after we reach peak production, less oil will be available, leading to shortages and price shocks. Forecasts put the peak in world crude oil production (about 90 million barrels per year) between the years 2020 and 2050.[6, 7] Oil production as we know it now is expected to end by about 2090 in the United States, and world production of oil should be nearly exhausted by 2100.[5-7] Some economists argue that we will never entirely run out of crude oil, because we will reach a point where finding it and extracting it will cost much more than it can sell for, and when that happens it will no longer be used as a fuel, but as a mineral to be made into comparatively expensive products.[8–11]

How much of the fossil-fuel resource is part of the reserve? Several decades ago, the known available reserve was about 1.6 trillion barrels. Today, the estimate is just over 3 trillion barrels.[6, 7] The increase is due primarily to discoveries in the Middle East, Venezuela, and Kazakhstan. Because so much of the world's oil is in the Middle East, oil revenues have flowed into that area, causing huge trade imbalances and many political consequences.

Two other sources of oil play a minor role: oil shale and tar sands. Both are sediments that contain low concentrations of oil, but because they are massive, in total they contain a lot of energy. The use of both is insignificant today, but tar sands could become important as oil from wells becomes scarce.

Oil shale is a fine-grained sedimentary rock containing organic matter (kerogen). When heated to 500°C (900°F), oil shale yields up to nearly 60 liters (14 gallons) of oil per ton of shale. The oil from shale is one of the so-called **synfuels** (from the words *synthetic* and *fuel*), liquid or gaseous fuels derived from solid fossil fuels. The best-known sources of oil shale in the United States are in the Green River formation that underlies large regions of Colorado, Utah, and Wyoming. Total identified world oil shale resources are estimated to be equal to about 3 trillion barrels of oil—about the same as estimated resources of standard oil. However, assessment of the feasibility of economic recovery and the environmental impact of recovery is not complete.[8]

Tar sands are sedimentary rocks or sands containing tar oil, asphalt, or bitumen. Petroleum cannot be recovered from tar sands by the usual commercial methods because the oil is too thick to flow easily. Oil in tar sands is recovered by first mining the sands, which is very difficult, and then washing the oil out with hot water. Some 75% of the world's known tar sands are in the Athabasca Tar Sands near Alberta, Canada. The total Canadian resource is about 2 trillion barrels, but we don't know how much of this will eventually be recovered. Today's production is 1.5 million barrels of synthetic crude oil per day, about 15% of North America's oil production.[9] The tar sand is extracted in a large open-pit mine, but the mining process is complicated by a fragile, naturally frozen environment that is difficult to restore. There is also a problem with waste disposal—the land surface after mining can be up to 20 meters (66 feet) higher than the original surface.

Today, for every four barrels of oil we consume, we are finding only one barrel.[5] However, this could improve in the future.[7] Recent studies suggest that about 20% more oil awaits discovery than was predicted a few years ago, and that there is more oil in known fields than we thought. An estimated 3 trillion barrels of crude oil may be recovered from remaining oil resources, while world consumption today is about 30 billion barrels per year (82 million barrels per day). Still, the new oil discovered in known fields will not significantly change the date when world production will peak and production will begin to decline.[5]

Natural Gas

Natural gas is considered a clean fuel—burning it produces fewer pollutants than does burning oil or coal, so it causes fewer environmental problems. As a result, it could serve as a transition fuel while we move away from oil and coal to alternative energy sources, such as solar power, wind power, and water power. However, despite new discoveries and the construction of pipelines, long-term projections for a steady supply of natural gas are uncertain. Considerable natural gas was recently discovered in the United States. There is an "energy rush" in Wyoming to recover shallow methane stored in coal, but it is controversial because poor-quality water may be released into the environment with the gas.[10, 11]

Methane hydrates could be a new source of natural gas if we can get at it. A little-known kind of fuel, methyl hydrates, was discovered in the oceans about 30 years ago. Formed by microbial digestion of organic matter in the seafloor sediments, methane hydrates are widespread in both the Pacific and Atlantic oceans and may contain twice as much energy as all the known natural gas, oil, and coal deposits on Earth.[12] But mining methane hydrates will be a difficult, since they occur at depths of 1,000 meters (3,000 feet) or more, and most drilling rigs cannot operate safely at these depths.

10.5 Coal

Coal is by far the world's most abundant—and most polluting—fossil fuel. Coal accounts for nearly 90% of the fossil-fuel reserves in the United States, and the total recoverable resource is sufficient for about 250 years at the current rate of use.[6] But coal is also the most polluting of the fossil fuels. Indeed, coal and nuclear fuels are the most polluting, period. Coal is the dirtiest fuel to burn because it has many impurities as a result of how it was formed.

Briefly, here's how coal was formed millions of years ago. Dead vegetation was buried in sediments, and these were crushed, heated, and transformed over millions of years into sedimentary rocks (Figure 10.4). The useful, burnable part of coal is the carbon stored in it, but the plants from which coal formed included many other chemical elements, and soil particles mixed with the dead plants add other impurities. The result is, environmentally, a pretty dirty product.

It seems unlikely that coal will be abandoned in the near future, because we have so much of it and we have spent so much time and money developing coal resources. The burning of coal produces about 50% of the electricity used and about 25% of the total energy consumed in the United States today.[13]

Coal is classified according to its energy and sulfur content. Hard, dense coal generally has a high en-

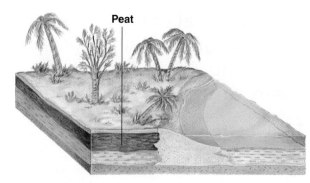

(a) **Coal swamps form.**

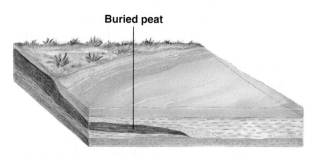

(b) **Rise in sea level buries swamp in sediment.**

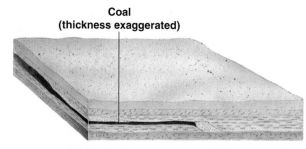

(c) **Compression of peat forms coal.**

■ **FIGURE 10.4**
How coal was made.
The diagram shows processes that transform buried plant debris (peat) into coal.

ergy content compared to softer, less dense types. The sulfur content of coal is important because low-sulfur coal emits less sulfur dioxide. But most low-sulfur coal in the United States is relatively low-energy coal found west of the Mississippi River. Power plants on the East Coast use high-sulfur coal from that region and treat it to lower its sulfur content and avoid excessive air pollution. Although it is expensive, treating coal to reduce pollution may be more economical than transporting low-sulfur coal from the western states.

10.6 The Environmental Effects of Extracting, Delivering, and Burning Coal

Strip Mining

We get coal from underground mines or from strip mining. Of all energy-extraction methods, strip mining has some of the worst effects. It involves opening up the surface of the soil and bedrock and digging the coal out from the surface. It leaves behind displaced soil open to erosion, deranged streams likely to flood, and acid water that drains from the mined land.

You may own the land, but not the minerals beneath the surface. In most of the United States, owning land does not mean that you own the rights to minerals below the surface. But many people do not know this. During the first half of the 20th century, mining companies that owned the mineral rights to coal let the landowners pay all the property taxes. But when the day came that they wanted to extract the coal, they would arrive unannounced and start bulldozing the land and any buildings in the way. Often they drove the farmers off the land, dug out the coal, and left the land cut open. Erosion, flooding, and acid mine drainage helped to destroy some of the greatest forests of the eastern United States, which grew in the Kentucky mountains.[14]

This method of extracting coal has caused environmental disasters. In October 2000 one of the worst such disasters in the history of mining in the Appalachian Mountains occurred in southeastern Kentucky. About 1 million cubic meters (250 million gallons) of toxic black coal sludge, produced when coal is processed, escaped from its reservoir and ended up flowing across people's yards and roads and into a stream of the Big Sandy River drainage. About 100 kilometers (65 miles) of the stream were severely contaminated, killing several hundred thousand fish and other life in the stream.

One of the largest and most controversial strip mines is at Black Mesa, near the Four Corners in Arizona (Figure 10.5). There the Hopi Indian Nation agreed to allow construction of a huge electric power generator that uses coal mined on this site, which includes lands with traditional spiritual value to the Hopi. Black Mesa is a case where the desires of the human spirit conflict with human materialism. The debate still rages while the coal continues to be mined, mixed with sacred water from Hopi lands, and made into a slurry that is piped to the power plant.

Coal dust and *acid mine drainage* are serious problems with strip mining. Acid mine drainage (produced when iron sulfide in coal reacts with oxygen to form sulfuric acid) is especially a problem in the eastern United States, where rainfall is abundant[15] (Figure 10.6). Mining also produces large amounts of coal dust that settles on towns and fields, further polluting the land and causing or worsening lung diseases, including asthma. Complaints by communities used to be ignored but are now getting more attention by state mining boards. As people become better educated about mining laws, they are more effective in reducing their risk of harm from mining, but much more needs to be done. In May 2002 a federal judge ordered the government to no longer allow mining companies to dump mining waste into streams and valleys. The decision upheld laws to protect our streams and rivers, but the ruling was overturned in January 2003.

■ **FIGURE 10.5**

The location of Black Mesa Coal.
(Source: Arizona Geological Survey).

■ **FIGURE 10.6**
Acid mine drainage affects streams large and small. Here the Wheeling watershed creek, part of a coal mining area since 1810, shows the effects of poor land use downstream from the mine (trash along the stream) and the barren rocks and pebbles, devoid of algae and plants because of the acid.

Meanwhile, more than half of U.S. coal comes from strip mining. Strip mining has steadily increased because it is cheaper and easier for the mining companies than underground mining.

Underground Mining

Underground mines provide about 40% of the coal mined in the U.S. (Figure 10.7). For a variety of reasons, many of them have been abandoned, particularly in eastern U.S. coalfields of the Appalachian Mountains. Underground coal mining is a risky profession. Besides the danger of collapse, explosion, and fire, respiratory illnesses such as black lung disease have killed or disabled many miners. Acid mine drainage from the mines and waste piles has polluted thousands of kilo-

meters of streams, and when coal mine tunnels collapse, the land above them drops, often leaving crater-shaped pits on the surface.

The Trouble with Coal

Although plentiful, coal is environmentally the least desirable fossil fuel. Coal-burning power plants in the United States are responsible for about two-thirds of the total emissions of sulfur dioxide, one-third of the nitrogen oxides, one-third of the carbon dioxide, and most of the mercury, as well as the release of other heavy metals. Clean Air Amendments of 1990 mandated that sulfur-dioxide emissions from coal-burning power plants be cut eventually by 70–90%, depending on the sulfur content of the coal, and that nitrogen oxide emissions be reduced by about 2 million metric tons per year.

What are the solutions to the coal problems? One is to stop using coal, but that is unlikely because it is so plentiful. Another is to find ways to convert coal to a purer form of carbon, perhaps even burn it in the underground mines, or produce a gas or liquid that can be purified and transported. Power companies are trying new technologies to reduce emissions.[16] One involves chemical and/or physical cleaning of coal. Another converts coal into syngas (discussed earlier), which is cleaner than coal but more polluting than natural gas. A third method uses new boiler designs to burn coal at a lower temperature, which reduces emissions of nitrogen oxides. "Scrubbing" (discussed in greater detail in Chapter 14 on air pollution) removes sulfur dioxides but produces sludge that has to be disposed of, which is a major problem. Another developing technology would remove more mercury from coal.

■ **FIGURE 10.7**
Digging coal down deep.

10.7 Environmental Effects of Extracting, Delivering, and Using Petroleum Products

Oil and natural gas burn cleaner than coal but are heavy polluters, too. Oil and gas are preferred fuels because coal is so polluting, and because it is less useful for many kinds of engines. But these fuels, too, cause a great deal of pollution. Burning gasoline in automobiles produces air pollution and smog. The effects of smog on vegetation and human health are well documented and are discussed in detail in Chapter 14. In addition, oil used in cars, trucks, and airplanes sometimes spills and soaks into the soil. Leaking oil and leaking underground gasoline tanks have caused pollution problems and expensive lawsuits, although it now seems that natural soil bacteria are capable of decomposing most oil.

Refineries, also, pollute. What comes out of the ground from a typical oil well is a thick substance that is a mixture of many chemicals, from very heavy tars to very light gasoline and natural gases. A refinery is basically a gigantic chemistry set that separates this "crude oil" into its components and can also convert one form of the crude oil into a more useful form, usually converting the heavier chemicals to lighter ones. Refineries have accidental spills and slow leaks of gasoline and other products from storage tanks and pipes. Over years of operation, large amounts of liquid hydrocarbons may be released, polluting soil and groundwater below the site.

The pollution continues during delivery. A famous example happened on March 24, 1989, when the supertanker *Exxon Valdez*, carrying 1.2 million barrels of crude oil, ran aground on Bligh Reef in Prince William Sound, Alaska, and broke open (Figure 10.8). The ship was full of Alaskan crude oil that had been delivered to it through the Trans-Alaska Pipeline, which itself is a controversial way of transporting oil. The oil poured out of ruptured tanks of the tanker at about 20,000 barrels per hour, spilling a total of about 250,000 barrels (11 million gallons) into the sound. An even bigger spill was avoided when the remainder of oil was off-loaded onto another vessel.

The spill killed thousands of fish, birds, and mammals—13% of the sound's harbor seals, 28% of the sea otters, and 645,000 seabirds died. Within three days, winds began spreading the huge oil slick so widely that there was no hope of containing it. Of the 11 million gallons of spilled oil, about 20% evaporated and 50% was deposited on the shoreline. Only 14% was collected by skimming and other waste recovery. The *Exxon Valdez* spill showed that the technology for dealing with oil spills was inadequate. The spill disrupted the lives of the people who live and work in the vicinity of Prince William Sound. Even after more than $3 billion was spent to clean up, few people were satisfied with the results.[17]

■ **FIGURE 10.8**

The *Exxon Valdez* tanker grounded and spilling oil in 1989 in Prince William Sound.

Long-term effects of large oil spills are probably not devastating. There is no evidence that the oceans' ecosystems are seriously threatened by oil spills.[25] Nevertheless, the effects can last several decades. Toxic levels of oil have been identified in salt marshes 20 years after a spill.[18, 19]

The *Exxon Valdez* spill led to the Oil Pollution Act of 1990 and new technology. More modern tankers are being built with double hulls designed to prevent or limit the release of oil in case of collision or grounding. We now also have new techniques to collect oil at sea, using floating barriers and skimmers (oil is lighter than water and so floats on water), but even the best methods are difficult to use in high winds and rough seas. Oil on beaches may be collected by spreading absorbent material, such as straw, waiting for the oil to soak in, and then collecting and disposing of the oily straw.

10.8 Three Basic Alternatives to Fossil Fuels: Solar, Geothermal, and Nuclear Energy

Solar energy is renewable and therefore sustainable—once used, it can be obtained again. Solar energy is stored in wind, water, tides, ocean currents, and organic matter, such as wood and animal dung (today called *biomass energy*). Scientists estimate that the amount of energy we could get from the sun is many times the amount of energy that the entire world is currently consuming. In fact, wind energy alone could provide enough to meet the world's current energy consumption. Although wind, water, and organic matter get their energy from the sun, we view them as separate alternative energy sources and therefore will discuss them separately.

Geothermal energy is heat from deep inside the Earth. Although the basic source of geothermal heat—Earth—is not going away, individual sources can be used up, and geothermal energy is therefore generally considered nonrenewable when developed on a large-scale, industrial level. However, heat trapped in the Earth's surface, or water from a well, or heat from the sun stored in soil and rock can be used to warm a house. This kind of Earth-based heat is renewable.

Nuclear energy is the most controversial option. It is called "nuclear energy" and used to be called "atomic energy" because it is stored in the nuclei of atoms. At least in theory, nuclear energy has three sources: conventional nuclear fission, breeder reactors, and fusion (which as yet is not practical). Nuclear energy from breeder reactors is considered renewable, but nuclear energy from conventional fission is not.

In choosing, we must assess their uses and their environmental effects. As we did in our discussion of traditional energy sources, we have to consider new alternative sources of energy in terms of how we would use them—for lighting; heating and cooling; moving objects; and information transfer/conversion—and in terms of how they could affect the environment.

10.9 Solar Energy: Two Types
Passive Solar Energy

People have long used passive solar energy. A classic Mediterranean house built by the ancient Greeks and Romans was U-shaped and faced south to collect and store sunlight for warming the house in winter. Deciduous trees (trees that drop their leaves in the fall) planted on the south side shaded the house in summer but let sunlight reach the house in the winter. Similarly, ancient American Indian dwellings—now famous ruins at places like Mesa Verde National Monument, in Arizona—were built into cliffs that face south and took advantage of the fact that stone and soil (the rock face and the building material) store heat well (Figure 10.10). Early European settlers on the American prairie built sod huts and sometimes dug cabins into the prairie sod because this kept the huts

Courtesy Powerlight Corporation

▪ **FIGURE 10.9**

Bavarian farm field with a solar-electric array.
In a farm field in Bavaria, Germany, sheep graze beneath an unusual crop: an array of black rectangles mounted on long metal tubes that rotate slowly during the day, following the sun like huge mechanical sunflowers (Figure 10.9). This field is part of the world's largest solar-electric installation, generating 10 megawatts of electricity on 62 acres. The installation was designed and built by an American company, PowerLight of California.

warmer in winter and cooler in summer, reducing the need for fuel. And when the Lewis and Clark expedition spent a winter in North Dakota, they built their cabins just below a south-facing bluff on the Missouri River, so the cabins were somewhat protected from the cold northern winds and gained some warmth from the winter sun.

Cheap and plentiful fossil fuels led us to ignore these methods later,[20] but they are becoming popular again. Today, systems that collect solar heat without using moving parts are known as *passive solar energy systems.*

Active Solar Energy Systems

Active solar energy systems either convert sunlight directly to electricity or they store the energy as heat and then use electric pumps to circulate air, water, or other fluids from solar collectors to a location where the heat is stored until used.

Solar collectors store the sun's energy to provide space heating and hot water. They are usually on rooftops and are commonly made of flat glass-covered plates over a black background where water or another liquid circulates through tubes. Sunlight enters the glass and is absorbed by the black background, which in turn heats the fluid circulating in the tubes (Figure 10.11).

Photovoltaics (PVCs) convert sunlight directly into electricity, as discussed in the case study about Bavaria. PVCs use thin layers of semiconductors (silicon or other materials) that produce an electric current when sunlight falls on them (Figure 10.12). (Even the chips in a computer, made of silicon oxides, will produce an electric current when sunlight falls on them. Some companies make solar PVCs from computer-part rejects.)

Solar-cell technology is advancing rapidly. Early versions could convert only 1–2% of sunlight falling on them to electricity, but today the most efficient ones convert 20%. The cells are modules encased in plastic or glass and can be combined to produce systems of various sizes, so their power output can be matched to the intended use. Photovoltaics is the world's fastest growing energy source, doubling every two years.

Environmental Effects of Using Solar Energy

Solar energy generally has relatively little environmental impact. One concern is that the manufacture of solar equipment uses a large variety of metals, glass, plastics, and fluids, some of which may cause environmental problems during manufacturing through accidental release of toxic materials. Some people also consider solar panels unattractive, but new materials and design are leading to roof tiles that look like normal roof tiles and protect the roof like normal roof tiles, but also produce solar electricity.

The Future of Solar Energy

Solar energy will likely become a major provider of the energy we use. In the future, the alternative energy source for many people and communities may well be lightweight PVCs in which the solar cells are mounted together to form modules. The growth (about 30% per year) and technology changes in PVCs suggest that in this century solar energy is likely to become a multigigawatt-per-year industry that will provide a significant portion of the energy we use.

MPI/Stringer/Getty Images, Inc

(a)

(b)   Courtesy Ansel Braseth

 **FIGURE 10.10**

Early use of passive solar energy.
(a) Ancient Indian houses at Mesa Verde, Arizona, were built against south-facing bedrock cliffs, so sunlight could warm them in the winter. (b) Claus Braseth's sod house in North Dakota.

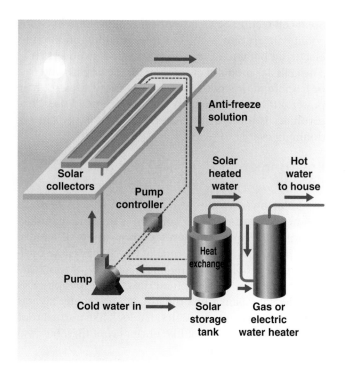

■ **FIGURE 10.11**

Diagram of a typical active solar heating system.
(*Source:* Active Solar Energy. Available at
http://www.newenergy.org/sesci/publications/
pamphlets/active.html. Accessed August 12, 2005.)

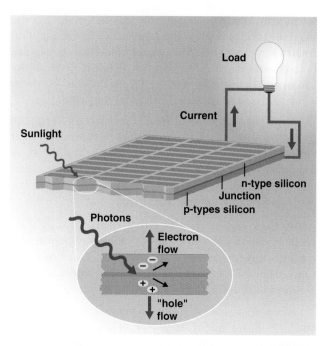

■ **FIGURE 10.12**

Diagram of a typical photovoltaic (PVC) sunlight-to-electricity system.
(*Source:* Why Invest in Solar? Available at
http://www.powerlight.com/solar/solar_basics.shtml.
Accessed August 12, 2005.)

Alternative Energy Sources: Bavaria Lights the Way

If fossil fuels are not our future, what are our choices? Here's one.

Germany isn't very hot and sunny, so how does this work? We tend to think that solar energy is reliably available only in places like Arizona or Florida or the Sahara Desert, and that it stops working on cloudy days. The sun seems a risky source of power for a country like Germany, famous not for sunshine but for high mountain peaks and beautiful winter sport resorts. Munich, Bavaria's major city, has an average January daytime temperature of 1°C (34°F), just above freezing, and an average January nighttime temperature of −5°C (23°F). Even summer is not all that warm—in August, the average daytime temperature is 23°C (73°F) and the average nighttime temperature is 9°C (48°F). Throughout the year, about one-third of the days are rainy. It is a climate somewhat like New York City's.

Wouldn't generating all of its energy this way take a lot of space? Germany's annual energy production is 548 million kilowatt hours. If its solar facility were scaled up, it could take over Germany's total energy production on 4,798 square miles. This may sound like a lot,

but it is only 3.5% of Germany's total land area and, more important, the solar facility could be on land that is already occupied. It could be on rooftops, above parking lots, and integrated with certain kinds of pasture and cropland.

The point is that solar energy is a real option. It is an alternative energy source that could be the solution to the world's energy problem.

10.10 Wind Power

The use of wind power, a form of solar energy like, solar power, has a long history. From early Chinese and Persian civilizations to the present, wind has propelled ships and driven windmills to grind grain and pump water. In the past, thousands of windmills in the western United States were used to pump water for ranches. Wind power is one of the fastest-growing sources of energy in the world, with its energy output doubling nearly every three years[21] (Figure 10.13).

Winds are a result of the sun's heating of Earth's surfaces. The different temperatures of the surface areas create air masses of different densities and heat. Abundant wind energy is often found in coastal areas and offshore as wind flows freely across oceans. Wind velocity also often

Courtesy of DOE/NREL. Photo by Todd Spink

■ **FIGURE 10.13**

A wind farm in the desert near Palm Springs, California.
Some criticize wind power as not fitting into the landscape. How well do you think these wind machines fit into
the local scene? Enough so that it seems okay? Or too ugly for the landscape?

increases over hilltops, and wind may be funneled through a mountain pass. In the United States, the regions with the greatest potential for wind energy are the Pacific Northwest coastal area, the coastal region of the northeastern United States, and a belt extending from northern Texas through the Rocky Mountain states and the Dakotas. Other good sites include mountain areas in North Carolina and the northern Coachella Valley in Southern California.

Wind energy is less reliable than solar energy, so must be stored. Although it holds large potential for energy, wind tends to be highly variable as to when, where, how long, and how strongly it blows. You may point out that the sun doesn't always shine either, but the fact is that Earth receives energy from the sun even on cloudy days without sunshine, whereas if the wind doesn't blow, it doesn't produce energy. For this reason, storage of wind energy is important, and is discussed later.

Europe is the undisputed leader in wind power today. World production of wind power in 2005 was about 47,300 megawatts (MW), of which about two-thirds was in Europe.[22] The U.S. produces about 6,700 MW, mostly in California, where wind farms were first developed, and Texas. Together these two states account for about one-half of U.S. wind power. It is believed that there is sufficient wind energy in Texas, South Dakota, and North Dakota to satisfy the electricity needs of the entire United States. Wind-power potential in Britain is more than twice Britain's present demand for electricity.[23]

Wind power holds great potential in the developing nations, with their huge populations and growing air-pollution problems. Today, China (the world's largest consumer of coal) produces about 75% of its electricity by burning coal. Many of the cities with the worst air pollution are in China, and this is driving a shift to cleaner energy. China gets only about 600 MW of wind energy today but is planning large projects with the goal of 20,000 MW by 2020. China could probably double its current capacity to generate electricity with wind alone! Other countries, including Colombia and several in Eastern Europe, are also planning new wind-energy projects.[23]

The rapid growth of wind power could make it a major energy supplier. Although wind now fills less than 1% of the world's demand for electricity, it is growing about 30% a year, more than ten times as fast as oil use. This suggests that wind could soon be a major supplier of power. One scenario suggests that it could supply 6% of the world's electricity in the coming decades and eventually could provide more energy than water power, which today supplies about 20% of the world's electricity.

Wind energy is already a big industry. It has created thousands of jobs and is also becoming a major investment opportunity. Technology is producing more efficient wind turbines that are lowering the price of wind power. One of the world's largest wind farms, on the Oregon–Washington border, produces nearly 300 MW at a cost of about 5 cents per kilowatt hour, which is competitive with electricity from burning natural gas.

Wind energy has some environmental disadvantages. Perhaps the most controversial effect of modern windmills is that large birds of prey and vultures collide with them and can be killed. Bats may also be vulnerable. In addition, many people consider modern wind farms, with hundreds of windmills, unattractive (Figure 10.13). This is a major concern in scenic coastal areas—for example, a project in the planning stage off the shore of Cape Cod is being hotly contested. All in all, however, wind energy has a relatively low environmental impact, especially in comparison to coal and nuclear energy.

10.11 Water power

Water power is a form of stored solar energy, as mentioned earlier. This is because the flow of water on Earth depends on the climate, which in turn depends on solar heating of the atmosphere. Rain is the result of sunlight heating and evaporating water, which turns into water vapor that condenses in the atmosphere and falls to the ground. The energy from the sun is stored in the water, and the energy of moving water has been used to do work since ancient times.

The faster the water moves, the more powerful it is. Elevation is important because water flowing downward from a great height has greater power. For this reason, today hydroelectric power plants use water that falls from reservoirs behind high dams (Figure 10.14). In the United States, hydroelectric plants generate about 80,000 MW of electricity—about 10% of the total electricity produced in the nation. Worldwide, water power provides about 19% of the electricity. In some countries, such as Norway and Canada, hydroelectric power plants produce most of the electricity. Water power has been used for centuries—in the 11th century there were more than 5,600 water mills in England alone![24] But it is in the last 200 years that people have built very large dams on many of the world's major rivers.

Water Power and the Environment

Water power is clean but has some negative environmental effects. Today, one of the great environmental advantages of water power is that it produces no greenhouse gases or other toxic substances—no radioactive waste, no sulfur oxides to pollute the air. So from a

Kathy Steen/iStockphoto

(a)

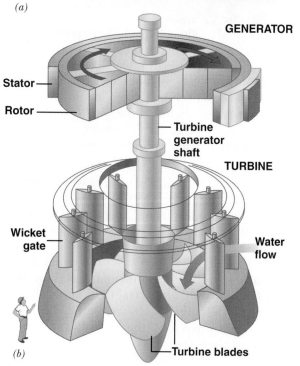

(b)

▪ **FIGURE 10.14**

(a) Hoover Dam, one of the world's great hydroelectric-power dams, is on the Colorado River about 30 miles from Las Vegas, Nevada. It provides electricity, water, and water recreation, and is one of the principal reasons that Las Vegas can exist in an otherwise desert environment. Built between 1933 and 1935, Hoover Dam was a huge construction project. Its generators, rated at 2,991,000 horsepower, produce 2 billion watts (2,080 megawatts) of electricity. Few sites remain on Earth where such huge dams could still be built. (b) Diagram of a hydropower turbine that converts the energy from falling water into electricity. Water flowing down an enclosed pipe spins a turbine that in turn spins a generator.

Courtesy United States Army Corps of Engineers

■ FIGURE 10.15
The Bonneville Dam.
This was the first of the great dams built on the Columbia River by the Bonneville Power Administration. The photo is an aerial view of the dam and the Columbia River.

physical and chemical point of view, it is an environmentally benign energy source. For centuries, water power was considered benign and extremely useful. In much of the twentieth century, many major hydroelectric dams were built (Figures 10.14 and 10.15). The environmental problems arise from effects on life—displacing people and their farms and towns, and threatening habitats of fish and other wildlife. In particular, large dams block the migration of some fish, such as salmon, and trap sediment that would otherwise reach the sea and replenish the sand on beaches. In addition, for a variety of reasons, many people do not want to turn wild rivers into a series of lakes. More water evaporates from the large surface areas of reservoirs, and the evaporative loss of water from reservoirs is even more significant in arid regions. One of the largest recent hydropower projects in the world, the Three Gorges project in China, is flooding areas that were farmed for centuries, displacing many people, taking away from the great landscape beauty of the region, and raising concerns about effects on biological diversity (Figure 10.16.)

These negative effects have led to a rising tide of opinion against construction of new dams, and even to the removal of existing ones, such as the now famous removal of Edwards Dam on the Kennebec River in Maine, the first major dam to be removed in the 20th-century in the United States.

The growth of large-scale water power will likely be limited for all these reasons, and because many good sites for dams already have one.

Tidal Power: Another Kind of Water Power

Tidal power is the force of ocean water flowing in and out with the tides. It can be used to turn waterwheels and turbines in much the same way that

freshwater produces water power, but the situation is more complex and obtaining power is more difficult. Tidal power can be traced back to 10th-century Britain,

(a)

Eddie Gerald/Alamy Images

(b)

■ FIGURE 10.16
Three Gorges Dam.
(a) Its location in China; (b) a photograph of the dam.

where tides were used to power coastal mills. However, only in a few places with favorable topography—such as the north coast of France, the Bay of Fundy in Canada, and the northeastern United States—are the tides strong enough to produce commercial electricity. The tides in the Bay of Fundy have a maximum range of about 15 meters (49 feet). A minimum range of about 8 meters (26 feet) appears necessary with current technology.

Dams and reservoirs harness tidal power. A dam is built across the entrance to a bay or estuary, creating a reservoir. As the tide rises (flood tide), water is initially prevented from entering the bay on the landward side of the dam. Then, when there is enough water on the ocean side (at high tide) to run the turbines, the dam is opened and water flows through it into the reservoir (the bay), which turns the blades of the turbines and generates electricity. When the bay/reservoir is filled, the dam is closed, stopping the flow and holding the water in the reservoir. When the tide falls (ebb tide), the water level in the reservoir is higher than in the ocean. The dam is then opened again to run the turbines backward (they are reversible), and electric power is produced as the water flows out of the reservoir and back into the ocean.

Tidal power has environmental impacts. The change in the hydrology that a dam causes in a bay or estuary can harm the vegetation and wildlife, and so can the periodic rapid filling and emptying of the bay as the dam opens and closes. The dam also restricts upstream and downstream passage of fish.

10.12 Biomass Energy

Biomass energy is a fancy term for solar energy stored in organic matter. This energy is a result of photosynthesis (see Chapters 3 and 4), so it is a kind of solar energy. Energy from the sun is fixed through photosynthesis and stored in organic matter. The longer-lasting organic matter is the woody tissue of plants. One of the major advantages of this kind of fuel is that it does not contribute to greenhouse gases. Since the carbon dioxide it releases was simply taken up from the atmosphere during photosynthesis, biomass fuel is a net zero in terms of greenhouse-gas production.

Biomass is the oldest fuel used by people. Our Pleistocene ancestors burned wood in caves to keep warm and to cook, and biomass continued to be a major source of energy throughout most of the history of civilization. When North America was first settled, there was more wood fuel than could be used. Forests often were cleared for agriculture by cutting through the bark all the way around the base of the trees to kill them and

then burning the forests. Wood remained the major fuel in the United States until the end of the 19th century, but burning wood become old-fashioned by the mid-20th century. With coal, oil, and gas plentiful, people burned wood in an open fireplace more for pleasure than for heat. Today, however, with other fuels becoming scarcer, there is renewed interest in the use of natural organic materials for fuel.

Over 1 billion people in the world still rely on wood for heat and cooking. Today, in developing countries, biomass provides about 35% of the total energy supply.[25] Energy from biomass can take several routes. One is direct burning of biomass either to produce electricity or to heat water and air. A second route is to heat biomass to form a gaseous fuel (gasification). A third is distillation or processing of biomass to produce *biofuels*, such as ethanol, methanol, and methane.[26]

Sources of Biomass Energy

While wood is the most widely used biomass fuel, there are many others. In less-developed, pastoral areas of some countries, cattle manure is burned for cooking. Peat, a form of compressed dead vegetation, serves as heating and cooking fuel in northern countries, such as Scotland, where it is abundant. The primary sources of biomass fuels in North America are forest products, agricultural residues, energy crops (see below), animal manure, and urban waste (Figure 10.17).

Some modern waste-processing facilities convert organic waste into methane. They use microorganisms to convert manure from livestock and other organic waste in specially designed "digestion" chambers to form methane, which is burned to produce electricity, or used in fuel cells or as fuel for tractors or other vehicles.

In some areas, crops are grown primarily to provide a fuel, not food. Brazil produces about 12 billion liters of ethanol per year from sugarcane, and in the United States farmers receive federal subsidies to grow crops to produce biomass fuels. But recent studies indicate that producing fuel from crops takes more fossil-fuel energy than we can obtain from these products. Ethanol made from corn grain requires 29% more fossil energy than the ethanol fuel can produce. Producing ethanol from wood is even less efficient, requiring 57% more fossil energy than the ethanol fuel produces.[27]

Converting waste into fuel is beneficial even without a net energy gain if that waste would otherwise pollute the environment. The wastes from ethanol distillation used to be dumped into rivers, causing water pollution. The wastes can now be treated to produce biogas and liquid fertilizers (recycled to sugarcane fields).[28] Methane and biogas can also be produced

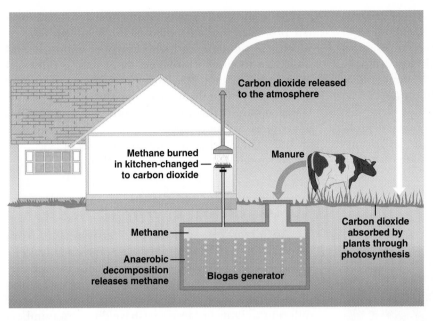

■ FIGURE 10.17
Diagram of a biogas system, from animal waste to electricity.
(Source: Re-energy.ca. Available at http://www.re-energy.ca/t-i_biomassbuild-1.shtml. Accessed July 5, 2005.)

ing habitat of the Indian lion (not to be confused with the Indian tiger).

In sum, should we turn to biomass fuels? In 2005, biomass fuel suddenly became very fashionable, and today it is being promoted by large agricultural corporations, some petroleum companies, and some government agencies. But it is still unclear where the production of biomass directly for fuel results in a net energy gain. Major scientific studies so far show that using modern industrial agricultural practices to produce a crop that will simply be burned as fuel requires more energy than we can obtain from burning it. At best, there may be a very slight gain, but one barely worth the investment. However, there are political pressures to promote this kind of fuel, as it provides a new kind of subsidy for big agriculture.

from urban waste in landfills and sewage at wastewater-treatment plants. In rural China, about 5 million small facilities are used to treat sewage. The original purpose was to reduce disease, but their potential as an energy source was soon recognized and used.[28] In Western Europe a number of countries use from one-third to one-half their municipal waste for energy production.

The United States has been slow to use urban waste as an energy source. However, today a number of U.S. facilities process waste to generate electricity or be used as fuel. California alone has 37 biomass-fueled power plants.[29] If all such power plants were operating at full capacity, about 15% of the country's waste, or 35 million metric tons per year, could be burned to extract energy.[30]

Biomass Energy and the Environment

The use of biomass fuels can pollute the air and degrade the land. It's true that biomass fuel does not contribute greenhouse gases to the atmosphere, and for most of us the smell of smoke from a single campfire is part of a pleasant outdoor experience. However, under certain weather conditions, woodsmoke from many campfires or chimneys in narrow valleys can lead to air pollution.

Using wood as fuel also puts added pressure on an already heavily used resource. A worldwide shortage of firewood is adversely affecting natural areas and endangered species. For example, the need for firewood has threatened the Gir Forest in India, the last remain-

10.13 Geothermal Energy

Geothermal energy is natural heat from the interior of the Earth. It can be converted to heat buildings and to produce steam for generating electricity. First used in Italy in 1904, today geothermal energy is generating electricity in 21 countries, including Russia, Japan, New Zealand, Iceland, Mexico, Ethiopia, Guatemala, El Salvador, the Philippines, and the United States. Total worldwide production is approaching 9,000 MW, which is double the amount in 1980 and equal to the energy produced by nine large modern coal-burning or nuclear power plants. Geothermal energy now supplies electricity to some 40 million people, at a cost in line with that of other energy sources.[30] It supplies 30% of El Salvador's total electricity consumption, but globally it accounts for less than 0.15% of the total energy supply.[28]

Is it renewable or nonrenewable? We would consider geothermal energy a nonrenewable energy source if we reached a point where we were using it up faster than the natural heat production within the Earth could replenish it. Right now we are using only a small fraction of the vast total resource base.

Groundwater is a low-heat source of geothermal energy. Because we usually tap high-heat sources, it may come as a surprise to learn that relatively low-temperature groundwater can be considered a source of geothermal energy. This is because the normal internal heat flow from Earth keeps the temperature of groundwater at a depth of 100 meters (320 feet) at about 13°C (55°F). This is warm compared with winter temperatures in much of the United States, so groundwater can help heat a house. And it is cool compared with summer temperatures, so it can be used for air-conditioning.

Geothermal energy pollutes. Environmental problems include on-site noise, emissions of gas, and disturbance of the land at drilling sites, disposal sites, roads and pipelines, and power plants. Geothermal development often produces thermal pollution from hot wastewater, which may be saline or highly corrosive and cause disposal and treatment problems. In Hawaii, geothermal power poses cultural problems. Hawaii's active volcanoes provide abundant heat near the surface, but native Hawaiians argue that the exploration and development of geothermal energy degrade tropical forests, and they are offended by using the "breath and water of Pele" (the volcano goddess) to make electricity (Figure 10.18).

Digital Vision/Getty Images

(a)

Courtesy Ormat Technologies, Inc.

(b)

▪ **FIGURE 10.18**
Hawaiian geothermal power plant.
(a) Energy from volcanoes on the big island of Hawaii produces steam that (b) generates electricity.

10.14 Nuclear Energy

An incredible amount of energy results from converting matter into energy. Some chemical elements do this by themselves. Their unstable **isotopes** decay spontaneously, releasing energy. (Isotopes are atoms of an element that are chemically the same but are of different weights because they have different numbers of neutrons. Some isotopes are stable and some are not.) All nuclear power plants make use of the enormous energy that results from converting matter into energy, following from Einstein's famous equation $E = mc^2$ (energy equals matter times the speed of light squared). We call this *radioactivity*, and we call elements that do this *radioactive*. Uranium and plutonium are two radioactive elements.

Nuclear power plants use these radioactive elements to generate electricity. In theory, there are three kinds of nuclear power plants: conventional nonbreeder fission; breeder fission; and fusion. However, only the first two exist—the third is still only a theoretical possibility. Most nuclear power reactors today are the conventional nonbreeder form, and these are only a short-term energy source because the necessary isotope, U-235, is rare.

At first, there was a lot of enthusiasm for nuclear energy. In 1953, Dwight D. Eisenhower, a popular World War II general and U.S. president, described an optimistic vision of the future of atomic energy, which he called "atoms for peace." The chairman of the Atomic Energy Commission, in a 1954 speech to science writers, predicted that the electricity produced by nuclear-powered generators would be so cheap, so nearly unlimited, and so clean that we would never need to meter it. More than 100 nuclear power plants were built in the next decades, and orders were placed for more than 100 more by 1978.

The public mood changed in the early 1980s. The primary reversal of opinion occurred three years after an accident at the Three Mile Island nuclear power plant in the U.S. and four years before the Chernobyl accident in the former Soviet Union (both discussed later in this chapter). No more nuclear reactors were ordered, and today there are still only 103 commercial nuclear power plants in the United States, producing about 20% of the nation's electricity.

Support for new nuclear plants is growing now, despite concerns. The rising cost of oil and natural gas, along with concern about global warming, is causing many people to rethink the value of nuclear energy. In a May 2003 survey, about 50% of U.S. adults interviewed agreed that we should build more nuclear power plants,[31] and the current fuel shortage has led to renewed interest in nuclear energy. Still, many people worry about issues of safety. To evaluate the risks and benefits of nuclear energy, we need to begin at the beginning, with an understanding of nuclear reactions.

Nonbreeder Reactors: Fission Reactors

The first human-controlled nuclear fission was demonstrated in 1942 by Italian physicist Enrico Fermi at the University of Chicago. This discovery of a way to split atoms and release their stored energy led to the use of nuclear energy not only to generate electricity for homes and industry but also to power submarines, aircraft carriers, and icebreaker ships.

On a weight-to-weight basis, nuclear fission produces much more energy than other sources. For example, 1 kilogram (2.2 pounds) of uranium oxide produces as much heat as about 16 metric tons of coal. The heat from nuclear fission makes steam to run turbines that generate electricity. A nuclear reactor has the same function as the boiler that produces heat in coal-burning or oil-burning power plants (Figure 10.19).

Three types, or *isotopes,* of uranium occur in nature: uranium-238, which accounts for about 99.3% of all natural uranium; uranium-235, which makes up about 0.7%; and uranium-234, which makes up about 0.005%. Uranium-235 is the only naturally occurring fissionable form, and therefore it is essential to the production of nuclear energy. Processing (called *enrichment*) increases the concentration of uranium-235 from 0.7% to about 3%, so it can be used as fuel for the fission reaction.

Fission reactors split uranium-235 atoms by neutron bombardment. The reaction produces more neutrons, fission fragments, and heat. The released neutrons strike other uranium-235 atoms, releasing more neutrons, fission fragments, and heat. The released neutrons are fast-moving and must be slowed down slightly, or *moderated,* to increase the probability of fission. In *light water reactors,* the kind most commonly used in the United States, ordinary water is used as the moderator. As the process continues, a chain reaction develops, with more and more uranium atoms splitting and more and more neutrons and heat being released.

Control rods keep the fission chain reaction stable. The main components of a reactor are the core (consisting of fuel and moderator), control rods, coolant, and reactor vessel. Control rods are made of materials that absorb the neutrons without nuclear fission. As the control rods are moved out of the core, the chain reaction increases; as they are moved into the core, the reaction slows. Full insertion of the control rods into the core stops the chain reaction.[32] The coolant removes the heat produced by the fission reaction.

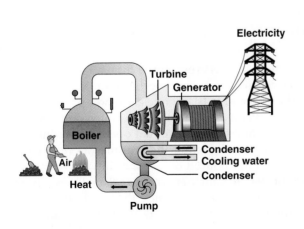

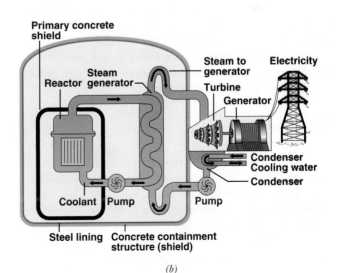

(a)   *(b)*

■ **FIGURE 10.19**

(a) A fossil-fuel power plant and (b) a nuclear power plant with a boiling water reactor.
Notice that the nuclear reactor has exactly the same function as the boiler in the fossil-fuel power plant. The coal-burning plant (a) is Ratcliffe-on-Saw, in Nottinghamshire, England. The nuclear plant (b) is in Leibstadt, Switzerland. (*Source:* American Nuclear Society, Nuclear Power and the Environment, 1973.)

An imbalance between heat buildup and cooling can be disastrous. All major nuclear accidents have occurred when something went wrong with the balance between heat removal by the coolant and heat buildup in the reactor core.[33] The core is enclosed in a heavy stainless-steel reactor vessel, and for extra safety the entire reactor is contained in a reinforced-concrete building. Even so, failure to maintain the balance between heat buildup and heat removal can cause a *meltdown*—a nuclear accident in which the nuclear fuel gets so hot that it forms a molten mass that breaches the containment of the reactor and contaminates the outside environment with radioactivity.

The nuclear industry now favors smaller, safer, less complex reactors. Large nuclear power plants, which produce about 1,000 MW of electricity, require an extensive set of pumps and backup equipment to make sure that adequate cooling is available to the reactor. Smaller reactors can be designed with cooling systems that work by gravity and thus are less vulnerable to pump failure in case of a power loss. Another approach is to design a fuel assembly that cannot hold enough fuel to reach the temperature that could cause a core meltdown.

A new reactor, the "pebble-bed reactor," may be more efficient. This design uses fuel elements called "pebbles" that are about the size of billiard balls. About 300,000 pebbles are loaded into a metal container shielded by a layer of graphite. About 100,000 nonfuel graphite pebbles are mixed in with the fuel pebbles to help control production of heat from the reactor. Fuel pebbles are fed into the core, continuously refueling the nuclear reaction. As a spent fuel pebble leaves the core, another is added. This is a safety feature, assuring just the right amount of fuel for optimal energy production. Pebble-bed reactors are expected to compete economically with the new generation of natural-gas power plants and be about 25% more efficient than present nuclear reactors.[34]

Breeder Reactors

Today's nuclear reactors use uranium very inefficiently. Only about 1% of the uranium provides electricity; the other 99% ends up as waste. Furthermore, uranium is a nonrenewable resource, and current reserves, if heavily exploited, will last only a few decades. Therefore, these reactors are part of the nuclear-waste problem (discussed later in this chapter) and are not a long-term solution to the energy problem.

Breeder reactors could make nuclear power sustainable for hundreds of years. Breeder reactors produce new nuclear fuel by transforming lower-grade uranium into fissionable material. Several thousand breeder reactors could supply about half the energy presently produced by fossil fuels for more than 1,000 years.[33] However, breeder reactors have two problems: Their waste is extremely radioactive and dangerous, and the fuels they produce can be used to make atom bombs. Also, fuel for the breeder reactors will have to be recycled, as reactor fuel must be replaced every few years. The recycling and disposal of wastes from breeder reactors remain an unsolved problem.

Fusion Reactors

Nuclear fusion is the source of energy in our sun and other stars. In contrast to fission, which involves splitting the nuclei of heavy elements (such as uranium), **fusion** involves combining (fusing) the nuclei of light elements (such as hydrogen) to form heavier ones (such as helium). As fusion occurs, heat energy is released.

Several conditions are necessary for fusion to take place. First, the temperature must be extremely high (about 100 million degrees Celsius). Second, the fuel elements must be incredibly dense. At the temperatures and pressures necessary for fusion, nearly all atoms are stripped of their electrons, forming a *plasma*—an electrically neutral material. Third, the plasma must be confined long enough to ensure that the energy released by the fusion reactions is greater than the energy supplied to maintain the plasma.[35, 36]

The potential energy from a fusion reactor power plant is nearly inexhaustible. However, no practical fusion reactor exists, and it is unclear whether one will ever be developed.

10.15 Environmental Problems of Nuclear Power

Three Mile Island: A Cooling Failure Leads to a Meltdown

The most serious event in the history of U.S. commercial nuclear power plants occurred on March 28, 1979, at the Three Mile Island nuclear power plant near Harrisburg, Pennsylvania.[37] A main water pump stopped running, and then a valve that opened to reduce pressure failed to close after the pressure was relieved, and cooling water poured out. As a result, the reactor core overheated, leading to a partial core meltdown. Intense radiation was released inside the plant, and some escaped into the atmosphere. By the third day after the accident, radiation levels near the site were high enough that the radiation dose a person there would have received would be six times as much as the average American receives in a year from natural radiation.

The state of Pennsylvania was unprepared to deal with the accident. There was no state bureau for

■ **FIGURE 10.20**
Chernobyl.
Where the largest nuclear power plant accident happened. (*Source:* Available at http://www.uic.com.au/graphics/Chernomap.gif. Accessed July 4, 2005.)

■ **FIGURE 10.21**
The Chernobyl nuclear reactor damaged by its explosion.
The center of the building in the foreground blew up and burned as a result of the nuclear reaction malfunction.

radiation help, and the state Department of Health did not have a single book on radiation medicine (the medical library had been dismantled two years earlier for budgetary reasons). One of the major impacts of the incident was fear, but there was no state office of mental health, and no staff member from the Department of Health was allowed to sit in on important discussions following the accident.[38]

This accident permanently changed the public view of nuclear power plants, changed the nuclear industry, and the Nuclear Regulatory Commission. The damaged reactor was permanently shut down and the radioactive materials removed to disposal sites.[39]

Chernobyl

Nuclear plants elsewhere were equally unprepared for a serious accident. This was dramatically illustrated on the morning of Monday, April 28, 1986, when workers at a nuclear power plant in Sweden measured alarmingly increased levels of radiation and searched frantically for the source near their plant. They soon concluded that it was not their plant that was leaking radiation. Rather, the radioactivity was coming from the Soviet Union on prevailing winds. Confronted, the Soviets announced that an accident had occurred two days earlier, on April 26, at a nuclear power plant at Chernobyl (Figures 10.20 and 10.21). This was the first notice to the world of the worst accident in the history of nuclear power generation.

At Chernobyl, too, it was a cooling-system failure. The system that supplied cooling waters for the reactor failed, causing the temperature of the reactor core to rise to over 3,000°C (about 5,400°F), melting the uranium fuel. Explosions removed the top of the building over the reactor, and the graphite surrounding the fuel rods in the core ignited. The fires produced a cloud of radioactive particles that rose high into the atmosphere. There were 237 confirmed cases of acute radiation sickness, and 31 people died of it.[40]

Many people in the Northern Hemisphere received radiation (Figure 10.22). Fortunately, except for the 30-kilometer (19-mile) zone surrounding Chernobyl, the global human exposure was relatively small. Even in Europe, where exposure was highest, it was considerably less than the natural radiation people receive in one year.[41] However, within the 30-kilometer radius of the plant, about 115,000 people were evacuated; and as many as 24,000 people were estimated to have received a very large radiation dose, as much radiation as the average person would receive naturally in 215 years, if people lived that long.

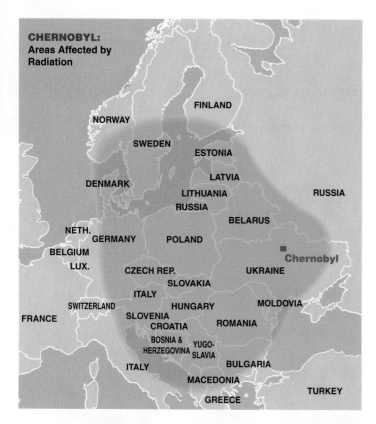

CHERNOBYL:
Areas Affected by
Radiation

■ **FIGURE 10.22**
Areas affected by radiation from the Chernobyl accident.
(*Source:* Available at http://history1900s.about.com/gi/
dynamic/offsite.htm?site=http%3A%2F%2Fwww.pathfinder.
com%2Ftime%2Fdaily%2Fchernobyl%2Fmaps.html
Accessed August 2, 2006.F.)

What were the health effects of Chernobyl?
Based on studies of Japanese atomic-bomb survivors, it
was expected that about 122 cases of leukemia would
occur from 1986 through 1998. Surprisingly, as of late
1998 there was no significant increase in cases of
leukemia, even among the most highly exposed people, although an increase in leukemia could still occur
in the future. However, the number of childhood thyroid cancer cases per year has risen steadily in Belarus,
Ukraine, and the Russian Federation (the countries
most affected by Chernobyl). Since the accident, a total of 1,036 cases of thyroid cancer cases have been diagnosed in children under 15. This is believed to be
linked to radiation from the accident, although other
factors, such as environmental pollution, may also play
a role. According to one estimate, Chernobyl will ultimately be responsible for several thousand deaths
worldwide.[42]

**Vegetation within 7 kilometers of the power
plant was killed or badly damaged.** Pine trees examined in 1990 around Chernobyl showed extensive tissue

damage and still contained radioactivity. The distance between annual rings (a measure of tree growth) showed
that the trees' growth rate had slowed since 1986.[43]

**Wildlife seem to be thriving but have genetic
mutations.** Scientists returning to the evacuated zone
in the mid-1990s were surprised to find thriving animal populations. With no people around, species such
as wild boar, moose, otters, waterfowl, and rodents
seemed to be enjoying a population boom. Indeed,
the population of wild boars had increased tenfold
since the evacuation. However, a study of gene mutations in voles (small mammals related to mice) in the
contaminated zone found more than five mutations
per animal, compared with hardly any (0.4 per animal) outside the zone. It is puzzling to scientists that
the high mutation rate has not crippled the animal
populations, but the benefit of excluding humans appears so far to outweigh the harm from radioactive
contamination.[44]

**Radioactive contamination remains in the areas
surrounding Chernobyl,** in the soil, vegetation, surface water, and groundwater, presenting a hazard to
plants and animals. The evacuation zone may be uninhabitable for a very long time unless some way is found
to remove the radioactivity.[45] Despite a government
warning, by 1987 over 100 people had returned to the
evacuation zone. However, by around 1995 fewer than
50 people still lived there, and by 2004 there were
fewer than 20. The city of Prypyat, 5 kilometers from
Chernobyl, is a "ghost city."

Could such an accident happen again? With several hundred reactors producing power in the world
today, the answer has to be yes. About ten accidents
have released radioactive particles during the past 34
years, and the probability of an accident increases with
every new reactor put into operation. According to the
U.S. Nuclear Regulatory Commission, the probability
of a large-scale core meltdown in any given year should
be no greater than 0.01% (one chance in 10,000).
However, using that guideline, if there were 1,500 nuclear reactors in the world (about four times the present number), a meltdown could be expected every
seven years.

Some Facts You Should Know About Radioactivity

**How long a radioactive substance remains dan-
gerous depends on its** *half-life* —the length of time
it takes for one-half of the original isotope to decay to
another form. Different isotopes have different half-
lives. Those with short half-lives remain dangerous for
only a short time, whereas those with long **half-life** can
contaminate the environment for a very long time.

Uranium-235 has a half-life of 700 million years, a very long time indeed! Radioactive carbon-14 has a half-life of 5,570 years, which is in the intermediate range, and radon-222 has a relatively short half-life of 3.8 days. Some have half-lives of only a fraction of a second.

There are three major kinds of nuclear radiation: alpha particles, beta particles, and gamma rays.

Alpha particles each have two protons and two neutrons (that is, an alpha particle is a helium nucleus) and have the greatest mass of the three types of radiation. Because alpha particles have a relatively high mass, they do not travel far—just a few centimeters in the air and a tiny fraction of a centimeter in living tissue. This makes alpha dangerous if ingested or inhaled, because essentially all of its energy is absorbed internally and can damage an individual's DNA and other cellular material.

Beta particles are electrons. These travel farther through air than the more massive alpha particles but can be blocked by even moderate shielding, such as a thin sheet of metal (aluminum foil) or a block of wood. Once again, this means it is dangerous to ingest or in-

hale beta particles, because your body will absorb most of their energy.

Gamma rays are the most penetrating type of radiation. They are similar to X-rays but more energetic and penetrating, so protection from gamma rays requires thick shielding, such as about a meter of concrete or several centimeters of lead. Still, they are generally somewhat safer to ingest than alpha or beta particles because most of the gamma energy passes out of your body.

Each radioactive isotope *(radioisotope)* has its own characteristic emissions: Some emit only one type of radiation, whereas others emit a mixture.

Danger exists in mining and using uranium, and in disposing of its wastes. Radioactive waste from uranium mines and mills can pollute the environment. In some instances, radioactive mine *tailings* have been used for foundation and building materials and have contaminated dwellings. (Tailings are leftover materials from mining and generally remain at the site.) Contamination also results from using uranium as a fuel. The waste materials produced every step of the way must be carefully handled and disposed of (Figure 10.23).

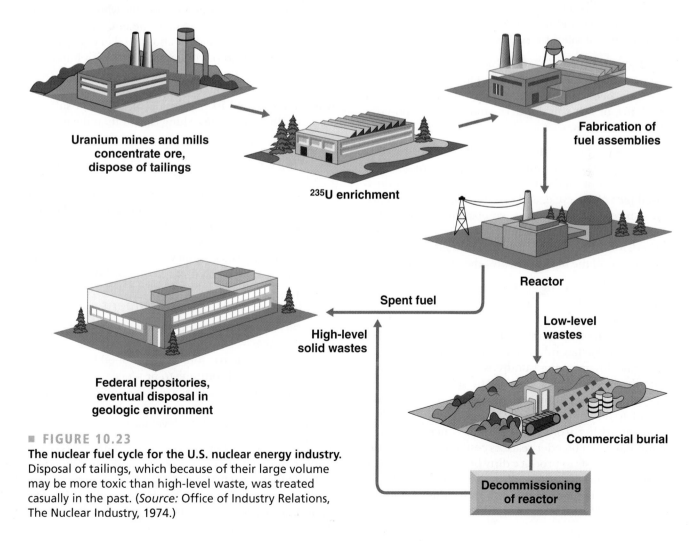

Uranium mines and mills concentrate ore, dispose of tailings

^{235}U enrichment

Fabrication of fuel assemblies

Reactor

Spent fuel

High-level solid wastes

Low-level wastes

Federal repositories, eventual disposal in geologic environment

Commercial burial

Decommissioning of reactor

■ **FIGURE 10.23**
The nuclear fuel cycle for the U.S. nuclear energy industry. Disposal of tailings, which because of their large volume may be more toxic than high-level waste, was treated casually in the past. (*Source:* Office of Industry Relations, The Nuclear Industry, 1974.)

Other Problems Associated with Nuclear Power Plants. We have already discussed the dangers of really big accidents at nuclear power plants. But even without an all-out disaster, there are problems associated with nuclear power plants.

To begin with, there is the question of where to put them. Site selection and construction of nuclear power plants in the United States are extremely controversial—nobody wants a nuclear power plant nearby. In addition, the environmental review process is extensive and expensive, often focused on the probability of such events as earthquakes and other hazards that could cause dangerous structural damage to the plant.

A second disadvantage is that they take longer to restart after a power outage. After the big blackout of 2003 (discussed later), coal plants could be brought back within six to eight hours, but the nuclear plants took as long as two days to restart.[46]

Then there is the problem of where to dispose of its radioactive waste. Just as no one wants a nuclear energy plant nearby, no one wants a nuclear-waste disposal facility nearby. Nuclear waste can remain hazardous for millions of years, and the public lacks confidence in our ability to store it safely for such a long time.

Small amounts of radioactivity have occasionally escaped into the environment. These accidents have generally been caused by human error, mechanical problems, or structural cracks, and have not posed a widespread threat.

Nuclear power plants have a limited lifetime of several decades. Decommissioning a plant (removing it from service) or modernizing a plant is another controversial part of the cycle, and one in which we don't have much experience. Contaminated machinery will need to be disposed of or safely stored to protect the environment. All told, the dismantling of decommissioned reactors may become one of the highest costs for the nuclear industry.[47]

There is also danger in supplying other nations with reactors. Terrorist activity and the possibility of irresponsible people in governments add risks that are not present in other forms of energy production. For example, Kazakhstan inherited a large nuclear-weapons testing facility, covering hundreds of square kilometers, from the former Soviet Union. The soil at several sites has "hot spots" of plutonium that present a serious problem of toxic contamination. There is also a security problem arising from international concern that the plutonium from breeder reactors could be collected and used by terrorists to produce dirty bombs (conventional explosives that disperse radioactive materials). There may even be enough plutonium to produce small nuclear bombs.[47]

10.16 How Are We Dealing with These Problems Today?
Radiation and Health

Most scientists agree that radiation can cause cancer. However, we don't yet know at what point exposure becomes a hazard to health. The effects of radioactivity are measured in units called *sieverts* (Sv). A dose of about 5 sieverts is deadly to 50% of people exposed to it. Exposure to 1 to 2 Sv is enough to cause health problems, including vomiting, fatigue, abortion of early pregnancies, and temporary sterility in males. The maximum allowed dose of radiation per year for workers in industry is 50-thousandths of an Sv (50 mSv), which is about 30 times the average natural background radiation received by people.[48, 49] For the general public, the maximum permissible annual dose (for infrequent exposure) is set in the United States at 5 mSv, which is about three times the annual natural background amount.[48] For continuous or frequent exposure, the limit for the general public is 1 mSv.

Most information about the effects of high doses of radiation comes from studies of people who survived the atomic-bomb attacks in Japan at the end of World War II. We also have information about workers in uranium mines, workers who painted watch dials with luminous paint containing radium, and people treated with radiation therapy for disease.[50] Workers in uranium mines who were exposed to high levels of radiation suffer a significantly higher rate of lung cancer than the general population.

Studies show a delay of 10–25 years between exposure and the onset of disease. Starting about 1917 in New Jersey, some 2,000 young women were employed painting watch dials with luminous paint. To keep a sharp point on their brushes, they licked them. By 1924, dentists in New Jersey were reporting cases of jaw rot; and within five years radium was known to be the cause. Many of the women died of anemia or bone cancer.[50]

Radioactive-Waste Management

Uranium mines and nuclear reactors produce radioactive waste. In the western United States, more than 20 million metric tons of abandoned tailings from uranium mines will continue to produce radiation for at least 100,000 years. Radioactive wastes may be grouped into three general categories: low-level waste, transuranic waste, and high-level waste.

Low-level radioactive waste contains low concentrations or quantities of radioactivity and is not supposed to be a significant environmental hazard if properly handled. However, large deposits of "low-level" radioactive waste contain a huge amount of dangerous materials. Low-level waste includes a wide variety of items, such as waste from chemical processing; solid or liquid plant waste, sludges, and acids; and slightly contaminated equipment, tools, plastic, glass, wood, and other materials.[51]

Low-level waste has been buried in near-surface areas where, it was believed, hydrologic and geologic conditions would keep the radioactivity from migrating.[51] However, monitoring shows that several of these U.S. sites have not provided adequate protection for the environment, and leaks of liquid waste have polluted groundwater. Of the original six burial sites, three had closed prematurely by 1979 due to leaks, financial problems, or loss of license. By 1995, only two government sites for low-level nuclear waste were still operating in the United States, one in Washington and the other in South Carolina. There is also a private facility in Utah run by Envirocare that accepts low-level waste. The public has strongly opposed construction of new burial sites, and there is continuing controversy as to whether low-level radioactive waste can be disposed of safely.[52]

High-level radioactive waste consists of commercial and military spent nuclear fuel; uranium and plutonium from military reprocessing; and other radioactive nuclear-weapons materials. It is extremely toxic, and finding ways to dispose of it has become urgent as the total volume of spent fuel accumulates. At present, in the United States, tens of thousands of metric tons of high-level waste are being stored at more than 100 sites in 40 states, and 72 of the sites are commercial nuclear reactors.[53, 54]

Serious problems have occurred where high-level waste is being stored. Current storage methods, including storage tanks, are at best a temporary solution, and eventually a better disposal program must be designed. Some scientists believe deep bedrock burial can best provide safe containment of high-level radioactive waste. Others have criticized proposals for long-term disposal of high-level radioactive waste underground.[55] A safe underground facility must have a low chance of earthquakes, slow movement of groundwater, and long flow paths to the surface to prevent the spread of radioactive materials dissolved in the groundwater.

Yucca Mountain, Nevada, is a controversial potential storage site. In 1978, the U.S. Department of Energy began studying Yucca Mountain as a possible site for the first long-term U.S. repository of high-level radioactive waste. In 2002 Congress voted to submit a license of application for Yucca Mountain to the Nuclear Regulatory Commission. In 2005 the Environmental Protection Agency (EPA) proposed a radiation standard for the Yucca Mountain site that would allow radiation exposures of 15 millirems near the site for the first 10,000 years after the repository opens and 350 millirems after that. (A standard chest X-ray is about 15 millirems.) Nevada's attorney general called the standard "the least stringent radiation protection standard in the world by far."[56] Nevada asked ten other states to join it in fighting against use of this site, while the federal government has continued to push for it[56] (Figures 10.24 and 10.25).

Some of the scientific questions at Yucca Mountain have concerned natural processes and hazards that might allow radioactive materials to escape, such as surface erosion, groundwater movement, earthquakes, and volcanic eruptions. One of the major questions is, how credible are geologic predictions covering several thousand to a few million years?[55] Unfortunately, there is no easy answer to this question, because geologic processes vary over both time and space. Climates change over long periods, as do areas of erosion, deposition, and groundwater activity. For example, large earthquakes even thousands of kilometers away may permanently change groundwater levels. The earthquake record for most of the United States extends back for only a few hundred years, so estimates of future earthquake activity are highly uncertain.

Bottom line: Geologists can suggest sites, but cannot offer guarantees. Policymakers (not geologists) need to evaluate the uncertainty of predictions in

Courtesy US Dept of Energy

■ **FIGURE 10.24**

Yucca Mountain, Nevada, the proposed location of a deep storage facility for high-level nuclear waste. This site has been extremely controversial. The U.S. Dept. of Energy first began studies of the Yucca Mountain's potential use for radioactive wastes in 1978, but the arguments for and against it still rage.

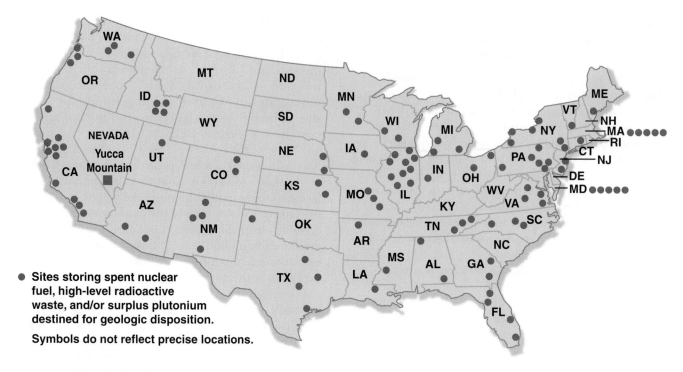

● Sites storing spent nuclear
fuel, high-level radioactive
waste, and/or surplus plutonium
destined for geologic disposition.

Symbols do not reflect precise locations.

▪ **FIGURE 10.25**
Map of the United States showing storage sites of radioactive waste.
(*Source:* U.S. Dept. of Energy. Available at http://www.ocrwm.doe.gov/newsroom/photos/images/
00199dc_012e_72dpi.jpg. Accessed July 5, 2005.)

light of pressing political, economic, and social concerns.[55,56] In the end, we must take great care to ensure that the best possible decisions are made on this important and controversial issue.

10.17 Energy: Storing it, Transporting it, Conserving it

Storing Energy

Not all energy sources provide energy 24/7 all year-round. Wind, water, and sunlight vary in output, so we must find ways to either store the energy or transport it to where it can be used when it is being produced. There are three ways to store energy at present: store it mechanically by moving matter; store it chemically; and store it electrically.

Mechanical storage includes pumping water uphill into a reservoir. Such "pump-storage facilities" are not a popular solution because they leave the landscape ugly—the reservoir is partly empty part of the day, leaving a big ring of barren soil like a ring around a bathtub (Figure 10.26). Another suggestion is a smaller-scale storage system that uses a heavy cylinder spinning at high speed to generate electricity. This has been proposed for cars, but it has some serious impracticalities, such as its weight and what centrifugal force would do if the spinning cylinder broke loose.

Chemical storage has possibilities. One form of chemical storage begins by passing an electric current through water to separate the water into hydrogen and oxygen. The oxygen is released to the atmosphere, but the hydrogen is stored to be burned later. Hydrogen is a high-quality fuel and is clean-burning—burning it just produces water. Hydrogen can be used easily in the same ways we normally use fossil fuels, such as to power automobile and truck engines and to heat water and buildings. Hydrogen can also be used in fuel cells that produce electricity by recombining hydrogen and oxygen to form water. Like natural gas, hydrogen can be transported in pipelines and stored in tanks, but it is very explosive, and to store it efficiently, it must be highly compressed. Even so, the idea of a hydrogen energy economy has been discussed for at least 40 years, and it is gaining popularity.

Iceland's economy may become the first to be based on hydrogen energy. Iceland has no fossil fuels but it does have enormous reserves of geothermal energy

Henry Westheim Photography/Alamy Images

■ **FIGURE 10.26**
Sun Moon Lake, a natural lake in Taiwan, has been converted into a pump-storage facility. This lake is famous for its scenic beauty, and its hotels are often used by honeymooners, but its beauty is marred by a ring of barren ground caused by the pump-storage activity.

that can be used to produce hydrogen for fuel cells. The most important step will be to create the necessary infrastructure for storage, transport, and fueling stations for hydrogen, which, as we said, is as flammable as gasoline.[57]

Once you have hydrogen a wealth of possibilities open up. Using systems something like the big oil refineries, only running them backward chemically, you can combine hydrogen with carbon to make a variety of fuels, such as alcohol, gasoline, and kerosene. This is a likely future. Even when the hydrogen is combined with carbon and the resulting fuel is burned, there is no net addition of carbon dioxide to the atmosphere, so there is no greenhouse effect—another important advantage of alternative energy.

Although batteries store electrical energy, they have serious environmental effects. They are made of heavy metals and acids or bases. Lead acid batteries, the kind in cars, are commonly used in solar houses. But the lead and sulfuric acid in them have posed serious disposal problems for years. The development of more efficient and environmentally friendly batteries is one of the greatest technological needs for the future of off-the-grid use of wind, solar, water, and tidal electric energy.

Transporting Energy

On the grid or off the grid, that is the question. No matter which energy source is used—conventional fuels or alternative sources—putting it on the grid is a major way to transport it. The really big power lines you see when you're traveling around the country are part of the grid, a huge, shared facility built and owned by both public and private entities and governments. Originally set up to respond to power emergencies, it is now used to distribute energy on an everyday basis.

The grid offers great potential for sharing energy from renewable sources. The grid has made it possible for energy generated from conventional sources in one part of the United States to be used in another, and the grid is beginning to transport energy from alternative energy, from solar and wind power. We can build solar energy power plants in the Southwest, where sunlight is often intense, and build wind farms in the Great Plains, Texas, the Northwest, and California, where the wind is strong and steady, and ship that energy to the Northeast, where it is needed.

A disadvantage is the potential for a huge blackout if the grid breaks down. The most recent event of this kind happened on Thursday, August 14, 2003. It was a hot day in much of eastern North America, and the high demand for electricity to run air conditioners overloaded the system. The overload began in the Midwest, and system after system automatically shut down to prevent damage to the equipment. As each system shut down, the overload on the others grew larger, causing more and more shutdowns. Soon 11 nuclear power plants in the United States and 11 in Canada went off-line, along with 80 fossil-fuel plants. The blackout extended from Detroit, Michigan, to the easternmost provinces of Canada and south to New York City. In Detroit, electric pumps ran the water system, so the city's fire department had trouble fighting fires. Major airports shut down, stranding passengers for days.[57]

How much of our electric power should be on the grid and how much off the grid? If you decide to install solar panels on your roof, this is a basic question you will have to answer. But the good news is that whatever method we use to generate electricity, it seems clear that there are adequate means to store and transport it, and this will not be a limit to the use of alternative energy.

Is hydrogen the answer? The pros and cons. There are many ways to transport energy. One is to make electricity, then use that electricity to dissociate water—that is, convert the water into free hydrogen and free oxygen—and then use the hydrogen as a fuel. The idea of a hydrogen energy economy has been talked about for decades. In addition to providing a way to transport energy in small containers rather than through an electrical grid, it has the advantage that many sources of energy can produce the same fuel. The primary disadvantage is that hydrogen is a light and explosive gas, and to transport it efficiently and safely it must be compressed and put into thick-walled containers. An alternative is to use the free hydrogen to make a small hydrocarbon compound, such as alcohol or gasoline, and transport this, as it is currently transported. The disadvantage is that each chemical step from hydrogen to another compound uses some energy and reduces the efficiency.

Fuel-cell-powered cars, which are frequently talked about, use hydrogen as a fuel and free oxygen as the oxidizing agent. Thus, a fuel-cell car is simply an efficient and clean way to transport energy. It is not a new source of energy, and ultimately the total pollution from a fuel-cell car depends on how polluting the original energy source was. There seems to be no doubt that as fossil fuels grow scarcer, some kind of hydrogen or small hydrocarbon will become the energy "currency" of the future.

Conserving Energy by Using It More Efficiently

The first rule is a lesson from the ancients: Minimize your need for energy. When you have to cut your own wood or collect dung to burn for heating and cooking, you quickly learn ways to make your fuel last longer—what today we call "energy conservation."

We can learn to use energy more efficiently. For example, *cogeneration* enables us to use waste heat from generating electricity—instead of simply releasing the heat into the environment, we use it to heat water. Total U.S. energy consumption leveled off in the early 1970s. Nevertheless, U.S. production of goods continued to grow. The reason we have had higher productivity with less energy use is that more industries are using cogeneration and more energy-efficient machinery.

▪ **FIGURE 10.27**
Green buildings.
Completed in spring 2004, China's first energy-efficient "green" building, which features a rooftop garden and solar panels, is the product of cooperation between the China Ministry of Science and Technology and the U.S. Department of Energy. Solar energy panels.

We can also design residential and office buildings that use sun and wind not only to decrease the use of fuels but also to create a more pleasant environment for living or working (Figure 10.27).[58]

Our modern way of life increases the use of energy for heating and cooling. Modern cities are warmer in summer than rural surroundings, in part because the dark, tarred streets and rooftops in cities, along with the lack of vegetation, raise a city's summer temperature an average of 5°F. The amount of increase was measured in five major U.S. cities—Los Angeles; Washington, DC; Phoenix and Tucson, Arizona; and Colorado Springs, Colorado—and it was found that energy demand increases 1–2% for each 1°F rise in daily maximum temperature.

Here's where vegetation enters the picture. If we planted just three more shade trees for one-half of the single-family homes in the United States, that would reduce late-afternoon air temperatures on a hot summer day by 5–10°F and reduce electricity use by 50 billion kilowatt hours per year—2% of the total U.S. annual electricity use (Figure 10.28).

The downside is that buildings constructed to conserve energy by having very tight and thick insulation are more likely to develop indoor air pollution, which is emerging as one of our most serious environmental problems (see Chapter 14). Another negative is that construction that incorporates environmental principles may be more expensive.

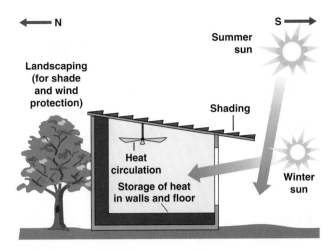

■ **FIGURE 10.28**

Trees cool a house in summer, reducing need for air-conditioning.

Until the Industrial Revolution, with its wealth of cheap energy, many societies built houses facing south and designed to collect heat from the sun in the winter. Deciduous trees planted on the south side allowed sunlight to warm the house in winter but shaded the house in summer.

Fuel-efficient automobiles have been developed. Hybrid (gasoline-electric) vehicles are getting 40–50 miles per gallon on the highway thanks to their smaller size and lighter-weight, more efficient engines. Unfortunately, about a quarter of the vehicles sold are SUVs and light trucks. Their large size not only increases their fuel requirements but also increases the amount of damage they can inflict in a collision with smaller vehicles. The number of serious accidents between cars and trucks in the U.S. has risen.

Some steps we can take to conserve energy: (1) Limit fuel used for daily travel by living closer to school or work and walking, bicycling, taking a bus or train, carpooling, or purchasing a hybrid vehicle; (2) turn off lights when leaving rooms and use compact fluorescent lamps (CFLs), which reduce electricity used for lighting by one-third; (3) take shorter showers to conserve hot water; (4) put on a sweater and turn down the thermostat in winter; (5) choose energy-efficient appliances and use them more efficiently (for example, a refrigerator placed beside a stove will need more energy to keep cool); and (6) switch to solar power in homes and offices.

Return to the Big Question

Can we assure a sustainable supply of energy?

Yes, the technology is at hand, and the most environmentally benign renewable, sustainable sources are solar and wind. Water power is also renewable but has more environmental effects. Geothermal energy can be sustainable for a long time but perhaps not indefinitely. While conventional nonbreeder nuclear-fission reactors are not sustainable, breeder reactors could provide energy for hundreds of years, and if a practical fusion reactor is ever invented, it could be a sustainable source of power. However, nuclear power plants are the most environmentally dangerous, while wind and solar energy cause the fewest environmental problems. And one of the most effective ways to ensure a sustainable supply of energy is by using energy more efficiently—getting more miles per gallon, more light per watt.

Summary

- Energy is the ability to do work, to move matter. Abundant energy is the foundation for much of what makes modern civilization pleasant and powerful, and allows us to do many things that most people throughout the history of our species were unable to do.

- Fossil fuels became the primary source of abundant energy in the 20th century and have made possible much that we enjoy about the modern world, but fossil fuels are also major sources of pollution. At some point, these fuels will run out, at least in an economically useful sense.

- Alternative energy sources are those that are not fossil fuels. They include energy from the sun (wind, direct solar, water, tidal, and biomass) and geothermal (deep Earth heat) and nuclear energy sources.

- Solar energy has the fewest undesirable environmental effects. Wind is a close second. Water power is clean in terms of pollution but alters aquatic habitats, affecting fish and wildlife and landscape beauty.

- Nuclear power is the most dangerous to the environment and to human health.

Key Terms

acid mine drainage

alternative energy source

conventional energy source

energy

fission

fossil fuels

fusion

geothermal energy

half-life

isotopes

nonrenewable energy

nuclear energy

nuclear fuel cycle

nuclear reactors

photovoltaic

radioactive decay

radioactive waste

radioisotope

renewable energy

reserve

resource

solar energy

strip mining

tidal power

work

Getting It Straight

1. Which energy resources used in centuries past are still useful today?

2. Why are petroleum products and coal called "fossil fuels"? Why are they considered to be nonrenewable resources?

3. How do the benefits and drawbacks of oil, natural gas, and coal compare?

4. What fossil fuel would you choose to use to (a) reduce local air pollution; (b) minimize global warming effects?

5. How is solar energy used to heat space and water, and to produce electricity?

6. How do the benefits and drawbacks of photovoltaic solar energy and wind energy compare?

7. Describe two ways that water can be used to generate electricity.

8. Burning biomass energy resources releases CO_2 into the atmosphere. Why then are these resources considered to be neutral in their impact on global warming?

9. How is a boiling water nuclear reactor like a coal-fired boiler? What are the benefits and drawbacks of nuclear energy?

10. If exposure to radiation is a natural phenomenon, why are we worried about it?

11. Why is hydrogen considered to be a way to store energy? Describe how energy can be stored in, and subsequently released from, hydrogen. What are the potential environmental impacts of this cycle?

12. List and describe several ways that the United States, your state, and you and your family can each conserve energy.

13. Which energy resources could be useful for each of the following? (a) transportation; (b) heating a building; (c) producing electricity; (d) lighting a building.

What Do You Think?

1. Under what conditions is it correct to say that an electric car does not pollute the local environment? The global environment?

2. Consider oil and coal. Which do you think is more damaging to the environment, and why? If you were a leader of the nation, what actions would you recommend to reduce environmental harm caused by these energy resources? What additional considerations would be important to consider?

3. Government and business leaders in your region propose a new electricity-generating power plant. What factors should they consider in deciding whether to select a coal-fired or nuclear plant design?

Pulling It All Together

1. Suppose you are the head of FEMA (the Federal Emergency Management Agency). What kind of energy system would you design to make available following a major catastrophe, such as Hurricane Katrina in New Orleans in 2005, or a major earthquake? Include in your discussion the sources of energy and methods of transmission and storage.

2. How would the energy system you designed in question 1 (above) compare with an energy system that would be best on a large scale to counter effects of global warming? Include in your discussion the sources of energy and methods of transmission and storage.

3. Design the least damaging and most sustainable energy system possible for a midwestern United States livestock and grain farm. How would this system differ from one you would design for an agricultural village in a remote, rural area of India? Include in your discussion the sources of energy and methods of transmission and storage.

Further Reading

Boyle, G. 2004. *Renewable energy: Power for a sustainable future.* New York: Oxford University Press.—An overview of the entire renewable energy field.

McDonough, & M. W., and Braungart. 2002. *Cradle to cradle: Remaking the way we make things.* North Point Press New York.

Perlin, J. 1999. *From space to Earth.* Ann Arbor, MI: Aatec Publications.—An interesting history about solar energy.

Wald, M. 2003. Dismantling nuclear reactors. *Scientific American,* March, pp. 60–69.—This is an in-depth discussion of steps in dismantling a nuclear power plant and some of the unforeseen difficulties.

James Phelps/IStockphoto

Water and Environment

Big Question

Can We Maintain Our Water Resources for Future Generations?

Learning Objectives

Although water is one of the most abundant resources on Earth, many important issues are involved in water management and pollution. After reading this chapter, you should understand . . .

- why sustainable water management will become more difficult as the demand for water increases;

- how a growing global water shortage is linked to our food supply;

- how dams, reservoirs, and canals affect the environment;

- why wetlands are important;

- what the major categories of water pollutants are;

- that polluted drinking water is a major problem in many places around the world;

- the various methods of wastewater treatment, and why some are environmentally preferable to others.

Case Study

The Colorado River: Water Resources Management, Water Pollution, and the Environment

The history of the Colorado River illustrates linkages among physical, biological, and social systems that are at the heart of environmental science. The Colorado originates in the Wind River Mountains of Wyoming and, in its long journey to the Gulf of California, flows through some of the most spectacular scenery in the world (see opening photograph and Figure 11.1a). About 800 years ago, Native Americans living in the Colorado River basin built a sophisticated water-distribution system. In the 1860s, settlers cleared the debris from these early canals and used them once again for irrigation.[1]

Considering its size, the Colorado River has only a modest flow. However, it is one of the most regulated and controversial bodies of water in the world. The Colorado River Compact of 1922 apportioned the river's total flow among various users, including seven U.S. states and Mexico, but allocated no water for environmental purposes. At that time, the concept of sustainable water management was not considered.

Today, Colorado River water only occasionally flows into the Gulf of California—it's stored in reservoirs created by dams and is used upstream. As a result, ecosystems of the lower river and delta, deprived of water and nutrients, have been damaged. The delta has shrunk to less than half its previous size, damaging fish populations and forcing some native people who depended on fishing to move away.

Whit Richardson/Aurora/Getty Images Inc.

(a)

(b)

■ **FIGURE 11.1**

Grand Canyon of the Colorado River.
(a) River rafters camping on a sandbar of the river, which the rafters call a beach. (b) Idealized diagram of how the sandbars may be produced by large water releases below the Glen Canyon Dam. At low flow sand is stored in the channel bottom (top). During release of water (middle) sand in suspension is moved to higher levels and deposited. After the flow has dropped the sand is left as sandbars on the side of the channel (bottom).
(*Source:* U.S. Geological Survey accessed 2/4/07 @water.usgs.gov. Controlled flooding of the Colorado River in Grand Canyon: The Rationale and Data-Collection Planned).

The complex issues of water management for the Colorado River illustrate major problems that are likely to be faced by other semiarid regions of the world in coming years: How shall we allocate scarce water resources? How can we best control water quality? How can we protect river ecosystems? There are no easy answers to these questions.

The two largest reservoirs on the river, Hoover Dam and Glen Canyon Dam, hold about 80% of the total water stored in the basin (Figure 11.2). With careful management, their total storage could meet user needs for several years. However, if there are several years of drought, maintaining a sufficient water supply for all users may not be possible.

The Glen Canyon Dam was completed in 1963. From a hydrologic viewpoint, the Colorado River has been changed by the dam. The river has been tamed. The higher flows have been reduced, the average flow has increased, and the flow changes often because of fluctuating needs to generate electrical power. Changing the hydrology of the river has also changed other aspects. For one thing, the rapids have changed, because large boulders delivered from tributaries can no longer be transported. Other changes include the distribution of sediments that form sandbars, called beaches by rafters (Figure 11.1a), and the vegetation near the water's edge.[2] The sandbars, which are valuable wildlife habitats, shrank in size and number following construction of the dam because sediment that would have moved downstream to nourish them was

trapped in the reservoir. All these changes affect the Grand Canyon, which is downstream from the dam.

A record snowmelt in the Rocky Mountains in June 1983 forced about three times the usual amount of water from the Glen Canyon Dam. The resulting floods scoured the riverbed and banks, releasing sediment that replenished sandbars and breaking off some vegetation that had taken root.[3] This release of water was beneficial to the river environment and highlighted the importance of floods in maintaining the system in a more natural state. Natural disturbances are a necessary part of the river ecosystem if it is to function on a sustainable basis.

As an experiment, about half the amount of water of the June 1983 flood was deliberately released for a week in 1996. Between March 26 and April 2, water was allowed to flow at full flood. The flow was reduced for the last two days of the experiment to redistribute the sand. The flood created 55 new sandbars and made 75% of the existing sandbars bigger than they had been. It also helped rejuvenate marshes and backwaters, which are critical habitats for native fish and some endangered species.[3] The process of how the sandbars are produced is illustrated on Figure 11.1b.

This experimental release of high flows of water marked a turning point in river management—it was the first time the U.S. government had opened the floodgates of a dam to improve a river's ecosystem. Although some scientists were concerned that the flooding was insufficient and too brief, the flood was hailed as a success. However, it will take some time to see how long the results last. It is hoped that what was learned will help to restore the environments and ecosystems of other rivers affected by dams.[3, 4]

Conflicts over the use and quality of Colorado River water have gone on for decades and extend beyond the river basin to urban centers and agricultural areas in California, Colorado, New Mexico, and Arizona. The need for water in these semiarid regions has led to the overuse of limited water supplies and the deterioration of water quality.

The Colorado River is just one illustration of a larger problem: Water is a critical, limited, renewable resource in many regions on Earth. This chapter discusses our water resources in terms of supply, use, pollution management, and sustainability.

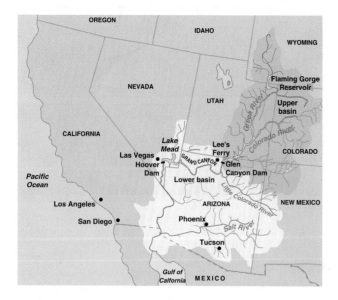

▪ **FIGURE 11.2**
The Colorado River basin.
The river begins in the mountains of Wyoming, flowing south to Arizona, west to near Las Vegas, Nevada, then south to the Gulf of California. Major dams and reservoirs in the lower river are Glen Canyon Dam upstream of the Grand Canyon, and Hoover Dam downstream of the canyon.

11.1 Water

Water is a unique liquid—without it, life as we know it is impossible. Consider the following:

Water has a high capacity to absorb and store heat. The capacity of water to hold heat has great significance for Earth's climate. Solar energy warms the oceans of the world, storing huge amounts of heat. The heat can be transferred to the atmosphere to develop hurricanes and other storms.

Water is the universal solvent. Because many natural waters are slightly acidic, they can dissolve a wide variety of compounds, from simple salts to minerals, including sodium chloride (common table salt) and calcium carbonate (calcite) in limestone rock.

Solid water is lighter than liquid water. Among the common compounds, water is the only one whose solid form (ice) is lighter than its liquid form. Water expands by about 8% when it freezes, becoming less dense. That is why ice floats. If ice were heavier than water, it would sink to the bottom of rivers and lakes, and they might freeze solid. In the ocean, sea ice would not form at the surface. As a result, the biosphere would be vastly different from what it is—and life, if it existed at all, would be very different.[5]

Sunlight can penetrate water to variable depths, permitting photosynthetic organisms (living things that require sunlight) to live below the surface.

A Brief Global Perspective

We face a growing global water shortage linked to our food supply. To better understand the problem, you will need to recall our discussion of the global hydrologic cycle in Chapter 3.

The main process in the cycle is the global transfer of water from the oceans to the atmosphere to the land and oceans and back to the atmosphere (Figure 11.3). The water cycle, while simple in concept, includes many different processes related to precipitation, evaporation,

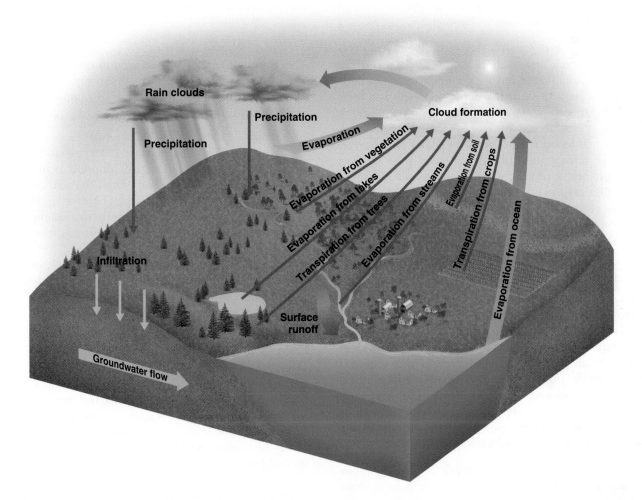

■ **FIGURE 11.3**
The hydrologic cycle, showing important processes and transfer of water. Most of the terms—such as *precipitation, runoff, infiltration,* and *groundwater*—are self-explanatory or already familiar to you. *Transpiration* is the release of water vapor into the atmosphere by vegetation. Trees, for example, take water from the soil, and release water from their leaves by transpiration. (*Source:* Modified from Council on Environment Quality and Department of State. The global 2000 report to the President. Vol. 2 Washington, DC.

☐ TABLE 11.1 THE WORLD'S WATER SUPPLY (SELECTED EXAMPLES)

Location	Surface Area (km²)	Water Volume (km³)	Percentage of Total Water	Estimated Average Residence Time of Water
Oceans	361,000,000	1,230,000,000	97.2	Thousands of years
Atmosphere	510,000,000	12,700	0.001	9 days
Rivers and streams	—	1,200	0.0001	2 weeks
Groundwater (shallow to depth of 0.8 km)	130,000,000	4,000,000	0.31	Hundreds to many thousands of years
Lakes (fresh water)	855,000	123,000	0.01	Ten of years
Ice caps and glaciers	28,200,000	28,600,000	2.15	Tens of thousands of years and longer

Source: U.S. Geological Survey.

transpiration (water vapor released into the atmosphere by plants), and infiltration and seepage of water into and out of the solid Earth. Table 11.1 lists the relative amounts of water in the major storage compartments of the cycle. Notice that more than 97% of Earth's water is in the oceans; the next-largest storage compartment—the ice caps and glaciers—accounts for another 2%.

Most of Earth's water is unusable for us. The oceans, ice caps, and glaciers account for more than 99% of the total water, but are generally unsuitable for human use because of their salinity (seawater) and their location. Only about 0.001% of the total water on Earth is in the atmosphere at any one time. However, this relatively small amount of water in the global water cycle, with an average atmosphere residence time of only about nine days, produces all our freshwater resources through the process of precipitation. In short, the amount of water for which all the people, plants, and animals on Earth's continents and islands compete is much less than 1% of the total.

Compared with other resources, water is used in very large quantities. In recent years, the total mass (or weight) of water used on Earth per year has been approximately 1,000 times the world's total production of minerals, including petroleum, coal, metal ores, and nonmetals.[6] Because of its great abundance, water is generally an inexpensive resource. In the semiarid southwestern United States, where there are large reservoirs on the Colorado River, government subsidies and water programs have kept the cost of water artificially low. The dams and reservoirs were built with the promise of greening the land with inexpensive water.

Can we avoid water shortages? Because the quantity and quality of water available at any particular time are highly variable, water shortages have occurred and will probably continue to occur with increasing frequency. Such shortages can lead to serious economic disruption and human suffering.[7] In the Middle East and northern Africa, competition for scarce water has also caused trou-

ble between countries, and future wars over water are a possibility. The U.S. Water Resources Council estimates that water use in the United States by the year 2020 may exceed surface water resources by 13%.[7] Therefore, an important question is, how can we best manage our water resources, use, and treatment to maintain adequate supplies?

Water Sources

There are two main sources: groundwater and surface water. In some places where freshwater is scarce, ocean water may be treated to remove the salt, a process known as **desalination.**

Groundwater, as the name implies, is water in the ground. The upper surface of the groundwater is called the *water table*, below which openings in soil or rock are filled with water. Sometimes the water table is so far underground that you have to drill a deep well to get water. Other times—for example, in a swamp—the water table is right on the surface and you'll get your feet wet walking through it.

An *aquifer* is an underground layer of rock, sand, or gravel that contains usable groundwater and releases it in significant amounts. Figure 11.4 shows the major features of a groundwater and surface water system.

Surface water consists of streams, rivers, and lakes. Streams may be classified as *effluent* or *influent* (Figure 11.4). In an effluent stream, groundwater seeping into the stream channel from below the surface maintains the stream's flow during the dry season. A stream that flows all year is called a *perennial* stream. An influent stream is above the water table everywhere along the channel and flows only in response to precipitation (rain or snow). Water from an influent stream seeps down into the subsurface. A stream that only flows during or shortly after rainfall or during snowmelt may be called ephemeral.

A stream may have both perennial and ephemeral reaches. A stream may have *intermittent reaches* that have a combination of influent and effluent flow, varying with the

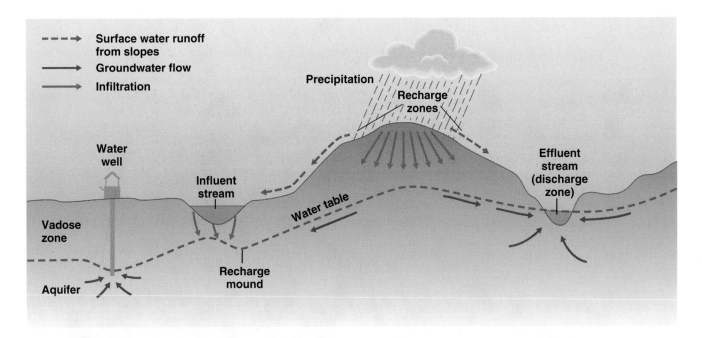

■ FIGURE 11.4
Groundwater and surface water flow system.
Precipitation infiltrates soil and rock or runs off at the surface. *Effluent* streams receive their low flow from groundwater seepage. *Influent* streams flow in response to surface runoff, and water seeps from the channel to recharge groundwater.

time of year. For example, streams flowing from the mountains to the sea in southern California often have reaches in the mountains that are perennial, supporting populations of trout or endangered southern steelhead, and further down stream intermittent reaches that change into ephemeral reaches. At the coast, these streams may receive groundwater and tidal flow from the ocean with a perennial lagoon.

Surface water and groundwater are parts of the same resource. Nearly all surface water environments, natural or human-made, such as reservoirs, have strong linkages with groundwater. For example, pumping groundwater from wells may reduce stream flow, lower lake levels, or change the quality of surface water. Lowering the groundwater level may turn a perennial stream into an intermittent or *influent* stream. Similarly, withdrawing surface water by diverting streams and rivers can deplete groundwater or change the quality of groundwater—for example, by increasing the concentrations of dissolved chemicals in the groundwater. Finally, pollution of groundwater may result in pollution of surface water, and vice versa.[8]

Figure 11.5 shows how surface water and groundwater interact in a semiarid urban and agricultural environment. Urban and agricultural runoff increase the amount of water in the reservoir. Pumping groundwater for agricultural and urban uses lowers the groundwater level. The

quality of both surface water and groundwater suffers from urban and agricultural runoff, which adds nutrients from fertilizers, oil from roads, and nutrients from treated wastewaters to streams and groundwater.

Desalination

Desalination is the process of turning seawater into freshwater. Seawater is about 3.5% salt, which means each cubic meter of seawater contains about 40 kilograms (88 pounds) of salt. Desalination, a technology to remove salt from water, is being used at about 15,000 plants around the world to lower the salt content of the water to about 0.05%, the level at which the water can be used as a freshwater resource. A large desalination plant can produce millions of gallons of water per day. It also produces a lot of salt that must be disposed of.

Desalinated water is getting less expensive, but still costs considerably more than traditional water supplies in the United States. Desalinated water has a *place value*, which means that the price rises quickly depending on how far the water must be moved from the plant. Moreover, because the various processes that remove the salt require large amounts of energy, the cost of the water is also tied to ever-increasing energy costs. For these reasons, desalination is used when other water sources are not available.

Desalination also has environmental impacts. When salty water is treated to remove the salt, a lot of salty

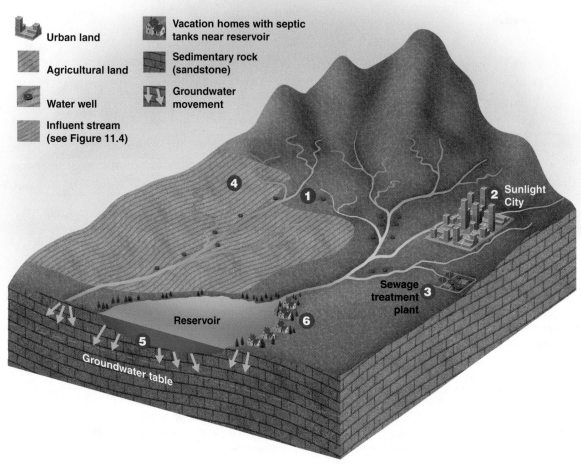

■ FIGURE 11.5

Some interactions between surface water and groundwater for a city in a semiarid environment near agricultural land and a reservoir.
(1) Water pumped from wells lowers the groundwater level. (2) Urbanization increases runoff to streams. (3) Sewage treatment discharges nutrient-rich waters to stream, groundwater, and reservoir. (4) Agriculture uses irrigation water from wells, and runoff to streams from fields contains nutrients from fertilizers. (5) Reservoirs allow water to seep into the ground, recharging groundwater. (6) Lakeside homes with septic systems add water to the soil and rocks and sometimes pollute water resources.

wastewater is produced. Discharging very salty water from a desalination plant into another body of water, such as a bay, may increase the bay's salinity and kill some plants and animals that can't tolerate more salt. The discharge from desalination plants may also cause wide variation in the salt content of local environments, which may damage ecosystems.

11.2 Water Supply

The water isn't always where you need it most.
The water supply anyplace on the land depends on several factors in the hydrologic cycle, including the rates

of precipitation, evaporation, and transpiration (the release of water vapor into the atmosphere by plants through pores in their leaves and stems), as well as stream flow and groundwater flow. The total average annual water yield (runoff) from Earth's rivers is approximately 47,000 cubic kilometers, but it is distributed far from equally (Table 11.2). Some runoff occurs in places where few people live, such as Antarctica, which produces about 5% of Earth's total runoff. South America, which includes the relatively uninhabited Amazon basin, provides about one-fourth of Earth's total runoff. North America's total runoff is about two-thirds that of South America, and, unfortunately, much of the North American runoff occurs in sparsely populated or uninhabited

□ TABLE 11.2 ANNUAL WATER BUDGETS FOR THE CONTINENTS[A]

Continental	Precipitation mm/yr	Precipitation km³	Evaporation mm/yr	Evaporation km³	Runoff km³/yr
North America	756	18,300	418	10,000	8,180
South America	1,600	28,400	910	16,200	12,200
Europe	790	8,290	507	5,320	2,970
Asia	740	32,200	416	18,100	14,100
Africa	740	22,300	587	17,700	4,600
Australia and Oceania	791	7,080	511	4,570	2,510
Antarctica	165	2,310	0	0	2,310
Earth (entire land area)	800	119,000	485	72,000	47,000[b]

[A] *Precipitation − evaporation = runoff.* [b] *Surface runoff is 44,800; groundwater runoff is 2,200.*
Source: I. A. Shiklomanov, "World Fresh Water Resources," in P. H. Gleick, ed., Water in Crisis (New York: Oxford University Press, 1993), pp. 3–12.

regions, particularly in the northern parts of Canada and Alaska.

How useful is the water that falls as precipitation? Approximately 10% of the water vapor passing over the United States every day falls as precipitation—rain, snow, hail, or sleet. Approximately two-thirds of the precipitation evaporates quickly or is transpired by vegetation. The remaining one-third enters the surface water or groundwater storage systems, flows to the oceans or across the nation's boundaries, is used by people, or evaporates from reservoirs. Due to natural variations in precipitation that cause either floods or droughts, only a portion of this water can be developed for intensive uses (only about 50% is considered available 95% of the time).[7]

How much water do people use? In the United States, the average person uses about 100 gallons a day. That's much more than people use in the rest of the world. Europeans use about one-half of that, and in some regions, such as Sub-Saharan Africa, people make do with just 5 gallons a day.

To put this in perspective, consider just the water in the Missouri River. In an average year, the water that flows down the Missouri River is enough to cover 25 million acres a foot deep—8.4 trillion gallons. If each American uses 100 gallons a day, the Missouri's flow is enough to supply domestic water and public water in the United States for about 230 million people. If we conserved water and reduced our per-capita use, the Missouri's flow could hypothetically be equivalent by volume to the water used for all the people in the U.S.

Groundwater is popular for drinking, but can be expensive. Nearly half the people in the U.S. use groundwater as their primary source of drinking water.

It accounts for approximately 20% of all water used in the United States. However, though the total amount of groundwater available in the United States is enormous, the high cost of pumping it limits the amount that can be economically obtained.[7]

Another problem is *overdraft*—**taking more groundwater out than is naturally replaced.** In many parts of the country, groundwater withdrawal from wells exceeds the natural inflow. We can think of water as a nonrenewable resource that is being *mined*. This can lead to a variety of problems, including damage to river ecosystems and land *subsidence* (cave-ins). Groundwater overdraft is a serious problem in the Texas–Oklahoma–High Plains area (which includes much of Kansas and Nebraska and parts of other states), as well as in California, Arizona, Nevada, New Mexico, and isolated areas of Louisiana, Mississippi, Arkansas, and the south Atlantic region.

In the Texas–Oklahoma–High Plains area, the overdraft amount per year is about equal to the natural flow of the Colorado River for the same period.[7] The Ogallala aquifer is the main groundwater resource in this area. Although the aquifer holds a tremendous amount of groundwater, it is being used in some areas at 20 times the rate at which it is being naturally replaced. The water table in many parts of the aquifer has declined in recent years (Figure 11.6), which has lowered the yields from wells and raised the costs of energy for pumping the water. The most severe water-depletion problems in the Ogallala aquifer today are in places where irrigation was first used in the 1940s. There is concern that eventually a significant portion of land that is now being irrigated will be returned to dryland farming as the available water is used up. Some cities and towns from Kansas to Texas are also facing water shortages as a result of overdraft.

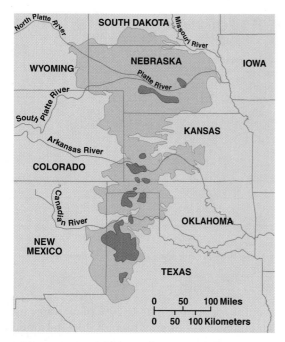

Water level changes from predevelopment since 1940

■ Declines of over 15 m (50 ft.)

▨ Changes between −15 m & +3 m (−50 ft. and +10 ft.)

■ Rises over 3 m (10 ft.)

■ **FIGURE 11.6**
Groundwater overdraft.
Changes in the groundwater level as a result of pumping in the Texas–Oklahoma–High Plains region. (*Source:* U.S. Geological Survey.)

11.3 Off-Stream and In-Stream Water Use

In discussing water use, it is important to distinguish between off-stream and in-stream uses.

Off-stream use refers to removing water from its source and using it elsewhere. Much of this water is returned to its source afterward. For example, water that is used as a coolant in industrial processes gets warm and may go to cooling ponds and then be discharged to a river, lake, or reservoir. In contrast, *consumptive use,* as the term implies, is an off-stream use in which water is consumed—by living things or by being used in industrial processes—and therefore is not returned to its source.[7]

In-stream use refers to using water right where it is, for such things as navigation, hydroelectric power generation, fish and wildlife habitats, and recreation. These multiple uses usually cause arguments among users because their needs often conflict. Fish and wildlife, for example, require certain water levels and flow rates for maximum biological productivity. These levels and rates will differ from those needed for generating hydroelectric power. In-stream uses of water for fish and wildlife will likely also conflict with shipping and boating. Figure 11.7 shows some of these conflicting demands on a graph. As can be seen, the preferred volume of water flowing per second (called *discharge*) is unchanging for navigation. Some fish, however, prefer higher flows in the spring for spawning, and recreational users prefer high flows in summer. Hydroelectric production can cause wide daily changes in discharge.

Another problem for in-stream use is off-stream use. That is, how much water can be diverted from a stream or river without damaging the stream's ecosystem. In the Pacific Northwest, diverting water for agriculture caused a large die-off of salmon in the Klamath River. Another example involved Mono Lake, the largest lake lying entirely in California. The lake has no outlets and is salty (lakes with no outlets become salty because water that evaporates from a lake leaves salts behind). The water that Mono Lake loses to evaporation is replenished by streams from the Sierra Nevada. However, the city of Los Angeles for years diverted the stream water for urban use, and lake levels dropped, endangering birds, brine shrimp, and brine flies. A program to save the lake resulted in a management plan that stopped or reduced water diversion, preserving the lake. The Aral Sea has not been so fortunate.

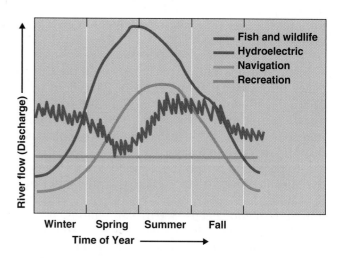

■ **FIGURE 11.7**
Water used in a stream.
In-stream water uses and optimal discharges (volume of water flowing per second) for each use. Discharge is the amount of water passing by a particular location and is measured in cubic meters per second. Obviously, all these needs cannot be met simultaneously.

The Aral Sea in Kazakhstan and Uzbekistan is a victim of diversion. It serves as a wake-up call about the environmental damage that can be caused by diverting water for agricultural purposes. Diversion of water from the two rivers that flow into the Aral Sea has transformed one of the world's largest inland bodies of water from a vibrant ecosystem into a dying sea.

Today, the Aral Sea's shoreline is surrounded by thousands of square kilometers of salt flats that formed as the sea's surface area shrank by about 40% over the past 40 years (Figure 11.8). The volume of water in the lake shrank by more than 50%, turning the Aral Sea into two separate bodies of water, and the salt content of the water increased, killing more fish, such as sturgeon, which are important to the region's economy. Dust raised by winds from the dry salt flats is causing a regional air-pollution problem, and the climate in the region has changed as the moderating effect of the sea has been reduced. Winters have grown colder and summers warmer. Fishing centers such as Muynak in the south and Aralsk to the north, which used to be on the shore,

are now many kilometers inland. Loss of fishing and the decline of tourism have damaged the local economy.[9]

The drying-up of the Aral Sea is considered to be one of the worst ecological disasters on Earth. But just when all seemed lost, there is new hope to restore at least part of it. The northern (small) body of water (Figure 11.8) was separated in 2005 from the southern (larger) part by a long dam, which looks like a long elevated gravel road. The dam has allowed water from the Syr Darya River to flow into the North Aral Sea, raising the water level and making it less salty. The ecology of that portion of the Aral Sea has improved, as has fishing. The southern portion, however, continues to dry up.

Transport of Water

Moving water to where it is needed is not a new idea. Ancient civilizations, including the Romans and Native Americans, built canals and aqueducts to transport water from distant rivers to where it was needed. Today, in many parts of the world, people are demanding that rivers supply water to agricultural and urban areas. As in the past, water is often moved long distances from areas with abundant rainfall or snow to areas that use a lot of water (usually agricultural areas). For instance, in California, two-thirds of the state's runoff occurs north of San Francisco, where there is a surplus of water. However, two-thirds of the water used in California is used south of San Francisco, where there isn't enough. In recent years, canals of the California Water Project have moved tremendous amounts of water from the northern to the southern part of the state, mostly for agricultural use but increasingly for urban uses as well.

New York City has imported water for more than 100 years. Water use and supply in New York City show a repeating pattern. Originally, local groundwater, streams, and the Hudson River itself were used. However, as the population increased and the land was paved over, surface waters were diverted to the sea rather than percolating into the soil to replenish groundwater. Furthermore, what water did infiltrate the soil was polluted by urban runoff. Water needs in New York exceeded local supply, and in 1842 the first large dam was built.

As the city grew, so did its water needs. As the city rapidly expanded from Manhattan to Long Island, its water needs increased. The shallow aquifers of Long Island were at first a source of drinking water, but this water was used faster than rainfall could replenish it. At the same time, the groundwater became contaminated with urban and agricultural pollutants and with saltwater seeping in underground from the ocean. A larger dam was built at Croton, some 30 miles from the city, in 1900, but further expansion of the population created the same pattern: initial use of groundwater, followed by pollution, salinization, overuse of the resource, and the

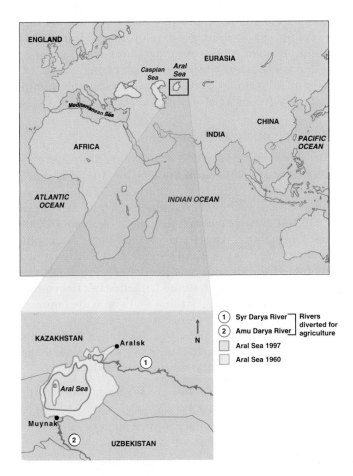

■ **FIGURE 11.8**

The Aral Sea is drying up and dying as a result of diversion of water for agriculture. [Courtesy of Philip P. Micklin.]

building of new, larger dams farther and farther upstate in forested areas.

Today, a lot of New York's water comes from upstate forests. The forests of the Catskill Mountains in upstate New York provide water to about 9 million people in the city.[10] The total contributing area in the forest is about 5,000 square kilometers (2,000 square miles), of which New York City owns less than 8%. The water from the Catskills was historically of very high quality and in fact was described as one of the largest municipal water supplies in the United States that did not require extensive treatment by filtering plants.

Catskills water can be filtered very effectively by natural processes. Water from rain and melting snow moves down through the soil into the rocks below. Some of this groundwater emerges to feed streams that flow into reservoirs and eventually through three large tunnels beneath the city and Manhattan Island. Two of the tunnels are over a century old, and one is still being completed. During its journey, the water undergoes a number of natural processes that treat and filter it. You can think of these as natural services that the Catskill forest ecosystem provides to the people of New York.

But these natural services may be overwhelmed by uncontrolled development in the watershed. A particular concern was runoff from buildings and streets, as well as seepage from septic systems that treat wastewater from homes and buildings, partly by allowing wastewater to seep through soil. At that time the Environmental Protection Agency warned the city that unless the water quality improved, New York City would have to build a water-treatment plant to filter the water. The estimated cost of such a facility is about $7 billion dollars, and the annual operating expense would be several hundred million dollars.[10]

New York City decided on sewage treatment rather than water treatment. Choosing to improve the water quality at the source, New Yorkers built a sewage-treatment plant upstate in the Catskill Mountains at a cost of about $2 billion. This seems very expensive, but it is only about one-third of the cost of building the treatment plant to filter water. The city chose to invest in the "natural capital" of the forest, hoping it would continue its natural service function of providing clean water. It is still a bit early to tell whether the gamble will work in the long term. The answer to that question will probably take several decades.[10]

The Catskill Mountain Forest Ecosystem offers other benefits as well. These benefits come directly from recreational activities, particularly trout fishing, which is a multibillion-dollar enterprise in upstate New York. Along with the trout fishermen come all sorts of people wanting to experience the Catskill Mountains through wildlife observation, bird watching, hiking, and winter sports.

How has the city managed to maintain good water quality? You may wonder how New York City has succeeded so far in maintaining high water quality when the city owns only about 8% of the land the water comes from. The answer is that the city has offered financial incentives to the area's farmers, homeowners, and others. Although the amount of money is not large, it is sufficient to encourage a sense of responsibility among the land owners, who are attempting to abide by the guidelines that help protect water quality.[10]

Still, transporting water to meet expanding needs has its limits. The case history of New York City's Catskill water points up the importance of valuing natural ecosystems for the functions they perform. With a little help, many of our ecosystems can provide a variety of services, helping to assure water quality, air quality, and many other necessities.[10] However, the cost of bringing water to large urban centers from far away, the increasing amounts needed by growing populations, and the competition for water will eventually limit the water supply of the city. As shortages develop, people will need to adopt stronger conservation measures, and the cost of water will rise. If the price goes high enough, more expensive sources may be developed—for example, pumping from deeper wells or using desalination.

Some Trends in Water Use

We're doing a better job of managing and conserving water. Today, in the U.S., we are using much more surface water than groundwater. Withdrawals of both surface water and groundwater increased between 1950 and 1980, reaching a maximum of about 430,000 million gallons a day. However, since 1980, water withdrawals have decreased and leveled off. It is encouraging that water withdrawals have decreased since 1980 even though the population of the United States has continued to grow. This suggests that we have made improvements in water management and water conservation.[11, 12] Trends in water withdrawals in the United States from 1960 to 1995 are shown in Figure 11.9. If you examine this graph, you will see that it suggests several things:

▪ The major uses of water are for irrigation and the thermoelectric industry.

▪ Agricultural use of water for irrigation began to level off around 1980.

▪ Water use by the thermoelectric industry and other industries peaked in 1980 and has fallen a bit.

▪ Use of water for public and rural supplies continued to increase in the period 1950–1995, presumably because of population growth.[12]

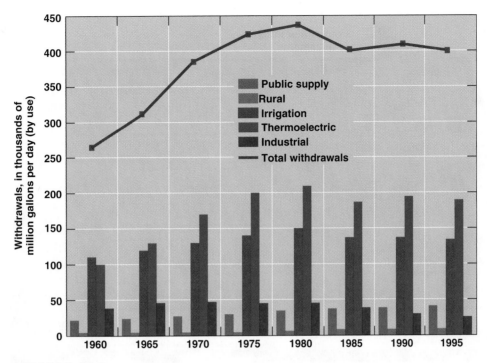

■ FIGURE 11.9

Trends in U.S. water withdrawals (fresh and saline) for the period 1960–1995.
[*Source:* W. B. Solley, R. P. Pierce, and H. A. Perlman. Estimated use of water in the United States in 1995. U.S. Geological Survey Circular 1200, 1998.]

11.4 Water Conservation

Water conservation is the careful use and protection of water resources. Conservation is an important part of sustainable water use. It is concerned with both the quantity of water used and its quality. Because the field of water conservation is changing rapidly, it is likely that a number of new approaches will reduce the total withdrawals of water for various purposes even though consumption will continue to increase.[6]

Agricultural Use

Agricultural use of water is key. Improved irrigation (Figure 11.10) could reduce agricultural use of water by 20–30%. Since agriculture is the biggest water user, this would be an enormous saving. Suggestions for agricultural conservation include the following:

■ Price agricultural water to encourage conservation (subsidizing water will encourage overuse).

■ Use lined or covered canals that reduce seepage and evaporation.

■ Use computer monitoring and schedule the release of water for maximum efficiency.

■ Irrigate when less water will be lost to evaporation, such as at night or in the early morning.

■ Use improved irrigation systems, such as sprinklers or drip irrigation.

■ Improve the soil to make it easier for the water to penetrate it and to minimize runoff. Where possible, use mulch to help retain water around plants.

■ Integrate the use of surface water and groundwater to more effectively use all of the water. That is, irrigate with surplus surface water when it is abundant, and also use surplus surface water to recharge groundwater aquifers by putting the surface water in specially designed infiltration ponds or injection wells. When surface water is in short supply, use more groundwater.

■ Encourage the development of crops that require less water or are more salt-tolerant, so that farmers won't need to flood irrigated land as often to remove accumulated salts in the soil.

Domestic Use

Domestic use of water is a small part of the total but often a big local problem. Domestic (that is, household) use of water accounts for only about 10% of total national water withdrawals. But because households are concentrated in urban areas, domestic water use can pose major local problems in areas where water is sometimes or often in short supply. Indeed, many urban areas in the United States are already experiencing

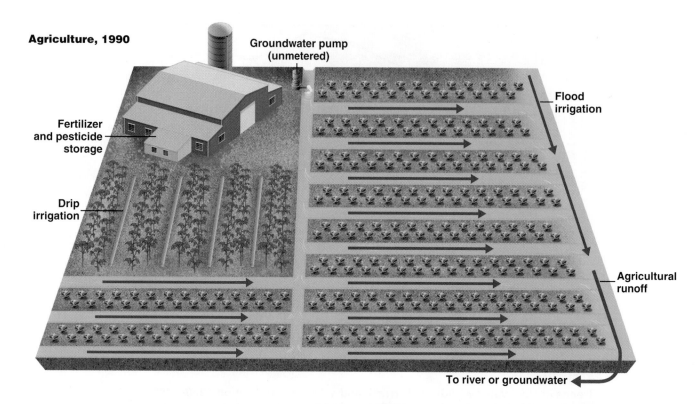

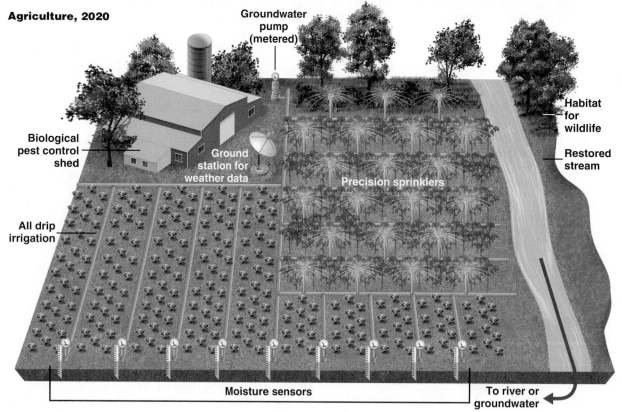

■ FIGURE 11.10

Comparison of agricultural practices in 1990 with what they might be by 2020.
The improvements call for a variety of agricultural procedures, from biological pest control to more efficient irrigation and the restoration of water resources and wildlife habitat. These practices are being used more often now and this trend is expected to continue. [*Source:* P. H. Gleick, P. Loh, S. V. Gomez, and J. Morrison. California water 2020, a sustainable vision. Oakland, CA: Pacific Institute for Studies in Development, Environment and Security, 1995.]

or will experience the impact of population growth on their water supply. For example:

- Southern California, in particular San Diego, is growing rapidly, and its water needs are quickly becoming greater than its local supplies. As a result, the city of San Diego has negotiated with farmers to the east in the Imperial Valley to purchase water for urban areas. The city is also building desalination plants and considering making dams taller so that more water can be stored for urban uses.

- In Denver, city officials, fearing future water shortages, are proposing strict water-conservation measures that include limits on the amount of water used for landscaping and the amount of grass that can be planted around new homes.

- Chicago, the seventh-fastest-growing area in the United States from 1990 to 2000, reported significant groundwater depletion after a recent drought.

- Tampa, Florida, fearing shortages of freshwater because of its continuing population growth, began operating a desalination plant in 2003 that produces about 25 million gallons of water a day.

- Atlanta, Georgia, the fourth-fastest-growing urban area in the United States from 1990 to 2000, is expecting increased demand on its water supplies as a result of population growth and is exploring ways to meet those demands.

- New York City declared a drought emergency in 2002 and placed water restrictions on its more than 9 million citizens.

What is clear from these examples is that while there is no overall shortage of water in the United States or the world, there are local and regional shortages, particularly in large, growing urban areas in the semiarid western and southwestern United States.[13]

What are some ways to use less water at home? In homes, most of the water use is in the bathroom and kitchen and for washing clothes. We can substantially reduce the amount of water used at home through the following measures:

- In semiarid regions, replace lawns with decorative gravels and native plants. Native plants require little if any watering, because they have evolved with the prevailing precipitation of the region.

- Use more efficient bathroom fixtures, such as low-flow toilets that use 1.6 gallons or less per flush rather than the standard 5 gallons, and low-flow showerheads that deliver less but sufficient water.

- Fix all leaks quickly. Dripping pipes, faucets, toilets, or garden hoses waste water. A small drip can waste several liters a day. Multiply this by millions of homes with a leak, and a large volume of water is lost.

- Use dishwashers and washing machines that use less water.

- Sweep sidewalks and driveways instead of hosing them.

- Wash cars at a car wash, where water may be reused.

- Consider using gray water (water from showers, bathtubs, sinks, and washing machines) to water plants. The gray water from washing machines is easiest to use, because it can be easily diverted before going down the drain.

- Water lawns and plants in the early morning, late afternoon, or at night to reduce evaporation.

- Use drip irrigation and mulch garden plants to keep the soil moist underneath.

- Plant drought-resistant vegetation that requires less water.

Water-pricing policies can help. Local water districts should adopt pricing policies that make water more expensive if a household uses more than a baseline amount that is determined by the number of people in a home and the size of the property.

Industry and Manufacturing Use

This is another area where there is room for improvement. For instance, the amount of water used for steam generation of electricity could be reduced up to 30% by using cooling towers that use less or no water. Manufacturing and industry could curb water withdrawals by increasing in-plant treatment and recycling of water and by developing new equipment and processes that require less water.[7]

Perception and Water Use

How people view their water supply affects how much they use. If water is abundant and inexpensive, we don't think much about it. If water is scarce or expensive, it is another matter. People in Tucson, Arizona, perceive the area as a desert (which it is) and view water as scarce. They therefore use gravel and a lot of native plants (cactus and other desert plants) in yards and gardens. Tucson's water supply is mostly from groundwater, which is being mined (used faster than it is being naturally replenished). Tucson also receives some Colorado River water, used for irrigation and industrial purposes.

The message is that we could all do with a little of Tucson's desert mentality. This is particularly true for those in large urban areas, such as Los Angeles and San Diego, in southern California.

11.5 Sustainability and Water Management

We can get along without a lot of things, but not without water. Water is essential to sustain life and to maintain ecological systems necessary for the survival of living things. As a result, water plays crucial roles in economic development, cultural values, and community well-being. Clearly then, managing water use for sustainability is vitally important.

Sustainable Water Use

What is it, and what does it require? **Sustainable water use** can be defined as the use of our water resources in a way that allows society to develop and flourish into an indefinite future. This means that we must find ways to use water without disturbing the various parts of the hydrologic cycle or the ecological systems that depend on it. Below are some general criteria for sustainable water use.[14]

▪ Develop water resources in sufficient volume to maintain human health and well-being.

▪ Provide sufficient water resources to guarantee the health and maintenance of ecosystems.

▪ Ensure minimum standards of water quality for the various users of water resources.

▪ Ensure that human actions do not damage or reduce long-term renewability of water resources.

▪ Promote the use of water-efficient technology and practice.

▪ Gradually eliminate water-pricing policies that encourage wasteful use of water.

Water Management and the Environment

Moving water from one area to another isn't easy. Many agricultural and urban areas require water to be delivered from nearby (and, in some cases, not-so-nearby) sources. Delivering the water requires a system for water storage and routing by way of canals and aqueducts from reservoirs. As a result, dams are built, wetlands may be modified, and rivers may be channelized to help control flooding.

Often, all of this creates a good deal of controversy. In the United States, we no longer embark on large water projects such as the construction of a dam and reservoir without environmental and public review. This involves input from a variety of government and public groups, who may have very different needs and concerns. Agricultural groups, for example, see water development as critical to their livelihood, while some other groups are primarily concerned with wildlife and wilderness preservation. Today, parties with special concerns about water issues are encouraged—and in some cases required—to meet and communicate their concerns and desires. We turn now to the focus of some of these concerns: wetlands, dams, channelization, and flooding.

11.6 Wetlands

What exactly are "**wetlands**"? Wetlands include salt marshes, swamps, bogs, prairie potholes, and vernal pools (shallow depressions that hold water at certain times of the year). Their common feature is that they are wet at least part of the year and therefore have a particular type of vegetation and soil. Figure 11.11 shows several types of wetlands.

(a)

Stephen Krasemann/Stone/Getty Images

(b)

Judy Foldetta/iStockphoto

(c)

Jim Brandenburg/Minden Pictures, Inc

▪ **FIGURE 11.11**
Several types of wetlands:
(a) aerial view of part of the Florida Everglades at a coastal site; (b) cypress swamp, water surface covered with a floating mat of duckweed, in northeastern Texas; and (c) aerial view of farmlands encroaching on prairie potholes in North Dakota.

Why do we need wetlands? Wetland ecosystems may provide a variety of natural services for other ecosystems and for people. For example:

■ Freshwater wetlands are a natural sponge for water. During high river flow, they store water, reducing downstream flooding. After a flood, they slowly release the stored water, nourishing low flows.

■ Many freshwater wetlands are important as areas of groundwater recharge (water seeps into the ground from a prairie pothole, for instance) or groundwater discharge (water seeps out of the ground in a marsh that is fed by springs).

■ Wetlands are among the primary nursery grounds for fish, shellfish, aquatic birds, and other animals. It has been estimated that as many as 45% of endangered animals and 26% of endangered plants either live in wetlands or depend on them for their existence.[15]

■ Wetlands are natural filters that help purify water—plants in wetlands trap sediment and toxins.

■ Wetlands are often highly productive and are places where many nutrients and chemicals are naturally cycled.

■ Coastal wetlands buffer inland areas from storms and high waves.

■ Wetlands are important storage sites for organic carbon. Carbon is stored in living plants, animals, and rich organic soils.

■ Wetlands are aesthetically pleasing to people and are favorite places for sightseeing, bird watching, and boating.

Freshwater wetlands are threatened in many areas. We lose 1% of our nation's total wetlands every two years, and 95% of this loss is freshwater wetlands. Wetlands such as prairie potholes in the U.S. Midwest and vernal pools in southern California are particularly vulnerable because their hydrology is poorly understood and establishing their wetland status is more difficult.[16] Over the past 200 years, over 50% of the wetlands in the United States have disappeared because they have been diked or drained for agricultural purposes or filled for urban or industrial development. Perhaps as much as 90% of the freshwater wetlands have disappeared.

Salt marshes have also suffered. Although most coastal marshes are now protected in the United States, the extensive salt marshes at many of the nation's major estuaries, where rivers entering the ocean widen and are influenced by tides, have been modified or lost. These include deltas and estuaries of major rivers such as the Mississippi, Potomac, Susquehanna (Chesapeake Bay), Delaware, and Hudson. The San Francisco Bay estuary, considered the estuary most modified by human activity in the United States today, has lost nearly all its marshlands to leveeing and filling[17] (Figure 11.12). Modifica-

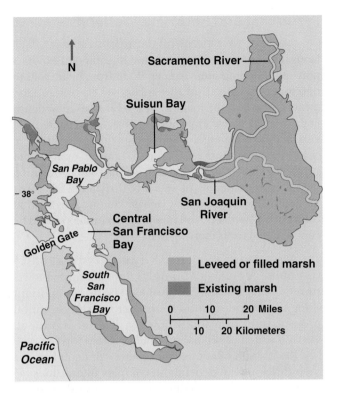

■ **FIGURE 11.12**
Loss of marshlands in the San Francisco Bay and estuary from about 1850 to the present. [*Sources:* T. J. Conomos, ed. *San Francisco, the urbanized estuary.* San Francisco: American Association for the Advancement of Science, 1979; F. H. Nichols, J. E. Cloern, S. N. Luoma, and D. H. Peterson. The modification of an estuary. *Science* 231: 567–573, 1986; © by the American Association for the Advancement of Science.]

tions result not only from filling and diking but also from loss of water. The freshwater inflow has been reduced by more than 50%, dramatically changing the hydrology of the bay in terms of flow characteristics and water quality. As a result of the modifications, the plants and animals in the bay have changed as habitats for fish and wildfowl have been eliminated.[17]

Redirecting the Mississippi is leading to loss of coastal wetlands. The Mississippi River delta includes some of the major coastal wetlands of the United States and the world. Historically, flooding of the Mississippi River maintained the coastal wetlands of southern Louisiana by delivering water, mineral sediments, and nutrients to the coastal environment. Today, levees line the lower Mississippi River, confining the river and directing floodwaters, mineral sediments, and nutrients into the Gulf of Mexico rather than into the coastal wetlands. Deprived of water, sediments, and nutrients in a coastal environment where the sea level is rising, the coastal wetlands are being lost.[18]

Preserving and Restoring Wetlands

We must offer incentives to wetlands owners. Wetlands hold potential for agricultural use, mineral exploitation, and building sites. To keep them from being sold to developers, we need to offer incentives to private land owners, who own the majority of several types of wetlands in the United States. Strategies for managing wetlands must also include planning to maintain quality and quantity of water that wetlands need in order to flourish, or at least to survive. Unfortunately, although we have laws governing the filling and draining of wetlands, the United States has no national wetlands policy. Debate continues as to what constitutes a "wetland" and how property owners should be compensated for preserving wetlands.[15, 19]

A related management issue is restoration of wetlands. A number of projects have attempted to restore wetlands, with mixed results. The most important thing to consider in most freshwater marsh restoration projects is the availability of water. If water is present, wetland soils and vegetation will likely develop.

Restoration of salt marshes is more difficult because their development depends on complex interactions between the hydrology, sediment supply, and vegetation. Careful studies of the relationship between the movement of sediment and the flow of water in salt marshes is providing important information that makes restoration of salt marshes and salt marsh vegetation more likely.

One idea is to construct wetlands to clean up agricultural runoff. This is being attempted in places that have a great deal of agricultural runoff. Wetlands have the natural ability to remove excess nutrients, break down pollutants, and cleanse water. However, constructing wetlands is difficult because, as we just noted, it involves linking complex ecological processes.[20]

Wetlands are being created in Florida to help restore the Everglades. The Everglades is a huge wetland ecosystem that functions as a wide, shallow river flowing south through southern Florida to the ocean. Fertilizers used in farm fields north of the Everglades make their way directly into the Everglades by way of agricultural runoff, disrupting the ecosystem. Phosphorus, in particular, causes undesirable changes in water quality and aquatic vegetation. The human-made wetlands are designed to prevent these nutrients from entering and damaging the Everglades.[21]

11.7 Dams and the Environment

Dams alter the environment in a number of important ways. Their effects include the following:

- Loss of land, cultural resources, and biological resources in the reservoir area.

- Storage of sediment behind the dam rather than allowing it to move downstream to coastal areas, where it previously supplied sand to beaches. The trapped sediment also reduces the reservoir's water-storage capacity, limiting the life of the reservoir.

- Fragmentation of river ecosystems through blocking upstream migration of fish.

- Downstream changes in hydrology and in sediment transport that change the entire river environment and the organisms that live there.

Some dams are being removed. For a variety of reasons—including displacement of people, loss of land, loss of wildlife, and undesirable changes to river ecology and hydrology—many people today are against turning our remaining rivers into a series of reservoirs with dams. In the United States, several dams, including the Edwards Dam near Augusta, Maine (Figure 11.13), have recently been removed. Removal of the Edwards Dam opened about 29 kilometers (18 miles) of river habitat to migrating fish, including Atlantic salmon, striped bass, shad, alewives, and Atlantic sturgeon. After the dam was removed, the Kennebec River came back to life as millions of fish migrated upstream for the first time in 160 years.[22] Other dams including several in Southern California and the Pacific Northwest are also being considered for removal because of the environmental damage they are causing.

■ **FIGURE 11.13**
Removal of a dam.
Location of Edwards Dam near Augusta, Maine.

In contrast, China recently constructed the world's largest dam. Three Gorges Dam on the Yangtze River (Figure 11.14) has drowned cities, farm fields, important archeological sites, and highly scenic gorges while displacing approximately 2 million people from their homes. In the river, habitat for endangered dolphins was damaged contributing to their extinction. On land, habitats are fragmented and isolated as mountaintops become islands in the reservoir. The dam is about 185 meters high and more than 1.6 kilometers wide, and produces a reservoir nearly 600 kilometers long (a distance greater than from Los Angeles to San Francisco). Raw sewage and industrial pollutants that are discharged into the river enter the reservoir, and there is concern that the reservoir will become seriously polluted. In addition, the Yangtze River has a high sediment load, and it is feared that the upstream end of the reservoir, where sediments will likely be deposited, will fill with sediment, damaging deep-water shipping harbors.

The dam may produce a false sense of security. It may encourage further development in flood-prone areas, which will be damaged or lost if the dam and reservoir are unable to hold back floods in the future. If this happens, loss of property and life from flooding may be greater than if the dam had not been built. Adding to this problem is the dam's location in a region where earthquakes and large landslides have been common in the past. If the dam fails, then a downstream city with a population of several million people might be submerged, with catastrophic loss of life.[23]

One positive is the dam's production of electricity. The giant dam and reservoir will be able to produce about 18,000 megawatts (MW) of electricity, the equivalent of about 18 large coal-burning power plants. This is big plus, since pollution from coal burning is a serious problem in China. However, some critics of the dam have pointed out that a series of dams on Yangtze River tributaries (smaller waterways that feed into the Yangtze) could have produced similar electric power without the potential for environmental damage to the main river.[24]

More and bigger dams? There is little doubt that if we continue to use water the way we do now, we will need additional dams and reservoirs, and some existing dams will be made taller to increase their water storage. Since there are few acceptable sites for new dams left, conflicts about the construction of additional dams and reservoirs are bound to occur.

11.8 Channelization and the Environment

What is channelization and why do it? Channelization of streams consists of straightening, deepening, widening, clearing, or lining existing stream channels. It is an engineering technique that has been used to control floods, drain land for agriculture and urban uses, control erosion, and improve navigation. The two most common goals of channelization projects are flood control and drainage improvement.

Channelization can sometimes harm the environment. Thousands of kilometers of streams in the United States have been modified by channelization, and all too often this has produced adverse environmental effects, such as the following:

- Turning a meandering stream with pools (deep, slow flow) and riffles (faster, shallow flow) into straight channels that are nearly all riffle flow, resulting in the loss of important fish habitats.
- Removal of vegetation along stream banks, which had provided wildlife habitats and shading of the water.
- Downstream flooding where the channelized flow ends, because the channelized section is larger and carries more floodwater than the natural channel downstream can carry without overflowing its banks.

Liu Liqun/Corbis Images

■ **FIGURE 11.14**
Three Gorges on the Yangtze River is a landscape of high scenic value. Shown here is the Wu Gorge, near Wushan, one of the gorges flooded by the water in the reservoir.

▪ Damage or loss of wetlands when their source of water is drained by channelization.

▪ Aesthetic degradation—channelized streams are much less attractive than natural streams.

Kissimmee River, Florida: A Case Study of Problems with Channelization

Turning a winding river into a straight ditch and back again. Channelization of the Kissimmee River started in 1962. After 9 years and $24 million of construction, the meandering river with many bends had been converted into a straight ditch 83 kilometers (about 50 miles) long. Unfortunately, the channelization not only failed to provide the expected flood protection but also damaged a valuable wildlife habitat, added to water-quality problems, and caused aesthetic degradation. As a result, in the 1990s, it was decided to return the river to its original meandering form.

Restoring the Kissimmee River has been no small task. Indeed, it is one of the most ambitious restoration projects ever attempted in the United States, and will be even more expensive than the channelization was. So far, several miles of nearly straight flood-control channel have been restored to a meandering channel with wetlands, returning ecosystems to a more natural state.

Not all channelization causes serious environmental degradation. In many cases, drainage projects are beneficial. Moreover, experience (such as the Kissimmee experience) is leading to improved channel design. Currently, more consideration is being given to the environmental aspects of channelization, and some projects are being designed with modified channels that behave more like natural streams.

11.9 Flooding

River flooding is the most widely experienced natural hazard in the world. A river and the flatland adjacent to it, known as the *floodplain*, together constitute a natural system. In most natural rivers, the water overflows the riverbanks onto the floodplain every year or so. This natural process has many benefits for the environment:

▪ Water and nutrients are stored on the floodplain.

▪ Deposits on the floodplain contribute to the formation of nutrient-rich soils.

▪ Wetlands on the floodplain provide an important habitat for many birds, animals, plants, and other living things.

▪ The floodplain functions as a natural greenbelt that is distinctly different from adjacent environments and provides environmental diversity.

Natural flooding is no problem unless people decide to live on floodplains. In the United States every year about 100 people lose their lives in floods, and damages to homes, businesses, and other structures on floodplains exceed $3 billion. The 1993 flood of the Mississippi River took about 50 lives and caused about $16 billion in damages (Figure 11.15). This loss of life, although terrible, is low compared with the toll in areas of the world that lack the sophisticated monitoring and warning systems the United States has for its rivers. For example, flooding associated with two cyclones that struck Bangladesh in 1970 and 1991 killed more than half a million people.

Urbanization and Flooding

What happens when you build towns and cities in a drainage basin? A drainage basin is the land area that contributes water to a particular stream system. Many small drainage basins have been urbanized—that is, people have decided to build streets, malls, and buildings there. In these drainage basins, it may be hard to even find the streams, which have been forced to flow through underground pipes or concrete channels.

Urbanization has had significant effects on basins and streams. With urbanization, much of the land is covered with buildings and pavement, which water cannot penetrate. As a result, rainwater quickly runs off the

Cameron Davidson/Comstock, Inc.

■ **FIGURE 11.15**
Mississippi River flood.
Failure of a levee in Illinois caused flooding in the town of Valmeyer.

artificial surfaces into storm sewers and then to streams. Because urbanization causes more water to run off, and to run off faster than it would normally, the risk of floods also increases. Urban areas experience more frequent and larger floods than do natural systems of similar size.

Flooding from urbanization can be reduced in several ways. One way is to use retention ponds or even parking lots to store runoff from storms. But the best way to lower flood risk is to limit urbanization of floodplains. Land-use planning or floodplain zoning that restricts development and building on floodplains is also the most desirable option from an environmental perspective. Ironically, the federal flood insurance program in the United States has had the opposite effect: by allowing repeated insurance coverage in hazardous areas, the program in effect encourages people to live on floodplains. The insurance program is being reevaluated, and repeated coverage may be denied. Instead, the government may use buyout programs for floodplain homes. After the 1993 flood, nearly 8,000 properties were acquired by the government in Illinois and Missouri, including the town of Valmeyer, Illinois.

It's a mistake to rely on dams, levees, and floodwalls to prevent flooding. These structures can fail. During the 1993 flood of the Mississippi, for example, about 70% of the levees failed (Figure 11.15). They simply were not designed to withstand a flood that lasted more than two months. After that flood, some communities relocated rather than rebuild on the floodplain. Urban development that already exists on floodplains will in most cases continue to be protected, but we should not repeat past mistakes with future development.

Hurricane Katrina in 2005 produced a storm surge (water driven inland by the storm) that flooded the city of New Orleans. Contributing factors were levee failures and removal of coastal wetlands that buffer inland areas from wave damage and advancing storm surge. Katrina was the most costly catastrophe in U.S. history causing about $80 billion in damages and claiming about 1,800 lives.

Not allowing more home-building on floodplains would be a good start. Unfortunately, this is not always done. Since the 1993 flood, about 28,000 new homes and other structures added $2.2 billion in new development in the St. Louis area on floodplain land that was underwater in 1993. The new development is supposedly protected by new, higher levees.[25]

11.10 Global Water Shortage Linked to Food Supply

Water shortages are occurring in many regions. In the past few years, we have begun to realize that isolated water shortages are apparently indicators of a global pat-

tern.[26] At numerous locations on Earth, both surface water and groundwater are being stressed and depleted:

■ Groundwater in the United States, China, India, Pakistan, Mexico, and many other countries is being used faster than it is being renewed.

■ Large bodies of water—for example, the Aral Sea—are drying up (as shown earlier in Figure 11.8).

■ Large rivers, including the Colorado in the United States and the Yellow River in China, do not deliver any water to the ocean in some seasons or years. The flow of others, such as the Nile in Africa, has been greatly reduced.

Water demand during the past half-century has tripled as the human population has more than doubled. In the next half-century, the human population is expected to swell by another 2–3 billion. Our increasing use of groundwater and surface water for irrigation has enabled us to produce more food—mostly crops such as rice, corn, and soybeans—but these water resources are being depleted and may not be sufficient to grow the crops to feed the estimated 8–9 billion people who will be inhabiting the planet by the year 2050.

To sustain our water resources, we must control human population growth. In this chapter, we have outlined a number of ways to conserve, manage, and sustain water. The good news is that a solution is possible—but it will take time, and we need to take steps now, before significant food shortages develop.

We will now shift our attention from water supply to the important topic of water pollution.

11.11 Water Pollution

Water pollution is anything that lowers the quality of water. In deciding whether a water supply is polluted, we generally consider several things: (1) what we intend to use the water for; (2) how different the water is from the norm; (3) its possible effects on public health; and (4) its ecological effects. From a public-health or ecological view, a pollutant is any biological, physical, or chemical substance that, in certain amounts, is known to be harmful to living organisms. Water pollutants include heavy metals, sediment, certain radioactive isotopes, heat, fecal coliform bacteria, phosphorus, nitrogen, sodium, and other useful (even necessary) elements, as well as certain harmful bacteria and viruses.

The lack of clean drinking water is a widespread problem. Today, the primary water-pollution problem in the world is the lack of clean, disease-free drinking water. Every year, several million people (mostly children in poor countries) die as a result of drought and water-borne diseases. In the past, epidemics of waterborne diseases, such as cholera, killed thousands of people in the United States. Fortunately, treating drinking water before it is used has largely eliminated such epidemics in the United States. However, this certainly is not the case worldwide. For example, an epidemic of cholera occurred in South America in the early 1990s, and outbreaks of waterborne diseases continue to be a threat, even in developed countries.

The quality of water determines its potential uses. The major uses for water today are agriculture, industrial processes, and domestic (household) supply. Water for household use must be free (within limits) of anything harmful to health, such as insecticides, pesticides, disease organisms, and heavy-metal concentrations. It also should taste good, be odorless, and not damage plumbing or household appliances. The quality of water required for industrial purposes varies widely. Some industrial processes may require distilled water. Others need water that is not highly corrosive and does not contain particles that could clog or otherwise damage equipment. Because most vegetation can tolerate a wide range of water quality, agricultural waters may vary widely in physical, chemical, and biological properties.[27] However, eating vegetables irrigated or washed with polluted water can make people sick.

Many processes and materials may pollute surface water or groundwater. Some of these are listed in Table 11.3. Table 11.4 lists categories and examples of water pollutants. All segments of society (urban, rural, industrial, agricultural, and military) contribute to the problem of water pollution. The major sources are runoff and leaks or seepage of pollutants into surface water or groundwater. Pollutants are also transported by air and deposited in water bodies.

The EPA sets limits for some water pollutants As part of our national drinking-water standards, the U.S. Environmental Protection Agency has set limits on the amounts of some pollutants that may be allowed in water. Because it is difficult to determine the effects of exposure to small amounts of pollutants, the EPA has set maximum concentration standards for only a small fraction of the more than 700 drinking-water pollutants. If the water contains more than the established limit, the water is considered unsatisfactory for a particular use.

More people usually means more pollutants as well as greater use of water resources.[28] As a result, we can expect that the quality of drinking water in some places will suffer in the near future. More than one-quarter of drinking-water systems in the United States have reported at least one violation of federal health standards.[29] Approximately 36 million people in the United States were recently supplied with water from systems that violated (at least once) federal drinking-water standards.[30]

There is a growing demand for bottled water, which may be treated water from a municipal supply, natural spring water, or even water that is thousands of years old, obtained from melting glacial ice. Bottled water has become a multibillion-dollar industry. A gallon of spring water or "specialty water" can be more expensive than a gallon of milk, but it tastes better than tap water to many people, and natural water sources sometimes do contain contaminants and pollutants and require treatment before use. Regulation and testing of bottled water are important parts of that industry's public relations and sales strategy. The industry needs to show that it offers a superior product.

☐ TABLE 11.3 SOME SOURCES AND PROCESSES OF WATER POLLUTION

Surface Water	Groundwater
Urban runoff (oil, chemicals, organic matter, etc.) (U, I, M)	Leaks from waste disposal sites (chemicals, radioactive materials, etc.) (I, M)
Agricultural runoff (oil, metals, fertilizers, pesticides, etc.) (A)	Leaks from buried tanks and pipes (gasoline, oil, etc.) (I, A, M)
Accidental spills of chemicals including oil (U, R, I, A, M)	Seepage from agricultural activities (nitrates, heavy metals, pesticides, herbicides, etc.) (A)
Radioactive materials (often involving truck or train accidents) (I, M)	Saltwater intrusion into coastal aquifers (U, R, I, M)
Runoff (solvents, chemicals, etc.) from industrial sites (factories, refineries, mines, etc.) (I, M)	Seepage from cesspools and septic systems (R)
Leaks from surface storage tanks or pipelines (gasoline, oil, etc.) (I, A, M)	Seepage from acid-rich water from mines (I)
Sediment from a variety of sources, including agricultural lands and construction sites (U, R, I, A, M)	Seepage from mine waste piles (I)
Air fallout (particles, pesticides, metals, etc.) into rivers, lakes, oceans (U, R, I, A, M)	Seepage of pesticides, herbicide and nutrients (U, R, A)
	Seepage from accidental spills (e.g., train or truck accidents) (I, M)
	Inadvertent seepage of solvents and other chemicals including radioactive materials from industrial sites or small businesses (I, M)

Key: U = urban; R = rural; I = industrial; A = agricultural; M = military.

☐ TABLE 11.4 CATEGORIES OF WATER POLLUTANTS

Pollutant Category	Examples of Sources	Comments
Dead organic matter	Raw sewage, agricultural waste, urban garbage	Produces biochemical oxygen demand and diseases.
Pathogens	Human and animal excrement and urine	Examples: Recent cholera epidemics in South America and Africa; See discussion of fecal coliform bacteria in Section 11.12
Organic chemicals	Agricultural use of pesticides and herbicides; industrial processes that produce dioxin (Chapter 8)	Potential to cause significant ecological damage and human health problems. Many of these chemicals pose hazardous-waste problems (Chapter 29).
Nutrients	Phosphorus and nitrogen from agricultural and urban land use (fertilizers) and wastewater from sewage treatment	Major cause of artificial eutrophication. Nitrates in groundwater and surface waters can cause pollution and damage to ecosystems and people.
Heavy metals	Agricultural, urban, and industrial use of mercury, lead, selenium, cadmium, and so on (Chapter 8)	Example: Mercury from industrial processes that is discharged into water (Chapter 8). Heavy metals can cause significant ecosystem damage and human health problems.
Acids	Sulfuric acid (H_2SO_4) from coal and some metal mines; industrial processes that dispose of acids improperly	Acid mine drainage is a major water pollution problem in many coal mining areas, damaging ecosystems and spoiling water resources.
Sediment	Runoff from construction sites, agricultural runoff, and natural erosion	Reduces water quality and results in loss of soil resources.
Heat (thermal pollution)	Warm to hot water from power plants and other industrial facilities	Causes ecosystem disruption (Chapter 8).
Radioactivity	Contamination by nuclear power industry, military, and natural sources	Often related to storage of radioactive waste. Health effects vigorously debated.

The following sections focus on several water pollutants to emphasize principles that apply to pollutants in general. We discuss other water pollutants (for example, heavy metals, organic chemicals, thermal pollution, and radioactive materials) elsewhere in this book.

11.12 Sources of Pollution

Biochemical Oxygen Demand (BOD)

BOD, biochemical oxygen demand, is the amount of oxygen consumed by microorganisms as they break down organic matter. In water-quality management, BOD is often measured and analyzed in a laboratory, using small water samples.

Streams and rivers carry organic waste. Dead organic matter—which produces BOD—enters streams and rivers from natural sources (such as dead leaves) and from human sources, particularly agricultural runoff and urban sewage. Approximately 33% of all BOD in streams comes from agriculture. However, urban areas—especially those with older sewer systems in which storm-water runoff and sewage share the same line—also considerably increase the BOD in streams. All streams have some capability to degrade organic waste, but when a stream is overloaded with oxygen-demanding waste, it overwhelms the stream's natural cleansing ability.

Waterborne Disease

In the United States, we tend not to think much about waterborne illness. Although historically epidemics of waterborne disease killed thousands of people in U.S. cities such as Chicago, public-health programs have largely eliminated such epidemics by treating drinking water to remove disease-carrying microorganisms and by not allowing sewage to contaminate drinking-water supplies. As we will see, however, North America is not immune to outbreaks of waterborne disease.

Fecal Coliform Bacteria

We use fecal coliform bacteria as an indicator of disease potential. Fecal coliform bacteria are normally found in the intestines and waste of all animals, including humans, and usually don't cause trouble. However, if we find fecal coliform bacteria in water, we know that there is the potential for waterborne diseases. The threshold used by the EPA for swimming water is not more than 200 cells of fecal coliform bacteria per 100 milliliters of water. Above that level, the water is considered unfit for swimming. Water that has any fecal coliform bacteria whatsoever is unsuitable for drinking.

One type, Escherichia coli (E. coli), has caused illness and death. Outbreaks have resulted from eating contaminated meat and drinking contaminated juices or water. There was an outbreak caused by contaminated meat at a popular fast-food chain in 1993. In 1998, 26 children became ill and one died after visiting a Georgia water park. In July 1998, the community of Alpine, Wyoming, suffered an outbreak of illness due to the presence of *E. coli* in the drinking-water supply.[31] In 2006 about 150 people in 23 states became sick and one person died from *E. coli* contamination caused by wild pigs foraging in a California spinich field. Clearly *E. coli* bacteria can be a threat to human health and must be carefully regulated.

The threat of disease causes thousands of warnings and beach closings each year in the United States. Advisories are often posted that swimming may be hazardous to health (Figure 11.16). In most cases, the pollutant is fecal coliform bacteria, which may indicate the presence of a specific disease-causing virus, such as hepatitis. Pollutants enter coastal waters from a variety of sources, including urban storm runoff, leaking sewer lines, overflowing sewage-treatment plants, and leaking septic systems from private homes. Urban storm runoff is particularly worrisome because it contains many pollutants, including pesticides and fertilizers from lawns and gardens, as well as engine oil, carwash detergents, animal waste, other waste from streets, and many kinds of urban trash. Coastal communities in many areas face potential loss of tourist income due to beach closings. As a result, increased research and testing of nearshore coastal waters are becoming more routine. Once the sources of pollutants are identified, management plans are designed to reduce the threat.

Nutrients

Land use is the source of two nutrients that pollute water: phosphorus and nitrogen. Streams in forested land have the lowest concentrations of

Robert Billstone/ iStockphoto

■ **FIGURE 11.16**
Coastal water pollution.
This beach in Southern California is occasionally closed as a result of contamination by bacteria.

phosphorus and nitrogen. Streams in urban areas have higher concentrations of these nutrients because of fertilizers, detergents, and products of sewage-treatment plants. Often, however, the highest concentrations of phosphorus and nitrogen are found in agricultural areas, where the sources are fertilized farm fields, feedlots, and farms that raise hogs, chickens, and turkeys. Over 90% of all the nitrogen added to the environment by human activity comes from agriculture.

Large industrial hog farms are particularly worrisome, as we learned from a North Carolina disaster in 1999. Hurricane Floyd struck North Carolina in September 1999, taking lives, flooding homes, and forcing some 48,000 people into emergency shelters. The storm had another effect as well: Floodwaters containing thousands of dead pigs, along with their feces and urine, flowed through schools, churches, homes, and businesses. The number of pig carcasses may have been as high as 30,000, and the stench was overwhelming. Several hundred million gallons of liquid pig waste—with its nitrogen nutrient load and accompanying BOD—ended up in flooded creeks, rivers, and wetlands. In all, something like 250 large commercial pig farms flooded out, drowning hogs whose floating carcasses had to be collected and disposed of (Figure 11.17).

North Carolina's hog population was nearly 10 million in 1997, making it the second-largest pig-farming state in the nation.[32] As the number of large commercial pig farms grew, the state allowed hog farmers to build automated and very confining farms housing hundreds or thousands of pigs. There were no restrictions on farm location, and many farms were constructed on floodplains. The North Carolina herd produced approximately 20 million tons of waste a year, which was flushed out into open, unlined lagoons about the size of football fields. Favorable regulations, along with the availability of inexpensive waste-disposal systems (the lagoons), were responsible for the tremendous growth of the pig population in North Carolina during the 1990s.

People, not a hurricane, caused this environmental disaster. The pig farmers blamed the hurricane, but the catastrophe was clearly caused by people, and it was not really a surprise. An early warning occurred in 1995, when a pig-waste lagoon failed, sending concentrated nutrient-and-BOD-rich waste down the New River, past the city of Jacksonville, and into the New River estuary. The environmental effects on marine life lasted about three months.

What can we learn from North Carolina's "Bay of Pigs"? The lesson is that we are vulnerable to environmental catastrophes caused by large-scale industrial

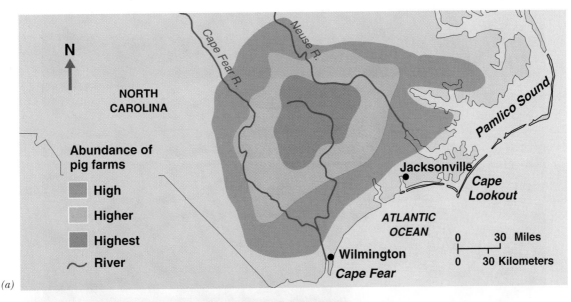

©AP/Wide World Photos

■ **FIGURE 11.17**

North Carolina's Bay of Pigs.
(a) Map of areas flooded by Hurricane Floyd in 1999 with relative abundance of pig farms. (b) Collecting dead pigs near Boulaville, North Carolina. The animals were drowned when floodwaters from the Cape Fear River inundated commercial pig farms.

agriculture. Economic growth and production of livestock must be carefully planned to anticipate problems, and waste-management facilities must be designed to prevent pollution of local streams, rivers, and estuaries.

Was the lesson learned in North Carolina? The pig farmers had big money and powerful friends in government. Incredible as it may seem, after the hurricane the farmers asked for $1 billion in grants to help repair and replace the pig facilities, including waste lagoons destroyed by the hurricane. Furthermore, they asked for exemptions from the Clean Water Act for a period of six months so that waste from the pig lagoons could be discharged directly into streams. This was not allowed.[33]

North Carolina's problem led to the formation of the "Hog Roundtable," a coalition of civic, health, and environmental groups whose goal was to control industrial-scale pig farming. Roundtable efforts, along with others, resulted in a mandate to phase out pig-waste lagoons and require buffers between pig farms and surface waters and wells. For example, some pig farms near

Chicago, Illinois, recycle hog waste by composting it with mulched grass, leaves, and wood chips from urban areas. This produces a rich, odorless humus that can be used as a fertilizer on farm fields or home gardens. The North Carolina coalition also halted construction of a slaughterhouse that would have allowed more pig farms to be established.

One of the problems of industrial agriculture is the nutrients that enter streams, rivers, ponds, lakes, and the ocean. We discuss the effects of these nutrients next.

Eutrophication

Cultural eutrophication is a **cultivation** process. When a body of water (such as a pond or lake) develops a high concentration of nutrients, such as nitrogen and phosphorus, the nutrients increase the growth of aquatic plants, bacteria, and algae. The algae may form surface mats (Figure 11.18), shading the water and reducing light to algae below the surface, greatly reducing photosynthesis. The bacteria and algae die, and as they decompose, the BOD increases, and oxygen in the water is

William E. Ferguson

■ **FIGURE 11.18**
Eutrophication.
Mats of dying green algae in a pond undergoing eutrophication.

consumed. If the oxygen is reduced too much, other organisms, such as fish, will die (Figure 11.19).

The fish do not die from phosphorus poisoning. If you added phosphorus to water in an aquarium where there were only fish and no algae or bacteria, the phosphorus would not affect the fish. The fish die from a lack of oxygen resulting from a chain of events that started with the input of phosphorus and affected the whole ecosystem. The unpleasant effects stem from the interactions among different species, the effects of the species on chemical elements in their environment, and the condition of the environment (the lake itself and the air above it). This is what we call an *ecosystem effect*.

Cultural eutrophication is not limited to lakes. Problems arising from artificial eutrophication of bodies of water are not restricted to lakes. In recent years, concern has grown about the flow of sewage from urban areas into tropical coastal waters, and about the effects of cultural eutrophication on coral reefs.[34, 35] For example, parts of the famous Great Barrier Reef of Australia, as well as some reefs around the Hawaiian Islands, are being damaged by eutrophication.[36, 37] The problem is that nutrient input stimulates algae growth, which covers and smothers the coral.

Nearshore environments of the East, West, and Gulf coasts of the U.S. and coasts near large urban areas worldwide are experiencing eutrophication prob-lems. Of particular concern are the so-called dead zones—wide areas of water containing so little oxygen that bottom-dwelling marine organisms cannot survive there.

The "dead zone" in the Gulf of Mexico, and how it got that way. Each summer, a dead zone develops off the nearshore environment of the Gulf of Mexico, south of Louisiana. The zone varies in size, but is about as large as the state of New Jersey. Within the zone, bottom water generally has low concentrations of dissolved oxygen. Shrimp and fish can swim away and seek oxygen elsewhere, but bottom dwellers such as crabs, snails, and other shellfish are killed.

The oxygen content is low because of cultural eutrophication. Nitrogen is believed to be the main cause of the dead zone, and it most likely gets there from one of the richest, most productive agricultural regions in the world—the Mississippi River drainage basin. The nitrogen from agricultural fertilizers causes the algae to bloom, and as the algae die, sink, and decompose, the oxygen in the water is depleted.

The level of nitrogen in the river water has leveled off, suggesting that the dead zone may have reached its maximum size. This gives us time to study the cultural-eutrophication problem and decide how best to reduce or eliminate it. We could partly reduce

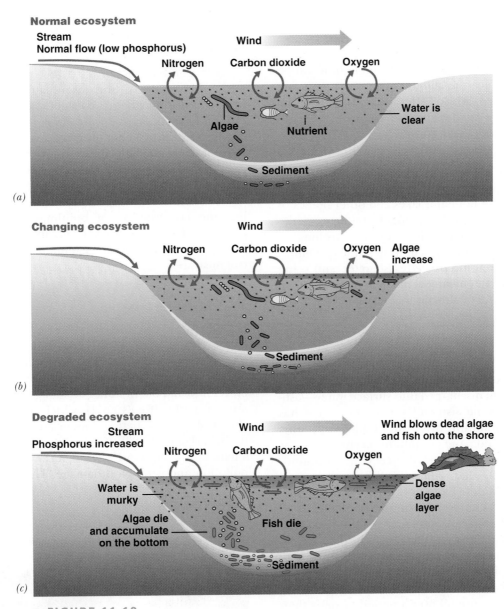

■ FIGURE 11.19
The eutrophication of a lake.
(a) In a low-nutrient lake, green algae is not abundant and the water is clear.
(b) Phosphorus is added to streams and enters the lake. Algae growth is stimulated, and a dense layer forms. (c) The algae layer becomes so dense that the algae at the bottom die. Bacteria feed on the dead algae and use up the oxygen. Finally, fish die from lack of oxygen.

the amount of nitrogen reaching the Gulf of Mexico via the Mississippi River by the following actions:[38]

■ Reduce the amount of nitrogen by using fertilizers more effectively and efficiently.

■ Restore and create river wetlands between farm fields and streams and rivers, particularly in areas known to contribute high amounts of nitrogen. The wetland plants will use the nitrogen, so less will enter the river.

■ Use nitrogen-reduction processes at wastewater-treatment plants for towns, cities, and industrial facilities.

Better agricultural practices could cut nitrogen in the Mississippi by up to 20%. This would require reducing the amount of fertilizers used by about 20%, which farmers say would harm productivity. Restoration and creation of river wetlands and *riparian* forests (forests along rivers and streams) could reduce the amount of nitrogen entering the river by up to 40%.[38]

There is no easy solution to cultural eutrophication in the "dead zone." Clearly, though, it will be necessary to reduce the amount of nitrogen entering the Gulf of Mexico. With a better understanding of the nitrogen cycle, we can design a management strategy to reduce or eliminate the dead zone.

What is the solution to cultural eutrophication in general? It is fairly straightforward: We need to make sure that high concentrations of nutrients from human sources do not enter lakes and other bodies of water. We can accomplish this goal by reducing pollutants, using phosphate-free detergents, controlling nitrogen-rich runoff from agricultural and urban lands, disposing of or reusing treated wastewater, and using more advanced water-treatment methods, such as special filters and chemical treatments that remove more of the nutrients.

Oil

Oil spills make headlines, routine discharge of oil does not. Oil discharged into surface water—usually in the ocean but also on land and in rivers—has caused major pollution problems. Tanker accidents have caused several large oil spills in recent years—for example, the spill into the Prince William Sound, Alaska, in 1989, that killed wildlife and spoiled beaches.[39] The Middle East war in 2006 resulted in destruction of oil facilities, releasing huge quantities of crude oil into the marine environment along the coast of Lebanon. The ultimate effects of the spills are expected to be an environmental disaster. However, although such spills make headlines and cause serious environmental problems, normal shipping activities probably release more oil over a period of years than is released by the occasional spill.

The long-term effects of large oil spills are uncertain. We know that the effects can last several decades. Toxic levels of oil have been identified in salt marshes 20 years after a spill.[40]

Sediment

By volume and mass, sediment is our greatest water pollutant. Sediment consisting of rock and mineral fragments (ranging from gravel particles to finer sand, silt, and clay) can produce a *sediment pollution* problem. In many areas, it chokes streams; fills lakes, reservoirs, ponds, canals, drainage ditches, and harbors; buries vegetation; and generally creates a nuisance that is difficult to remove. Sediment pollution is a twofold

problem: It results from erosion, which depletes an important land resource (soil), and it lowers the quality of the water resource it enters.[41]

Many human activities affect runoff, erosion, and sedimentation. Streams in naturally forested or wooded areas may be nearly stable, without excessive erosion or sedimentation. However, converting forestland to agriculture generally increases runoff, sediment yield, and erosion. Soil-conservation practices on farms can minimize soil loss but can't eliminate it. Converting farmland, forestland, and countryside into highly urbanized land has even more dramatic effects. Excavating and construction of buildings and roads can produce large quantities of sediment. Fortunately, sediment and soil erosion can be minimized by on-site control measures.

Acid Mine Drainage

Water with a high concentration of sulfuric acid drains from mines—mostly coal mines but also metal mines (copper, lead, and zinc). The sulfuric acid is produced when surface water or shallow groundwater moves into and out of mines or tailings that contain the mineral pyrite (fool's gold), which is iron sulfide. A chemical reaction between water and pyrite produces the acid (Figure 11.20). If the acid-rich water runs off to a natural stream, pond, or lake, it can cause significant pollution and ecological damage. The acidic water is

John Cancalosi/DRK Photo

▪ **FIGURE 11.20**
Acid mine drainage.
Acid drainage from an abandoned mine is entering a small stream channel and polluting the surface waters. This mine is in the mountains of southwestern Colorado.

toxic to the plants and animals of an aquatic ecosystem; it damages biological productivity, and fish and other aquatic life may die. The acidic water can also seep into and pollute groundwater.

Acid mine drainage is a significant water-pollution problem in Wyoming, Indiana, Illinois, Kentucky, Tennessee, Missouri, Kansas, and Oklahoma, and it is probably the most significant water-pollution problem in West Virginia, Maryland, Pennsylvania, Ohio, and Colorado. The total impact is considerable because thousands of miles of streams have been damaged.

11.13 Surface-Water Pollution

How does it happen? Flowing water has a natural ability to remove or dilute harmful substances. However, when too much of an undesirable substance enters a body of water, that natural ability is overwhelmed, and the water becomes polluted.

Pollutants are categorized as coming from point or non-point sources. **Point sources** are distinct and confined, such as pipes from industrial or municipal sites that empty into streams or rivers (Figure 11.21). In general, point source pollutants from industries are controlled through on-site treatment or disposal and are regulated by permit. Municipal point sources are also regulated by permit. **Non-point sources**, such as runoff from an urban area, are less distinct, occur irregularly, and are influenced by such things as land use, climate, hydrology, topography, native vegetation, and geology. Common urban non-point sources include runoff from streets or fields; such runoff contains all sorts of pollutants, from heavy metals to chemicals and sediment. Rural sources of non-point pollution are generally associated with agriculture, mining, or forestry. Non-point sources are difficult to monitor and control.

Three approaches to dealing with surface-water pollution: (1) reduce the sources; (2) treat the water to remove pollutants; or (3) convert the pollutants to forms that can be disposed of safely. Which approach to use depends on the circumstances. Reduction at the source is the environmentally preferable way to deal with pollutants. For example, air-cooling towers, rather than water-cooling towers, may be used to prevent heat from power plants from causing thermal pollution of water. Water treatment is used for a variety of pollution problems and includes chlorination to kill microorganisms, such as harmful bacteria, and filtering to remove sediment and heavy metals.

Cities are undoing the damage they've done to their rivers. Many large cities in the United States—such as Boston, Miami, Cleveland, Detroit, Chicago, Portland, and Los Angeles—grew on the banks of rivers, but the cities often damaged their rivers with pollution and concrete. Today, there are grassroots movements all around the country to restore urban rivers and adjacent lands as greenbelts, parks, and scenic areas. One example is the Cuyahoga River in Cleveland, Ohio, which by 1969 was so polluted that sparks from a train ignited oil-soaked wood in the river, setting the surface of the river on fire! The burning of an American river became a symbol for a growing environmental consciousness. Today, from Cleveland to Akron, the Cuyahoga River is a beautiful greenbelt (Figure 11.22). The greenbelt changed this part of the river from a sewer into a valuable public resource and focal point for economic and environmental renewal.[42] However, in downtown Cleveland and Akron, the river remains an industrial stream, and parts remain polluted.

David Woodfall/DRK Photo

■ **FIGURE 11.21**
Point-source pollution.
This pipe is a point source of chemical pollution entering a river in England from an industrial site.

©AP/Wide World Photos

■ **FIGURE 11.22**
Recovery of a river that once burned.
The Cuyahoga River (lower left) flows toward
Cleveland, Ohio, and Lake Erie. A trail along the Ohio
and Erie Canal (lower right) is in the Cuyahoga National
Park. The skyline is that of industrial Cleveland.

11.14 Groundwater Pollution

Groundwater pollution differs in several ways from surface-water pollution. Groundwater has a low concentration
of oxygen, a situation that kills microorganisms that require oxygen-rich environments but may provide a happy
home for microorganisms that live in environments that
require little oxygen. Also, the breakdown of pollutants
that occurs in the soil and in material a meter or so below
the surface does not occur readily in groundwater. Finally,
groundwater moves through channels that are very small
and variable, so it usually moves slowly and has only limited opportunity to disperse or dilute pollutants.

Groundwater can be easily polluted by several
sources (see Table 11.3), and the pollutants, even
though they are toxic, may be difficult to recognize.
Only a small portion of the groundwater in the United
States today is known to be seriously contaminated; but
the problem is worsening as the swelling human popu-

lation places increasing pressure on water resources. Atlantic City, New Jersey, and Miami, Florida, are two eastern cities threatened by polluted groundwater that is
slowly migrating toward their wells.

Principles of Groundwater Pollution: An Example

Old, leaking underground gasoline tanks have
polluted water and soil. Buried gasoline tanks belonging to automobile service stations are a widespread
environmental problem that no one thought very much
about until only a few years ago. Underground tanks are
now strictly regulated. Meanwhile, thousands of the old,
leaking tanks have been removed, and the surrounding
soil and groundwater have been treated to remove the
gasoline. Cleanup can be very expensive, involving removal and disposal of soil (as a hazardous waste) and
treatment of the water using a process known as *vapor
extraction*. Treatment may also be accomplished underground by microorganisms that degrade (consume) the
gasoline. This is known as *bioremediation* and is less expensive than removal, disposal, and vapor extraction.

Pollution from buried gasoline tanks highlights
some important points about groundwater pollutants:

■ Some pollutants, such as gasoline, are lighter than water and thus float on the groundwater.

■ Some pollutants are heavier than water and sink or
move downward through groundwater. Examples of
sinkers include some particulates and cleaning solvents. Pollutants that sink may become concentrated
deep in groundwater aquifers.

■ Some pollutants have multiple phases: liquid, vapor,
and dissolved. Dissolved pollutants chemically combine
with the groundwater, much as salt dissolves into water.

■ Because cleanup or treatment of water pollutants in
groundwater is very expensive, and because undetected
or untreated pollutants may harm living things and the
environment, the emphasis should be on preventing
pollutants from entering groundwater in the first place.

Another Example: Long Island, New York

How urbanization has led to groundwater pollution. Two counties on Long Island (Nassau and Suf-
folk), with a population of several million people,
depend entirely on groundwater. The most serious
groundwater problem on Long Island is shallow-aquifer
pollution stemming from urbanization. Sources of pollution in Nassau County include urban runoff; household sewage from cesspools and septic tanks; salt used to
de-ice highways; fertilizers; pesticides; and industrial and
solid waste. These pollutants enter surface waters and
then migrate downward, especially in areas where
groundwater levels are declining.[43] Landfills for municipal solid waste have been a significant source of shallow-

aquifer pollution on Long Island because pollutants (garbage) placed on sandy soil over shallow groundwater quickly enter the water. For this reason, most Long Island landfills were closed in the last three decades.

11.15 Water Treatment

In the U.S., water for domestic use comes from surface water and groundwater. Although some groundwater is of high quality and needs little or no treatment, most sources are treated to conform with national drinking-water standards.

Water is first stored, then filtered and treated. Before treatment, water is usually stored in reservoirs or special ponds to allow solids, such as fine sediment and organic matter, to settle out. The water is then run through a water plant, where it is filtered and chlorinated before being distributed to homes. Even so, many people filter their tap water or buy bottled water for drinking and cooking. As a result, production of bottled water has become a multibillion-dollar industry.[30]

The drinking water in the United States is among the safest in the world. Treating water with chlorine has nearly eliminated waterborne diseases, such as typhoid and cholera, which previously caused widespread suffering and death in the developed world and still do in many parts of the world. However, we need to know much more about the long-term effects of exposure to low concentrations of toxins in our drinking water. Although the water in the United States is much safer than it was 100 years ago, low-level contamination (lower than the amount considered dangerous) with organic chemicals and heavy metals is a concern that requires continued research and evaluation.

11.16 Wastewater Treatment

Water used for industrial and municipal purposes is often degraded during use by the addition of solids, salts, nutrients, bacteria, and other material that uses up oxygen. In the United States, the law requires that these waters be treated before being released back into the environment. Wastewater treatment, or sewage treatment, costs many billions of dollars per year in the United States, and the cost continues to rise. Wastewater treatment is big business.

Conventional methods include septic tanks in rural areas and treatment plants in cities. Newer approaches include applying wastewater to the land and wastewater renovation and reuse. We discuss the conventional methods first.

Septic-Tank Disposal Systems

Many rural areas have no central sewage system or wastewater-treatment facilities. As a result, individual septic-tank disposal systems, not connected to sewer systems, continue to be an important way to get rid of sewage in rural and suburban areas. Because not all land is suitable for the installation of a septic-tank disposal system, the law requires an evaluation of each site before issuing a permit. For this reason, before purchasing property in areas where a septic system is necessary, a buyer should make sure the site is satisfactory for such a system.

How does a septic-tank disposal system work? The basic parts of a septic-tank disposal system are shown in Figure 11.23. The sewer line from the house leads to an underground septic tank in the yard. The tank is designed to separate solids from liquid, digest (biochemically change) and store organic matter for a

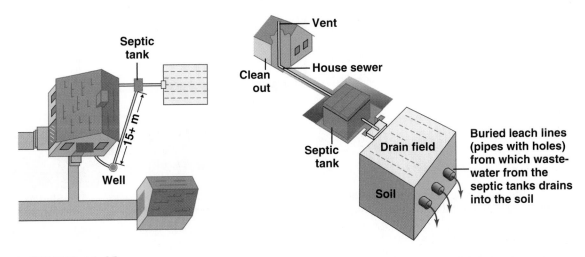

■ **FIGURE 11.23**
Septic-tank sewage disposal system and location of the drain field with respect to the house and the well. [*Source:* Based on data from Indiana State Board of Health.]

while, and allow the clarified liquid to discharge into a drain field from a system of pipes through which the treated sewage seeps into the surrounding soil. As the wastewater moves through the soil, it is further treated by the natural processes of oxidation and filtering. By the time the water reaches any freshwater supply, it should be safe for other uses.

Sewage drain fields may fail for several reasons. The most common causes are (1) failure to pump out the septic tank when it is full of solids, and (2) poor soil drainage, which allows the wastewater to rise to the surface in wet weather. When a septic-tank drain field fails, pollution of groundwater and surface water may result. Solutions to septic-tank system problems include putting septic systems in well-drained soils, making sure systems are large enough, and maintaining them properly.

Treatment Plants

In urban areas, wastewater is treated at specially designed plants that accept sewage from homes, businesses, and industrial sites. The raw sewage is delivered to the plant through a network of sewer pipes. After it is treated, the wastewater is discharged into surface water—river, lake, or ocean—or, in a few cases, used for another purpose, such as crop irrigation. The main purpose of standard treatment plants is to break down and reduce the BOD and kill bacteria with chlorine. A simplified diagram of a wastewater treatment plant is shown in Figure 11.24.

Wastewater treatment has three categories: primary, secondary, and advanced. *Primary treatment* removes 30–40% of the BOD from the wastewater, mainly in the form of suspended solids and organic matter, by passing the waste through a series of screens and settling tanks.[44]

Secondary treatment uses bacteria to digest about 90% of the BOD that entered the plant in the sewage. The sludge from the digester is dried and disposed of in a landfill or applied to improve soil. In some instances, treatment plants in urban and industrial areas contain many pollutants, such as heavy metals, that are not removed in the treatment process. Sludge from these plants is too polluted to use to improve the soil and must be disposed of.

Advanced treatment is used when treatment beyond primary and secondary is needed. Some additional pollutants can be removed by adding more treatment steps.

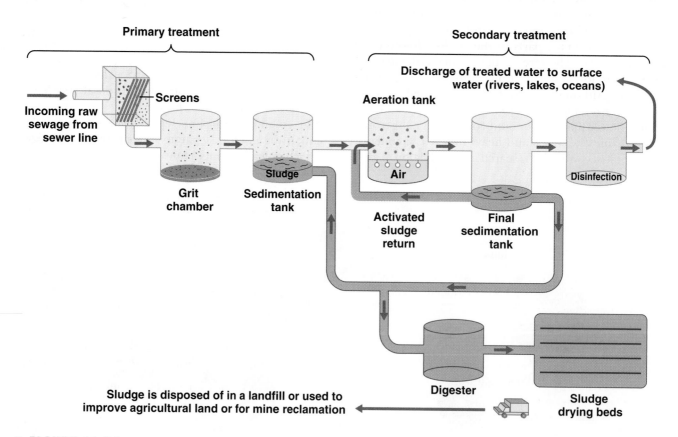

■ **FIGURE 11.24**
Wastewater treatment plant.
The diagram shows the sewage-treatment process from primary to secondary treatment.

For example, phosphates and nitrates, organic chemicals, and heavy metals can be removed by specially designed sand filters and carbon filters, and with the help of some chemicals. Treated water is then discharged into surface water or may be used to irrigate agricultural or municipal land, such as golf courses, city parks, and the grounds surrounding wastewater-treatment plants.

Primary and secondary treatment are required by federal law for all municipal plants in the United States, but some plants may temporarily be exempt from full secondary treatment if installing secondary treatment facilities poses too great a financial burden. Where secondary treatment is not sufficient to protect the quality of the surface water into which the treated water is discharged—for example, a river with endangered fish species that must be protected—advanced treatment may be required.[44]

Boston Harbor: Cleaning Up a National Treasure

The city of Boston is steeped in early American history. The names of Samuel Adams and Paul Revere immediately come to mind when considering the late 1700s, when the colonies were struggling to obtain freedom from Britain. In 1773, a group of patriots led by Samuel Adams boarded three British ships and dumped their cargo of tea into Boston Harbor to protest an unfair tax on tea. The event came to be known as the Boston Tea Party.

More than tea has been dumped into the harbor since 1773. The tea dumped overboard by the patriots did not pollute the harbor, but the growing city and the dumping of all sorts of waste eventually did. For over 200 years, Boston Harbor was a disposal site for sewage, treated wastewater, and water contaminated from sewer overflows into Massachusetts Bay. Late in the 20th century, court orders demanded that measures be taken to clean up the harbor.

It was decided to dump waste farther out in the bay. Although there is strong tidal action between the harbor and the bay, it takes about a week to flush water in the harbor out to sea. A study of the bay suggested that disposing of waste farther offshore, where water is deeper and currents are stronger, would dilute the pollutants and lower the pollution in Boston Harbor.

Moving waste disposal farther offshore is a step in the right direction, but dilution by itself cannot solve the urban waste management problem. Even pollutants dumped farther offshore will eventually accumulate and cause environmental damage. In short, any long-term solution must include reducing the amount of pollutants at their source. The Boston Regional Sewage Treatment Plan called for a new treatment plant designed to significantly reduce the amounts of pollutants discharged into the bay. The new Deer Island sewage treatment plant now collects and treats wastewater from 43 greater Boston communities. Moving the sewage outfall offshore, when combined with source reduction of pollutants, is a positive example of what can be done to better manage our waste and reduce environmental problems.[45]

Land Application of Wastewater: An Old Practice Made Cleaner

The practice of applying wastewater to the land is based on the belief that waste is simply a resource out of place. Land application of untreated human waste was practiced for hundreds if not thousands of years before the development of wastewater-treatment plants. However uncontrolled land application caused water pollution problems and disease. Today we are much more careful with land application of wastewater.

Wastewater is being applied successfully to wetlands at a variety of locations.[46–48] Natural or human-made wetlands can be effective in treating the following kinds of water:

- Municipal wastewater (BOD, pathogens, phosphorus, nitrate, suspended solids, metals).
- Storm-water runoff (metals, nitrate, BOD, pesticides, oils).
- Industrial wastewater (metals, acids, oils, solvents).
- Agricultural wastewater and runoff (BOD, nitrate, pesticides, suspended solids).
- Mining waters (metals, acidic water, sulfates).
- Groundwater seeping from landfills (BOD, metals, oils, pesticides).

Using wetlands to treat wastewater is economically attractive. Some communities can't afford to purchase traditional wastewater-treatment plants. Others—such as the city of Arcata in northern California—has for many years used wetlands as part of their wastewater-treatment system. Arcata's wastewater comes mostly from homes, with minor inputs from the numerous lumber and plywood plants. It is treated by standard primary and secondary methods, then chlorinated and dechlorinated before being discharged into Humboldt Bay.[46]

The state of Louisiana also uses its wetlands to treat wastewater. Louisiana, with its coastal wetlands, is a leader in developing advanced treatment using wetlands after secondary treatment (Figure 11.25). Applying wastewater rich in nitrogen and phosphorus to coastal wetlands increases the production of wetland plants and improves the quality of the water as these nutrients are used by the plants. When the plants die, the organic material from their stems, leaves, and roots

Courtesy John Day, Louisiana State University

(a)   (b)   (c)

■ **FIGURE 11.25**
Using wetlands to treat wastewater.
(a) Pointe au Chene Swamp, three miles south of Thibodaux, Louisiana, receives wastewater; (b) one of the pipes delivering wastewater; and (c) ecologists doing fieldwork at Pointe au Chene Swamp to evaluate the wetland.

accumulate, raising the level of the wetland, which partially offsets wetland loss due to the rise in sea level.[49] There are significant economic savings in applying treated wastewater to wetlands, because the financial investment needed is small compared with the cost of advanced treatment at conventional treatment plants.[48, 50]

©Integrated Water System, Inc.

■ **FIGURE 11.26**
Artificial wetlands can treat agricultural wastewater.
Photograph of artificial wetlands project for treatment of agricultural wastewater at Avondale, Arizona, showing wetlands integrated with housing development (lower left).

In arid regions, wetlands can be constructed to treat poor-quality water. For example, Avondale, Arizona, near Phoenix, constructed a wetland facility in a residential community to treat agricultural wastewater (Figure 11.26). Designed to treat several million gallons of water a day, the artificial wetlands contain naturally occurring plants and bacteria that reduce the nitrate level in the water to below the maximum allowable level. After treatment, the water flows by pipe to a recharge basin on the nearby Agua Fria River, where it seeps into the ground to become a groundwater resource. The wetland facility cost about $11 million to create, about half the cost of a more traditional treatment facility.

Conclusion: Using wetlands to treat wastewater is a practical solution to improving water quality in small, widely scattered communities in the coastal zone, and constructing wetlands can serve the same purpose in arid regions. As water-quality standards are tightened, wetland wastewater treatment will become an effective alternative that is less expensive than traditional treatment.[49, 50]

Reuse of Treated Wastewater

Treated wastewater can be piped directly from a treatment plant to the next user.

In most cases, the water is reused in industry, agriculture, or to irrigate golf courses, institutional grounds (such as university campuses), and parks. Direct water reuse is growing rapidly and is already the norm for industrial processes at many factories. In Las Vegas, Nevada, new resort hotels that use a great deal of water for fountains, rivers, canals, and lakes are required to treat wastewater and reuse it (Figure 11.27). Because of people's negative attitudes about consuming treated wastewater, very little direct reuse of water is planned for human consumption (except in emergencies).

11.17 Water Pollution and Environmental Law

Environmental law deals with conservation, use of natural resources, and pollution. It is very important as we debate environmental issues and make decisions about how best to protect our environment. In the United States, laws at the federal, state, and local levels address these issues.

Federal laws to protect water resources go back to the Refuse Act of 1899, which was enacted to protect streams, rivers, and lakes from pollution caused by navigation. Table 11.5 lists major federal laws that have had an important effect on water quality. The purpose of many of these laws was to clean up or treat pollution or treat wastewater, but there has also been a focus on

Prof. Ed Keller

■ **FIGURE 11.27**
Reusing water.
Water reuse at a Las Vegas, Nevada, resort hotel.

preventing pollutants from entering water in the first place. Prevention has the advantage of avoiding environmental damage and costly cleanup.

The mid-1990s was a time of controversy about U.S. water pollution. In 1994 Congress tried to rewrite major environmental laws, including the Clean Water Act, to give industry more flexibility in choosing how to

☐ TABLE 11.5 FEDERAL WATER LEGISLATION

Date	Law	Comment
1899	Refuse Act	Protects navigable water from pollution.
1956	Federal Water and Pollution Control Act	Enhances the quality of water resources and prevents, controls, and abates water pollution.
1958	Fish and Wildlife Coordination Act	Water resources projects such as dams, power plants, and flood control must coordinate with U.S. Fish and Wildlife Service to enact wildlife conservation measures.
1969	National Environmental Policy Act	Requires environmental impact statement prior to federal actions (development) that significantly affect the quality of the environment. Included are dams and reservoirs, channelization, power plants, bridges, and so on.
1970	Water Quality Improvement Act	Expands power of 1956 act through control of oil pollution and hazardous pollutants and provides for research and development to eliminate pollution in Great Lakes and acid mine drainage.
1972 (amended in 1977)	Federal Water Pollution Control Act (Clean Water Act)	Purpose is to clean up nation's water. Provides billions of dollars in federal grants for sewage-treatment plants. Encourages innovative technology, including alternative water treatment methods and aquifer recharge of wastewater.
1974	Federal Safe Drinking Water Act	Aims to provide all Americans with safe drinking water. Sets contaminant levels for dangerous substances and pathogens.
1980	Comprehensive Environmental Response, Compensation, and Liability Act	Established revolving fund (Superfund) to clean up hazardous-waste disposal sites, reducing groundwater pollution.
1984	Hazardous and Solid Waste Amendments to the Resource Conservation and Recovery Act	Regulates underground gasoline storage tanks. Reduces potential for gasoline to pollute groundwater.
1987	Water Quality Act	Established national policy to control nonpoint sources of water pollution. Important in development of state management plants to control nonpoint water pollution sources.

comply with regulations concerning water pollution. Industry interests favored proposed new regulations that, in their opinion, would be more cost-effective without causing further harm to the environment. Environmentalists, on the other hand, viewed the attempts to rewrite the Clean Water Act as a giant step backward in the nation's fight to clean up our water resources.

Apparently, Congress misread the public's values on this issue. Survey after survey has shown that there is strong support for a clean environment in the United States, and that people are willing to pay to have clean air and clean water. Congress has continued to debate changes in environmental laws, but little has been resolved.[51]

Return to the Big Question:

Can we maintain our water resources for future generations?

Our water resources can be maintained for future generations by applying known principles of sustainable water use. It is particularly important to be sure our actions do not damage or reduce the long-term sustainable use of water. This will involve continued water conservation by promoting water-efficient technology and practice. A large amount of groundwater is used today, and it is a challenge to keep it pollution-free and renewable. We need water policy to reduce and eliminate pollutants before they enter the groundwater system, where pollution abatement is difficult and expensive. We also need to carefully evaluate groundwater resources and how and when new water enters the system. We must then take care not to reduce a groundwater resource over a period of years faster than it is replenished by natural processes. Even with careful conservation and protection, some regions will likely continue to experience water shortages, especially in drought cycles. Therefore, planning for sustainable use of water must take into account the natural cycles of water resources and include contingency and emergency plans for the times when water resources become naturally scarce. This may mean setting aside some deep wells as reserves to be used only during shortages, or using water desalination plants for a coastal city during times of need.

Summary

- Although it is one of the most abundant and important renewable resources on Earth, more than 99% of Earth's water is unavailable or unsuitable for human use.

- During the next several decades, water consumption will increase due to greater demands from a growing population and industry.

- Taking water from streams leaves less water for fish and wildlife habitats and navigation, and may therefore cause conflicts.

- Our growing global water shortage is linked to population growth and agricultural use of water to feed the world's people. Thus, controlling population growth and using water more efficiently in agriculture will have the greatest effect on sustainable water use.

- Wetlands serve a variety of functions that benefit other ecosystems and people.

- Flooding is perhaps the most universal natural hazard, and urbanization increases both the frequency and severity of flooding. The best remedy is to avoid building on floodplains.

- The primary water-pollution problem in the world today is the lack of safe drinking water.

- Major categories of water pollutants include disease-causing organisms, dead organic material, heavy metals, organic chemicals, acids, and sediment.

- Pollutants may come from point sources, such as pipes that discharge into a body of water, or non-point sources, such as runoff.

- Cultural eutrophication of water occurs when our activities increase the concentration of such nutrients as phosphorus and nitrogen, which may cause a population explosion of photosynthetic bacteria. The decay of dead bacteria uses up oxygen in the water, causing fish to die.

- Sediment pollution is a twofold problem: soil is lost through erosion, and water quality suffers.

■ Acidic water draining from mines pollutes streams and other bodies of water, damaging aquatic ecosystems and degrading water quality.

■ Conventional treatment plants use primary, secondary, and occasionally advanced water treatment. In some locations, natural ecosystems, such as wetlands and soils, are being used as part of the treatment process.

■ Cleanup and treatment of polluted surface water and groundwater are expensive and may not be completely successful. Furthermore, environmental damage may occur before a pollution problem is noticed and treated. Therefore, we should continue to focus on preventing pollutants from entering water, which is a goal of much water-quality legislation.

Key Terms

acid mine drainage
cultural eutrophication
cultivation
desalination
environmental law
groundwater
in-stream use

non-point sources
off-stream use
point sources
sustainable water use
wastewater treatment
water conservation
wetlands

Getting It Straight

1. If water is one of our most abundant resources, why are we concerned about its availability in the future?

2. Give some examples of in-stream and off-stream uses of water. Why is in-stream use controversial?

3. What are some important environmental problems related to groundwater use?

4. How might your community better manage its water resources?

5. What are some of the major ways that dams affect the environment? How might these be minimized?

6. How can the flood hazard be minimized?

7. Do you think outbreaks of waterborne diseases will be more common or less common in the future? Why? And where are outbreaks most likely to occur?

8. What is *water pollution*, and what are the major processes that contribute to water pollution?

9. Compare point and non-point sources of water pollution. Which is easier to treat, and why?

10. What is the twofold effect of sediment pollution?

11. Describe the major steps in wastewater treatment (primary, secondary, advanced). Can natural ecosystems perform any of these functions? Which ones?

12. Why is acid water that drains from coal mines an important environmental problem?

13. What is cultural eutrophication, and what can we do about it?

14. Do you think our water supply is vulnerable to terrorist attacks? If so, how could the threat be minimized?

What Do You Think?

1. How can we ease or eliminate the growing global water shortage? Do you agree that the shortage is related to our food supply? Why?

2. In a city along an ocean coast, rare waterbirds inhabit a pond that is part of a sewage-treatment plant. How could this have happened? Is the water in the sewage pond polluted? Consider this question from the birds' point of view and from your own point of view.

3. How might a drought impact the water supply within the two largest reservoirs along the Colorado River, Hoover Dam and Glen Canyon Dam?

4. Is the price of water cheap, and has that encouraged people to use more water than what they actually need?

Pulling It All Together

1. Do future generations of people in the United States have to be concerned about water supply? How would they feel about possibly consuming cleaned/treated gray water?

2. Why is so much of the Earth's water unsuitable? What are ways we might be able to consider when thinking about expanding our usable water supply? Is conservation the best method for improving our current water supply?

3. What human-induced activities are contributing to the contamination of surface and ground water sources? What can we begin to do to reduce contamination of these water sources?

4. Strategize a domestic water conservation plan for your home, apartment, or dorm. Identify three ways in which water could be conserved or recycled for use.

5. Wetlands are an important part of ecosystem sustainability. Why are wetlands needed and how can we preserve and restore them throughout the United States?

Further Reading

Borner, H., ed. 1994. *Pesticides in ground and surface water.* Vol. 9 of *Chemistry of Plant Protection.* New York: Springer-Verlag.—Essays on the fate and effects of pesticides in surface water and groundwater, including methods to minimize water pollution from pesticides.

Gleick, P. H. 2003. Global freshwater resources: Soft-path solutions for the 21st century. *Science* 302:1524–1528.

Gleick, P. H. 2000. *The world's water 2000–2001.* Washington, DC: Island Press.

Graf, W. L. 1985. *The Colorado River.* Resource Publications in Geography. Washington, DC: Association of American Geographers.—A good summary of the Colorado River water situation.

Hester, R. E., and R. M. Harrison, eds. 1996. *Agricultural chemicals and the environment.* Cambridge: Royal Society of Chemistry, Information Services.—A good source for information about the impact of agriculture on the environment, including eutrophication and the impact of chemicals on water quality.

Newman, M. C. 1995. *Quantitative methods in aquatic ecotoxicology.* Chelsea, MI: Lewis Publishers.—Up-to-date text on fate, effects, and measurement of pollutants in aquatic ecosystems.

Twort, A. C., F. M. Law, F. W. Crowley, and D. D. Ratnayaka. 1994. *Water supply,* 4th ed. London: Edward Arnold.— Good coverage of water topics from basic hydrology to water chemistry, its use, management, and treatment.

Wheeler, B. D., S. C. Shaw, W. J. Fojt, and R. A. Robertson. 1995. *Restoration of temperate wetlands.* New York: Wiley.— Discussions of wetland restoration around the world.

12

Oceans and Environment

Big Question

Can We Learn to Manage the Oceans' Resources?

Learning Objectives

The oceans cover 70% of Earth's surface and contain much of its life. They also contain some of the least-known ecosystems and habitats on our planet. The ocean not only is an environment in itself but also affects the rest of Earth's surface and thus affects all living things (see the discussion of climate in Chapter 13). After reading this chapter you should understand that . . .

- although there are many fish in the oceans, we have not found a way to harvest them sustainably;

- that overfishing, pollution, and—especially near shore and on coral reefs—habitat destruction are the major human causes of declining fish populations;

- a liquid environment is a constantly changing environment, so marine organisms must be highly adaptable;

- even under the best of circumstances, the abundance of marine populations will fluctuate, and our use of these resources must take such changes into account.

Case Study

Shrimp, Mangroves, and Pickup Trucks

Maitri Visetak owns a small plot of land along the coast of southern Thailand and wanted to improve life for his family. In the early 1990s, Mr. Visetak began farming shrimp in two small ponds (Figure 12.1a). Within two years, he saved enough money to buy two pickup trucks—in Thailand, these are a clear indication of financial success. By then, though, his ponds were contaminated with shrimp waste, antibiotics, fertilizers, and pesticides. Shrimp could no longer live in the ponds. And there was an even more widespread effect: Pollutants escaping from

Frans Lanting/Minden Pictures, Inc.

(a)

(b)

Tom Bean

▪ **FIGURE 12.1**
(a) Shrimp farms such as this one threaten the survival of mangrove forests. (b) Mangroves on the banks of Indian River, Isle of Dominica, West Indies. Mangrove trees grow in coastal wetlands. Their specialized roots can survive immersion in ocean water at high tide and exposure to the drying sun at low tide. Swamps formed by mangroves provide habitat for many kinds of ocean life and are important for commercial fisheries in many parts of the world.

the ponds threatened survival of the area's mangrove trees (Figure 12.1b). Like thousands of other shrimp farmers in Southeast Asia, India, Africa, and Latin America, Mr. Visetak considered abandoning these ponds and moving on to others.

Maitri Visetak is trying to feed his family in the best way he knows how, but, along with thousands of other shrimp farmers in the world, he is unwittingly contributing to the destruction of coastal mangroves, one of the world's valuable ecosystems. The growing demand for shrimp as a luxury food and the overfishing of wild shrimp fueled growth of the world market for farmed shrimp from a $1.5 billion industry 30 years ago to an $8 billion business today. Along the way, half of the world's mangrove forests have been destroyed and with them a major source of food for local human populations and breeding grounds for much of the tropical world's sea life.

The United Nations Environment Program has estimated that one-fourth of the destruction of mangroves can be traced to shrimp farming. Environmentalists have become alarmed, and in many areas local people have staged protests against shrimp farming. With the world's population expected to increase from 6.2 billion to 9 billion by the middle of the 21st century, concern about the world's mangrove forests is growing.[1–6]

Maitri Visetak's story is a clear illustration of what overexploitation has done to almost all the fisheries in the oceans. Sometimes people do it simply to survive, and sometimes people do it out of greed and lack of caring. Even when they care, they often fail because of poor management. In any case, populations of the major commercially valuable species of fish have crashed, one after another.

Ironically, probably no other living resource has been subjected to so much formal scientific management, and yet no other living resource has been so badly managed worldwide. Understanding why we keep destroying the ocean's living resources and what we can do to improve the way we manage these resources is the primary purpose of this chapter.

Earth is the Water Planet and the Ocean Planet (Figure 12.2). Oceans are full of fascinating creatures of many different kinds, from sharks to huge tube worms. Oceans also have a great variety of geological processes, from underwater volcanoes to rifts in the tectonic plates, and have many kinds of currents.

Space Science and Engineering Center, University of Wisconsin-Madison

Earth is the Ocean Planet, as seen in a satellite image of the Pacific Ocean side of our planet.
This infrared image of Earth was taken by the GOES 6 satellite on September 21, 1986.

12.1 Lots of Fish in the Sea: World Fish Production

Scientists estimate 27,000 species of fish and shellfish live in the oceans. People catch many of these species for food, but only a few kinds provide most of the food: Anchovies, herrings, and sardines provide almost 20%.

Fish are the main source of animal protein for about 1 billion people[7]—one of every six people in the world—and about 35 million people make their living by fishing or by **aquaculture** (the farming of things that live in water). At the beginning of the 21st century, the annual world fish catch was worth $81 billion, and international trade in fish amounted to $55 billion.[8]

The world's total fish harvest has increased greatly since the mid-20th century. The total ocean harvest was about 40 million metric tons (MT) annually in the early 1960s, almost doubling to the high 60 millions and low 70 millions in the first years of the present century. The world's total fish catch from all sources increased from just over 100 million MT in early 1990s to 132 million MT in 2003.[9] (Figure 12.3).

Much of the increase in the world's total fish catch is from aquaculture—fish farming more than doubled between 1992 and 2001, from about 15 million

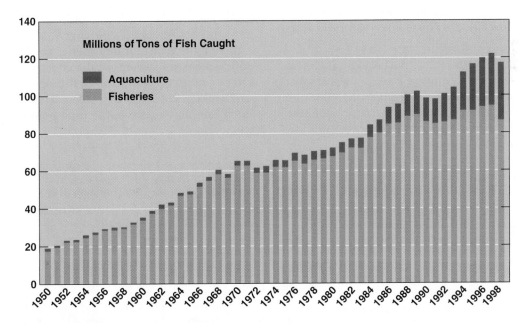

The world fish catch.
The world's catch of ocean fish has increased greatly, from 20 million metric tons in the middle of the 20th century to 80 million MT recently (four times the 1950s catch). One question is, how big an ocean catch can the fish of the world sustain—that is, how many can the fish replace? To date, few, if any, ocean fisheries have been sustainable. [*Source:* http://www.iffo.org.uk/tech/alaska.htm. Accessed June 4, 2005.]

MT to more than 37 million MT. Aquaculture presently provides more than 20% of all fish harvested, up from 15% in 1992.[10]

How big an ocean catch can the fish of the world sustain—that is, how many can the fish replace? If they can't replace as many as are being caught, the catch is not sustainable and the fishery is not sustainable. To date, few, if any, ocean fisheries have been sustainable.

12.2 The World's Fisheries Are in Trouble: The Decline of Fish Populations

Fishing is an international trade, but a few countries dominate. Japan, China, Russia, Chile, and the United States are among the world's major fisheries.[11] The world's ocean fisheries are among the worst managed of all living resources, and are, in general, in big trouble.

Which fish are the most threatened? Species suffering declines include codfish, flatfishes, tuna, swordfish, sharks, skates, and rays.[12, 13] The North Atlantic, whose George's Bank and Grand Banks have for centuries provided some of the world's largest fish harvests, is suffering. The Atlantic codfish catch was 3.7 million MT in 1957, peaked at 7.1 million MT in 1970, declined to 1 million MT in 2000.[14] European scientists recently called for a total ban on cod fishing in the North Atlantic, and the European Union came close to accepting this. Instead, however, they cut the allowable catch by 65% for North Sea cod for the years 2004 and 2005.[15]

Sharks—who needs them? Surprisingly, the great white shark—the shark of the movie *Jaws* and the shark that people most fear—is endangered (Figure 12.4). But who would care? Well, actually, despite their reputation, the great whites kill only about six people per year worldwide, and in 2004 only two people in the United States died from attacks by any shark species[16] (Figure 12.5). Compare this to the yearly worldwide average of 50,000 people killed by snake bites and 500 by elephants, or compare it to the U.S. average of 66 people killed each year by lightning and more than 42,000 killed every year in auto accidents.

Why is the shark endangered? One reason is that the great white's reputation has made it a target, because there is a market for its teeth and other parts. Chinese cuisine also sometimes uses sharks for shark fin soup, and sharks are also caught inadvertently by fishermen seeking other species.

Why should we care? One of the primary concerns about the great white is that it is a top predator in some marine food webs,

▪ **FIGURE 12.4**
The great white shark.

and its elimination might have undesirable effects on the ecosystems in which it functions. Also, those who want to preserve Earth's biological diversity are concerned about this species, as there are about 19 other species of shark that are threatened or endangered (there are a total of 400 shark species).

Understandably, it's getting harder to catch fish. Although the total marine fisheries catch has increased during the past half-century, the effort required to catch fish has increased as well. More fishing boats with better and better gear search the oceans. That is why the total catch can increase while the total population of a fish species declines.

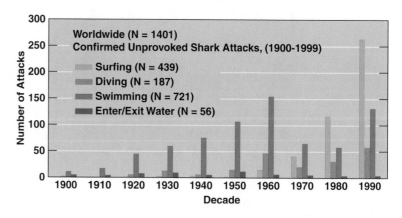

▪ **FIGURE 12.5**
Shark attacks on people in the 20th century.
A total of 1,401 shark attacks were recorded worldwide in the 20th century. Since surfing became popular after World War II, most attacks are on surfers.

(a)

(b)

■ **FIGURE 12.6**
Some modern commercial fishing methods.
(a) Trawling using large nets; (b) workers on a factory ship.

How do we know that fish populations are declining? One source of evidence is the number of marine fish caught with lines and hooks. The catch rate has generally fallen from 6–12 fish per 100 hooks—the typical catch rate before a fish population began to be exploited—to 0.5–2.0 fish per 100 hooks just 10 years later. This amounts to a decline of about 80% in only 15 years, suggesting that fishing depletes fish quickly. Many of the fish that people eat are predators, and on fishing grounds the biomass of large predatory fish appears to be only about 10% of what it was before fishing became an industry. These changes indicate that the biomass of most major commercial fish has declined greatly. In short, we are mining, not sustaining, these living resources.

Why are fish populations declining? One reason many species of fish are declining is improved fishing technology, as well as an increase in the number of ships involved (Figure 12.6). Scallops in the western Pacific show a typical harvest pattern. The catch started very low in 1964 at 200 MT, increased rapidly to 10,000 MT in 1975, declined by more than half by the 1980s, increased to about 10,000 MT in 1992, and then fell to 3,000 MT in 2000.[17] The catch of tuna and their relatives peaked in the early 1990s at about 730,000 MT and fell to 680,000 MT in 2000, a decline of 14%.

Chesapeake Bay—one of the worst fishery disasters in the U.S. Chesapeake Bay, America's largest estuary, was one of the world's great fisheries, famous for oysters and crabs and as the breeding and spawning ground for bluefish, sea bass, and many other commercially valuable species. The bay, 200 miles long and 30 miles wide, drains an area of more than 165,000 square kilometers from New York State to Maryland and is fed by 48 large rivers and 100 small ones.

One difficulty is that food chains in the bay are complex (Figure 12.7). Typical of marine food chains, the food chain of the bluefish that spawns and breeds in Chesapeake Bay is linked to a number of other species,

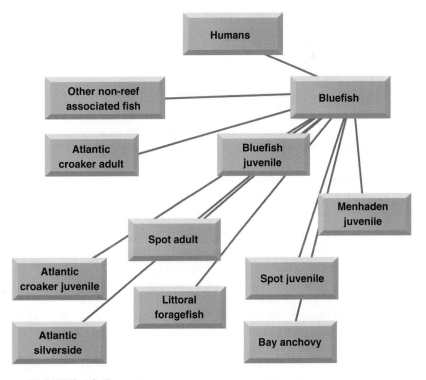

■ **FIGURE 12.7**
Food chain of the bluefish in Chesapeake Bay.
[*Source:* Chesapeake Bay Foundation.]

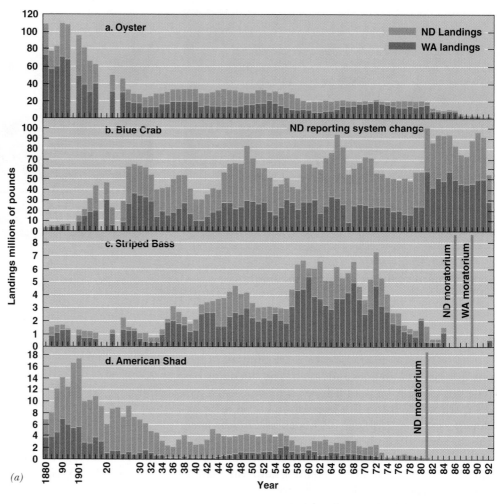

(a) **Fish catch in the Chesapeake Bay.** Oysters have declined dramatically. [*Source: The Chesapeake Bay Foundation.*] (b) Map of Chesapeake Bay Estuary. [*Source: U.S. Geological Survey. The Chesapeake Bay: Geologic product of rising sea level, 1998.*]

each requiring its own habitat within the space and depending on processes that require varying amounts of space and time.

Chesapeake Bay has a number of other problems as well. It is affected by many events on the land that surrounds it—runoff from farms, including chicken and turkey farms, that is highly polluted with fertilizers and pesticides; introductions of exotic species; and the alteration of habitats from fishing and the development of shoreline homes. Add to these problems the varied salinity of the bay's waters—freshwater inlets from rivers and streams, seawater from the Atlantic, and brackish water resulting from the mixture of these.

How do we figure this all out and fix it? Just determining which of these factors, if any, are responsible for a major change in the abundance of any fish species is difficult enough, let alone finding a solution that is affordable and doesn't mean job losses for the bay's fisheries employees. Chesapeake Bay's problems are at the limit of what environmental sciences can deal with at this time. Scientific theory remains inadequate, as do observations, especially of fish abundance (Figure 12.8).

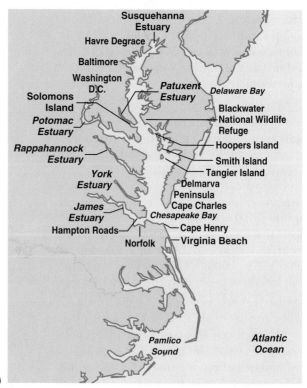

(b)

Why has science-based management failed to conserve fisheries? First, management has been based largely on the logistic growth curve (Figure 12.9), whose problems we discussed in Chapter 2. This curve assumes that an environment remains unchanged except when we harvest resources. The theory gave rise to a calculation called the **maximum sustainable yield.** This term, now famous among fishermen and fisheries scientists, refers to the maximum amount of any biological resource that can be harvested indefinitely. Traditional fisheries equations and computer programs calculated that the maximum sustainable yield of a fish population occurs when the population reaches exactly one-half of its **carrying capacity.** Unfortunately, the oceans are never constant and fish populations do not grow according to this curve.

Second, fisheries are an open resource, meaning that in international waters the numbers of fish that may be harvested can be limited only by international treaties, which are not tightly binding. Open resources offer ample opportunity for unregulated or illegal harvest, or harvests that violate agreements. (We discuss this in more detail in Chapter 18 on environmental economics.)

Another problem is exploitation of a new fishery before scientific assessment. This happens frequently, and as a result the fish are depleted by the time any reliable information about them is available.

Also, some fishing gear is destructive to the habitat. Ground-trawling equipment destroys the ocean floor, ruining habitat for both the target fish and its food. Long-line fishing causes the death of sea turtles and other nontarget surface animals. Large tuna nets have killed many dolphins that are also hunting the tuna.

And that's not all. There are also qualities of the oceans as habitat, and qualities of fish populations and their ecosystems, that make forecasts and harvest management difficult. For example, the harvest of large predators raises questions about ocean ecological communities, such as whether these large predators play an important role in controlling the abundance of other species.

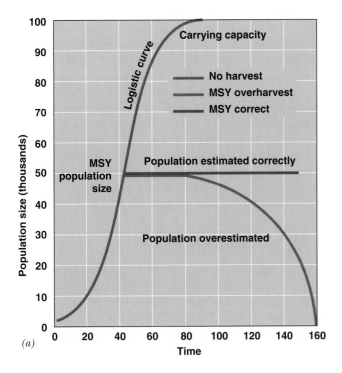

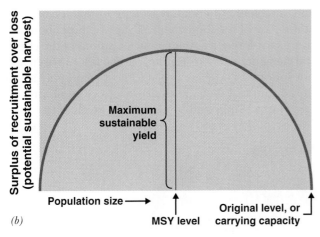

■ **FIGURE 12.9**

The harvest of many fish has been based on this smooth S-shaped growth curve.

This curve, called the *logistic*, assumes that a population grows smoothly from a small number to its carrying capacity, and that its environment is forever constant. Based on this curve, the maximum sustainable yield (MSY) of a fish population occurs when the population reaches exactly one-half of its carrying capacity. Unfortunately, the oceans are never constant and fish populations do not grow according to this curve. (a) Shows the logistic growth over time. (b) Shows logistic growth rate as a function of population size.

12.3 An Ocean Is Many Habitats and Ecosystems

The oceans are actually not just one habitat and ecosystem. Their habitats and ecosystems include **intertidal** areas (which flood and drain twice a day). These in turn include saltwater marshes, sandy shores and rocky shores, and mangrove swamps. Then there are *continental shelves* (the underwater edges of the continents), *upwellings* (currents rising from the depths of the ocean on continental shelves); open oceans (called the **pelagic** zones); coral reefs; the deeps (called the **benthos**); and *hydrothermal vents* (geysers on the seafloor, created where tectonic plates are moving apart, spewing super-hot, mineral-rich water into the cold ocean depths).

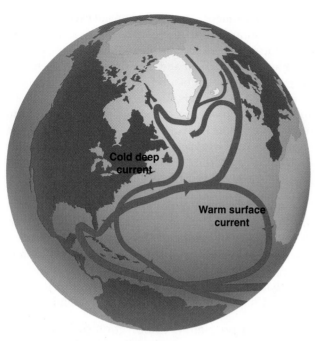

■ **FIGURE 12.10**
Major ocean currents.
[*Source:* Woods Hole Oceanographic Institute (WHOI).]

12.4 Ocean Currents

The oceans are warmed by the sun in the tropics. These warmed waters expand and flow north (in the Northern Hemisphere) along the surface, bringing their heat to northern latitudes. As they reach the North, these warm waters give up their heat and, as they cool, condense, become heavier, and sink, returning in the depths to the tropics. This global circulation is called the *ocean conveyor,* and it is important to life in the oceans and on the land (Figure 12.10).

The Gulf Stream is one of the most famous of the upper warm currents. It is like a saltwater river flowing on the surface of the Atlantic Ocean from the southeastern coast of North America northeast across the Atlantic toward Europe. The Gulf Stream provides warmth to Northern Europe—indeed, it was long believed that the Gulf Stream was the major source of warmth for Europe, making Western Europe warm enough for human habitation. However, recent investigations show that Western Europe is kept warm just as the West Coast of North America is—the ocean waters hold heat much better than land, and moderate the climate of the western coasts of mid- and high-latitude continents. This is why Seattle is warmer than similar towns and cities of similar latitudes in Siberia, and why Madrid, although it is at about the same latitude as New York, has a much warmer climate. The Gulf Stream is only part of this process, but these warm currents at the surface have major effects on ocean life, as we shall see.

12.5 Where Are the Fish?

Most of the fish we eat come from continental shelves. Commercial fisheries are concentrated in relatively few areas of the world's oceans (Figure 12.11). Continental shelves, which make up only 10% of the oceans, provide more than 90% of the fishery harvest, and half of the world's fish catch are made in upwellings.[18]

Fish are abundant where their food is abundant and, ultimately, where there is high production of algae at the base of the food chain. Algae are most abundant in areas with relatively high concentrations of the chemical elements necessary for life, particularly nitrogen and phosphorus. These areas occur most commonly along the continental shelf, particularly in regions of wind-induced upwellings and sometimes quite close to shore.

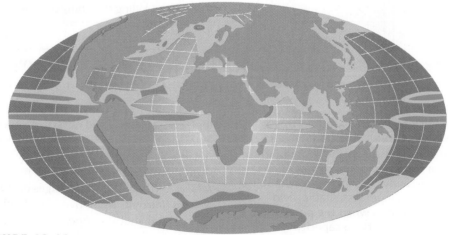

■ **FIGURE 12.11**
The world's major fisheries.
Red areas are major fisheries; the darker the red, the greater the harvest and the more important the fishery. Most major fisheries occur in areas of ocean upwellings—places where rising currents bring nutrient-rich waters from the depths of the ocean. Upwellings tend to occur near continents.

The intertidal habitat is teeming with life. These near-shore areas, which are exposed to air at low tide and ocean waters at high tide, include saltwater shrub and grass marshes and mangrove swamps (as in the opening case study). The constant movement of the water carries nutrients into and out of these areas, which are therefore usually rich with living things and important to people as a direct source of food and as a spawning and breeding ground for many important foods. As a result, intertidal areas are major economic resources. Large algae are found here, from giant kelp of temperate and cold waters to algae of coral reefs in the tropics. Birds and shellfish are usually abundant and are economically important.

Near-shore habitats are most susceptible to pollution from land sources, including rivers and streams. These are often heavily polluted by human activities, because major cities and civilizations tend to develop at the mouths of major rivers and along productive intertidal coastlines. In addition, because they are used heavily for recreation, near-shore areas are considerably altered by people. Some of the oldest environmental laws concern the rights to use resources of the intertidal and near shore, and major legal conflicts continue today about access to intertidal areas and the harvesting of their biological resources.

Disturbances are frequent in the intertidal, making it essential for living things to adapt to variation and extreme conditions. Some of the most extreme variations in environmental conditions occur here, including not only the daily changes in sea level with the tides, but also seasonal changes in tide levels, and ocean storms. The ability to adapt to these disturbances is critical to survival in the intertidal area. Consider, for example, a barnacle or mussel that twice daily experiences a change from a cool or cold saltwater environment to direct exposure to bright sunlight and the highly oxygenated atmosphere. Imagine lying there all day with those changes occurring, from one kind of harsh environment to another.

The upper open ocean (the pelagic region) often has low biological diversity. This region includes vast areas that tend to be low in nitrogen and phosphorus, and are thereby chemical deserts with low productivity and low biological diversity. Surprisingly, even few kinds of algae do well there. Many species of large animals occur, but not in abundance.

The benthos (the ocean bottom) remains relatively little known. The primary input of food is dead organic matter that falls from above. The waters are too dark for photosynthesis, so no algae grow there.

Hydrothermal vents occur in a few locations in the deep ocean. Plate tectonic processes create openings in the ocean floor where hot water with a high concentration of sulfur compounds is released. These sulfur compounds provide an energy basis for chemosynthetic bacteria, which are the base of the food chain for strange creatures such as giant clams, large worms, and other unusual life forms. Water pressure at the vent is high, and temperatures range from the boiling point in waters of vents to the frigid (about 4°C) waters of the deep ocean. Some scientists believe that the vents may have been the location where life on Earth originated.

12.6 Salmon, Anchovies, and Upwellings

Managing ocean fisheries is difficult. This is partly because fish tend to have complicated life cycles, are part of complex food webs (see Chapter 4), and many species use a variety of habitats. In sum, many factors can influence fish populations.

Salmon illustrate the problems of managing ocean fisheries (Figure 12.12). King salmon (also called chinook salmon) start life in small freshwater streams in a woodland region in the Pacific Northwest. When they are about a year old, they swim downstream to larger and larger bodies of water, eventually reaching an estuary, where they stop for a while to adjust to life in saltwater. Then they enter the ocean, where they spend three to five years feeding and fattening up. After this, they return to the stream where they were born, to *spawn* (lay their eggs) and die.

Bill Brooks/Masterfile

■ **FIGURE 12.12**

King salmon swimming upstream on their way to spawn.

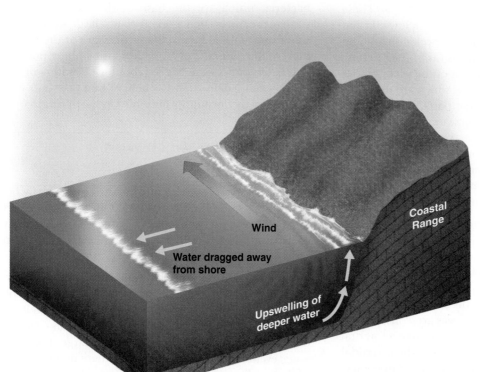

▪ **FIGURE 12.13**
Upwellings are caused by winds pushing surface waters away from a shoreline. Deep waters rise to replace those that have moved away.

As a result, many factors can influence salmon birth and death rates. One factor is the condition of the forests along the stretch of the stream where they spawn. Another factor is the amount of water flowing in the stream where they are born, which varies according to the rainfall and temperature of the past year. Also important are conditions of all land ecosystems along their entire pathway to the ocean—today including farms and cities. On these streams, dead trees form log dams, creating a complex habitat important for the salmon, especially as habitats for the small animals they eat, such as mayfly larvae. Forest conditions can influence salmon, and so can forest disturbances, such as forest fires, logging, and any other land-clearing. Farming, with its use of fertilizers and pesticides and general erosion of soil that gets into the streams, can also affect salmon.

Even rock affects salmon. Salmon lay their eggs in gravel beds on the bottom of shallow streams and are very particular about the size of the bed, the size of the gravel, and other conditions of the gravel. The gravel comes from the bedrock slopes of mountains near the streams and lands in the streams when a headwall fails and the rock crumbles. The spawning beds can be affected by long-term geological events that affect the quality of the bedrock, the steepness of the slopes, and the climate that helps degrade the rock.

Upwelling ocean currents can affect the smaller fish that the salmon eat. Upwellings (Figure 12.13)

are important to many fish (as well as to ocean birds that feed on fish), and some scientists believe that these vertical currents are especially important to salmon. Upwellings bring nutrients, especially phosphorus, from the deep sediments and waters of the ocean to the surface, where they become available to photosynthetic algae. Upwelling regions are therefore highly productive, and many of the world's major fisheries are in upwelling areas (Figure 12.14). Among the most famous is the up-

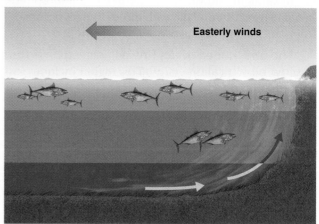

▪ Mixed layer

▪ Thermocline

▪ **FIGURE 12.14**
Upwellings make fish happy by bringing nutrients such as phosphorus to the surface, fertilizing the growth of algae, the base of food chains.

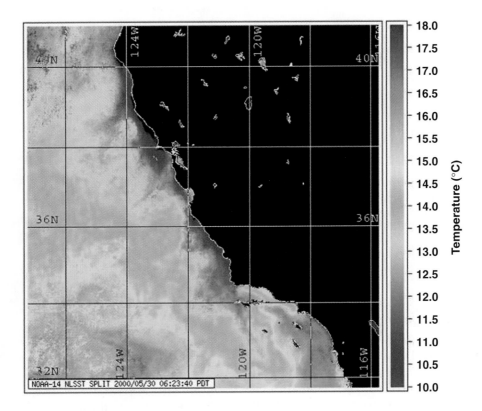

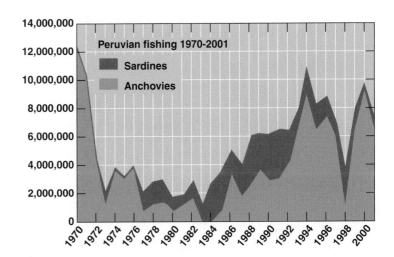

■ **FIGURE 12.15**

Effects of upwellings, as shown in this satellite image.
Water temperatures measured by a satellite from space show upwelling currents as blue along the west coast of South America (the land appears in black). Blue means the water is cold, because these waters have just come up from the deep ocean. The surface waters become warm as they move out to the west, becoming greenish and yellow in this image. [*Source:* http://oceanexplorer.noaa.gov/explorations/02quest/background/upwelling/media/fig2_map.html. Accessed June 4, 2005.]

welling area in the Pacific Ocean off Peru in South America, where anchovies thrive on the upwelling food web. There was a time when this anchovy fishery was the largest commercial fishery in the world. In 1970, 8 million tons of anchovies were caught, but the catch dropped to 2 million tons 2 years later and has continued to decline, and was below 1 million tons in 2002.

Prevailing winds create upwellings by pushing surface water away from the coast. Water flows upward to replace the water that moved away (Figure 12.15). But these so-called prevailing winds do not always prevail. They are the result of climatic conditions that can change, and sometimes the winds fail, and the upwellings with them. The failing of the upwellings off Peru are called *El Niño events*, and they have large-scale climatic effects. During El Niño, fish populations decline (Figure 12.16), and the birds that feed on those fish suffer too, seeking food elsewhere temporarily or dying.

■ **FIGURE 12.16**

Ups and downs of the anchovy catch off Peru.
[*Source:* http://www.iffo.org.uk/tech/alaska.htm. Accessed June 5, 2005.]

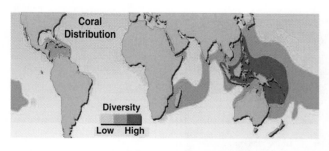

■ **FIGURE 12.17**
Coral reefs at risk and their relative biodiversity.
[*Source:* http://maps.grida.no/go/graphic/
distribution_of_coral_mangrove_and_seagrass_diversity.
Philippe Rekacewicz, UNEP/GRID-arendal.]

12.7 Coral Reefs: A Special Problem

Coral reefs are among the most loved and most studied of all marine habitats. People use them for vaca-
tions and for commercial fishing, and there are island
nations that are essentially just clusters of coral reef islands. These reefs are restricted to warm waters and
therefore low latitudes (Figure 12.17). They are a biological product, built up by a variety of organisms that
make hard limestone-like structures, the coral, within
which they live.

The trouble is, we love them to death. We end
up damaging them, and as a result many coral reefs
are in trouble. By the beginning of the 21st century,
more than one-quarter of known coral reefs had been
lost. Massive bleaching, the result of climatic warming
in 1998, destroyed 16% of the known coral reefs. It is
estimated that another 22% have been destroyed by
human actions—11% lost because of pollution from
sediments and nutrients, and a similar percentage lost
from overfishing, mining of sand and rock, and development of reefs for recreation and tourism.[19] See Figure 12.18, for example.

More than half of the remaining coral reefs are
considered at risk.[20] Those especially at risk from human activities are in the Philippines and Indonesia in
the South Pacific, Tanzania and the Comoros along the
East African coast, and the Lesser Antilles in the
Caribbean. There are more than 400 coral reef parks
and reserves, but most are very small.[21]

12.8 We Pollute the Oceans Too, Which Gets Fish in Trouble

A lot of our wastes end up in the oceans, including biological waste and its embedded disease organisms, toxic chemical wastes (such as oil from spills), chemical fertilizers, and

Peggy Chen/iStockphoto

■ **FIGURE 12.18**
Bora Bora, one of the most famous resort islands in
the world, is a coral reef that is part of Tahiti's island
chain. The reef is heavily developed for tourism.

physical waste, like plastics (Figure 12.19). Because of
pollution, shellfish today sometimes contain organisms
that produce diseases, such as polio and hepatitis. In
the United States, at least 20% of the nation's commercial shellfish beds have at times been closed
(mostly temporarily) because of pollution. Beaches
and bays have been closed to recreational use (again,
mostly temporarily) for the same reason. Ocean dumping of organic wastes has created lifeless zones—such
as occurred offshore from New York City when sewage
was dumped into the ocean.[22, 23]

Marine pollution affects oceanic life in a variety
of ways. It may cause death or retarded growth, reduced vitality, and lower reproductivity of marine organisms. It can also cause major changes in habitats,
both physically and chemically, such as eutrophication
from nutrient-rich waste in shallow waters of estuaries,
bays, and parts of the continental shelf, resulting in the
overgrowth and death of algae, which may wash up and
pollute coastal areas.[24]

Plastics in the Ocean

Plastics have become a special ocean pollution
problem. Vast quantities of plastic are used in the thou-

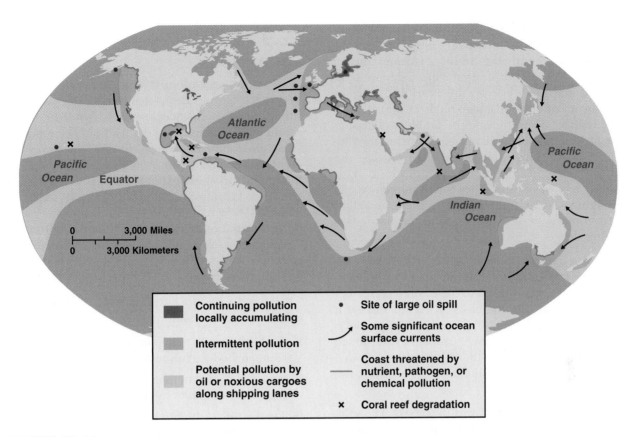

■ FIGURE 12.19
Ocean pollution of the world.
Notice that the areas of continuing and locally accumulating pollution, as well as the areas with intermittent pollution, are in near-shore environments. [*Source:* Modified from the Council on Environmental Quality, Environmental Trends, 1981, with additional data from A. P. McGinn. Safeguarding the health of the oceans. WorldWatch Paper 145. Washington, DC: WorldWatch Institute, 1999, pp. 22–23.]

sands of products we use every day—ranging from food-and-beverage containers to toys, plumbing, and auto interiors—and for decades, oceans have been the ultimate dump for many. Some are dumped from passing ships by passengers. Others are dropped as litter along beaches and carried out to sea by the tides. Still others flow down rivers and streams, sometimes dumped there on purpose, sometimes carried there by storm-water sewerage pipes, and sometimes just washed into the ocean by rain from the shore.

What happens to these plastics when they land in the ocean? Once in the ocean, many of them float with ocean currents and tend to accumulate in places of *convergent currents*—that is, places where currents running in different directions come together. Convergent currents in the Pacific (Figure 12.20) have a whirlpool-like action that concentrates debris near the center of these zones. One such zone of convergence is north of the equator, where the northwestern Hawaiian Islands are located. These islands are very remote, and most people would think of them as unspoiled and

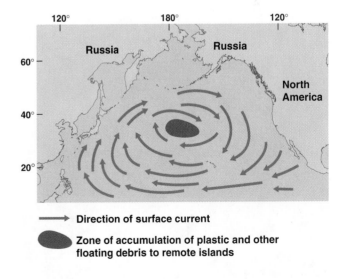

→ **Direction of surface current**

⬭ **Zone of accumulation of plastic and other floating debris to remote islands**

■ FIGURE 12.20
General circulation of the North Pacific Ocean.
Arrows show direction of currents. Notice the tightening clockwise spiral pattern that carries floating debris to remote islands.

pristine. Yet there are literally hundreds of tons of plastics and other types of human debris on these islands. Indeed, recently the National Oceanographic and Atmospheric Administration collected more than 80 tons of marine debris on the Pearl and Hermes atolls. The marine scientist Jean-Michel Cousteau and colleagues have been studying the problem of plastics on the northwestern Hawaiian Islands, including Midway Island and Kure Atoll. They reported that the beaches of some of the islands and atolls look like a "recycling bin" of plastics. They found numerous cigarette lighters, some with fuel still in them, as well as caps from plastic bottles and all kinds of plastic toys and other debris.

Wildlife live among the debris. The islands are home to sea turtles, monk seals, and a variety of birds, including albatross. Attracted to the colored plastic but not knowing what it is, the birds pick it up and eat it (Figure 12.21). Plastic rings from a variety of products have also been found around the snouts of seals, which ultimately then starve to death, and are also swallowed by sea turtles. In some areas, the carcasses of albatrosses litter the shorelines.

What can we do about this problem? The solution to the problem of plastics in the ocean is to be more conscious of recycling plastic products to ensure that they do not enter the marine environment. Collecting the plastic items on beaches where they accumulate is a step in the right direction, but keeping it from landing on the beaches in the first place is even better.

Courtesy of Cynthia Vanderlip/Algalita Marine Research Foundation

▪ **FIGURE 12.21**
An albatross killed on a remote Pacific island by ingesting a large amount of plastic and other debris.

12.9 Can We Make Ocean Fisheries Sustainable?

The first problem is that much of the open ocean is a "commons"—that is, it is international territory, used by many nations (in Chapter 17, on environmental economics, we discuss how a commons can affect environmental issues). In 1977 the U.S. government, concerned about overfishing in U.S. waters by foreign factory ships, extended the nation's coastal waters from 12 miles to 200 miles (from 19 to 322 kilometers). Many nations did the same, and the move to push international waters 325 kilometers from their coasts has turned some fisheries from completely open common resources to national resources open only to domestic fishermen.

How do we go about managing fisheries? In fisheries, there are four main management options:

▪ Establish a total catch quota for the entire fishery and allow anybody to fish until that total is reached.

▪ Issue a restricted number of licenses but allow each licensed fisherman to catch many fish.

▪ Tax the catch (the fish brought in) or the effort (the cost of ships, fuel, and other essential items).

▪ Allocate fishing rights—that is, assign each fisherman a quota, which can be transferred or sold.

With Option One, the fishery is closed when the total catch quota is reached. Whales, Pacific halibut, tropical tuna, and anchovies have been regulated this way. However, when this option was used in Alaska, all the halibut were caught in a few days, and restaurants no longer had halibut for most of the year. This led to a change in policy: The total-catch approach was replaced by the sale of licenses.

Total-catch quotas encourage bigger fishing fleets. Although regulating the total catch can be done in a way that helps the fish, it tends to increase the number of fishermen and the capacity of vessels so that a fleet can catch as many fish as possible before a fishery is closed. The end result is a hardship for fishermen. Recent economic analysis suggests that Option Three, taxing the catch, can work to the best advantage of fishermen and fish. Similar results are achieved by Option Four, allocating a transferable and salable quota to each fisherman.

Which method works best depends on the fishery and the fishermen. Indeed, no matter what environmental resource you attempt to manage, deciding which management method will make the best use of the resource will depend on the specific characteristics of both the resource and the users. The tools of economics can help in making the decision.

Suppose you went into fishing as a business and expected a reasonable increase in that business in the first 20 years. As we said earlier, when the world's ocean catch of fish doubled in the 20 years between 1960 and 1980, that was an average annual increase of 3.7%. From a business point of view, even assuming you can sell all fish you catch, that is not rapid sales growth. But it is a heavy burden on a living resource.

Sometimes attempts to improve the economics of fishing have unexpected results. For example, to encourage domestic fishermen, the National Marine Fisheries Service provided loan guarantees for replacing older vessels and equipment with newer boats that had high-tech equipment for locating fish. However, during this same period, demand for fish increased as Americans became more concerned about cholesterol in red meat. Consequently, the number of fishing boats, the number of days at sea, and fishing efficiency increased sharply, and as much as 50–60% of the populations of some species were landed each year.

Harvest quotas were tried, but proved unpopular. In 1982, the New England Fisheries Management Council attempted to enforce harvest quotas but rescinded the order under pressure from the commercial fishing industry. In the wake of a bitter controversy that resulted from this decision, as well as further declines in fish populations, the council later prohibited fishing at certain times and in certain areas, mandated minimum net sizes, and set quotas on the catch.

In 1992, the council adopted a more ambitious plan. Its goal was to cut the fishing effort in half by 1997—but without the use of quotas and without the removal of current fishermen. The council decided to issue a limited number of fishing permits, limit the number of days at sea, and use high-tech monitoring equipment to ensure compliance.

The National Marine Fisheries Service prefers individual transferable quotas (see Option Four above). Called ITQs, these permits allow boat owners to harvest a fixed amount of fish each year or to lease, sell, or bequeath the permits to others. Although ITQs have been successful in several U.S. fisheries and in some other countries, some small operators in New England fear that large corporations will buy up permits and dominate the industry.

Marine Sanctuaries

Marine sanctuaries might help restore fish populations. In 1972 the U.S. government passed the National Marine Sanctuaries Act, which allowed the Department of Commerce to designate and manage marine sites that had conservation, ecological, recreational, historical, aesthetic, scientific, or educational value. No fishing nor habitat modifications would be allowed in sanctuaries set up for conservation. Since that time 14 marine sanctuaries have been established, including: one in the Florida Keys to protect coral reefs; others at the Channel Islands, Monterey Bay, Gulf of the Farallones, and Cordell Bank in California to protect fish and shellfish habitats; Flower Garden Banks, a coral oasis in a sea of oil rigs in the Gulf of Mexico; a sanctuary in the Hawaiian Islands for humpback whales; one off the coast of North Carolina to protect the sunken wreck of the *Monitor*, the Civil War ironclad ship; and one in the Great Lakes to protect sunken ships as archeological sites.

The most recent addition to the program, a sanctuary off the coast of the state of Washington, adjacent to Olympic National Park, contains some of the world's richest fish and shellfish habitats (Figure 12.22). This Olympic Coast National Marine Sanctuary is also used by 29 species of marine mammals, and has one of the largest seabird colonies in the United States.

Sanctuaries can be invaluable in helping populations recover. The idea is to set aside areas within territorial waters where a nation's laws apply, in hopes that recovered populations will not only increase within the sanctuaries but also provide a breeding ground that can help repopulate areas where fishing is going on. Other nations have established similar protected marine areas, using a variety of names. Such areas seem to be a necessary part of the solution—though not the entire solution—to the great declines in fish populations.

The fact is, sustainable harvesting of wild biological resources usually isn't profitable. Few wild biological resources can sustain a harvest that is big enough to meet even low requirements for a growing business (as we discuss in greater detail in Chapter 18). The United States found this out with harvests of white pine in Michigan, where 19 million acres of white pine—believed to be a resource that could never be depleted—were clear-cut and are no longer sustainable. We learned this lesson also from the fate of the bison, discussed in an earlier chapter, and it is true for whales as well. There have been a few exceptions, such as the several hundred years of fur trading by the Hudson's Bay Company in northern Canada. But past experience suggests that economically profitable sustainability is unlikely for most wild populations.

With that in mind, we can turn to farming fish—aquaculture. This has been an important source of food in China for centuries and is an increasingly important food source worldwide.

Altrendo/Getty Images, inc.

(a)

■ **FIGURE 12.22**
Marine sanctuaries.
(a) An island within the Olympic Coast National Marine Sanctuary, and (b) a map showing that this sanctuary is at the northwestern tip of the Olympic Peninsula of the state of Washington. [*Source: Olympic Coast Marine Sanctuary. Available at http://sequim.com/ocmns/ocmns.html. Accessed September 12, 2005.*]

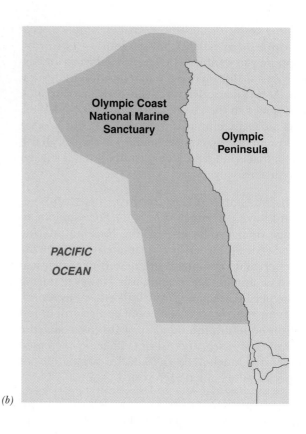

(b)

Aquaculture and *Mariculture*

The farming of food in both oceans and freshwater is growing rapidly and could be one of the major solutions to providing nutritional quality. Popular freshwater aquacultural species include carp, tilapia, catfish (in the southern and midwestern United States), and trout (United States). But most farmed fish and shellfish are marine species. Popular **mariculture** species include oysters, mussels (France, Spain, and Southeast Asia), shrimp, yellowtail (important in Japan), eels (China), salmon (Norway and the United States), plaice, sole, the Southeast Asian milkfish (Great Britain), and sturgeon (Ukraine).

Aquaculture is not a new idea. Although relatively new in the United States, aquaculture has a long history elsewhere, especially in China, where it is an ancient practice that can be traced back to a treatise on fish culture written by Fan Li in 475 B.C.[25] In China, at least 50 species are grown, including finfish, shrimp, crab, other shellfish, sea turtles, and sea cucumbers (a marine animal, not a vegetable). In the Szechuan area of China, fish are farmed in more than 100,000 hectares (about 250,000 acres) of flooded rice fields. On a per-area basis, aquaculture can be extremely productive, especially because flowing water brings food into the pond or enclosure from outside.

Mariculture has grown rapidly and will likely continue to do so, as illustrated by the increase in mariculture of blue mussels (Figure 12.23). Mariculture of abalone

and oysters is also increasing. In the United States and Canada, for example, researchers are seeking ways to attract the young, swimming stages of these shellfish to areas where they can be conveniently grown and harvested.

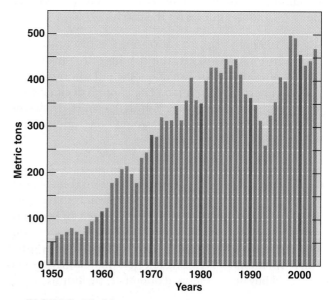

■ **FIGURE 12.23**
World production of blue mussels, a species favored in Europe and grown there. These mussels can attract a price as high as $1,700 per ton, and in recent years production has reached 500,000 tons, a potential value of $850 million!

Oysters and mussels are grown on rafts that are lowered into the ocean, a common practice in the Atlantic Ocean in Portugal and in the Mediterranean in such nations as France. These shellfish are filter feeders, obtaining food from water that moves past them. Because a small raft is exposed to a large volume of water and thus a large volume of food, rafts can be extremely productive. Mussels grown on rafts in bays of Galicia, Spain, produce 300 metric tons per hectare, whereas public harvesting grounds of wild shellfish in the United States yield only about 10 kilograms per hectare.[26] Oysters and mussels are also grown on artificial pilings in the intertidal zone in the state of Washington.

Unfortunately, aquaculture can create its own environmental problems. One of the best-known environmental issues involves Atlantic salmon, the major species of salmon grown as a crop. In 2000, Atlantic salmon were put on the endangered-species list. One of the explanations suggested for the problems of this species is extensive mariculture of Atlantic salmon in such places as the coast of Maine. Observers have raised two concerns. First, salmon excrement and excess food provided for the farmed salmon pollute the habitat of the wild salmon. Second, breeding between native salmon and farmed salmon that are not native to a locale may lead to genetic strains that are less fit for a particular stream.

12.10 Conservation of Whales and Other Marine Mammals

It's easier to save a species than to both harvest *and* save it. Fisheries biologists have managed marine mammals by using primarily the same concepts (such as the logistic growth curve) that have not worked very well in fisheries management. But although marine mammals live in the same, highly variable ocean as ocean fish, the goal of the Marine Mammal Protection Act, enacted by the United States in 1972, is an **optimum sustainable population** (OSP) rather than a **maximum sustainable yield**. An OSP means the largest population that can be sustained indefinitely without negatively affecting the ability of that population or its ecosystem to continue to support that same number. Because the primary goal in managing marine mammals has been to conserve species rather than maximize their yield, the results have tended to be better.

Fossil records show that all marine mammals originally lived on land. During the last 80 million years, several separate groups of mammals returned to the oceans and became adapted to marine life. The degree of adaptation depends on how long ago a group returned to the water. Each group of marine mammals shows a different degree of adaptation to ocean life. Understandably, those that began the transition longest ago have had time to become more fully adapted to life in the ocean. Some marine mammals—such as dolphins, porpoises, and great whales—are so well adapted that they can complete their entire life cycle in the oceans and cannot move on land. Others, such as seals and sea lions, spend part of their time on shore.

There are two major categories of whales: baleen and toothed. Figure 12.24 shows a toothed whale and a baleen whale. Sperm whales are the only great whales that are toothed. The rest of the toothed

Francois Gohier/Photo Researchers Inc.

Seapics.com

(a)

(b)

■ **FIGURE 12.24**
Photographs of (a) a sperm whale and (b) a blue whale. Sperm whales are the only great whales that have true teeth. The blue whale and other baleen whales have modified teeth that act as filters.

group are smaller whales, dolphins, and porpoises. The great whales in the baleen group, such as the blue whale, have highly modified teeth that look like giant combs and act as water filters. Baleen whales feed by filtering ocean plankton.

Drawings of whales date back to 2200 B.C.[27] Eskimos used whales for food and clothing as long ago as 1500 B.C. In the 9th century, whaling by Norwegians was reported by travelers, whose accounts were written down in the court of the English king Alfred. The earliest whale hunters killed these huge mammals from the shore or from small boats near shore, but gradually whale hunters ventured farther out. In the 11th and 12th centuries, Basques hunted the Atlantic right whale from open boats in the Bay of Biscay, off the western coast of France. The boats returned to land once the whale hunt was finished, and the catch was processed on shore.

Eventually, whaling became *pelagic* — whalers remained at sea for long periods, hunting the open ocean and processing their catch on board as well. This was made possible by the invention of furnaces and boilers for extracting whale oil at sea. Thus, pelagic whaling was a product of the Industrial Revolution. With this invention, whaling grew as an industry. American fleets developed in New England in the 18th century, and by the 19th century the United States dominated the industry, providing most of the whaling ships and even more of the crews.[28, 29]

Whales provided many 19th-century products. Whale oil was used for cooking, lubrication, and lamps. Whales provided the main ingredients for the base of perfumes. The modified, elongated teeth of baleen whales are flexible and springy and were used for corset stays and other products before the invention of inexpensive steel springs. But actually, although 19th-century whaling ships were made famous by such novels as *Moby Dick*, more whales were killed in the 20th century.

The decline in most species of whales is a global environmental issue. Conservation of whales has been a concern for many years. Attempts to control whaling began with the League of Nations in 1924. The first agreement, the Convention for the Regulation of Whaling, was signed by 21 countries in 1931. In 1946, a conference in Washington, D.C., initiated the International Whaling Commission (IWC), and in 1982 the IWC called a temporary halt to commercial whaling. Currently, 12 of approximately 80 species of whales are protected.[30]

The IWC has played a major role in almost eliminating commercial whaling. Since the formation of the IWC, no species has become extinct, the total take of whales has decreased, and harvesting of species that are considered endangered has ceased. Some of the great whales remain rare. The largest, the blue whale, appears to have recovered somewhat but remains endangered, numbering somewhere around 1,000.[31] Gray whales are now relatively abundant, numbering about 26,000.[32] Global climate change, pollution, and ozone depletion now pose greater risks to whale populations than does whaling.[33]

The establishment of the IWC was a landmark in wildlife conservation. It was one of the first major attempts by a group of nations to agree on a reasonable harvest of a biological resource. The annual meeting of the IWC has become a forum for discussing international conservation, working out basic concepts of maximum and optimum sustainable yields, and formulating a scientific basis for commercial harvesting. The IWC demonstrates that even an informal commission whose decisions are accepted voluntarily by nations can be a powerful force for conservation. We now realize that management policies for marine mammals must be expanded to include ecosystem concepts and the understanding that populations interact in complex ways.

Dolphins and Other Small Whales

Many of the small whales are killed accidentally. Among the many species of small "whales" (cetaceans) are dolphins and porpoises, more than 40 species of which have been hunted commercially or have been killed inadvertently (not on purpose) by other fishing efforts.[34] A classic case is the inadvertent catch of the spinner, spotted, and common dolphins of the eastern Pacific. Because these fish-eating mammals often feed with yellowfin tuna, a major commercial fish, more than 100,000 of these dolphins have been netted and killed inadvertently in recent years.[35]

Groups are working together to solve this problem. The U.S. Marine Mammal Commission and commercial fishermen have cooperated in seeking ways to reduce dolphin deaths. Research into dolphin behavior helped in the design of new netting procedures that trap far fewer dolphins. The attempt to reduce dolphin mortality illustrates cooperation among fishermen, conservationists, and government agencies and also highlights the role of scientific research in managing renewable resources.

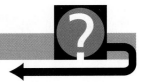

Return to the Big Question

Can we learn to manage the oceans' resources?

Probably the ocean will always be only a partially manageable system, because it is so large and complex, and because there is so much interaction between the ocean and the atmosphere. But we could do a much better job of sustaining our ocean resources if we could (1) get away from the old beliefs about how fish populations grow, (2) stop dumping plastics, other sediments, and toxic chemicals into our rivers and streams and thereby into the oceans, and (3) work out international agreements that all the nations of the world would abide by. Obviously, none of these is easy.

Summary

- Oceans cover 70% of Earth's surface and contain many kinds of ecosystems and species.
- Marine fish are an important food, and world harvests of fish are large, but the fish populations on which the harvests depend are easily exploited, generally declining, and difficult to restore.
- The main reasons that fish populations are shrinking are overharvesting, pollution, and habitat destruction.
- We desperately need new approaches to forecasting acceptable harvests and workable international agreements to limit catch. This is a major environmental challenge, needing solutions within the next decade.

Key Terms

aquaculture
benthos
carrying capacity
intertidal

mariculture
maximum sustainable yield
optimum sustainable population
pelagic

Getting It Straight

1. What can we do to reduce the deaths of seabirds from waste plastics?

2. How will the growth of China's human population change commercial fisheries in the Atlantic Ocean?

3. Which fish are most threatened in our oceans? What can we do to begin to help sustain the aquatic environments on Earth?

4. Why is it becoming more difficult to catch fish in the United States? Why are fish populations declining?

5. What are some environmental problems affecting Chesapeake Bay?

6. Where do most fish we eat come from?

7. Which aquatic habitats are most at risk for destruction? Explain why.

8. Are humans putting coral reefs in danger? Explain.

9. Identify three types of pollution that have become a concern for aquatic/ocean environments.

10. Can managing fisheries help improve aquatic life and maintain aquatic environments?

11. Are marine sanctuaries making a difference in the fight to protect our aquatic life and aquatic environments?

What Do You Think?

1. Design a shrimp farm that would not pollute the nearby ocean waters. (Refer to the opening case study.)

2. What would be an optimum sustainable population of great white sharks? Consider the effects of sharks on people, the roles of sharks in ocean ecosystems, and the commercial harvest of sharks for food. Is it possible to combine these three concerns and achieve a single optimum population of sharks?

3. Should all ocean fish harvested for food be from aquaculture? Present an argument for or against this idea.

4. Could the species that live in and near deep-sea hydrothermal vents become a major source of food? Explain.

Pulling It All Together

1. It has been suggested that if the Gulf Stream stopped flowing, Europe would become very cold, perhaps experiencing another ice age. Discuss what else might help keep western Europe warm. Refer to the climates of cities in the Pacific Northwest of the United States and Canada.

2. What is necessary to make the commercial harvest of oysters in the Chesapeake Bay sustainable? Making use of available reference sources, consider whether this will ever be possible, given the multiple uses of the bay and its large watershed.

3. The International Whaling Commission is an organization of nations interested in the problem of harvesting whales. Only a few countries harvest whales today, including Japan, Norway, and (because of treaties with the Eskimo) the United States. Those harvesting whales today claim that it is important for cultural reasons, and that sustaining human cultures is as important or more important than sustaining whales. Using available resources, present an argument for or against the continued harvest of whales.

Further Reading

Ellis, Richard. 2004. *The Empty Ocean.* Washington, DC: Island Press.—More stories about species in the oceans and what has happened to them because of harvesting, pollution, and habitat destruction.

Russell, Dick. 2005. *Striper Wars: An American Fish Story.* Washington, DC: Island Press.—A book about the history of the decline and recovery of the striped bass, typical of the problems with many marine fish.

Thurman, Harold V., and Alan P. Trujillo. 2001. *Essentials of Oceanography.* New York: Prentice Hall.—An introduction to the physical and chemical ocean, as well as some more information on biological oceanography.

Digital Vision/Getty Images, Inc.

13

Earth's Atmosphere and Climate

Big Question

Global Warming Is Happening: What Part Do We Play?

Learning Objectives

Global warming is a difficult and controversial subject, but it boils down to a set of questions:

- Is the temperature rising?
- If so, why? What causes the climate to change, and what may be causing it to get warmer?
- Why are greenhouse gases increasing in the atmosphere?
- What is the greenhouse effect, and how does it warm the Earth's surface?
- Are human activities at least part of the cause of the recent global warming? A significant part?

- Why should we care if global warming is occuring? What will be its effects, and how damaging will they be?
- Can we do anything to stop it, or slow it, or keep it from getting much hotter?
- Can we do anything to lessen the effects of global warming?

In attempting to answer these questions, we will use scientific analysis. By the time you finish this chapter, you should understand the answers and their scientific bases.

Polar bears are the largest carnivore in North America; they can reach 2.5 meters (about 8 feet) in length and weigh over 700 kilograms (about as much as a three-quarter-ton truck). Today, some claim that polar bears may be in trouble in western Hudson Bay, and the early breakup of sea ice each spring is thought to be the problem. Sea ice everywhere has thinned by as much as 40% over the past 40 years and now covers 10% less area than it used to, presumably in response to global warming. Some of global warming's most significant impacts on wildlife are occurring in the Arctic, because temperature changes are more dramatic there than at lower latitudes.

For polar bears in the western Hudson Bay, sea ice is a critical habitat for hunting seals (Figure 13.1). In the spring, polar bears prey on the seals to fatten up before the annual melting of the sea ice. After the ice melts, the bears move to land, where they go on a fast that may last for months. Because pregnant female bears fast for up to eight months, they need a large reserve of fat to carry, care for, and feed their cubs until they can again return to the ice to feed.

Since 1981, when studies on polar bears in the western Hudson Bay began, the bears there have weighed less than average and have given birth to fewer cubs. Biologists have established a link between the decline in polar bears and the earlier breakup of sea ice. If the trend continues, there will be fewer and fewer bears. Another consequence is that bears will be forced onto land earlier, where dangerous contact with people is more likely.[1,2]

Our short story of polar bears in the Hudson Bay suggests that climate change can cause serious problems in the biosphere.

Johnny Johnson/DRK Photo

▪ **FIGURE 13.1**
Polar bear moving through thin ice while hunting for seals in Hudson Bay.

13.1 Is the Global Temperature Rising?

Yes, it is—as we learned in the opening Case Study and it's rising globally[3,4]. And here is more evidence, not only of rising temperatures but also that the temperature is always changing, at every time scale and over long periods, as shown in Figure 13.2. In the last 2 million years, glacial ages (ice ages) have come and gone. The periods between glacial ages, called **interglacial,** have been warm, but today's temperatures are the warmest known during the last 2 million years.

Has the temperature risen steadily? The climate began to warm around 1850 but began to cool again in the 1940s (Figure 13.2e). The temperature leveled off in the 1950s, then dropped further during the 1960s. After

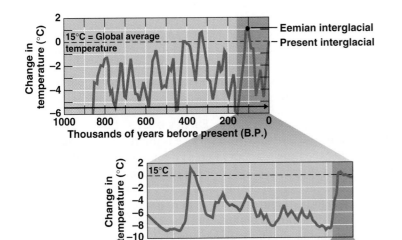

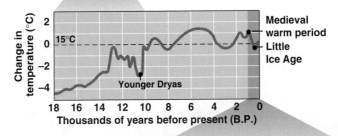

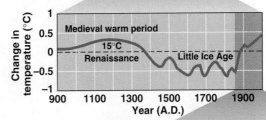

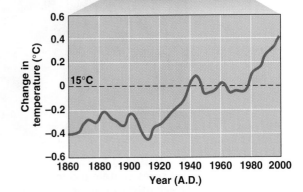

■ **FIGURE 13.2**

Earth's changing climate.
Each graph shows changes in Earth's average temperature, beginning with the last 800,000 years, with each subsequent graph zooming in on a shorter time period. In every case—at every time scale—Earth's temperature varies. Also, temperatures in the past 100,000 years are among the warmest in the past million, and the average temperature is continuing to rise. [*Sources:* UCAR/DIES. Science capsule, changes in the temperature of the Earth. *Earth Quest* 5(1), Spring 1991; Houghton, J. T., G. L. Jenkins, and J. J. Ephranns, eds. *Climate change: The science of climate change.* Cambridge: Cambridge University Press, 1996; U.K. Meteorological Office. *Climate change and its impacts: A global perspective,* 1997.]

that, the average temperature climbed steadily through the 1990s. In the last 100 years, the global average annual temperature has risen approximately 0.6°C (1°F).[5-7] The 1990s were the warmest decade in the 142 years that temperatures have been recorded, and in the last 1,000 years according to geologic data.[8-10]

It has not only gotten warmer but it has also gotten warmer faster. Since the mid-1970s, the temperature has risen about three times faster than in the preceding 100 years. The ten warmest years have all occurred since 1990, and the five warmest since 1997. The warmest year on record was 2005, with 1995 second and 2002 third.

In the United States, 2003 was cooler and wetter than average in much of the eastern part of the country, and warmer and drier in much of the western part. Ten western states were much warmer than average: New Mexico had its warmest year on record. Alaska was warmer in all four seasons—indeed, 2003 was one of the five warmest years since Alaska began taking measurements in 1918. In 2003, Europe experienced summer heat waves, with the warmest seasonal temperatures ever recorded in Spain, France, Switzerland, and Germany. About 15,000 people died in heat waves in Paris during the summer. Heat and drought contributed to severe wildfires in Australia, southern California, and British Columbia.

The overall picture supports the view that global warming is occurring. A year or two of high temperatures, drought, heat waves, and wildfires does not by it-self indicate longer-term global warming. However, the persistent overall trend of rising temperatures over three decades is compelling evidence that global warming is real and happening.

13.2 What Causes Climate Change of Any Sort, and What Is Making It Get Warmer?

Studies of ice cores (Figure 13.3) and other geologic records that allow reconstruction of temperature show that the average temperature of Earth's surface has varied over time, as we saw in Figure 13.2. What could have caused that?

The answer lies in Earth's energy balance. The temperature of Earth's surface is the result of a balance between the amount of energy coming in from the sun and the amount of energy radiated out by the Earth. This energy exchange is referred to as Earth's energy balance (Figure 13.4). When energy from the sun reaches the top of the atmosphere, 30% of it is reflected immediately. Of the remaining 70%, approximately 25% is absorbed in the upper atmosphere and 45% reaches Earth and is absorbed at the surface.

An important concept here is **albedo**, which is the percentage of light received by a surface that is reflected and scattered—the rest of the energy is absorbed by the surface. The albedo of a surface has important

Roger Ressmeyer/Corbis Images

■ **FIGURE 13.3**
Ice cores contain a climate record.
A scientist examines a glacial ice core stored in a freezer.

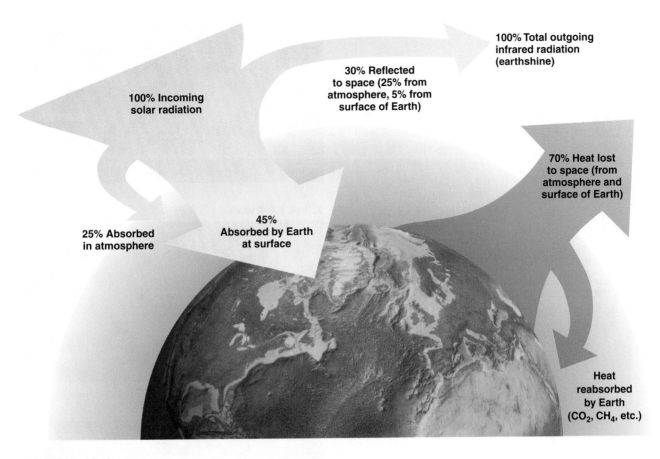

■ FIGURE 13.4
Earth's energy balance.
[*Source:* Modified from N. L. Pruitt, L. S. Underwood, and W. Surver. *Bio inquiry.* New York: Wiley, 2002.]

effects. This is why people who live in desert climates wear white, which has a higher albedo—reflects more of the light—than black. Similarly, snow reflects much of the light that falls on it—it has a high albedo—while tropical forests have a low albedo, absorbing much of the light. There is positive feedback from albedo. Snow accumulates where it is cold, but because snow has a high albedo, it reflects most of the light coming in and keeps the surface cold. A lot of snow cools an area and can increase the amount of precipitation that falls as snow, creating positive feedback. Some typical albedo is shown in Figure 13.5.

Energy may take a complex path from Earth's surface to space. As the energy from the sun is absorbed by plants, soils, rocks, water, and other surface materials, these become warmer and re-radiate the energy as infrared radiation, which is eventually lost to space from Earth's upper **atmosphere**. Part of the infrared radiation is reabsorbed by gases in the atmosphere, including carbon dioxide and methane, known as *greenhouse gases* (discussed later in the chapter). Some of this is, in turn, radiated back to the surface, warming

the Earth. But ultimately it, too, is lost to space as heat, completing Earth's energy balance.

The warmer an object, the more energy it radiates. Earth, like all physical objects, reacts to an increase in incoming energy by warming up, and the warmer it gets, the more energy its surface radiates. In addition, the warmer it gets, the shorter the dominant wavelengths of the energy it radiates. The very hot sun radiates mostly in infrared (invisible) wavelengths and visible wavelengths. Earth radiates mainly in the infrared.

Variation in the Sun's Energy May Be One Reason for Climate Changes

The sun's energy seems to have varied over the ages. Scientists have figured out how intense sunlight was in the past by looking at atoms emitted from the sun that landed on glaciers and were buried in the ice.[11] Ice cores taken from glaciers reveal that during a medieval period, around A.D. 1100–1300, the amount of solar energy reaching Earth was relatively high compared with today. These cores also suggest that the sun radiated less energy during the 14th century, which coincided with

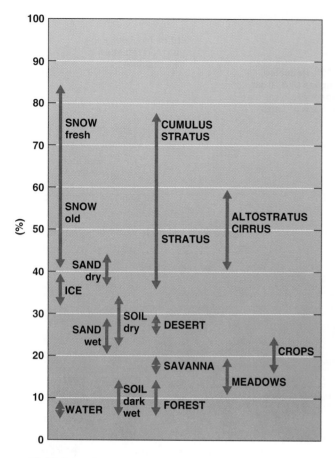

■ **FIGURE 13.5**

Typical ranges of albedo for common surfaces.
Interestingly, fresh snow reflects much more light
than sand, but ice does not, and water has an albedo
as low as that of forests.

the beginning of a cold period, called the Little Ice Age,
that lasted from about 1450 to 1850 (see Figures 13.2
and 13.3). Thus, we conclude that the variability of solar
energy may well explain some of the changes in Earth's
climate during the past 1,000 years. The effect, however,
seems relatively small.[12]

Milankovich Cycles Are Another Possible Explanation

Some temperature cycles stem from Earth's wobble in an elliptical orbit. The spinning Earth wobbles
like a top on its elliptical (not perfectly circular) orbit
around the sun. Its wobble and its elliptical orbit make
it impossible for Earth to keep a constant position in relation to the sun. Scientist Milutin Milankovitch, realizing that Earth's position relative to the sun is a factor in
how much sunlight reaches Earth, suggested that this
could account for variations in climate (Figure 13.6).
This climate effect could occur because most of the land

1. Now

(a)

2. In c. 5,250 years

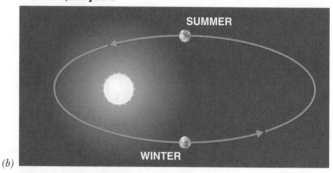

(b)

3. In c. 10,500 years

(c)

■ **FIGURE 13.6**

Milankovitch cycles.
Earth revolves around the sun in an elliptical orbit. It also
rotates—spins like a top—causing our night and day. But
like a top, it wobbles slightly. The combination of the
wobble and the elliptical orbit appear to affect how cold
winters are and how warm summers are. Winters are
different when the Northern Hemisphere faces away from
the sun at the time it is closest to the sun and when it is
farthest from the sun. The Milankovitch theory is that this
causes ice ages and interglacial ages. *[Source:
http://www.ldeo.columbia.edu/~polsen/nbcp/cmintro.html.
Accessed September 24, 2005.]* (a) In this orbit, Earth
receives 20–30% more sunlight when nearest the sun (listed
as "winter" here, for the Northern Hemisphere's season)
than at the most distant point (listed as "summer" here).
Similarly, in (c), "summer" in the Northern Hemisphere
occurs when the Earth is closest to the sun, and receives
20–30% more sunlight than at the most distant point, listed
as "winter" for the Northern Hemisphere.

on Earth is in the Northern Hemisphere, so summers and winters act differently in the Northern Hemisphere than in the Southern Hemisphere. He also showed that variations in Earth's orbit around the sun followed a cycle of approximately 100,000 years, which corresponds with the major glacial and interglacial episodes. Curiously, ice ages are promoted when there are warmer winters in the Northern Hemisphere, where most of the land is, because the warmer winter air produces more snowfall, which builds up on the land, and the cooler summers are less likely to melt all the snow.[13]

In addition, Milankovitch suggested that climatic cycles of about 40,000 and 20,000 years could also result from changes in Earth's tilt and wobble. Milankovitch cycles may account for many, but not all, of the large variations in Earth's climate. One way to think about these cycles is that they set up conditions where an ice age is either likely or unlikely. When it is likely, other factors come into play to determine whether an ice age actually happens.

Volcanoes Can Alter Climate

Volcanic eruptions cool the climate in two ways: The dust that they release into the atmosphere reflects sunlight back out into space, so less sunlight reaches Earth, and the dust's smaller particles provide surfaces for water to condense on, forming clouds that also reflect incoming solar energy away from Earth.[14] How long a particle of dust stays in the atmosphere depends on its size. The smallest particles, called aerosols, are less than 10 microns across and remain in the atmosphere for a long time, bounced around by the air molecules.

The eruption of Mount Pinatubo in the Philippines in 1991 illustrates volcanic effects on climate. The eruption sent volcanic ash up 30 kilometers (19 miles) into the stratosphere. The aerosol cloud of ash, containing 20 million tons of sulfur dioxide, remained in the atmosphere, circling Earth for several years. These particles of ash and sulfur dioxide scattered incoming solar radiation, resulting in a slight cooling of the global climate during 1991 and 1992.[15]

Volcanic eruptions are believed to have contributed to the cooling in the Little Ice Age, from about 1450 to 1850[16] (see Figure 13.2).

Dust from Our Own Activities Also Cools the Climate

Our aerosol emissions have reduced global warming. Aerosol emissions from human activities have increased since the Industrial Revolution (see the discussion of air pollution in Chapter 14). These air-pollution particles have reduced the amount of sunlight reaching Earth today by as much as 10% in some regions, and reduced global warming due to greenhouse gases by an estimated 50%. This is known as **global dimming.**[17]

Variations in Ocean Currents May Affect the Climate

Ocean currents together with prevailing winds can warm or cool our planet. Variations in ocean currents, along with variations in the atmosphere's prevailing winds, may make Earth's climate fundamentally unstable, resulting in surprisingly rapid changes. One study suggests that during the past 4,000 years there has been a cycle of about 1,500 years that may help explain a medieval warm period that peaked around A.D. 1100, followed by the colder Little Ice Age mentioned earlier (from about A.D. 1450 to 1850). Some scientists suggest that the present warming trend is just another natural warming cycle of the same length, and thus can be expected to continue until around the year 2400.[18] If this is correct, then any warming caused by human activities would be added to a system that is already slowly warming.[19]

The *Gulf Stream* is part of a major ocean circulation that affects climate. The strong northward movement of upper warm waters of the **Gulf Stream** in the Atlantic Ocean passes near the coast of Florida and flows to Europe. There it plays a part in keeping Europe warm. When the stream arrives near Greenland, its waters cool, become denser, and begin to sink to the bottom.[20] The cold, deep current then flows southward, then eastward, and finally northward in the Pacific Ocean. Upwelling in the north Pacific starts the warm shallow current again. The flow in this system is huge (20 million cubic meters per second, about equal to 100 Amazon Rivers). If this giant current were to shut down, there would be an effect on Europe's climate.[21]

El Niño: A Special Climate Phenomenon Linked to Ocean Currents

A curious and historically important climate change linked to variations in ocean currents is the Southern Oscillation, known informally as **El Niño.** From the time of early Spanish settlement of the west coast of South America, people observed a strange event that occurred about every seven years. Usually starting around Christmas (hence the Spanish name El Niño, referring to the little Christ child), the ocean waters would warm up, fishing would become poor, and seabirds would disappear. The explanation (which you may recall from our discussion in Chapter 12 about oceans) is that under normal conditions there are strong vertical, rising currents, called upwellings, off the shore of Peru.

Upwellings cool the surface water and help to support bird and fish populations. The upwellings bring cold water up from the depths, along with important nutrients that promote the growth of algae (the base of the food chain) and thus produce lots of fish.

Seabirds feed on those fish and live in great numbers, nesting on small islands just offshore.

El Niño occurs when upwellings slow or cease—that is, the cold vertical currents become weak or stop rising altogether. As a result, nutrients decline, algae grow poorly, and so do the fish, which either die, fail to reproduce, or move away. The seabirds, too, either leave or die.

Scientists learned that upwellings are driven by the prevailing winds. The direction of the winds is influenced by the topography of the land and the ocean. When these winds fail, El Niño. occurs. Scientists now understand that the Southern Oscillation is a seesawing pattern. Under normal conditions, air pressure is high in the *eastern* tropical Pacific and low in the *western* tropical Pacific. When this pattern reverses, the winds and upwellings off Peru fail, and El Niño begins. Modern satellite remote sensing of the ocean surface temperature shows the warm ocean waters forming a long extension into the midst of the Pacific Ocean from Peru during El Niño (Figure 13.7). When the normal pattern reappears, El Niño ends.[22]

El Niño is important to our discussion of climate change for two reasons. First, it turns out that El Niño is actually a global event, involving changes in rainfall, temperature, and frequencies of storms in many parts of the world. Second, one of the forecasts about global warming is that El Niño is likely to become more common and more intense as Earth's average temperature rises.

Past events now thought to have been linked to El Niño include the great flood of 1993 on Mississippi River (which inundated St. Louis and many other towns); California floods in 1995; drought conditions in South America, Africa, and Australia, and resulting large brush fires in Australia; and a period in which serious storms, such as hurricanes, were infrequent in the North Atlantic, sparing Florida, Georgia, and other parts of the eastern United States and the Gulf of Mexico from serious storm damage.[23]

Elsewhere, according to a U.S. National Academy of Sciences report, "Severe El Niño events have resulted in a few thousand deaths worldwide, left thousands of people homeless, and caused billions of dollars in damage" (Figure 13.8). Yet during that period, "residents on the northeastern seaboard of the United States could credit Elño with milder-than-normal winters (and lower heating bills) and relatively benign hurricane seasons.".[24]

Scientists are still not sure what is the first cause of El Niño, but they do understand the various linked events, and have proposed a number of theories to explain them. They believe that the comings and goings of upwellings have something to do with the way the com-

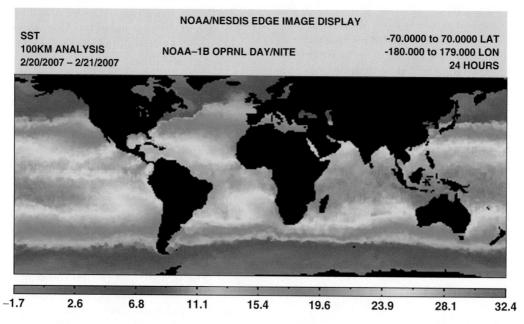

■ **FIGURE 13.7**

Cold water and chemical nutrients may cease rising during El Niño. Blue is cold water; red is the warmest water; green is cool; yellow warm.
Satellite remote sensing of ocean temperature during El Niño shows warm water reaching westward from Peru far into the Pacific Ocean. In normal years, upwellings keep the surface water west of Peru cold, more like the rest of the Pacific. Blue shows cold water; red is the warmest water. Green is cool; yellow is warm [*Source:* Available at http:/envisat.esa.int/live/envisat_live_09.htm. Accessed February 21, 2007.]

©Vince Streano/Corbis

■ **FIGURE 13.8**
Results of the 1998 El Niño event.
Flooded area in Lakeport, California.

paratively dense ocean water interacts with the much lighter, less dense gases of the atmosphere. This interaction is complex because both the ocean and the atmosphere are fluid systems that are highly dynamic, always moving, and always changing, and changes in each affect the other. El Niño became stronger and more frequent in the 1990s, suggesting a link to the global temperature rise, but that is still an open scientific question.

The bottom line is (1) that El Niño is a frequent natural climate change with many effects on the environment and on people, and (2) that global warming may increase El Niño's frequency, with many effects on our lives.

13.3 What Is the Greenhouse Effect, and How Does It Warm Earth's Surface?

Earlier we introduced the idea of the Earth's energy balance. Here, to understand what greenhouse gases do, we add that the temperature at or near the surface of Earth is determined by four main factors:

1. the amount of sunlight Earth receives
2. the amount of sunlight Earth reflects
3. retention of heat by the atmosphere
4. evaporation and condensation of water vapor

Playing Ping-Pong with infrared rays. The sun gives off a wide range of electromagnetic radiation, over many wavelengths. As we noted earlier, most of the sun's energy reaches Earth as visible and infrared radiation. This sunlight warms both Earth's atmo-

sphere and surface. The surface reflects this energy back to the atmosphere,[25] but certain gases in Earth's atmosphere send it back again, making Earth warmer than it otherwise would be.

What are "greenhouse gases"? Earth's atmosphere contains many gases. The major ones are nitrogen (78%), oxygen (21%), argon (0.9%), and carbon dioxide (0.03%). The atmosphere also contains trace amounts of methane, ozone, hydrogen sulfide, carbon monoxide, oxides of nitrogen and sulfur, hydrocarbons, chlorofluorocarbons (CFCs), and various particulates, including aerosols (small particles). Water vapor is also present in the lower few kilometers of atmosphere. Each kind of gas absorbs and radiates heat and light in its own way. The ones that are especially good at absorbing infrared radiation are called **greenhouse gases.** These include water vapor, carbon dioxide, methane, nitrogen oxides, ozone, and chlorofluorocarbons (CFCs).

Greenhouse gases trap heat much as panes of glass do in a greenhouse. That's why the result is called the **greenhouse effect** (although, actually, the process by which the heat is trapped is not exactly the same as in a greenhouse). The greenhouse effect is a natural phenomenon that has occurred for hundreds of millions of years on Earth, as well as on other planets (see later). Water vapor and small drops of water are a much larger percentage of our atmosphere than the other greenhouse gases (about 1–4% of the content of Earth's atmosphere is water, though this varies from time to time and place to place). Water therefore accounts for much of the "greenhouse effect" (Figure 13.9).

This raises an interesting question. Carbon dioxide (CO_2) is only 0.037% of the atmosphere, less than four-one-hundredths of the amount of water in the lower atmosphere. How can such a small amount of this gas have a noticeable effect on temperature? The answer is not simple. In part, it is because CO_2 freezes (as dry ice) at a much lower temperature than water, and therefore CO_2 vapor exists at a higher altitude than water vapor. Infrared radiation absorbed and re-emitted upward by water can be absorbed and radiated downward or upward by CO_2, so CO_2 has a last chance at what happens to the heat radiation, so to speak. Part of the reason a small concentration of CO_2 can have such a large effect is that it absorbs in a region of infrared light that water transmits.

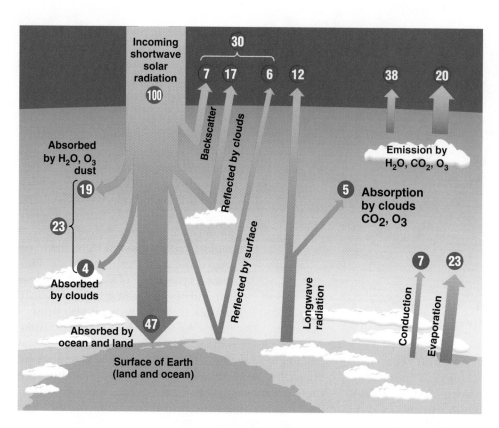

■ **FIGURE 13.9**
The greenhouse effect.
Incoming visible solar radiation is absorbed by the Earth's surface and re-emitted as infrared radiation. Most of the re-emitted infrared radiation is absorbed by the atmosphere, leading to the greenhouse effect.

The greenhouse effect keeps Earth warmer than it would otherwise be. Scientists have compared the **energy budget** of Earth with greenhouse gases to a hypothetical identical planet without those gases, and calculate that greenhouse gases keep Earth 33°C warmer. In addition, infrared radiation from the atmosphere due to the greenhouse effect keeps variations in Earth's surface temperature from day to night relatively small. Without this effect, the land would cool much more at night and warm faster during the day.

Greenhouse gases produce a greenhouse effect on other planets, too. Calculations made for our two neighbors, Venus and Mars (Figure 13.10), show this.

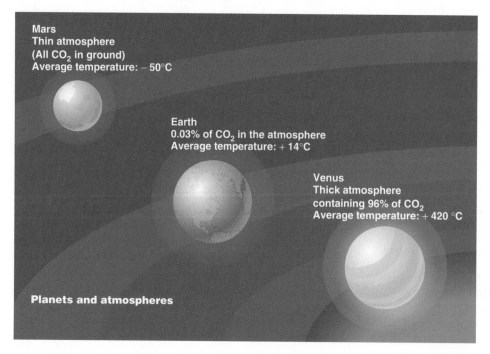

■ **FIGURE 13.10**
The greenhouse effect on Mars, Earth, and Venus warms each planet.
Our two neighbors are even warmer than we are because their atmospheres contain much more carbon dioxide, a major greenhouse gas. Life has removed most of the carbon dioxide from the Earth's atmosphere.

The atmospheres of these planets are mostly carbon dioxide, with a small percentage (about 1%) of nitrogen, whereas on Earth, living things have removed most of the carbon dioxide. A comparison of the energy budgets of Mars and Venus with their present atmosphere and with an Earth-like atmosphere shows that they are much warmer now than they would be if they had our atmosphere.

13.4 Greenhouse Gases Are Increasing, and We Are Part of the Reason

The major greenhouse gases, other than water, have increased in concentration in the Earth's atmosphere since the beginning of the Industrial Revolution, which suggests that the increases may be related to human activities. Carbon dioxide contributes 50–60% of the **anthropogenic** (human-caused) greenhouse effect. The rest of the anthropogenic effect comes from trace gases, the most important of which are the CFCs and methane. These trace gases contribute 27–45% of the anthropogenic greenhouse effect, and they have accumulated in the atmosphere much faster than carbon dioxide. Here are the facts about each of these gases.

Carbon Dioxide

CO_2 has been increasing in the atmosphere for some time (Figure 13.11). The idea that the increase may be linked to our activities is not new. It was first suggested by a scientist early in the 19th century, shortly after the beginning of the Industrial Revolution with its major burning of coal. He speculated that carbon dioxide emitted by burning coal could increase the concentration of carbon dioxide in the atmosphere, thereby increasing the greenhouse effect and warming Earth's surface. But few people believed him or even knew about his calculations. The idea came back again in the 1930s, when a scientist named Guy Stewart Callendar calculated the amount of carbon dioxide that could have been added to the atmosphere from the burning of fossil fuels since the start of the Industrial Revolution, and also made calculations about the effects of this increase on Earth's temperature. His work, too, was mostly ignored.

As the 20th century wore on, the idea gained attention. A small group of scientists, mainly climatologists and ecologists, began to agree that global warming was possible and in fact likely, and they started to gather evidence about the concentration of greenhouse gases in the atmosphere. In 1957—designated an "international geophysical year," in which nations cooperated to study Earth as a planet—an observatory was established at 11,000 feet on Mauna Loa Volcano in Hawaii. There, far above and away from most local industrial and agricultural activities, measurements of carbon dioxide in the atmosphere began and have continued ever since. These and other long-term measurements of atmospheric carbon dioxide at other locations since then show that the concentration of

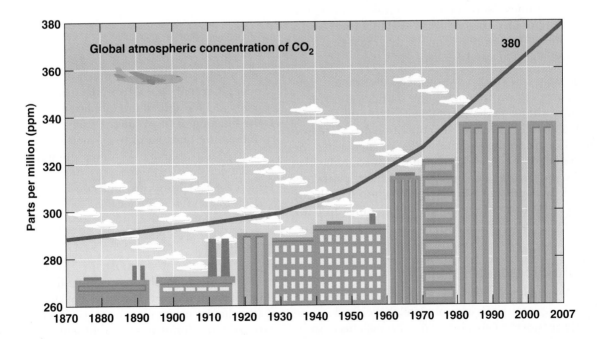

■ **FIGURE 13.11**
Carbon dioxide has increased in the atmosphere since the 19th century.
Today, the concentration is about 380 parts per million.

carbon dioxide in the atmosphere has risen 32%, from 280 ppm (parts per million) at the beginning of the Industrial Revolution to 380 ppm today.

Scientists now can estimate the carbon dioxide concentration in ancient atmosphere by measuring the concentration in air bubbles trapped in polar ice sheets. Measurements of CO_2 trapped in air bubbles in the Antarctic ice sheet suggest that over the last 160,000 years, the atmospheric concentration of CO_2 varied from about 200 to 300 ppm.[26] Currently, the rate of increase is about 0.5% per year. If growth continues at this rate, the concentration of CO_2 will double before the end of the 22nd century.

People add to atmospheric carbon dioxide in several ways: by burning fossil fuels, by burning wood, and through certain major changes in land use—such as cutting down forests and removing prairies—that destroy the living vegetation and the organic matter in soils.

Methane

The methane concentration more than doubled in the past 200 years and is believed to account for 12–20% of the human-caused greenhouse effect.[27] Methane is produced by certain kinds of bacteria that cannot live where there is oxygen. They therefore exist in three major habitats: in the parts of wetlands that are so saturated with water that they lack oxygen; in the intestines of cattle and other ruminants; and in the intestines of termites. Methane is also released during the processing of fossil fuels (it is a fossil fuel itself, and a good one at that) and by landfills that create environments similar to natural wetlands.

People can add to atmospheric methane in several ways: (1) by increasing the number and size of habitats where methane-producing bacteria live (including the land area used for cultivating rice, since lowland rice grows in wetlands); (2) by increasing the number of domesticated ruminants—cows and the like; (3) by processing and burning fossil fuels; (4) by destroying wetlands and thereby releasing stored methane; and (5) by increasing the size of landfills and the amount of organic matter stored in them.

Chlorofluorocarbons (CFCs)

We used to think these artificial compounds were harmless. CFCs were originally developed for use in refrigerators but are also used as propellants in spray cans and as an industrial cleaning agent. Like DDT, CFCs were long thought to be benign, since they are

nontoxic, nonexplosive, and relatively inert chemically (meaning they don't combine easily or at all with other chemicals to form new substances). But then we discovered that CFCs are greenhouse gases and destroy ozone in the upper atmosphere.

CFCs in the atmosphere have been increasing about 5% per year, due to both deliberate release and accidental leaks of sizable amounts. It has been estimated that about 15–25% of the anthropogenic (human-caused) greenhouse effect may be related to CFCs.[28]

Use of CFCs as propellants was banned in the United States in 1978. But although CFCs are no longer used in spray cans in many countries, they have not yet been banned worldwide. In 1987, 24 countries signed a treaty, the Montreal Protocol, that included an agreement to reduce and eventually eliminate production of CFCs, and accelerate development of alternative chemicals. Because of the treaty, production of CFCs was nearly phased out by 2000. If CFCs had not been regulated by the Montreal Protocol, they would have become the major contributor to the anthropogenic greenhouse effect by the early 1990s.[29]

Reduced emissions are evidently responsible for the recent decrease in the concentration of atmospheric CFCs. However, not all countries signed the treaty, and illegal production and use of CFCs continues in some countries.

In sum, CFCs are entirely a human product, and therefore their effects on global warming are entirely the result of human activities.

Nitrous Oxide

Atmospheric nitrous oxide (N_2O) is increasing and likely contributes as much as 5% of the anthropogenic greenhouse effect.[30] Human sources of nitrous oxide include agriculture (application of fertilizers) and the burning of fossil fuels. Using less fertilizer and reducing the burning of fossil fuels would reduce emissions of nitrous oxide. However, this gas also has a long residence time, so even if we reduced or at least did not increase N_2O emissions, high concentrations would persist for at least several decades.[27]

Ozone

Ozone is a greenhouse gas, in addition to its role in the lower atmosphere as an undesirable pollutant, and its role high in the atmosphere blocking ultraviolet light (see the discussion of ozone in Chapter 14 on air pollution). However, it has been difficult to determine the percentage of the greenhouse effect due to ozone, and this is still a matter under discussion among scientists.[31]

13.5 Would It Really Be So Serious If Earth Warmed Up a Bit?

Yes, if changes in the atmosphere affect living things and habitats. There are two steps in forecasting what these effects might be. First, we need to forecast changes in the climate and the physical conditions of the oceans and land surfaces. Second, using these forecasts, we forecast how species and ecosystems will respond.

Computer models are the major scientific tool in the first step. Computer models of the atmosphere, or of the oceans and atmosphere as a coupled system, view the atmosphere and the ocean three-dimensionally. A unit within one of these models is a boxlike compartment touching others (Figure 13.12). Gases and energy flow from one compartment to the surrounding ones. There are inputs into the atmosphere of energy and gases from the land and ocean surfaces. The models contain hundreds and hundreds of compartments and require considerable computer power to run. A handful of these have been developed at major climate-research institutes around the world, and the forecasts from these models provide the basis for scientific statements of what might happen.

What are these models telling us so far? According to the forecasts from such models, global warming will increase the average temperature of the Earth's surface by 1.5 to 4.5°C from 1990 to 2100 with 3°C being the most likely.[3,4] The models vary in their projections about changes in the amount of rain and snow, but generally agree that soils will dry out in many places as warmer temperatures cause water to evaporate faster. It also seems that the greatest temperature increases will take place in polar regions (Figure 13.13). As ice melts, the sea level will rise, icy habitats will shrink, Earth's climate will change, and the world's food supply will be threatened by changes in the water supply.

What Will Be the Effects of a Rising Sea Level?

The sea level has been rising naturally since the end of the last ice age, in part because the melting ice adds to the ocean waters and in part because water expands as it warms.[32] The sea level is now more than 120 meters higher than at the peak of the last ice age, and satellite measurements indicate that it is rising about 3 millimeters (about an eighth of an inch) a year.[4] Scientists expect that the rate will increase but that it will total less than 1 meter higher than today by 2100.

A rising sea level increases the damage from major storms, including flooding of low-lying coastal areas, entire islands, and possibly even entire island nations. It could increase beach erosion, making buildings and other structures in the coastal zone more vulnerable to damage, as happened in recent hurricanes. It could also cause a landward migration of estuaries and salt marshes, and lead to loss of coastal wetlands.[30] The people and wetlands at risk from a 44-centimeter rise in sea level are shown by region in Figure 13.14.

How Will Global Warming Affect the World's Climate?

While the glaciers are melting, Antarctica's central ice cap has been growing. Strange as it may seem, this is consistent with the predictions of global climate models and therefore appears consistent with global warming. The ice sheets are growing because, for complicated reasons, more snow falls on Antarctica as Earth warms.[33] Recent satellite measurements suggest that the east Antarctic ice sheet increased in mass by about 50 billion tons per year from 1992 to 2003.

Changes in the amount and location of ice will affect Earth's climate. Solar energy that would have been reflected back into the atmosphere by sea ice will instead be absorbed by the dark water. This is part of what is termed **polar amplification**.[34]

Global warming will also change the frequency and intensity of storms, because hurricanes and typhoons derive their intensity from warm ocean water.

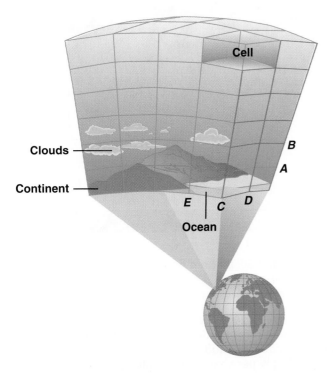

■ **FIGURE 13.12**
A view of the atmosphere in one of the global computer models.

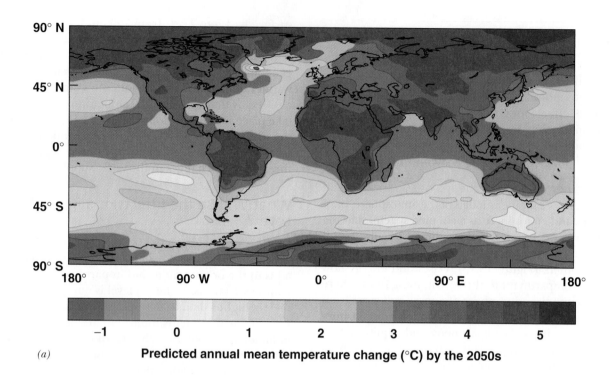

(a) **Predicted annual mean temperature change (°C) by the 2050s**

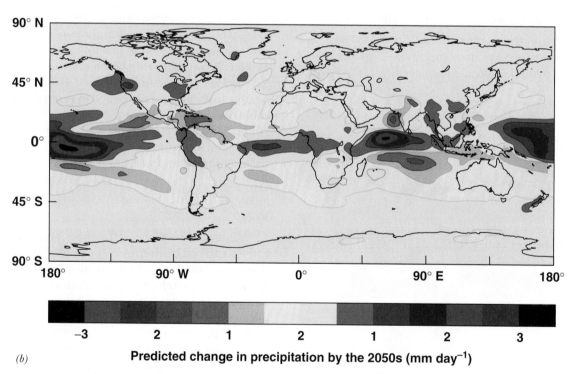

(b) **Predicted change in precipitation by the 2050s (mm day^{-1})**

■ **FIGURE 13.13**

Forecasts suggest that temperatures are rising.
(a) Projected changes in annual temperatures from today to the 2050s. Notice that changes are greatest at the polar regions. (b) Projected changes in annual precipitation from today to the 2050s indicate that the greatest increases will be near the equator. Models that predicted these changes assume that greenhouse gases will increase about 1% per year. [*Source:* Met Office, Hadley Center for Climate Prediction and Research, in R. T. Watson, presentation at the Sixth Conference of the Parties of the United Nations Framework Convention on Climate Change, Intergovernment Panel on Climate Change, November 13, 2000. Available at www.ipcc.ch/press/sp-COPG.htm. Accessed December 1, 2000.]

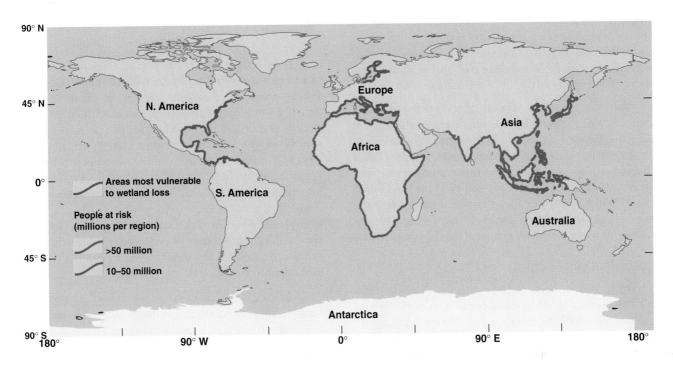

■ **FIGURE 13.14**
People and wetlands at risk from a 44-cm rise in sea level by the 2080s.
This map assumes that coastal flood protection will remain as it is today. [*Source:* Modified from R. Nicholls, Middlesex University in the U.K. Met Office, in R. T. Watson, presentation at the Sixth Conference of the Parties of the United Nations Framework Convention on Climate Change, Intergovernment Panel on Climate Change, November 13, 2000. Available at www.ipcc.ch/press/sp-COPG.htm. Accessed December 1, 2000.]

This became clear in the summer of 2005, when very warm waters in the Gulf of Mexico made hurricanes Katrina and Rita into two of the most intense and destructive ever. Over the past several decades, the worldwide frequency of hurricanes has not changed, but their intensity has increased. Two cyclones that hit highly populated Bangladesh in the last 25 years killed more than 400,000 people and caused over $1.6 billion in property damage.

The double impact of a rising sea level and more frequent and powerful cyclones and other tropical disturbances would have devastating effects on people in developing countries. Approximately half the human population lives in low-lying coastal areas and will suffer from these increasingly intense storms, as happened in 2005.

Greenhouse warming is expected to cause other climate changes, including wetter winters, hotter and drier summers, and an increased possibility of droughts in the northern temperate latitudes.[35]

Agriculture

Global warming may seriously affect the world's food supply, by impacting agriculture in a variety of ways. Although some of the climate models forecast that precipitation may increase in some areas, the overall forecast is that the amount of water evaporated from warmed soils and vegetation will be greater than the amount added by rain and snow. As a result, in general, groundwater and soil water will decrease. Since a reliable water supply is necessary for agriculture, this will put greater pressure on water demand.

The best agricultural areas may no longer be in North America. North American agriculture is especially fortunate in having some of the best agricultural soils in the world and some of the best climates for agriculture. But with global warming, those good climates for farming will move north to the less fertile Canadian shield. As a result, world production, especially of small grains, could decline, or at least not increase rapidly enough to meet the needs of the world's growing human population.

Winter snowpacks will store less water. Winter snowpacks in mountains are an important way that water is stored and then becomes available, through natural stream flow, for summer agriculture. But global warming will decrease the depths of snowpacks near major agricultural regions, heightening the demand for water for farming.

Lowering of Water Tables and Reservoirs Could Cause Serious Shortages

If, as projected, global warming leads to a drying out of soils, then, in general, stream and river flow will be less than in the past, and less water will be stored in underground aquifers. Add to this problem that as the average temperature rises, plants require more water, so agricultural irrigation needs will increase. This will be happening at the same time that the human population increases, putting more demands on water.

Water use in some regions is already unsustainable. Many parts of the world, including areas in the United States that depend on such sources as the Colorado River, are already using more water than is sustainable. Global warming can only make matters worse. Mark Twain said more than a century ago that in the West (of the United States) "whiskey is for drinking and water is for fighting over." Those "water fights" are likely to increase. (We discussed our water supply in greater detail in Chapter 11.)

Biological and Ecological Changes

The biosphere will change as a result of damage to ecosystems, from tropical coral reefs to birds and bears in the Arctic. As the ice melts, frozen habitats will shrink, affecting animals that depend on these habitats, such as penguins and polar bears (see the Chapter opening Case Study).

Conservationists are also concerned about the declining numbers of black guillemots on Cooper Island, Alaska (Figure 13.15). Rising temperatures in the 1990s caused the sea ice to recede farther from Cooper Island each spring. The ice receded before the black guillemot chicks were mature enough to survive on their own. Parent birds feed on Arctic cod found under the sea ice, then return to the nest to feed their chicks. For the parents to accomplish this, the sea ice must be less than about 30 kilometers (18 miles) away from the nest, but in recent years the ice in the spring has been receding as much as 250 kilometers (156 miles) from the island before the chicks are able to leave the nests. As a result, the black guillemots on the island have lost an important source of food, and their numbers have shrunk.

The fate of black guillemots on Cooper Island depends on future springtime weather. Too warm, and the birds may disappear. Too cold, and there may be too few snow-free days for breeding, and in this case, too, they will disappear.[36]

Spring now arrives up to two weeks earlier than it did three decades ago, and this is having ecological effects.[37] An endangered species of woodpecker in North Carolina is laying eggs about a week earlier than it did two decades ago. Marmots in the mountains of Colorado are waking from winter hibernation about 6 weeks earlier than they did 17 years ago. Some cherry trees in Washington, D.C., are blooming about a month earlier than they did 50 years ago. Robins are arriving in Wisconsin a few days earlier than they did just 10 years ago. Mexican jays are breeding earlier in the Chiricahua Mountains, according to a 30-year study. One theory is that these birds have adjusted to earlier spring temperatures by breeding earlier so that their young will arrive when food sources, including insects, are plentiful.

Early arrival of spring can stress some species, changing communities of organisms. Rapid change may require adaptions that some species can't make. Earlier-arriving birds may compete for food with other birds that migrate later, when days are longer. Some plants may flower earlier, then be damaged by spring snowstorms.[38] Wild relatives of domesticated plants exist in remnants of their original habitats, some of which have become ecological islands. As the climate warms, these remaining habitats may no longer be suitable for these wild plants, but the wild strains are important for food production because they provide genetic diversity that is valuable in developing new hybrids to combat diseases and adapt to new climate conditions.

Some species are changing their geographic ranges, moving northward in the Northern Hemisphere. Subalpine forests are moving to higher elevations in the Olympic Mountains of Washington State (Figure 13.15). Edith's checkerspot butterflies have been moving north and to higher elevations during the last century, apparently in response to global warming in their western North American habitat. As early as 1989, Edith's checkerspot butterflies in some high meadows of the Sequoia National Forest in California died, not from poisoning or disease but because the snowpack melted early and the warm temperatures caused the butterflies to emerge before the nectar-rich plants they feed on were available.

Sachem skipper butterflies have been expanding their range from northern California into Oregon and southeastern Washington (Figure 13.15). This has been going on over a 50-year period, coinciding with gradual global warming. Each move northward has occurred during an unusually warm summer.

Many other species are also moving northward (Figure 13.15). Snails, sea stars, and other intertidal organisms have been shifting northward in Monterey Bay as a result of rising shoreline water temperatures over the past 60 years. Mexican voles (small rodents with short tails and stout bodies) have expanded their range during the past 100 years into northeastern Arizona and Colorado from the south and west. Studies suggest that a number of other mammal species are also shifting northward to adjust to changes in climate and habitat. Zone-tailed hawks have gradually moved northward over a 50-year period, as have a number of other bird species of the Southwest and Mexico, such as common blackhawks, whippoorwills, and brown-crested flycatchers.

Migration of Species Can Spread Diseases

There has been a shift in the range of mosquitoes that carry diseases, including malaria and dengue fever.

1. Black guillemots, sea birds living in the Arctic environment, have been declining in numbers as sea ice receded earlier each spring.

2. Edith's checkerspot butterflies are moving north and to higher elevations.

3. Sachem skipper butterflies have been expanding their range from northern California into Oregon and southeastern Washington.

4. Snails, sea stars, and other intertidal organisms have been shifting northward in Monterey Bay.

5. Mexican jays are breeding earlier in the Chiricahua Mountains.

6. Mexican voles (small rodents with short tails and stout bodies) have expanded their range during the past 100 years into northeastern Arizona and Colorado from the south and west.

7. Zone-tailed hawks have gradually moved northward.

8. Prairie grasses are on the decline in the central plains.

9. Fire ants are expanding their range in all directions from the Deep South as far north as North Carolina.

10. Palm trees along Florida's west coast are dying a result of exposure to salt water.

11. Polar bears in western Hudson Bay are losing weight and bearing fewer cubs.

12. Caribou populations are declining in the high arctic environment of Canada.

13. Mosquitoes carrying dengue fever (a severe flu-like viral illness) are moving to higher elevations in Mexico.

14. Coral reefs in the Florida Keys, Bermuda, and the Caribbean are declining as a result of bleaching.

15. Subalpine forests are moving to higher elevations in the Olympic Mountains of Washington.

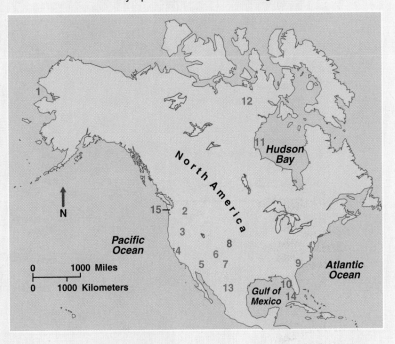

■ FIGURE 13.15
Selected examples of how rising global temperatures are shifting the range of plants and animals in North America.
[*Sources:* Modified from S. Levy, "Wildlife on the Hot Seat," *National Wildlife 38*, no. 5(2000):20–27; and N. Holmes, "Has Anyone Checked the Weather [Map]?" *Amicus Journal 21*, no. 4(2000):50–51.]

Mosquitoes that carry dengue fever (a severe viral illness that may cause fatal internal bleeding) are moving to higher elevations in Mexico. Malaria, which causes chills and fever, is particularly worrisome, as it kills about 3,000 people per day, mostly children. Global warming is projected to increase the land area where malaria can be transmitted (Figure 13.16). Today, that area is home to 45% of the world's population. With global warming, malaria could threaten 60% of the world's population.[39, 40]

West Nile virus is a good example of how global warming can spread disease. The emergence of West Nile virus in New York City in 1999 illustrates the linkages between physical, biological, and social systems— the essence of environmental science. We don't know how the West Nile virus arrived in North America, but we do know that its spread from mosquitoes to birds to people was aided by a warm winter followed by a dry spring and a hot, wet summer. These conditions are symptoms of global warming. This is how it works:[40]

■ During a mild winter, more mosquitoes survive in sewers as well as in still water in various locations, such as ponds, waste cans, and abandoned tires.

■ During a dry spring, surface-water sites, such as small ponds, get smaller, concentrating both birds and mosquitoes at the water sites.

■ Mosquitoes infected with the virus bite uninfected birds, passing the virus on to them.

■ Infected birds are bitten by uninfected mosquitoes, passing the virus to more mosquitoes.

■ Hot summer months, with their warm air and heavy rains, cause the mosquito population to mature and grow rapidly. More mosquitoes become infected and pass on the virus to birds and, eventually, to people.

Endangered Species

What will happen to species that can neither migrate nor adapt? Consider the Kirtland's warbler, whose

Malaria Plasmodium vivax

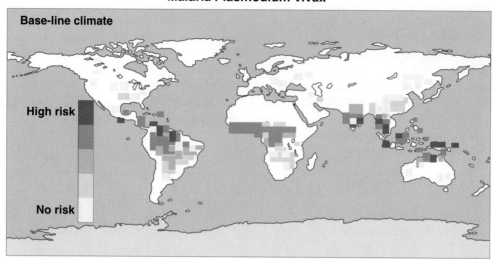

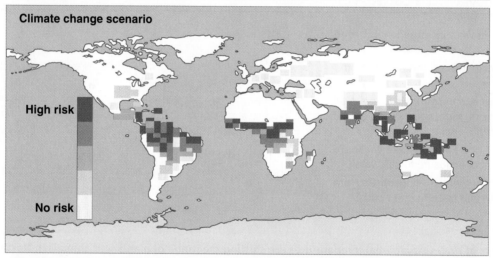

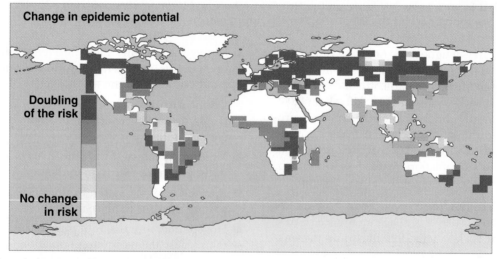

▪ **FIGURE 13.16**

The forecast for malaria epidemics.

The forecast shows that the potential for malaria epidemics could change as a result of global warming. [*Source:* Martens, P., et al. Potential impacts of climate change on malaria risk. *Environmental Health Perspectives* 103(5):458–464, 1995.]

story was told in Chapter 5 (Figure 13.17). Can this small, endangered bird of North America survive global warming, or will it become extinct? The species has long been of interest and concern to conservationists, ornithologists, and people who just enjoy the outdoors. Remember that in 1951 the Kirtland's warbler became the first songbird in the United States to have a complete census. The census found about 400 nesting males, but just 20 years later their numbers had been cut in half.

Kirtland's warblers require a very specific habitat. As we noted earlier, they are known to nest only in young jack-pine woodlands, and the jack pine is a "fire species"—that is, it can sustain itself only where there are periodic forest fires.

To save these birds, 38,000 acres were set aside in Michigan. There, prescribed burning was introduced, based on planning by the Audubon Society, the U.S. Fish and Wildlife Service, and the State of Michigan Department of Natural Resources. But computer simulation of forests in this region shows that jack pine will not be able to grow there in global warming climates. The warbler only nests on jack pine in certain kinds of sandy soils found only in southern Michigan, and appears unable to move its range northward, and so the birds will have no place to nest, and therefore are likely to go extinct.[41-43] This is working. In 2006 there were 1478 singing males.

A scarcity of suitable habitats could cause many extinctions, especially because habitats are no longer continuous stretches of land but are broken up by cities and suburbs, farms, industries, and other human uses of the land. As the climate changes, a species may attempt to move to a more suitable area. Some will be able to do so, but others will not. The efforts of some species to relocate may be restricted by environmental disruptions, such as habitat loss or fragmentation of habitat from land-use change. Those that require a very specific type of habitat may simply be unable to find it anywhere else.[44] The good news is that species can adjust better to changes that are slow, so if we could slow the rate of climate change, the undesirable biological effects would be fewer.

13.6 Can We Do Anything to Slow the Temperature Rise?

We may be able to slow it down, but time is running out. There are three approaches:

■ **Reduce our production and release of greenhouse gases.** In this approach, the primary direct action is to reduce the world's use of fossil fuels and the destruction of organic matter in forests, grasslands, peatlands, marshes, and their soils. As Chapter 10 on energy explains, alternative energy sources can replace fossil fuels, but there is a limited time to act. We can

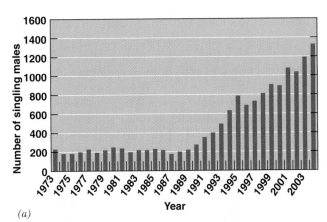

(a)

(b)

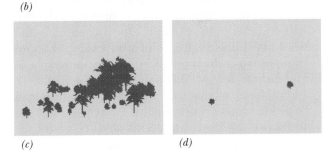

(c)   *(d)*

■ **FIGURE 13.17**

The Kirtland's warbler is threatened by global warming.
(a) The Kirtland's warbler is an endangered-species success story, with this endangered species increasing, as indicated by the known number of singing males. (b) This species nests only in young jack-pine forests on certain kinds of sandy soils, which restricts its breeding to just a few areas in Michigan. (c) Computer forecasting for the growth of jack pine in these habitats under 20th-century conditions are contrasted with (d) the expected failure of these trees to grow under a global-warming climate by the mid-21st century.
[*Sources:* Range map: http://www.globalchange.umich.edu/webprojects/group5/Warbler_files/warbler_intro.htm. http://www.michigan.gov/dar/0,1607,7-153-10370_12145_12202-32591—,00. html (c) and (d) www.naturestudy.org Daniel B. Botkin, and Botkin D. B. Forest Dynamics: An Ecological Model, Oxford University Press, N. Y. 1993].

also use fossil fuels more efficiently—for example, by using mass transit much more than we do now, and by using more fuel-efficient automobiles, such as hybrids. But we are unlikely to see an overall decline in the world's use of fossil fuels, especially as the economies of formerly undeveloped nations, particularly China and India, become increasingly prosperous and increase their use of fossil fuels for automobiles, construction, and industrialization.

Burning forests to convert land to agriculture accounts for about 20% of human emissions of carbon dioxide into the atmosphere. Therefore, burning less and protecting the world's forests would help reduce the rate of global warming.

▪ **Find ways to sequester (store) greenhouse gases.** In this second approach, we can promote biological uptake and storage of carbon dioxide. Reforestation (which is happening in the eastern United States) will help because young, fast-growing trees take up and store carbon dioxide. Underground storage of carbon in plant roots and in dead organic matter in soil could reduce the amount of carbon dioxide that would otherwise enter the atmosphere.

Carbon may also be stored deep in the earth in geologic reservoirs, as, for example, in depleted oil and gas fields. A carbon sequester experiment is being conducted beneath the North Sea.

There are also some rather far-out suggestions. For example, the southern ocean adjacent to Antarctica is one of the world's great centers of biological production. But the algae in those oceans seem to be limited by the availability of iron. One suggestion is that we dump huge quantities of iron oxide dust into these oceans, which would increase photosynthesis and carbon storage. Although the algae do not store carbon for a long time, food chains in the ocean lead to production of much organic matter that sinks to the bottom of the ocean and is stored.

▪ **Take actions that cool the climate.** Some scientists have suggested that we could cool the climate by injecting sulfur oxides into the upper atmosphere, just as volcanic eruptions do. The problem with this and with similar suggestions—such as adding large amounts of iron oxides to the southern oceans—is that they are global experiments, they are fooling with the entire biosphere, a system that we little understand. We don't know all the effects of these actions, and there may be large undesirable ones. Tinkering with the biosphere without understanding the implications of what we do is not a wise path to take.

What Has Been Done So Far to Mitigate Global Warming?

We took the first steps in 1988. The attempt to control emissions of greenhouse gases, particularly carbon dioxide, began at a major scientific conference on the issue of global warming in 1988 in Toronto, Canada. At that meeting, scientists recommended a 20% reduction in carbon dioxide emissions by 2005. The meeting was a catalyst for scientists and other concerned people to work with politicians in formulating international agreements to reduce emissions of greenhouse gases. Although at that time many uncertainties remained concerning global warming, the prevailing attitude was that it was advisable to be conservative and reduce emissions before problems became apparent.

A second important step was taken in 1992, at the Earth Summit in Rio de Janeiro, Brazil, where a general blueprint for reducing global emissions was suggested. Some in the United States, however, objected that the reductions in CO_2 emissions would be too costly. Furthermore, agreements from the Earth Summit did not include legally binding limits. Following the meetings in Rio de Janeiro, governments worked to strengthen a climate-control treaty that included specific limits on the amounts of greenhouse gases that could be admitted into the atmosphere by each industrialized country.

In December 1997, legally binding emission limits were discussed in Kyoto, Japan, but specific aspects of the agreement divided the delegates. The United States eventually agreed to cut emissions to about 7% below 1990 levels. However, that was far less than the reductions suggested by scientists, who recommended reductions of 60–80% below 1990 levels. In fact, after the conference, it was realized that emissions of CO_2 in 2010 would likely be about 30% higher than the 1990 emissions.

We have run out of excuses. Every time an environmental solution to a problem has been suggested, there have been those who said it couldn't be done. But it has been proved again and again that it can be done. Advances in alternative renewable energy appear sufficient for society to meet goals for reducing CO_2 emissions without jeopardizing our economic future.

What is being done in the United States? The United States, with 5% of the world's population, emits about 20% of the atmospheric carbon dioxide. It is encouraging that the U.S. government has acknowledged that reducing emissions of CO_2 into the atmosphere is an important goal. Providing economic incentives to improve and design new energy-efficient technologies and alternative energy sources would be a positive step. However, the U.S. Congress has been slow to act or to fund necessary studies, and the United States, in the Hague meetings of late 2000, refused to honor reductions in emissions of CO_2 that were agreed to at Kyoto in 1997.

In March 2001, the new administration in Washington disappointed European allies by announcing that the United States would not abide by the Kyoto Accord. More than 180 countries meeting in Bonn, Germany, in

July 2001 adopted the Kyoto Accord despite the U.S. rejection of the agreement, and it went into effect in early 2005.

Why is the Kyoto Accord so important? It's true that the agreement is far from popular with all participants, and the goals it sets for total reductions in CO_2 emissions are not sufficient to reach the ultimate goal of eliminating potentially damaging effects of global warming. Even so, the Kyoto Accord is important because it is a first attempt to unite the people of Earth in seeking ways to reduce and eventually reverse human-induced climate change.

13.7 Can We Do Anything to Alleviate the Effects of Global Warming?

Some of the effects of global warming will be so massive that we will be unable to do much to counter them. But there are some things we will be able to do, especially regarding biological effects.

- **We can move species to new habitats.** Not waiting for the slow natural rate of migration will lower the number of extinctions.
- **We can establish new nature preserves in areas that provide appropriate habitats** for species as climate changes.
- **We can establish wildlife corridors among these preserves** so that species can travel as needed to suitable climates.
- **We can develop new strains of crops** that will improve yields in the new climates.

A likely adjustment will be learning to live with the changes—a warmer climate, more variability in weather patterns, changes in the biosphere, and a higher sea level. If the changes occur relatively slowly and total mean global warming is less than 2°C, then learning to live with new conditions may be possible. Indeed, in some cases the changes will offer opportunities. However, it is equally likely that surprises, including unexpected problems, will emerge. A 2°C change is at the low end of predicted global warming. With increases greater than 2°C, potential consequences to people increase significantly.

Return to the Big Question

Global Warming Is Happening: What Part Do We Play?

The strong correlation between the rise in the Earth's temperatures and the amount of greenhouse gases added to the atmosphere from burning fossil fuel strongly suggests that we are playing an important role in global warming. Much of the other evidence discussed in this chapter adds further support to the belief that people are playing a role in global warming. The great majority of scientists today believe that this is the case.

Summary

- The atmosphere, a layer of gases that envelops Earth, is a dynamic system that is constantly changing. A great number of complex chemical reactions take place in the atmosphere, and atmospheric circulation produces the world's weather and climates.
- Major climate changes have occurred during the past 2 million years, with periodic appearances and retreats of glaciers. During the past 1,500 years, several warming and cooling trends have affected people. During the past 100 years, the mean global annual temperature has apparently risen about 0.5°C.

- Water vapor and several other gases, including carbon dioxide, tend to warm Earth's atmosphere through the greenhouse effect. Most of the greenhouse effect is produced by water vapor, but more than half of the human-induced (anthropogenic) greenhouse effect is from carbon dioxide. Greenhouse gases occur naturally in the atmosphere, but since the Industrial Revolution, human activity has added substantial amounts of them to the atmosphere, especially carbon dioxide.
- Climate models suggest that in the next few decades carbon dioxide will reach a level that is double its

▪ preindustrial level, and the mean global temperature may rise by 1°–2°C. Total warming during the 21st century may range from 1.5°C to 4.5°C.

▪ We can reduce CO_2 emissions by conserving energy, sequestering carbon, and using alternative energy sources.

▪ Major effects of global warming include (1) changes in climate patterns and the frequency and intensity of storms, (2) a rise in sea level, and (3) changes in the biosphere.

▪ Changes in the biosphere will include shifts in where specific plants and animals live and the spread of diseases such as malaria to higher elevations.

▪ Adjustments to global warming include (1) attempts to mitigate warming by reducing emissions of carbon dioxide or by storing it to keep it out of the atmosphere and (2) learning to live with the changes. It seems likely that people will choose to learn to live with change, but if climatic change is too rapid, there may not be enough time for this.

Key Terms

albedo
atmosphere
climate
climatic change
El Niño

global warming
greenhouse effect
greenhouse gases
Gulf Stream
interglacial periods

Getting It Straight

1. Why can a small amount of methane in the atmosphere have a big effect on the air temperature, when there is so much water in the atmosphere, and water is also a major greenhouse gas?

2. What are the four major greenhouse gases?

3. Does the greenhouse effect impact other planets?

4. What are the effects of the rising sea levels on land areas?

5. What three human activities are most likely to contribute to increased atmospheric carbon dioxide levels?

6. What is methane? And how is it produced?

7. When were propellant chlorofluorocarbons (CFCs) banned from use in the U.S.?

8. Why have CFC levels been increasing 5% per year?

9. What are the biological and ecological changes associated with global warming?

10. What is the primary cause of climate change of any kind?

11. The percentage of light received by a surface that is reflected and scattered is called _____.

12. How do volcanic eruptions cool the climate? When is El Niño most likely to occur?

13. What is the greenhouse effect? How do greenhouse gases affect the Earth's atmosphere or climate?

What Do You Think?

1. You have built a greenhouse in your town. Discuss what you would do to try to keep the temperature constant inside day and night, winter and summer.

2. Long, long ago, the modern continents of Australia, India, South America, Africa, and Antarctica existed together as one huge landmass, today called "Gondwana" or "Gondwanaland." This formed about 650 million years ago and lasted until about 130 million years ago.

a. Would there have been ice ages on Gondwana? Under what conditions?

b. If Earth now had this one supercontinent, would we be facing global warming?

3. Would a very, very large oil spill have a greater effect on climate if it occurred in Alaska or in Venezuela? Explain.

4. Where you live, would you expect it to become cloudier or sunnier, or stay about the same as it is now, when global warming occurs? Explain your answer. Discuss how you would go about finding out the answer as a science project.

5. Bowhead whales live in cold Arctic waters. Humpback whales breed in such places as the waters around Hawaii. How might the effects of global warming differ in these two species? How might they be the same?

6. Is El Niño a global event? Will it become more intense as the Earth's average temperature rises? Explain.

Pulling It All Together

1. Some scientists have suggested that we counter global warming by adding large amounts of sulfur dioxide particles to the atmosphere—these small dust particles will reflect sunlight, cooling Earth's surface. Present an argument for or against this idea.

2. Do you think the major destructive hurricanes in the American Gulf Coast in 2005 were the result of global warming? What are the arguments for and against this idea?

3. Do you believe that human induced global warming exists? Is the Earth really warming up or is it a weather trend that will eventually end?

Further Reading

Anthes, A. R. 1992. *Meteorology*, 6th ed. New York: Macmillan.—A short text providing a good overview of basic meteorology and atmospheric processes.

Fay, J. A., and Golumb, D.S. 2002. *Energy and environment.* New York: Oxford University Press.—See Chapter 10 on global warming.

Gore, Al. 2006. *An Incovenient Truth: The Planetary Emergency of Global Warming and What We Can Do About It.* Rodale Books NY, NY. Book accompanies an accademy award winning documentary film about global warming and its potential impacts. Also suggests solutions to minimize warming.

IPCC. 2007. *The Intergovernmental Panel on Climate Change scientific Assessment.* New York: Oxford University Press.—A detailed scientific review and assessment of global warming.

Leggett, Jeremy K. 2001. *The carbon war: Global warming and the end of the oil era.* London: Routledge.—A petroleum geologist, who became an advisor to Greenpeace, followed the many meetings around the world negotiating about global warming, and relates them here in something resembling a mystery story.

Schneider, Stephen H., and Terry L. Root. 2002. *Wildlife responses to climate change.* Washington, DC: Island Press.—The best book available about case histories regarding wildlife and climate change.

Steve Allen /Photo Researchers, Inc.

14

Air Pollution and Environment

Big Question

Why Is Air Pollution in Cities Still Such a Big Problem?

Learning Objectives

Since people first used fire, but particularly since the Industrial Revolution, the atmosphere has been a place where gaseous and particulate wastes are dumped. Problems arise when the amount of waste entering the atmosphere in an area is too great for the atmosphere to disperse or break down. This chapter discusses first outdoor air pollution and then indoor air pollution, and ways to control both kinds. After reading this chapter, you should understand . . .

■ what the major categories and sources of air pollutants are;

■ why human activities that pollute the air, combined with meteorological conditions, may overwhelm the atmosphere's natural ability to remove wastes;

■ what acid rain is, and how we can minimize it;

■ why an international agreement to reduce stratospheric ozone depletion is an environmental success story;

■ what air-quality standards are, and why they are important;

■ why indoor air pollution is one of our most serious environmental health problems.

Case Study

An Olympic Success Story

When the summer Olympic Games came to Georgia in 1996 (see Figure 14.1) there was concern because Atlanta's air quality in the summer is among the worst in the United States. Atlanta is a proud southern city and wanted to project a positive image with the very best environment for athletic competition. The city also wanted to reduce traffic congestion so the many visitors could get to the Games. To accomplish these goals, Atlanta made a major effort that included (1) improved public transit; (2) closure of the downtown area to private automobiles; and (3) encouragement for people to work from home or at hours when traffic wasn't a problem.[1]

These measures reduced air pollutants and congestion, allowing spectators to get to the Games and providing cleaner air for the athletes. Local hospitals also reported that during the Games the number of children between the ages of 1 and 16 who went to hospitals for asthma attacks dropped significantly.

This success story demonstrates that reducing air pollution in our cities can lead to health benefits. Most of this chapter will be concerned with urban air pollution, but we will also discuss the serious problem of indoor air pollution, acid rain from burning fossil fuels, and ozone depletion in the upper atmosphere that starts with emission of chemicals in the lower atmosphere.

Simon Bruty/Getty Images, Inc.

■ **FIGURE 14.1**
The 1996 Olympic Games in Atlanta Georgia had a successful program to reduce air pollution.

14.1 A Brief History of Air Pollution

People have long recognized the existence of atmospheric pollutants. Leonardo da Vinci wrote in 1550 that trees caused a blue haze. What he had observed was a natural *photochemical smog* (smog produced by chemicals in the presence of light) from hydrocarbons given off by living trees. This haze, whose cause is still not completely understood, gave rise to the name Smoky Mountains for the range in the southeastern United States. Acid rain was first described in the 17th century, and by the 18th century it was known that smog and acid rain damaged plants in London.

Nature itself produces air pollution. In fact, except for sulfur and nitrogen oxides, nature produces more air pollutants than we do (see Table 14.1). Examples include the following:[2,3]

▪ Sulfur dioxide from volcanic eruptions. For example, volcanic activity on the island of Hawaii releases pollutants that react in the atmosphere to produce volcanic smog called "vog," which can be a health hazard and cause local acid rain.

▪ Hydrogen sulfide from geysers and hot springs and from biological decay in bogs and marshes.

▪ Ozone in the lower atmosphere as a result of unstable meteorological conditions, such as violent thunderstorms.

▪ A variety of particles from wildfires and windstorms.

▪ Natural hydrocarbon "seeps," such as the La Brea Tar Pits in Los Angeles.

However, our own activities cause a large portion of our air pollution. In urban areas, it is the air pollution that we ourselves produce that is most abundant and leads to the most severe problems for our health. The atmosphere is the fastest-moving part of the global environment, and we therefore have always viewed it as a quick way to get rid of unwanted substances, such as smoke. Indeed, ever since we first used fire, the atmosphere has been a sink for waste disposal. As a result, in the United States today, about 150 million metric tons of pollutants per year enter the atmosphere from human-related processes. If these pollutants were widely and equally distributed in the atmosphere, the concentration would be only a few parts per million by weight. Unfortunately, pollutants tend to be produced, released, and concentrated locally or regionally—for example, in large cities.

The problem grew with the Industrial Revolution in the 18th century. The word *smog* was introduced by a physician at a public-health conference in 1905. The term referred to poor air quality resulting from a mixture of smoke and fog.

The "Donora Fog" in Donora, Pennsylvania, led to increasing research on air pollution in the United States. The Donora event occurred in 1948 and remains the worst industrial air-pollution incident in U.S. history, causing 20 deaths and 5,000 illnesses. Pollutants including sulfur dioxide, carbon monoxide, and heavy metals from the Donora Zinc Works smelting plant and other sources were trapped by weather conditions in a narrow valley. The "Donora fog" lasted about three days, until pollutants were washed out and dispersed by rainstorms.

The London smog crisis followed in 1952, killing several thousand people, and 46 years later fires in Indonesia (1997–1998) caused severe air pollution. The Indonesian fires caused several hundred deaths and made millions of people sick.[2]

Legislation has helped, but more needs to be done. Today in the United States and other countries, legislation to reduce emission of air pollutants has been successful, but chronic exposure to high levels of air pollutants continues to contribute to illnesses that kill people around the world.

☐ TABLE 14.1 MAJOR NATURAL AND HUMAN-PRODUCED AIR POLLUTANTS

Air Pollutants	Emissions (% of total)		Major Sources of Human-Produced Components	Percent
	Natural	Human-Produced		
Particulates	85	15	Fugitive (mostly dust)	85
			Industrial processes	7
			Combustion of fuels (stationary sources)	8
Sulfur oxides (SO_X)	50	50	Combustion of fuels (stationary sources, mostly coal)	84
			Industrial processes	9
Carbon monoxide (CO)	91	9	Transportation (automobiles)	54
Nitrogen dioxide (NO_2)		Nearly all	Transportation (mostly automobiles)	37
			Combustion of fuels (stationary sources, mostly natural gas and coal)	38
Ozone (O_3)	A secondary pollutant derived from reactions with sunlight, NO_2, and oxygen (O_2)		Concentration present depends on reactions in lower atmosphere involving hydrocarbons and thus automobile exhaust	
Hydrocarbons (HC)	84	16	Transportation (automobiles)	27
			Industrial processes	7

What are the chances of another air-pollution health disaster? Unfortunately, the chances are all too good, given the tremendous amount of air pollution in some large cities. Beijing, for example, might be a candidate. The city uses an immense amount of coal, and coughing is such a common problem that residents often refer to it as the "Beijing cough." Another likely candidate is Mexico City, which has one of the worst air-pollution problems in the world today.

14.2 General Effects of Air Pollution

Air pollution affects many aspects of our environment: its visual qualities, vegetation, animals, soils, water quality, natural and human-made structures, and the environment's effects on our health. Air pollution affects visual resources by discoloring the atmosphere and by reducing visual range and clarity.

In many cities, it is a significant factor in the human death rate. For example, in Athens, Greece, the number of deaths is estimated to be several times higher on days when the air is heavily polluted. The most polluted air in the United States is found in the Los Angeles urban area, where millions of people are exposed to unhealthy air. It is estimated that as many as 150 million people live in areas of the United States where exposure to air pollution contributes to lung disease, which causes more than 300,000 deaths per year. Air pollution in the United States is directly responsible for annual health costs of about $50 billion.[4] In China, whose large cities have serious air-pollution problems, mostly from burning coal, the health cost may be about $100 billion per year by 2020 (Figure 14.2).

Air pollution's effects on people depend on several factors, including the dose or concentration and individual susceptibility. Some of the primary effects include toxic poisoning, cancer, birth defects, eye irritation, respiratory problems (including aggravation of chronic diseases, such as asthma and emphysema), and increased susceptibility to viral infections and heart disease. Air Pollutants that are known to cause cancer or other serious health problems are called **air toxins**. Examples include ammonia, chlorine gases, and hydrogen sulfide.

Many air pollutants have *synergistic effects*—that is, the combined effects of two pollutants are greater than the sum of their separate effects. For example, sulfate may attach to small coal dust particles in the air and be inhaled deep into lung tissue. The combination of sulfate and coal dust does more damage to the lungs than the two pollutants would do acting separately. This phenomenon has obvious health consequences: Consider joggers breathing sulfate and coal dust particles deep into their lungs as they run through city streets.

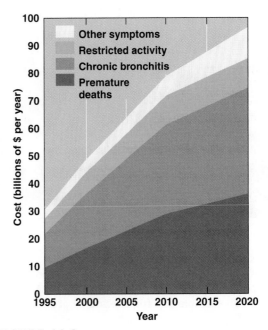

■ **FIGURE 14.2**
Air pollution's annual cost to human health in China.
[*Source:* Modified from World Bank. *Clear water, blue skies: China's environment in the new century.* World Bank, 1997.]

14.3 Primary and Secondary Pollutants, Natural and Human

Air pollutants occur either as gases or as particulate matter (PM). PM pollutants are very small particles (1–10% the diameter of a human hair) of solid or liquid substances and may be organic or inorganic. The major gaseous pollutants include sulfur dioxide, nitrogen oxides, carbon monoxide, ozone, hydrogen sulfide, hydrogen fluoride, and volatile organic compounds (VOCs) such as hydrocarbons.

Primary air pollutants are emitted directly into the air. They include particulates, sulfur dioxide, carbon monoxide, nitrogen oxides, and hydrocarbons.

Secondary pollutants are produced by reactions between primary pollutants and normal atmospheric compounds. For example, ozone forms over urban areas when primary pollutants react with sunlight and natural atmospheric gases. Thus, ozone is a secondary pollutant.

14.4 Major Air Pollutants: Where Do They Come From and What Do They Do?

Stationary sources have a relatively fixed location and include *point sources, fugitive sources,* and *area sources.*[3]

■ *Point sources* emit pollutants from one or more controllable sites, such as smokestacks at power plants (Figure 14.3a).

F. Hoffman/The Image Works

(a)

■ **FIGURE 14.3**
Point source.
(a) This steel mill in Beijing, China, is a major point source of air pollution. (b) Burning sugarcane fields, Maui, Hawaii—an example of a fugitive source of air pollution.

Courtesy Ed Keller

(b)

■ *Fugitive sources* include dirt roads, construction sites, farmlands, surface mines, and other exposed areas where fire or wind can inject material into the air (Figure 14.3b).

■ *Area sources* include urban areas, heavily industrialized areas, agricultural areas sprayed with herbicides and pesticides, and similar, well-defined areas.

Mobile sources emit pollutants while moving from place to place. These include automobiles, trucks, buses, aircraft, ships, and trains (Figure 14.4).[3]

The major air pollutants and their sources are summarized in Table 14.1, which also shows the percentage

Jules Bucher/ Photo Researchers

■ **FIGURE 14.4**
Mobile source.
Exhaust from older diesel buses such as this one in Thailand are a major source of air pollution.

of each that is produced by nature and the percentage produced by our own activities—called *anthropogenic* sources (from the Greek *anthro,* meaning "man").

Sulfur Dioxide

Sulfur dioxide is produced equally by people and nature. The major human source is the burning of fossil fuels, mostly coal in power plants (see Table 14.1 and Chapter 10). A variety of industrial processes—ranging from petroleum refining to the production of paper, cement, and aluminum—are also major sources.[3, 5] When this colorless, odorless gas enters the atmosphere, it can be converted into fine sulfate particles that end up on land and water. Depending on how much of it is present and in what form (see Chapter 8), it can corrode paint and metals and cause injury or death to people and other animals, severely damaging their lungs. It is also an important precursor to acid rain (discussed later).[3, 5] U.S. emission rates of sulfur dioxide peaked at about 32 million tons in the early 1970s and have dropped about 50% since then, to about 16 million tons, thanks to effective emission controls (see Figure 14.5).

Nitrogen Oxides

Two forms are important to us. Nitrogen oxides are emitted mainly in two forms, nitric oxide (NO) and nitrogen dioxide (NO_2), and only these two forms are subject to emission regulations. The more important of the two is NO_2, a yellow-brown to reddish-brown gas. Both are major contributors to smog, and NO_2 is a major con-

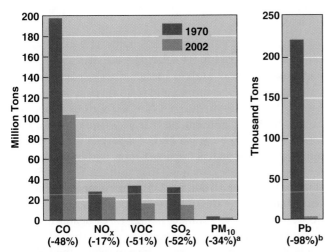

^a Based on 1985 emission estimates. Emission estimates prior to 1985 are uncertain.

^b Values for lead are based on 2001 data; 2002 data for lead are not yet available.

■ **FIGURE 14.5**

U.S. emissions of six major air pollutants in 1970 compared with 2002 (carbon monoxide, nitrogen oxides, volatile organic compounds, sulfur dioxide, particulate matter, and lead). Notice there have been significant reductions. [*Source:* U.S. Environmental Protection Agency 2002 highlights. Available at www.epa/gov. Accessed March 24, 2004.]

tributor to acid rain. Nearly all nitrogen dioxide comes from human sources, particularly from automobiles and from power plants that burn fossil fuels. They can cause irritation of the eyes, nose, throat, and lungs and increased susceptibility to viral infections, including influenza (which can lead to bronchitis and pneumonia).[3] They may also hinder plant growth. However, when the oxides are converted to nitrates and are deposited on the soil, they can promote plant growth. U.S. emission rates have fallen about 17% since 1970 (see Figure 14.5).

Carbon Monoxide (CO)

This colorless, odorless gas can be deadly. About 90% of carbon monoxide comes from natural sources. The other 10% comes mainly from fires, automobiles, and other sources of incomplete burning of organic compounds. Carbon monoxide is highly toxic to people and other animals. Its effects can range from dizziness and headaches to death, depending on the dose and the altitude (the effects tend to be worse at higher altitudes, where there is less oxygen). CO is particularly hazardous to people with heart disease, anemia, or respiratory disease. It may also cause birth defects.[3] Many people have been accidentally asphyxiated by carbon monoxide from burning fuel in campers, tents, and houses. CO detectors (like smoke detectors) are now commonly used to warn

people if CO in a building reaches a dangerous level. Emissions of CO in the United States peaked in the early 1970s at about 200 million metric tons but fell 50%, to about 100 million metric tons, by 2002, thanks largely to cleaner-burning automobile engines (see Figure 14.5).

Ozone

Ozone is a photochemical oxidant that plays two roles. Atmospheric interactions of nitrogen dioxide and sunlight produce *photochemical oxidants*. Ozone is the most common of these. It is a colorless gas with a slightly sweet odor. (Photochemical oxidants that occur with photochemical smog are discussed later in this chapter.) Ozone (O_3) is a form of oxygen whose molecules have three atoms of oxygen rather than the normal two. The molecules release their third oxygen atoms readily, so ozone oxidizes or burns things more readily than normal oxygen does. Ozone is sometimes used to sterilize—for example, bubbling ozone gas through water kills bacteria and other microorganisms in the water. And the ozone layer in the stratosphere (the upper atmosphere) protects us from ultraviolet rays of the sun. But in the lower atmosphere, ozone can do harm.

Ozone is a secondary pollutant, produced on bright, sunny days in areas where there is a lot of primary pollution. The chemicals that produce ozone and other oxidants come from burning fossil fuels, especially in automobile engines, and from industrial processes that produce nitrogen dioxide. Because there are so many automobiles and primary pollutants that with sunlight produce ozone, the health standard for ozone is exceeded more frequently than any other in U.S. urban areas.[6, 7]

Effects of ozone depend in part on the dose and include damage to plants and animals as well as to materials such as rubber, paint, and textiles. At low concentrations, ozone can slow plant growth. At higher concentrations, it kills leaf tissue and may kill the whole plant. The death of white pine trees along highways in New England is believed to be due partly to ozone pollution. Ozone can damage the eyes and respiratory systems of people and other animals. Millions of Americans are often exposed to ozone levels that damage lungs and airways. Eventually, scars and lesions may form in the airways, and the lungs become less elastic and more susceptible to bacterial infection. Even young, healthy people may not be able to breathe normally, especially on days when pollution is heavy.[7]

Volatile Organic Compounds (VOCs)

VOCs include organic compounds used as solvents in industrial processes, such as dry cleaning, degreasing, and graphic arts. Hydrocarbons—compounds of hydrogen and carbon—are one group of VOCs. They include methane, butane, and propane. Some of the many hydrocarbons in urban air react with sunlight to produce photochemical smog. Many are toxic to plants and animals, and some may be turned into harmful compounds by complex chemical changes that occur in the atmosphere.

Globally, only about 15% of hydrocarbon emissions are from human activities. In the United States, however, nearly half of hydrocarbons entering the atmosphere come from anthropogenic sources. The largest human source of hydrocarbons in the U.S. is automobiles. Anthropogenic sources are particularly abundant in urban regions, but in some southeastern U.S. cities, natural emissions probably exceed those from automobiles and other human sources.[4] VOC emissions peaked in the early 1970s and have since dropped 50%, due to the government's emission standards for automobiles (Figure 14.5).

Particulate Matter

Particulates are particles of matter suspended in the air we breathe. Particles are classified by their size. PM10 describes particles less than 10 micrometers (10 μm) in diameter, and PM 2.5 describes particles less than 2.5 μm in diameter (a human hair is about 100 μm in diameter). PM2.5 particles are particularly important because they are easily inhaled into the lungs, where they may be absorbed into the bloodstream. Farming adds a lot of particulate matter to the atmosphere, as do windstorms and volcanic eruptions. Nearly all industrial processes, including those that burn fossil fuels, release particulates into the atmosphere. Much of it is easily visible as smoke, soot, or dust, but some of it is not easily seen. Particulates include airborne asbestos particles and small particles of heavy metals, such as arsenic, copper, lead, and

zinc, which are usually emitted from industrial facilities, such as smelters.

Particulate matter threatens human health, ecosystems, and the biosphere. Particles are everywhere, but high concentrations and/or specific types of particles—asbestos particles, for example, and very fine particles that are easily inhaled into the lungs—can be a serious danger to human health. Measurements of *TSPs—total suspended particulates*—tend to be much higher in large cities in developing countries, such as Mexico, China, and India, than in developed countries, such as Japan and the United States (Figure 14.6). But even in the U.S. particulate air pollution contributes to the death of 60,000 people annually.[8] Indeed, recent studies estimate that 2–9% of deaths in cities are linked to particulate pollution, and that the risk of dying is about 15–25% higher in cities that have the highest levels of fine particulate pollution.[9]

Particulates are linked to both lung cancer and bronchitis and are especially hazardous to the elderly and to people with respiratory problems. There is a direct relationship between particulate pollution and increased hospital admissions for respiratory distress. Modern industrial processes have greatly increased total suspended particulates in Earth's atmosphere. These particulates block sunlight and may cause changes in climate and have lasting effects on the biosphere. However, particulate emissions have been reduced by about one-third (34%) since 1970 (see Figure 14.5).

Hydrogen Sulfide

Hydrogen sulfide is a highly toxic corrosive gas that smells like rotten eggs. It is produced from natural sources such as geysers, swamps, and bogs, and from

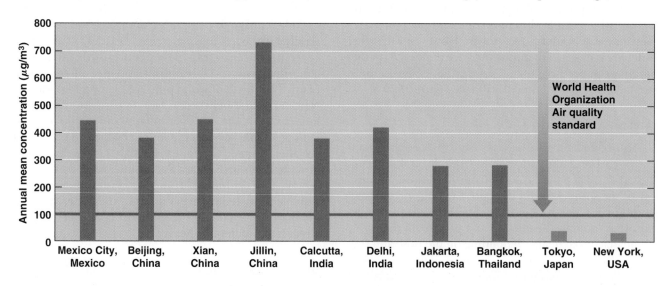

■ **FIGURE 14.6**
Total suspended particulates (TSP) for several large cities in developing countries (blue) and developed countries (green).
The value of 100 μg/m³ is the air-quality standard set by the World Health Organization. [*Source:* Modified from R. T. Watson. Intergovernment Panel on Climate Change. Presentation at the Sixth Conference of Parties to the United Nations Framework Convention on Climate Change, November 13, 2000, Figure 20.]

human sources such as industrial sites that produce petroleum or smelt metals. Hydrogen sulfide can damage plants and cause health problems ranging from illness to death for humans and other animals.[5]

Hydrogen Fluoride

Hydrogen fluoride is a gaseous pollutant released by some industrial activity, such as aluminum production, coal gasification, and the burning of coal in power plants. Hydrogen fluoride is extremely toxic—even a low concentration may cause problems for plants and animals. It can be dangerous to grazing animals because some forage plants can become toxic when exposed to this gas.[3]

Other Hazardous Gases

Dangerous gases come from a variety of sources. It is a rare month when the newspapers do not carry a story of a truck or train accident that releases toxic chemicals in a gaseous form into the atmosphere. People are often evacuated until the leak is stopped or the gas has dispersed to a nontoxic level. Chlorine gases and a variety of other materials used in chemical and agricultural processes may be involved.

Sewage-treatment plants are another source of gaseous air pollution. Urban sewers deliver a variety of organic chemicals to treatment plants. These chemicals—which include paint thinner, industrial solvents, chloroform, and methyl chloride—are not usually removed in treatment plants. In fact, the treatment processes actually enable the chemicals to evaporate into the atmosphere, where we can inhale them. Many of the chemicals are toxic or are suspected of being carcinogens, and some are so toxic that extreme care must be taken to make sure they do not enter the environment.

A pesticide was the source of a disaster in India. On December 3, 1984, a toxic liquid from a pesticide plant leaked, vaporized, and formed a deadly cloud of methyl isocyanate over a 64-square-kilometer area of Bhopal, India, killing more than 2,000 people and injuring more than 15,000. Methyl isocyanate is used in a common pesticide known in the U.S. as Sevin, as well as two other insecticides used in India. An industrial plant in West Virginia also makes the chemical, and small leaks that did not lead to major accidents occurred there before and after the catastrophic accident in Bhopal.

In our urban world, hazardous chemicals should not be stored near population centers. In addition, chemical plants must have reliable accident-prevention equipment and personnel trained to spot potential problems.

Lead

Lead really gets around. It is an important part of automobile batteries and many other industrial products. When lead is added to gasoline, it helps protect engines and enhances fuel consumption. Lead in gasoline (still used in some countries) is released into the air in vehicle exhaust. This is how lead has come to be spread widely around the world, reaching high levels in soils and waters along roadways. Once released, lead particulates can be transported by air and water. When the particulates are taken up by plants through the soil or deposited directly on leaves, lead enters terrestrial food chains. When lead is carried by streams and rivers, deposited in quiet waters, or transported to oceans or lakes, it is taken up by aquatic organisms and enters aquatic food chains.

Lead has even been found in glacial ice in Greenland. The concentration of lead in Greenland glaciers was essentially zero in A.D. 800 and began to creep up with the beginning of the Industrial Revolution in the mid-18th century. It increased steadily from 1750 until about 1950, then began to increase rapidly with the rapid growth of leaded gasoline. The accumulation of lead in the Greenland ice illustrates that our use of heavy metals in the 20th century eventually affected the entire biosphere.

Nearly all gasoline is now lead-free in the U.S., Canada, and much of Europe. In the United States, lead emissions have been reduced by about 98% since the early 1970s (Figure 14.5). Reducing and eventually eliminating lead in gasoline is a good start in lowering the levels of anthropogenic lead in the biosphere.

14.5 Urban Air Pollution

Air pollution varies among regions of the world and even within the U.S. For example, in the Los Angeles basin and many U.S. cities, nitrogen oxides and hydrocarbons are especially troublesome because they combine in the presence of sunlight to form photochemical *smog* (Figure 14.7). Most of the nitrogen oxides

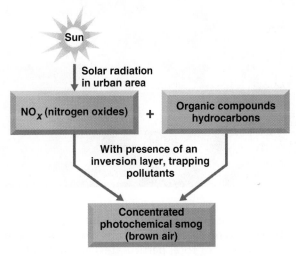

■ **FIGURE 14.7**
How photochemical smog is produced.

and hydrocarbons come from automobiles (mobile sources). In other parts of the U.S., such as Ohio and the Great Lakes region, additional air pollution comes from sulfur dioxide and particulates from industry and coal-burning power plants (point sources). This produces gray air called *sulfurous smog.*

Air pollution also varies with the time of year. For example, particulates are a problem in hot, dry months, when wildfires are likely, and during months when the wind blows across the desert. Smog, too, is mostly a problem in the summer, when there is a lot of sunlight.

The brown haze over Las Vegas, Nevada, is due mostly to particulates—in this case, naturally occurring particles from the desert environment. Las Vegas in the 1990s and into the 21st century is one of the fastest-growing urban areas in the United States. The population of Clark County, which includes Las Vegas, increased from less than 300,000 in 1970 to over 1.5 million in 2005. Las Vegas also has some of the most polluted air from particulates in the southwestern United States (Figure 14.8). About 60% of the dust comes from new construction sites, dirt roads, and vacant land. The rest is from natural and other sources.

Topography and meteorology are important factors. Pollution can develop wherever many sources emit pollutants over a wide area. Whatever the source—automobiles in Los Angeles or woodburning stoves in Vermont—whether air pollution develops depends on *topography* (the natural and human-made features of the

land) and *meteorology* (the atmosphere, including the weather). These factors determine the rate at which pollutants are carried away from their sources and converted into harmless compounds in the air. When pollution is produced faster than it can be degraded and carried off, dangerous conditions can develop. But meteorological conditions can determine whether air pollution is a nuisance or a major health problem.[10]

Atmospheric inversion can cause a serious pollution event, leading to increases in illnesses and deaths. In the lower atmosphere, such events can develop over a period of days. An **atmospheric inversion** occurs when warmer air lies above cooler air. It poses a particular problem when it limits air circulation below, resulting in a stagnant air mass. Figure 14.9 shows two types of atmospheric inversion that may contribute to air-pollution problems. In the upper diagram, which is somewhat similar to the situation in the Los Angeles area, descending warm air forms a semipermanent inversion layer. Because topographical features—the mountains—act as a barrier to the pollution, polluted air tends to move up canyons, where it is trapped. This air pollution occurs primarily in summer and fall.

The lower part of Figure 14.9 shows another type of inversion in a valley with relatively cool air overlain by warm air. This type of inversion can occur when cloud cover associated with a stagnant air mass develops over an urban area. Incoming solar radiation is blocked by the clouds, which reflect and absorb some of the solar energy and are warmed. Near Earth's surface, the air cools. If there is moisture in the air (humidity), then as the air cools, the water vapor condenses and fog may form. Because the air is cold, people burn more fuel to heat their homes and workplaces, so more pollutants are released into the atmosphere. As long as the stagnant conditions exist, the pollutants build up. This was what caused the 1952 London smog event that killed about 4,000 people.

Cities surrounded by hills or mountains fare the worst. Cities situated in a valley or topographic bowl surrounded by mountains are more susceptible to smog problems than are cities in open plains. The surrounding mountains and temperature inversions prevent pollutants from being carried off by winds and weather systems. Los Angeles, for example, has mountains surrounding part of the urban area, and lies within a region where the air lingers, allowing pollutants to build up.

Automobile use and solar radiation are factors in photochemical smog. As we said earlier, the reactions that occur in the development of photochemical smog are complex and involve both nitrogen oxides and organic compounds (hydrocarbons). The development of photochemical smog is directly related to automobile use. As the smog develops, visibility may be greatly reduced (Figure 14.10) as light is scattered by the pollutants.

▪ **FIGURE 14.8**
Las Vegas haze from particulate pollution.
Sources of particulates include construction sites and dirt roads (60% of total particulates) and natural and other sources (40%).

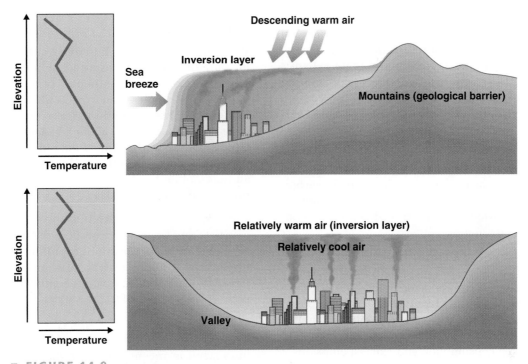

■ FIGURE 14.9

Two causes of an atmospheric inversion: warmer air over cooler air , which may aggravate air-pollution problems.

Reducing Urban Air Pollution at Its Source

The best strategy is to keep pollutants from entering the atmosphere in the first place. For both stationary and mobile sources of air pollutants, the most reasonable strategies are to reduce, collect, capture, or retain pollutants before they enter the atmosphere. From an environmental viewpoint, reducing pollution through energy efficiency and conservation (for exam-ple, by burning less fuel) has clear advantages over all other approaches. Here, we discuss pollution control for selected air pollutants.

Automobiles

Pollution-control measures for automobiles re-duce some urban pollutants, such as carbon monox-ide, nitrogen oxides, and hydrocarbons. Control of these

(a) *(b)*

■ FIGURE 14.10

The city of Los Angeles, California, on (a) a clear day and (b) a smoggy day.

materials will also limit ozone formation in the lower atmosphere, since ozone forms through reactions with nitrogen oxides and hydrocarbons in the presence of sunlight. Nitrogen oxides from automobile exhausts are controlled by recirculating exhaust gas, diluting the air-to-fuel mixture that is burned in the engine. Exhaust recirculation to reduce nitrogen-oxide emissions has been common practice in the U.S. for more than 20 years.

The catalytic converter is the device most often used to reduce carbon monoxide and hydrocarbon emissions from automobiles. It converts carbon monoxide to carbon dioxide and converts hydrocarbons to carbon dioxide and water.

Replacing carburetors with computerized fuel injection helped a lot. As government regulations controlling emissions became stricter, it became difficult to meet the new standards without the aid of computer-controlled engine systems. Computer-controlled fuel injection began to replace carburetors in the 1980s and has lowered fuel consumption and exhaust emissions.[11]

Do U.S. automobile emission regulations really reduce pollutants? Critics point out that pollutants may be relatively low when a car is new, but that many people do not take care of their cars well enough to ensure that the emission-control devices continue to work. Some people even disconnect smog-control devices. Evidence suggests that these devices tend to become less efficient each year after purchase.

Some suggest effluent fees instead of emission controls as the primary method of regulating air pollution from automobiles in the United States.[12] Under this scheme, vehicles would be tested each year for emission control, and fees would be assessed on the basis of test results. The fees would give people an incentive to buy cars that pollute less, and annual inspections would ensure that pollution-control devices were properly maintained. Although there is considerable controversy about mandatory inspections, they are common in a number of areas and will likely increase as air-pollution abatement becomes essential.

Automobiles that burn less gasoline pollute less. Other measures include (1) developing cleaner automobile fuels through the use of fuel additives and reformulation; (2) requiring new cars to use less fuel; and (3) encouraging the use of cars with electric engines and hybrid cars that have both an electric engine and an internal combustion engine.

With the discussion of specific types of air pollution and urban air pollution behind us, we now consider the topics of acid rain and ozone depletion. These are two serious air-pollution problems that involve pollutants in a more complex way than direct emission.

14.6 Acid Rain

We've all heard about it, but what exactly is it? We use a measure called *pH* to measure acidity and to classify substances as acid or alkaline (neutral or not acid). All rainfall is slightly acidic—water reacts with carbon dioxide in the atmosphere to produce weak carbonic acid. Thus, pure rainfall has a pH of about 5.6. **Acid rain** is any precipitation—not only rain, snow, and fog, but also dry particulate deposits—with a pH below 5.6. The acid in an automobile battery has a pH of about 1, which is highly acidic (Figure 14.11).

Acid rain results from burning fossil fuels. It occurs near and far from areas where the burning of fossil fuels creates major emissions of sulfur dioxide and nitrogen oxides. Other acids are also involved—for example, hydrochloric acid, which is emitted from coal-fired power plants (Figure 14.12).

Today, acid rain is a global problem. Human-induced acid rain likely began with the Industrial Revolution, but in recent decades it has gained more and more attention, and today it is a major environmental problem affecting all industrial countries. In the United

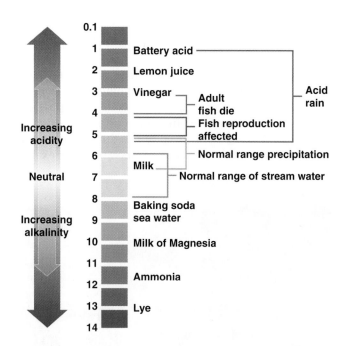

▪ **FIGURE 14.11**

The pH scale shows the levels of acidity in various fluids.
The scale ranges from less than 1 to 14, with 7 being neutral. pHs lower than 7 are acidic, while pHs greater than 7 are alkaline (basic). Acid rain can be very acidic and can be harmful to the environment. [*Source:* http://ga.water.usgs.gov/edu/phdiagram.html. Accessed Aug 12, 2005.]

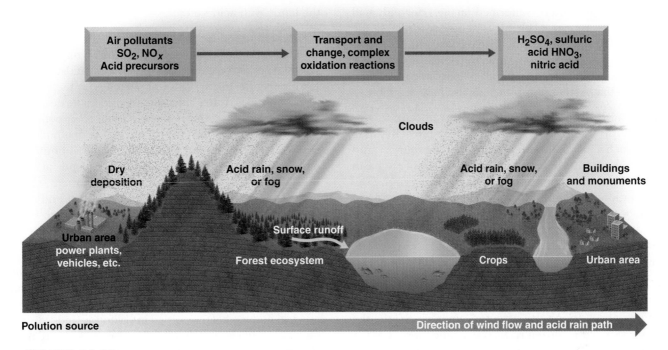

Air pollutants SO$_2$, NO$_x$ Acid precursors → **Transport and change, complex oxidation reactions** → **H$_2$SO$_4$, sulfuric acid HNO$_3$, nitric acid**

Clouds

Dry deposition

Acid rain, snow, or fog

Acid rain, snow, or fog

Buildings and monuments

Urban area power plants, vehicles, etc.

Surface runoff

Forest ecosystem

Crops

Urban area

Polution source

Direction of wind flow and acid rain path

■ **FIGURE 14.12**
Formation of acid rain.
Idealized diagram showing selected aspects of acid rain formation and paths.

States, nearly all of the eastern states are affected, as well as West Coast urban centers such as Seattle, San Francisco, and Los Angeles. The problem is also of great concern in Canada, Germany, Scandinavia, and Great Britain. Developing countries that rely heavily on coal, such as China, are facing serious acid-rain problems as well.

Some have tried to ease the local effects by using taller emission stacks. Taller stacks reduce local concentrations of air pollutants but worsen regional effects by spreading pollution more widely. For example, problems with acid precipitation in Canada can be traced to emissions of sulfur dioxide and other pollutants in the Ohio Valley. Tall stacks increase the length of time pollutants remain in the atmosphere (residence time) from 1 to 2 days to 10 to 14 days, because pollutants enter the atmosphere higher up, where mixing and transport by wind are more effective.[13]

Effects depend on geology, climate, types of vegetation, and soil composition. Areas most sensitive to acid rain have bedrock and soil that cannot buffer acid input. Calcium carbonate, which is present in many soils and rocks (limestone), is an important natural buffer against acid rain. Hydrogen in acid reacts with calcium carbonate, neutralizing acid. Thus, areas less likely to suffer damage from acid rain are those whose bedrock contains limestone or other carbonate material or whose soils contain calcium carbonate. Soils without

a buffer may lose their fertility when exposed to acid rain, either because nutrients are leached out by acid water or because the acid in the soil releases elements that are toxic to plants.

Acid rain hurts forest ecosystems. It has long been suspected that acid precipitation, wet or dry, harms trees. Studies in Germany led scientists to conclude that acid rain and other air pollution killed thousands of acres of evergreen trees in Bavaria. Similar studies in the Appalachian Mountains of Vermont (where many soils are naturally acidic) suggest that in some locations half the red spruce trees have died in recent years, due partly to acid rain and fog. As trees weaken and die, there is less habitat and food for wildlife. Bare trees allow more light through to the forest floor, changing the temperature and water content of soil at the surface. This affects what grows on the forest floor and what lives in the soil. Thus, the entire forest ecosystem is affected by acid rain. This is an example of the principle of environmental unity introduced in Chapter 1.

Lake ecosystems suffer, too. In recent years, fish have disappeared from lakes in Scandinavia, and the problem has been traced to acid rain from industrial processes in other countries, particularly Germany and Great Britain. Acid rain affects lake ecosystems in two ways. First, it disrupts the life processes of fish, amphibians, and other aquatic species in ways that limit their growth or cause death. (For example, crayfish produce fewer eggs and often malformed larvae.) Second, acid rain dissolves chemical elements necessary for life in the lake. The dissolved elements leave the lake with water outflow and are lost. Without these nutrients, algae do not grow, animals that feed on the algae have little to eat, and animals that feed on these animals also have less food.

Acid rain causes heavy metals to end up in water we drink and fish we eat. Acid rain leaches metals—such as aluminum, lead, mercury, and calcium—from the soils and rocks in a drainage basin and discharges them into rivers and lakes. High concentrations of aluminum are particularly damaging to fish, because the metal can clog their gills and cause suffocation. The heavy metals may pose health hazards to birds and other animals, including people, who eat the fish. Drinking water taken from acidic lakes may also have high concentrations of toxic metals.

Thousands of rivers and lakes in the United States and Canada are affected. In Nova Scotia, the water in at least a dozen rivers is so acidic for part of the year that it can no longer support healthy populations of Atlantic salmon. In the northeastern United States, about 200 lakes in the Adirondacks are no longer able to support fish; and thousands more are slowly losing the battle with acid rain.

Acid rain also damages building materials, such as steel, galvanized steel, paint, plastics, cement, masonry, and several types of rock, especially limestone, sandstone, and marble (Figure 14.13). Classical buildings on the Acropolis in Athens and other cities show considerable decay that accelerated in the 20th century as a result of air pollution. The problem has grown to such an extent that buildings require restoration, and statues and other monuments must have protective coatings, replaced frequently, at a cost of billions of dollars a year. Particularly important statues have been moved to protective glass containers, and replicas stand in their place outdoors for tourists to view.[14]

In the United States, the problem is coast to coast (Figure 14.14). Cities along the eastern seaboard are more susceptible to acid rain today because emissions of sulfur dioxide and nitrogen oxide are more abundant there and the regional wind pattern is east and north. However, acid precipitation has been recorded in California. Indeed, acid fog in Los Angeles may have a pH as low as 3—more than 10 times as acidic as the average acid rain in the eastern United States. In contrast to acid rain, acid fog forms near the ground when water vapor mixes with pollutants and

Don & Pat Valenti/Stone/Getty Images

■ **FIGURE 14.13**
Damage from acid rain.
Damage to a statue in Chicago from acid rain (left) and the same statue following restoration (right).

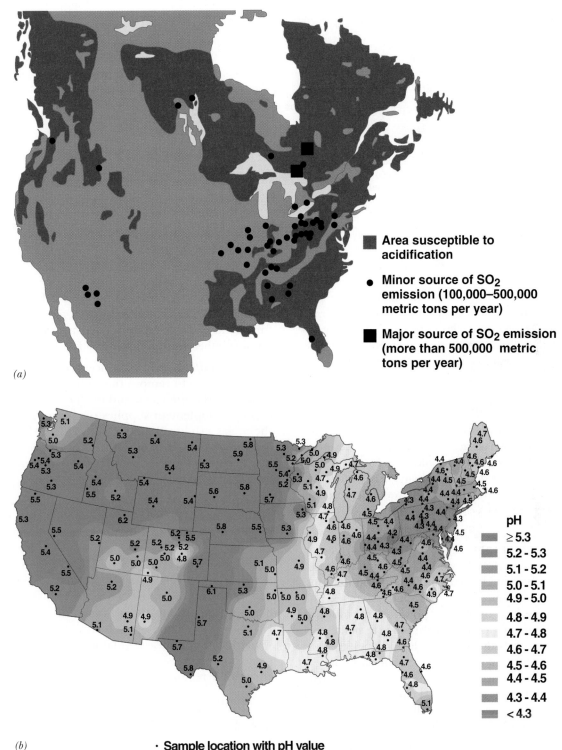

(a)

Area susceptible to acidification

Minor source of SO$_2$ emission (100,000–500,000 metric tons per year)

Major source of SO$_2$ emission (more than 500,000 metric tons per year)

(b)

pH
- ≥ 5.3
- 5.2 - 5.3
- 5.1 - 5.2
- 5.0 - 5.1
- 4.9 - 5.0
- 4.8 - 4.9
- 4.7 - 4.8
- 4.6 - 4.7
- 4.5 - 4.6
- 4.4 - 4.5
- 4.3 - 4.4
- < 4.3

· **Sample location with pH value**

■ **FIGURE 14.14**

Areas sensitive to acid rain: (a) Areas in Canada and the United States that are sensitive to acid rain. (b) pH of precipitation over the United States in 2000. Notice the relationship between the numerous sources of sulfur dioxide (SO$_2$) in the eastern U.S. [in (a) and low (more acidic) pH values]. [*Sources*: (a) How many more lakes have to die? *Canada Today* 12(2), 1981; (b) National Atmospheric Deposition Program/National Trends Network, 2001.]

▪ **FIGURE 14.15**
How a scrubber works.
The scrubber removes sulfur oxides from the gases emitted by tall stacks.

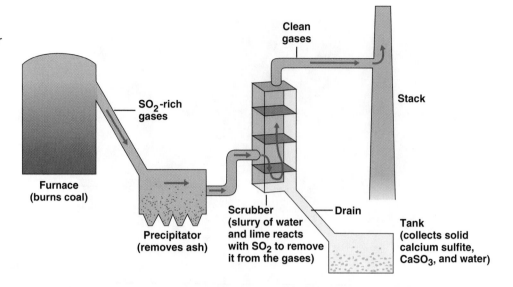

turns into an acid, which condenses around very fine particles of smog. If the air is sufficiently humid, a fog may form. When the fog eventually disappears, it can leave behind nearly pure drops of sulfuric acid. People can inhale tiny particles containing the acid deep into their lungs—a considerable health hazard.

Control of Acid Rain

We are struggling to find solutions for this problem. One solution to lake acidification is to neutralize the acid in a lake by adding lime periodically, as has been done in New York State, Sweden, and Ontario. This solution is not satisfactory over a long period, however, because it is expensive and requires a continuing effort.

The real solution is to decrease emissions of sulfur dioxide and nitrogen oxides. The best strategy is to increase energy efficiency and conservation measures that result in burning less coal and using nonpolluting alternative energy sources. Another strategy is to use pollution-abatement technology at power plants to lower emissions of air pollutants.[15] Technology to make coal burn more cleanly is already available. Although the cost of removing sulfur makes fuel more expensive, the expense must be balanced against the long-term consequences of burning sulfur-rich coal. In the United States, sulfur-dioxide emissions have been reduced about 50% since 1970 by the following methods.

Switching from high-sulfur coal to low-sulfur coal: This seems an obvious solution to reducing emissions of sulfur dioxide. Unfortunately, however, most low-sulfur coal in the United States is in the western part of the country, whereas most coal is burned in the East. Thus, transportation is an issue; and use of low-sulfur coal is a solution only where it can be done economically.

Cleaning up high-sulfur coal by washing it: This involves washing finely ground coal with water. But although wash-

ing is effective in removing nonorganic sulfur from minerals such as pyrite, it is ineffective for removing organic sulfur bound up with carbonaceous material. Cleanup by washing is therefore limited, and it is also expensive.

Coal gasification: This process converts high-sulfur coal into a gas to remove the sulfur. The gas obtained from coal is quite clean and can be transported fairly easily to supplement supplies of natural gas. The synthetic gas produced from coal is still fairly expensive compared with gas from other sources, but the price may become more competitive in the future.

"Scrubbers" are being used at power plants. Sulfur-dioxide emissions from stationary sources, such as power plants, can be reduced by removing the oxides from the gases in the stack before they reach the atmosphere. Perhaps the most highly developed technology for the cleaning of gases in tall stacks is **scrubbing** (Figure 14.15). The technology to scrub sulfur dioxide and other pollutants at power plants was developed in the 1970s in the United States in response to the passage of the Clean Air Act. However, the technology was not initially used in the United States because regulators allowed plants to disperse pollutants using tall smokestacks instead. This increased the regional acid-rain problem.

14.7 Ozone Depletion in the Stratosphere

Too much ozone causes trouble down here, and too little is a problem up there. Ozone occurs in the stratosphere about 30 km (19 mi) above Earth's surface. When it is near the surface, it is a secondary pollutant that can cause serious health problems. However, in the stratosphere, the so-called **ozone shield** protects us from the sun's harmful ultraviolet radiation. Roughly 99% of the harmful radiation is screened out in the ozone shield (Figure 14.16).

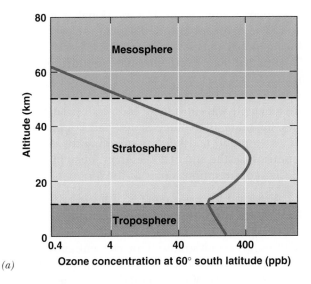

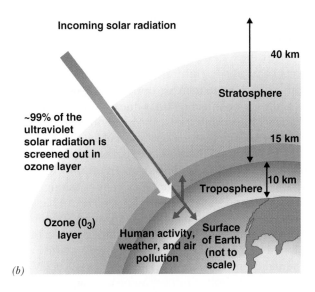

(a) *(b)*

■ **FIGURE 14.16**

Atmospheric structure and the ozone shield: (a) Structure of the atmosphere and ozone concentration. (b) Reduction of the most potentially damaging ultraviolet radiation by ozone in the stratosphere. [*Source*: Ozone concentrations modified from R. T. Watson. Atmospheric ozone. In J. G. Titus, ed., *Effects of change in stratospheric ozone and global climate*, Vol. 1, Overview. Washington, DC: U.S. Environmental Protection Agency, p. 70.]

The story of ozone depletion starts with emissions of chlorofluorocarbons (CFCs) and to a lesser extent nitrous oxide and methylbromide near the surface of the Earth. CFCs were produced and used for many years as the gas in refrigeration and air conditioners and as propellants for aerosol spray cans and production of Styrofoam. CFC emissions were about a million tons per year in the later part of the 20th century.

That CFCs were harming the ozone layer was suggested in 1974, by Mario Molina and F. Sherwood Rowland.[16] The idea received a tremendous amount of exposure in newspapers and on television and was heatedly debated by scientists, by companies that were producing CFCs, and by other interested parties. The public was concerned because such everyday products as shaving cream, hair spray, deodorants, paints, and insecticides were packaged in spray cans that contained CFCs as a propellant. That these products could threaten their health and the environment caused many Americans to write to their senators and representatives and to buy fewer products containing CFCs.[17]

The major features of the Molina and Rowland hypothesis are as follows:[18]

■ CFCs emitted in the lower atmosphere by human activity are stable and unreactive, and therefore have a very long residence time (about 100 years). The only known significant sink for CFCs is perhaps soil, which evidently does remove a small but unknown amount of CFC from the atmosphere at Earth's surface.[19]

■ Because CFCs have a long residence time in the lower atmosphere and because the lower atmosphere is very

fluid, the CFCs eventually wander upward and enter the stratosphere. Once they get above most of the stratospheric ozone, they may be destroyed by solar ultraviolet radiation. This releases chlorine, a highly reactive atom.

■ The reactive chlorine may enter into complex chain reactions that deplete ozone in the stratosphere.

■ Most depletion is in polar areas and the so-called *ozone hole*, a region of ozone depletion that was first identified over Antarctica.

■ The depletion of ozone causes an increase in the amount of ultraviolet B (UVB) radiation that reaches Earth's surface. UVB is a cause of human skin cancers and is also thought to be harmful to the human immune system.

The use of CFCs as a refrigerant has increased dramatically, especially in developing countries, such as China. However, CFCs as aerosol propellants are no longer a problem—this use was banned in a number of countries in the late 1970s.[18]

Ozone depletion can have several serious environmental effects, including damage to food chains on land and in the oceans and damage to human health. A 1% decrease in ozone can cause an increase of 1–2% in UVB radiation and an increase of 2% in the incidence of skin cancer. As a result, the incidence of skin cancer has increased around the world. UV radiation can also cause cataracts (an eye disease in which the lens becomes clouded and vision is impaired) and may damage the human immune system as well, resulting in higher numbers of a variety of diseases. Finally, a variety of environmental pollutants in the air and water could have synergistic effects, increasing potential health risks of exposure to ultraviolet radiation.[20, 21]

Controls on CFCs should lower the incidence of skin cancer, but not right away. The incidence of skin cancer is projected to increase until about 2060, then decline as the ozone shield recovers due to controls on CFC emissions.[22] Meanwhile, we've changed our habits when it comes to the sun. For years, people seeking that tanned, healthy look sunned themselves in parks and backyards, at pools and beaches, and even on city rooftops ("Tar Beach"). Today, hats and sunblock lotions, along with makeup and even shampoos with sunscreen, are big sellers all year-round. Newspapers in the United States provide readers with the Ultraviolet (UV) Index (Table 14.2) developed by the National Weather Service and the Environmental Protection Agency. The index predicts UV intensity on a scale from 1 to 11+. Some news agencies also use the index to recommend the level of sunblock.

Here are some ways to reduce your risk of skin cancer and other skin damage (such as wrinkles) from UV exposure:

▪ Try to avoid exposure to the sun between 10:00 a.m. and 4:00 p.m., when solar radiation is most intense.

▪ When possible, remain in the shade.

▪ Use a sunscreen with sun protection factor (SPF) of at least 30 (and remember to put more on, since it becomes less effective with time of exposure).

▪ Wear a wide-brimmed hat and, where possible, a full-length cover-up of tightly woven fabric (for example, loose-fitting light cotton).

▪ Wear UV-protective sunglasses.

▪ Avoid tanning salons and sunlamps.

▪ Consult the UV Index before going out.

Reducing Ozone Depletion: An Environmental Success Story

The signing of the Montreal Protocol in September 1987 was a diplomatic achievement of monumental proportions. Twenty-seven nations signed the

☐ TABLE 14.2 ULTRAVIOLET (UV) INDEX FOR HUMAN EXPOSURE

Exposure Category	UV Index	Comment
Low	<2	Sunblock recommended for all exposure
Moderate	3–5	Sunburn can occur quickly
High	6–7	Potentially hazardous
Very high	8–10	Potentially very hazardous
Extreme	11+	Potentially very hazardous

Source: *Modified after U.S. Environmental Protection Agency 2004 (with the National Weather Service). Available at www.epa.gov. Accessed June 16, 2004.*
Note: *At moderate exposure to UV, sunburn can occur quickly; at high exposure, fair-skinned people may burn in 10 minutes or less of exposure.*

agreement originally, and 119 additional nations signed later. The protocol outlined a plan for eventually reducing global emissions of CFCs to 50% of 1986 levels. The original plan was to eliminate production of CFCs by 1999, but because of scientific evidence that the ozone layer was being depleted faster than predicted, the timetable was shortened. Most industrialized countries, including the United States, had stopped production by the end of 1995; the deadline for developing countries was the end of 2005. An eventual phase-out of all CFC consumption is part of the Montreal Protocol.

Already, the increase in CFC emissions has slowed, thanks to the protocol and other agreements and amendments. Assessments in London (1990) and Copenhagen (1992) suggest that stratospheric concentrations of CFCs will shrink to pre-1980 levels by about the year 2050. Nevertheless, because of the long residence time of CFCs in the stratosphere, ozone depletion from CFCs will likely continue for many years to come.[23, 24] That's the bad news. The good news is that ozone levels in the stratosphere will slowly increase in the next few decades.

We are seeking safe and effective substitutes for CFCs. Two substitutes being experimented with today are *hydrofluorocarbons (HFCs)* and *hydrochlorofluorocarbons (HCFCs)*. These chemicals are controversial and more expensive than CFCs but do have advantages.

The advantage of HFCs is that they do not contain chlorine. They do contain fluorine, though; and when fluorine atoms are released into the stratosphere, they participate in reactions similar to those of chlorine. Thus, they can cause ozone depletion. However, this may not be a significant problem with fluorine, because it is about 1,000 times less efficient in those reactions.[18, 25] Also, some blends of HFCs have little or no ozone-depleting potential.[26]

HCFCs contain hydrogen in place of chlorine. They can be broken down in the lower atmosphere, but they can cause ozone depletion if they do reach the stratosphere before being broken down. Although their atmospheric lifetime is much shorter than that of the CFCs, when HCFCs are used in tremendous quantities, they do cause ozone depletion.[18] HCFCs are at best a temporary solution measure until substitutes that do not cause ozone depletion are available. HCFCs will be phased out by 2030.

Until the ozone layer recovers, we will have to adapt to UV radiation levels. As we have seen, depletion of stratospheric ozone will be a story of gradual recovery by the mid-21st century.[23, 27] Meanwhile, people will have to learn to live with higher levels of exposure to ultraviolet radiation. In the long term, achieving sustainability of stratospheric ozone will require careful management of human-produced chemicals that damage the ozone layer.

Our discussion of acid rain and ozone depletion has emphasized some of the complexity of air pollution. We will now consider the serious problem of indoor air pollution.

14.8 Indoor Air Pollution

Indoor air pollution began when we started building and heating shelters for protection from the elements. An autopsy of a 4th-century Native American woman, frozen shortly after death, revealed that she had black lung disease from breathing very polluted air over many years. The pollutants included particles from lamps that burned seal and whale blubber.[28] This same disease has long been recognized as a major health hazard for underground coal miners and has been called "coal miners' disease" (Figure 14.17). By the mid-1970s, black lung disease was estimated to be responsible for about 4,000 deaths a year in the United States.[29]

Sources and Concentrations of Indoor Air Pollution

Potential sources of indoor air pollution are incredibly varied (see Figures 14.18 and 14.19). Indoor air pollutants can arise from both human activities and

■ **FIGURE 14.17**
Dirty air in mines.
Coal miners in Eastern Europe covered with lung-damaging coal dust.

1. Heating, ventilation, and air-conditioning systems may be sources of indoor air pollutants, including molds and bacteria, if filters and equipment are not maintained properly. Gas and oil furnaces release carbon monoxide, nitrogen dioxide, and particles.

2. Restrooms may be sources of a variety of indoor air pollutants, including second-hand smoke, molds, and fungi resulting from humid conditions.

3. Furniture and carpets often contain toxic chemicals (formaldehyde, organic solvents, asbestos), which may be released over time in buildings.

4. Coffee machines, fax machines, computers, and printers can release particles and chemicals, including ozone, which is highly oxidizing.

5. Pesticides can contaminate buildings with cancer-causing chemicals.

6. Fresh air intake that is poorly located—for example, above a loading dock or first-floor restaurant exhaust fan—can bring in air pollutants.

7. People who smoke indoors, perhaps in offices or restaurants, and people who smoke outside buildings, particularly near open or revolving doors, may cause indoor pollution when the smoke is drawn into and up through the building.

8. Remodeling, painting, and other such activities often bring a variety of chemicals and materials into a building. The fumes may enter the heating, ventilation, and air-conditioning system, causing widespread pollution.

9. A variety of cleaning products and solvents used in offices and other parts of buildings contain harmful chemicals, whose fumes may circulate throughout a building.

10. People can increase carbon dioxide levels, emit bioeffluents (gas from the digestive system), and spread bacterial and viral contaminants.

11. Loading docks can be sources of organic pollution from garbage containers, and of particulates and carbon monoxide from vehicles.

12. Radon gas can seep into a building from soil, and molds can grow when dampness enters foundations and rises up the walls.

13. Dust mites and molds can live in carpets and other indoor places.

14. Pollen can come from indoor and outdoor sources.

■ **FIGURE 14.18**
Some Potential Sources of Indoor Air Pollution.

(a)   *(b)*

▪ **FIGURE 14.19**

Dust mites and pollen: (a) This dust mite (magnified about 140 times) is an eight-legged relative of spiders. It feeds on human skin in household dust. It lives in materials such as fabrics on furniture. Dead dust mites and their excrement can produce allergic reactions and asthma attacks in some people. (b) Microscopic pollen grains that in large amounts may be visible as a brown or yellow powder. The pollen here are dandelion and horse chestnut.

natural processes. Several sources have made news in recent years, so the public is especially aware of them. They are described in the following list:

▪ **Environmental tobacco smoke** (secondhand smoke) is the most hazardous common indoor air pollutant, linked to over 40,000 deaths per year in the U.S. (mostly from heart disease and lung cancer).

▪ *Legionella pneumophila,* a bacterium that normally lives in pond water, causes "Legionnaires' disease," a sometimes fatal type of pneumonia, when inhaled. This disease is usually spread by air-conditioning equipment, which harbors the bacteria in pools of stagnant water in air ducts and filters.[30] However, one epidemic occurred in a hospital as a result of contamination from an adjacent construction site.

▪ **Some molds (fungal growths)** in buildings release toxic spores. Inhaling them over a period of time can cause chronic inflammation and scarring of lungs, as well as other serious respiratory illnesses that are painful, disabling, and sometimes fatal.[30] It is believed that molds may be responsible for as many as half of all health complaints resulting from indoor air environments.

▪ **Radon gas** seeps up naturally from soils and rocks below buildings and is thought to be the second most common cause of lung cancer.

▪ **Pesticides** used in buildings to control ants, flies, fleas, moths, and rodents may be toxic to people as well.

▪ **Some varieties of asbestos,** used as an insulating and fireproofing material in homes, schools, and offices, cause a particular type of lung cancer (see Chapter 8).

▪ **Formaldehyde** used in some foam insulation, as a binder in particleboard and wood paneling, and in

many other materials found in homes and offices can emit a gas into the indoor environment. Some mobile homes have high concentrations of formaldehyde gas because products containing the chemical were used in their construction (wood paneling, for example).

▪ **Dust mites and pollen** irritate the respiratory system, nose, eyes, and skin of people who are sensitive to them.

Sick Buildings

Many buildings—including schools—can make you sick. People today spend 70–90% of their time indoors (in homes, workplaces, automobiles, restaurants, and so forth), but we only recently have begun to study indoor pollution and how it affects our health. The World Health Organization has estimated that as many as one in three workers may be working in a building that causes them to become sick (Figure 14.20). As many as 20% of public schools in the United States have problems with indoor air quality. The U.S. Environmental Protection Agency (EPA) considers indoor air pollution one of the most significant environmental health hazards people face in the modern workplace.[30]

An entire building can be considered sick because of environmental problems. There are two types of sick buildings:

▪ Buildings with identifiable problems, such as toxic molds or bacteria known to cause diseases. Such diseases are known as *building-related illnesses* (BRI).

▪ Buildings with **sick building syndrome** (SBS), where the symptoms people report cannot be traced to any one cause. One common aspect of SBS is that often no specific disease or cause is easily identified.[31]

■ **FIGURE 14.20**
Sick building.
This modern art museum, which opened in 1998 in Stockholm, Sweden, closed for repairs in 2002 due to building-related illness. Employees in the kitchen and bookstore complained of a variety of symptoms, including persistent coughs, headaches, and difficulty in breathing. The indoor air pollution is thought to be partly due to toxic mold, resulting from problems with the ventilation system.

So how do we decide whether it's a case of sick building syndrome? Not just one but a number of people in the building report health problems that they believe are linked to the amount of time they spend in the building. Their complaints may range from funny odors to more serious symptoms, such as headaches, dizziness, nausea, and so forth. A number of people in the building may have contracted a disease such as cancer.

It is often hard to identify what may be making occupants ill. Sometimes health problems can be traced to poor management and low worker morale rather than exposure to toxins in the building. When a building's occupants report health problems and a study of the building finds no cause, a number of things may be happening:[31]

■ The problem stems from not one contaminant but the combined effects of a number of contaminants in the building.

■ Environmental stress from a source other than air quality—such as noise, high or low humidity, poor lighting, or overheating—is responsible.

■ Employment-related stress—such as poor relations between labor and management, poor morale, or overcrowding—may be causing the symptoms reported.

■ Other, unknown factors may be responsible. For example, pollutants or toxins may be present but not identified.

■ Finally, the problem could be the combined effect of various aspects of some or all of these factors.

Air pollutants are more concentrated indoors than outdoors. Not only are many products and processes used in our homes and workplaces sources of pollution, but common air pollutants are often more highly concentrated indoors than they are outside. For example, carbon monoxide, particulates, nitrogen dioxide, radon, and carbon dioxide are generally found in much higher concentrations indoors than outdoors.

Why are concentrations generally greater indoors? One obvious reason is that there are so many potential indoor sources of pollutants. Another reason is simply that the steps we have taken to conserve energy in homes and other buildings have caused pollutants to be trapped inside. Two of the best ways to conserve energy in homes and other buildings are to increase the insulation and keep out more of the outside air. Caulking, weather-stripping, and windows that do not open do reduce energy consumption, but also tend to affect the indoor air quality by reducing natural ventilation. An important function of ventilation is that it replaces the indoor air with outdoor air, in which the concentrations of pollutants are generally much lower. With less natural ventilation, we must depend more on the ventilation systems that are part of heating and air-conditioning systems.

Faulty systems can be a source of trouble. No matter which type of heating, ventilation, and air-conditioning system is used, it must be designed to meet the needs of the building, properly installed, and correctly maintained and operated.[31] If filters become plugged or contaminated with fungi, bacteria, or other potentially infectious agents, serious problems can result.[31, 32]

Symptoms of Indoor Air Pollution

People react in different ways to pollutants. Some people are particularly susceptible to indoor air pollution, but even in the same environment, the symptoms they report may vary. And in some cases, their symptoms stem from factors other than air pollution. Some chemical pollutants can cause nosebleeds, chronic sinus infections, headaches, and irritation of the skin, eyes, nose, and throat. More serious problems include loss of balance and memory, chronic fatigue, difficulty speaking, and allergic reactions, including asthma.

For example, chlorine tablets, often used to disinfect swimming pools and hot tubs, cause coughing and shortness of breath if dust from the tablets is inhaled. Other pollutants cause dizziness or nausea. Exposure to low concentrations of carbon monoxide causes shortness of breath, but high concentrations are extremely toxic and sometimes fatal.[33]

Symptoms can come on slowly or quickly, and be short-term or long-term. Because of long lag times between exposure and disease, it may be difficult to establish relationships between a particular indoor air environment and disease in an individual.

Two Particularly Important Indoor Pollutants

Environmental tobacco smoke (ETS). This comes from two sources: smoke exhaled by smokers, and smoke emitted from burning tobacco in cigarettes, cigars, or pipes. People who are exposed to ETS (or secondhand smoke) are referred to as *passive smokers*.[34]

ETS is the most widely known hazardous indoor air pollutant. It is hazardous for the following reasons:[34, 35]

- Tobacco smoke contains several thousand chemicals, many of which are irritants and about 40 of which are *carcinogens* (substances that can cause cancer).

- Studies of nonsmoking people exposed to ETS found their airway functions comparable to those of people who smoke up to ten cigarettes a day. They suffer more illnesses (such as coughs, eye irritation, and colds) and lose more work time than people who are not exposed to ETS.

- In the United States, about 3,000 deaths from lung cancer and 40,000 deaths from heart disease each year are thought to be associated with ETS.

The number of U.S. smokers has declined, but there are still about 40 million. The rate is higher in the developing world, where health warnings are few or nonexistent. Smoking is a hard habit to break, because tobacco contains nicotine, a highly addictive substance. Nevertheless, education and social pressure have worked to some extent to influence some thoughtful people to quit smoking and encourage others to quit.

Radon gas. Unlike ETS, radon gas is a problem not of our making. It has become apparent only within the past few decades that *radon gas* may be a significant environmental health problem in the United States.[36] And one of the interesting things about the radon gas hazard is that it comes from natural processes, not from human activities. Radon is a naturally occurring radioactive gas that is colorless, odorless, and tasteless. It was discovered in 1900 by a German chemist named Ernest Dorn.

In the early 1900s, people thought radon was good for you. During this period, bathing in radon water became a health fad, and many products containing radium (and radon) hit the market. These included chocolate candies, bread, and toothpaste. As recently as 1953, a contraceptive jelly containing radium was marketed in the United States.[36]

Radon enters buildings from the rocks and soil underneath. The concentration of radon gas that reaches the surface and can enter our dwellings depends on the concentration of radon in the rocks and soil, as well as how efficiently it is transferred from there to the surface. In some regions of the United States, the bedrock has an above-average natural concentration of uranium. A large area that includes parts of Pennsylvania, New Jersey, and New York is now famous for high concentrations of radon gas. Areas with elevated concentrations of radon have also been identified in a number of other states, including Florida, Illinois, New Mexico, South Dakota, North Dakota, Washington, and California.

Radon gas enters homes and other buildings in three main ways (Figure 14.21):

- It migrates up from soil and rock into basements and lower floors.

- Dissolved in groundwater, it is pumped into wells and then into homes.

- Materials contaminated with radon, such as building blocks, are used in construction.

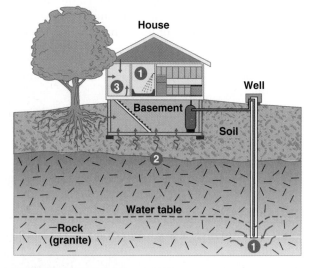

■ **FIGURE 14.21**
How radon can enter homes.
(1) Radon in groundwater enters a well and goes to a house, where it is used for dishwashing, showers, and other purposes. (2) Radon gas in rocks and soil migrates into a basement through cracks in the foundation and pores in construction. (3) Radon gas is emitted from construction materials used in building a house.
[*Source:* U.S. Environmental Protection Agency.]

How many homes have elevated radon levels? The EPA estimates that about 7% of U.S. homes have elevated radon levels and recommends that all homes and schools be tested.[37] The test is simple and inexpensive.

Controlling Indoor Air Pollution

There are strong financial incentives to improve the work environment. As much as $250 billion per year might be saved by decreasing illnesses and increasing productivity.[30] A good starting point would be passing environmental legislation requiring minimum indoor air-quality standards. This should include increasing the inflow of fresh air through ventilation. In Europe, systems of filters and pumps in many office buildings circulate air three times more frequently than U.S. buildings typically do. Many building codes in Europe require that workers have access to fresh air and natural light. Unfortunately for U.S. workers, no such codes exist in the United States, and many buildings use central air-conditioning with windows permanently sealed.[30]

Air-conditioning isn't a cure-all. You might think that heating, ventilating, and air-conditioning systems, when operating properly and well maintained, will ensure good indoor air quality. Unfortunately, these systems are not designed to maintain all aspects of air quality. Ventilation is one control strategy when faced with high concentrations of any indoor air pollutant, including radon. Other strategies include source removal, source modification, and air cleaning.[33]

Some combination of strategies may be the best approach. One of the principal ways to control the quality of indoor air is by diluting it with fresh outdoor air via a ventilating air-conditioning system and windows that can be opened. Outside air is brought in and mixed with air in the return flow system from the building; the air is filtered, heated or cooled, and supplied to the building. Various types of air-cleaning systems for residential and nonresidential buildings are available to reduce potential pollutants, such as particles, vapors, and gases. These systems can be installed as part of the heating, ventilation, and air-conditioning system or as stand-alone appliances.[33]

The good news about radon problems is that they are often easy to fix. There are well-known ways to lower the concentration of radon gas in buildings. The simplest step is to find the places where it is getting in and seal them. However, this is often not sufficient; so additional ventilation, using fans and other devices, may be necessary. Increased ventilation is the primary remedy for radon problems. If these steps are not successful, a venting system can be constructed.[38, 39]

Education plays an important role. Education gives people the knowledge necessary to make intelligent decisions. These decisions may be about exposure to chemicals, such as paints and solvents, and about strategies to avoid potentially hazardous conditions in the home and workplace,[33] such as deciding not to install unvented or poorly vented appliances. A surprising (and tragic) number of people are killed each year by carbon monoxide poisoning due to poor ventilation in homes, campers, and tents. Educated people are also more aware of their legal rights in terms of product liability and safety.

14.9 Air Pollution Legislation, Standards, and Index of Air Quality

Disastrous smog events in Donora, Pennsylvania, in 1948 and London, England, in 1952 prompted legislation to reduce air pollution in both England and the United States.

Clean Air Act Amendments of 1990

The Clean Air Act Amendments of 1990 were enacted by the U.S. Congress to address acid rain, toxic emissions, ozone depletion, and automobile exhaust. In dealing with acid rain, the amendments set limits on the amount of sulfur dioxide that can be released by utility companies burning coal. The goal was to reduce such emissions by about 50%, to 10 million tons a year, by 2000, and this goal was achieved (see Figure 14.5). The legislation provides incentives for utility companies by allowing them to buy and sell the right to pollute.[40] The total amount of pollution allowed is divided among a certain number of permits. Utilities with clean power plants that don't need the permits can sell them to those that do need them.

Environmentalists, too, can buy these permits to keep them from being bought by utility companies. This forces the companies to use more vigorous pollution-control technology. Buying of permits by environmentalists, however, has not been a major factor. As the permits are bought and sold, they take on economic value, and polluters begin to view polluting as an expensive way to do business.[41]

The Clean Air Amendments also deal with ozone depletion in the stratosphere. The goal is to end production of all chlorofluorocarbons (CFCs) and other chlorine chemicals by 2030.[42]

The legislation seeks to reduce urban smog through stricter emission controls on automobiles, and requirements for cleaner-burning fuels. These requirements raise the cost of automobile fuels and the price of new automobiles.

Ambient Air-Quality Standards

Air-quality standards work with emission standards to control air pollution. Many countries including the United States have developed air-quality standards, for carbon monoxide, nitrogen dioxide, ozone, sulfur dioxide and particulates.

Air Quality Index (AQI)

The U.S. uses the Air Quality Index to describe air pollution on a given day (Table 14.3). In urban areas, the day's air quality may be reported as good, moderate, unhealthy for sensitive groups, unhealthy, very unhealthy, or hazardous, corresponding to a color code of the **Air Quality Index**. The AQI is based on measurements of five major pollutants: particulate matter, sulfur dioxide, carbon monoxide, ozone, and nitrogen dioxide. An AQI value greater than 100 is unhealthy. In most U.S. cities, the AQI ranges between 0 and 100 and exceeds that range only a few times a year. However, cities with serious air-pollution problems may exceed an AQI of 100 many times a year. In a typical year, AQI values above 200 for all U.S. sites are rare, and those above 300 are very rare. By contrast, large, densely population urban areas outside the United States that have many uncontrolled sources of pollution frequently exceed an AQI value of 200.

During a pollution episode, ozone levels are monitored hourly, and a smog episode begins if the primary National Ambient Air Quality Standard (NAAQS) of 0.12 ppm of ozone (1-hour average) is exceeded. This measure corresponds to unhealthy air with an AQI of 100–300 (Table 14.3). An air pollution "alert" is issued if the AQI exceeds 200. An air pollution "warning" is issued if the AQI exceeds 300, the point at which air quality is hazardous to all people. If the AQI exceeds 400, an "air-pollution emergency" is declared, and people are urged to remain indoors and not exert themselves. Driving private automobiles may be prohibited, and industry may be required to reduce emissions to a minimum during the episode.

14.10 The Cost of Reducing Air Pollution

The cost of reducing air pollution varies widely among industries. For example, it would cost a utility that burns fossil fuel only a few hundred dollars for each additional ton of particulates it removed. For an aluminum plant, the cost to remove an additional ton of particulates may be as much as several thousand dollars.[10] For this reason,

☐ TABLE 14.3 AIR QUALITY INDEX (AQI) AND HEALTH CONDITIONS

Index Values	Descriptor Action Level (AQI)[a]	Cautionary Statement	General Adverse Health Effects	
0–50	Good	None	None	None
51–100	Moderate	Unusually sensitive people should consider limiting prolonged outdoor exertion.	Very few symptoms[b] for the most susceptible people.[c]	None
101–150	Unhealthy for sensitive groups	Active children and adults, and people with respiratory disease, such as asthma, should limit prolonged outdoor exertion.	Mild aggravation of symptoms in susceptible people, few symptoms for healthy people.	None
151–199	Unhealthy	Active children and adults, and people with respiratory disease, such as asthma, should avoid prolonged outdoor exertion; everyone else, especially children, should limit prolonged outdoor exertion.	Mild aggravation of symptoms in susceptible people, irritation symptoms for healthy people.	None
200–300	Very unhealthy	Active children and adults, and people with respiratory disease, such as asthma, should avoid outdoor exertion; everyone else, especially children, should limit outdoor exertion.	Significant aggravation of symptoms in susceptible people, widespread symptoms in healthy people.	Alert (200+)
Over 300	Hazardous warning	Everyone should avoid outdoor exertion.	300–400: Widespread symptoms in healthy people	(300+)
	Emergency		400–500: Premature onset of some diseases.	
			Over 500: Premature death of ill and elderly people; healthy people experience symptoms that affect normal activity.	(400+)

[a] Triggers preventive action by state or local officials.   [b] Symptoms include eye, nose, and throat irritation; chest pain; breathing difficulty.
[c] Susceptible people are young, old, and ill people and people with lung or heart disease.
AQI 51–100: Health advisories for susceptible individuals.
AQI 101–150: Health advisories for all.
AQI 151–200: Health advisories for all.
AQI 200+: Health advisories for all; triggers an alert; activities that cause pollution might be restricted.
AQI 300: Health advisories to all; triggers a warning; probably would require power plant operations to be reduced and carpooling to be used.
AQI 400+: Health advisories for all; triggers an emergency; cessation of most industrial and commercial activities, including power plants; nearly all private use of vehicles prohibited.
Source: U.S. Environmental Protection Agency.

some economists would argue that it is wise to raise the standards for utilities and relax them, or at least not raise them, for aluminum plants. This should lead to more cost-efficient pollution control while maintaining good air quality. However, the geographic locations of various facilities will determine what trade-offs are possible.[40, 41]

Consider also the law of diminishing returns. In sum, eventually a point is reached at which the cost of each additional gain is very high in relation to the additional benefit. Because of this and other economic factors, some argue that penalizing polluters with fees or taxes might make more economic sense than trying to evaluate the costs and benefits of enforcing standards. Another approach is to issue vouchers that allow businesses to release a certain total amount of pollution in a region. These vouchers are bought and sold on the open market. All these economic alternatives are controversial and may be objectionable to people who believe that polluters should not be allowed to buy their way out of doing what is socially responsible—that is, not polluting our air.

Some of the variables are hard to put a value on. But we do know the following:

■ With increasing air-pollution controls, the cost for technology to control air pollution increases.

■ As the controls for air pollution increase, the loss from pollution damages decreases.

■ The total cost of air pollution is the cost of pollution control plus the cost of environmental damage from the pollution.

Determining the cost of the damage is the bigger problem. The cost of pollution-control technology is fairly well known. It is much more difficult to determine the loss from pollution damages, particularly when we consider health problems and damage to vegetation, including food crops. Air pollution may cause or aggravate chronic respiratory diseases in people (especially the young and the elderly), at a very high cost. A recent study of the annual cost of air pollution in the Los Angeles basin is 1,600 lives and about $10 billion.[42] Air pollution also leads to loss of revenue in places such as Los Angeles and Mexico City because many people choose not to visit areas known to have serious air-pollution problems.

Even if we can't calculate all benefits and costs, reducing pollution is worthwhile. In the United States, the National Ambient Air Quality Standards set a minimum acceptable air-quality level. However, as discussed, it is also a good idea to consider alternatives, such as charging fees or taxes for emissions. If the charges are carefully determined and emissions are closely monitored, the charges should provide an incentive for industries to invest in air-pollution control. The end result would be better air quality.[43, 44]

14.11 What Lies Ahead for Air Pollution?

Air-quality problems are not just in urban areas—pollution travels. The North Slope of Alaska, a vast strip of land about 200 kilometers (125 miles) wide, is considered one of the last unspoiled wilderness areas left on Earth. It seems logical that air in such a place would be pristine, except perhaps near areas where petroleum is being developed. However, studies suggest that the North Slope has an air-pollution problem that originates in Eastern Europe and Eurasia! Apparently, pollutants from burning fossil fuels in Eurasia are transported by the jet stream, possibly at speeds of more than 250 miles an hour, over the North Pole and eventually to the North Slope of Alaska. There, the air mass slows, stagnates, and produces what is known as the "Arctic haze." The pollution level has been compared with that of some eastern U.S. cities, such as Boston.[45]

Another global event occurred in the spring of 2001, when a white haze consisting of dust from Mongolia and industrial particulate pollutants arrived in North America. The haze affected one-fourth of the United States and could be seen from Canada to Mexico. The particulates were close enough to the ground to cause respiratory problems for people. In the United States, pollution levels from the haze alone were as high as two-thirds of federal health limits. The haze demonstrates that pollution from Asia is carried by winds across the Pacific Ocean.

The optimistic view is that urban air quality will continue to improve as it has in the past 35 years, because we know so much about the sources of air pollution and have developed effective ways to reduce it. The pessimistic view is that in spite of this knowledge, population pressures and economics will dictate what happens in many parts of the world, and the result will be poorer air quality in many places.

A mixture of optimistic and pessimistic points of view is likely. In the 21st century, the air quality in large urban areas in developing countries may get worse even as attempts are made to improve the situation, because population and economic factors will likely outweigh pollution abatement. Large urban areas in developed and more affluent countries (particularly the United States) may well continue to experience improved air quality in coming years.

Los Angeles is an example of trends in developed countries. As we have discussed, this region has the worst air quality in the United States. However, Los Angeles is coming to grips with the problem. The people studying air pollution in the region now understand that pollution abatement will take massive efforts, unlike the limited approach of past strategies. Los Angeles is considering a new, controversial air-quality plan involving the entire urban region and including the following features:[45, 46]

- strategies to discourage automobile use and reduce the number of cars;
- stricter emission controls for automobiles;
- a requirement for a certain number of zero-pollutant automobiles (electric cars) and hybrid cars with fuel cell and gasoline engines;
- a requirement that more gasoline be reformulated to burn cleaner;
- improvements in public transportation and incentives for people to use it;
- mandatory carpooling;
- stricter controls on industrial and household activities that are known to contribute to air pollution.

At the household level, for example, common materials such as paints and solvents will be reformulated so that their fumes will cause less air pollution. Eventually, certain equipment, such as gasoline-powered lawn mowers that contribute to air pollution, may be banned.

Southern California's air quality is better but is still the nation's worst. Since the 1950s, the peak level of ozone (considered one of the best indicators of air pollution) has declined from about 0.68 to 0.3 parts per million even though during this period the population nearly tripled and the number of motor vehicles quadrupled.[47] However, even if all the controls we have talked about are implemented, air quality will continue to be a problem in coming decades, particularly if the urban population continues to increase.

Many other U.S. cities also have poor air quality a significant part of the year. Based on the criterion of 30 days per year of unhealthy air due to ozone pollution, many millions of Americans live in cities where hazardous air pollution exists. Cities with high ozone levels include Riverside, California; Houston, Texas; Baltimore, Maryland; Charlotte, North Carolina; and Atlanta, Georgia. U.S. cities with the cleanest air in the United States include Bellingham, Washington; Cedar Rapids, Iowa; Colorado Springs, Colorado; and Des Moines, Iowa.[7] However, with the exception of the Pacific Northwest, no U.S. region is free from ozone pollution and its adverse health effects.

The future will likely bring some success stories and some near or actual tragedies. What is apparent is that urban air pollution is important to people, and many urban areas are drawing up ambitious plans to control it. Whether these plans are put into action will depend on global, regional, and local economies (reducing air pollution is expensive); population growth (more people means more air pollution); international cooperation (air pollutants travel across international borders); and whether we make air pollution a priority.

Return to the Big Question

Why is urban air pollution still such a big environmental problem?

The quality of urban air in the United States has improved dramatically in recent decades due to strict regulations and enforcement of pollution-abatement laws and regulations. This does not mean that our urban air is always healthy to breathe, but that it is not as bad as it used to be. Many U.S. cities still have a number of days each year when the air quality is unhealthy. It is estimated that about half the people in the United States live in areas where exposure to air pollution contributes to lung disease that causes several hundred thousand premature deaths every year. Health costs are tens of billions of dollars. In other parts of the world, air pollution is worse and health effects are even more severe.

It is apparent that air pollution in our cities is still a big problem and that further improving the situation will be challenging. This is because we have already made the easy big fixes by controlling emissions from our power plants and automobiles. Additional pollution abatement will be more expensive and will not lead to improvements as big as those from earlier strategies. Finally, many more people are moving to large urban areas, increasing the population in those areas and the amount of air pollutants admitted into the atmosphere from automobiles even though today's automobiles release far fewer pollutants that those only a few decades ago.

Summary

- Every year, about 150 million metric tons of primary pollutants enter the atmosphere above the United States from processes related to human activity. Considering the size of the atmosphere, these pollutants would not be a major problem if they were evenly distributed. Unfortunately, they generally are not. They are concentrated in urban areas and other areas where the air naturally lingers.

- The two main types of pollution sources are stationary and mobile. Stationary sources have a relatively fixed position and include point sources, area sources, and fugitive sources.

- The two main groups of air pollutants are primary and secondary. Primary pollutants are emitted directly into the air: They include particulates, sulfur dioxide, carbon monoxide, nitrogen oxides, and hydrocarbons. Secondary pollutants are produced by reactions between primary pollutants and other atmospheric compounds. A good example is ozone, which forms over urban areas through photochemical reactions between primary pollutants and natural atmospheric gases.

- Air toxins are those pollutants that are known to cause cancer or other serious health problems.

- The major air pollutants affect the visual quality of the environment, vegetation, animals, soil, water quality, natural and artificial structures, and human health.

- Burning fossil fuels releases sulfur and nitrogen oxides into the atmosphere and creates acid rain. Environmental effects of acid rain include loss of fish and other life in lakes, damage to trees and other plants, leaching of nutrients from soils, and damage to statues and buildings in urban areas.

- There are two major types of smog: photochemical and sulfurous. Each type brings particular environmental problems that vary with geographic region, time of year, and local urban conditions.

- International agreements to reduce stratospheric ozone depletion are an environmental success story.

- Indoor air pollution has been with us for thousands of years, since people first built structures and burned fuel indoors. It is one of our most serious environmental health problems.

- Concentrations of indoor air pollutants are generally greater than concentrations of the same pollutants outdoors.

- Sources of indoor air pollution may include the materials we use to construct our buildings, the furnishings we put in them, and the types of equipment we use for heating and cooling, as well as natural processes that allow gases to seep into buildings.

- There are two basic types of sick buildings: those with an identifiable problem, such as mold or bacteria, and those with "sick building syndrome," in which people's symptoms can't be traced to any one cause.

- In urban areas, meteorological conditions greatly affect whether polluted air is a problem. In particular, temperature inversion layers that restrict circulation in the lower atmosphere may lead to pollution events.

- Efforts to reduce air pollution in urban regions center on automobiles, buses, and other vehicles, because they release most of the pollutants that enter the urban atmosphere.

- Emissions of air pollutants in the United States are decreasing, but they remain a significant health hazard in many cities. In developing countries, air pollution in large urban centers is often a serious problem.

- Air quality in U.S. urban areas is usually reported in terms of whether the quality is good, moderate, unhealthy for sensitive groups, unhealthy, very unhealthy, or hazardous. These levels are defined by the Air Quality Index (AQI).

- Establishing a minimum total cost is a compromise between the cost of controlling pollutants and the value placed on losses or damages from pollution. If more controls are needed to lower pollution to a more acceptable level, additional costs are incurred. Beyond a certain point, these costs can begin to increase more rapidly than the benefits.

Key Terms

acid rain
air quality index
air-quality standards
air toxins
atmospheric inversion
indoor air pollution

ozone shield
primary pollutants
scrubbing
secondary pollutants
sick building syndrome (SBS)
smog

Getting It Straight

1. What is the difference between point and non-point sources of air pollution? Which type is easier to manage?

2. What are the differences between primary and secondary pollutants?

3. Why is it so difficult to establish national air-quality standards?

4. What are some of the common sources of air pollutants where you live, work, or attend classes?

5. What air pollutants does nature produce?

6. When did the air pollution crisis begin in the United States?

7. What effects does air pollution have on the following? Explain.

   a. People
   b. Buildings
   c. Plants

8. List examples of several primary air pollutants.

9. List examples of several secondary air pollutants.

10. Define the following and give an example of a pollutant from each source.
    a. Point source
    b. Fugitive source
    c. Area source

11. Can acid rain be controlled? How might you control the pollution problem?

12. Has the success in decline of use of chlorofluorocarbons (CFCs) been noticed in our air quality?

What Do You Think?

1. Why do we have air-pollution problems when the amount of pollutants released into the air is a very small fraction of the total material in the atmosphere?

2. Why is acid rain a major environmental problem, and how can it be minimized?

3. Why will air-pollution control strategies in developed countries probably be much different in terms of methods, process, and results from air-pollution control strategies in developing countries?

4. In a highly technological society, is it possible to have 100% clean air? Is it likely?

5. Do you agree or disagree that most of the people in the world will adjust to ozone depletion by doing little (doing nothing is an adjustment) to decrease their personal exposure to ultraviolet radiation?

6. What do you think about sick building syndrome? If you were a manager at a large corporation and a number of your employees complained of illness with similar mysterious symptoms, what would you do? Play the role of the administrator and develop a plan to look at the potential problem.

Pulling It All Together

1. How good is the Air Quality Index? How might we evaluate its usefulness? Do you think the index will change in the future? If so, what are the likely changes?

2. Develop a research plan to assess the indoor air quality in your local library. How might that plan differ from a similar assessment of the science buildings on a college campus?

3. Develop a plan to study the potential radon hazard in your community. Where would you start? How would you gather data, and so on? If your community has undergone extensive testing already, review the results and decide whether further testing is necessary.

Further Reading

Boubel, R. W., D. L. Fox, D. B. Turner, and A. C. Stern. 1994. *Fundamentals of air pollution*, 3rd ed. New York: Academic.—A thorough book covering the sources, mechanisms, effects, and control of air pollution.

Brenner, D. J. 1989. *Radon: risk and remedy*. New York: W. H. Freeman.—A wonderful book concerning the hazard of radon gas. It covers everything from the history of the problem to what was happening in 1989, as well as solutions, and is highly recommended.

Brooks, B. O., and W. F. Davis. 1992. *Understanding indoor air quality*. Ann Arbor, MI: CRC Press.—A comprehensive evaluation of indoor air pollution. It discusses most of the sources of indoor air pollutants, as well as health effects and controls.

Rowland, F. S. 1990. Stratospheric ozone depletion by chlorofluorocarbons. *AMBIO* 19(6–7):281–292.—An excellent summary of stratospheric ozone depletion that discusses some of the major issues.

Toon, O. B., and R. P. Turco. 1991. Polar stratospheric clouds and ozone depletion. *Scientific American* 246(6):68–74.—An article that provides valuable information concerning polar stratospheric clouds and their importance in ozone depletion. It offers a good explanation of the formation of polar stratospheric clouds and the chemistry that occurs there.

Wang, L. 2002 (February). Paving out pollution. *Scientific American*, p. 20.—Discussion of an innovative approach to reducing air pollution.

Calvin Larson/Photo Researchers, Inc.

15

Minerals and Environment

Big Question

Is It Possible to Use Nonrenewable Mineral Resources Sustainably?

Learning Objectives

Modern society depends on the availability of mineral resources, which can be considered a nonrenewable heritage from the geologic past. After reading this chapter, you should understand . . .

- that the standard of living in modern society is related in part to the availability of minerals;
- the processes responsible for the distribution of mineral deposits;
- the differences between mineral resources and reserves;
- what factors control the environmental impact of mineral exploitation;

- how wastes generated from the use of mineral resources affect the environment;
- what social impacts result from mineral exploitation.
- how sustainability may be linked to use of non-renewable minerals.

Case Study

Fossil Trace Golf Club, a Story of Successful Mine Reclamation

The city of Golden, Colorado, has an award-winning golf course on land that was for about 100 years an open-pit mine (quarry) excavated in limestone rock. From clay layers between limestone beds, the mine produced clay for making bricks. Over the life of the mine, the clay was used as a building material at many sites, including prominent buildings in the Denver area, such as the Colorado Governor's Mansion. The site included not only unsightly mining pits with limestone walls but also a landfill for waste disposal. However, it had spectacular views of the foothills and the Rocky Mountains.

Today the limestone cliffs with their exposed plant and dinosaur fossils have been transformed into golf greens, fairways, and a driving range (Figure 15.1). The name of the course reflects its geologic her-itage, and the project includes trails to fossil locations. It also includes channels, constructed wetlands, and three lakes that store runoff of floodwater, helping to protect Golden from flash floods. The reclamation project started with a grassroots movement by the people of Golden to have a public golf course. The reclamation is now a money-maker for the city, and demonstrates that mining sites not only can be reclaimed but also can be transformed into valuable property.

Fossil Trace Golf Club is a unique instance of mine reclamation; but each renovation is somewhat unique, based on local conditions—physical, hydrological, and biological. This chapter discusses the origin of mineral deposits, as well as environmental consequences of mineral development.

Courtesy of Fossil Trace Golf Course

■ **FIGURE 15.1**
Mine to golf course.
At Fossil Trace Golf Club, in Golden, Colorado, the rock faces are part of a clay mine.

15.1 The Importance of Minerals to Society

Modern society absolutely depends on the availability of mineral resources—without them, our society as we know it would collapse.[1] Many mineral products are found in a typical American home (Table 15.1). The iPod, Blackberry, cell phone, and personal computer are all made of minerals, including metals, nonmetals, and petroleum. And consider your breakfast this morning: You probably drank from a glass made primarily of sand, ate from dishes made of clay, flavored your food with salt mined from Earth, ate fruit grown with the aid of fertilizers such as potassium carbonate (potash) and phosphorus, used utensils made from stainless steel, which comes from processing iron ore and other minerals, and read the news on your computer made of plastic (from petroleum) and metals such as iron, copper, silver, and gold.

A society's standard of living rises with the availability of minerals. The availability of mineral resources is one measure of the wealth of a society. Those who have been successful in locating and extracting or importing and using minerals have grown and prospered. Without mineral resources to grow food, construct buildings and roads, and manufacture everything from computers to televisions and automobiles, modern technological civilization as we know it would not be possible. In maintaining our standard of living in the United States, every person requires about ten tons of nonfuel minerals per year.[2]

Minerals are our nonrenewable heritage from the geologic past. Nonrenewable mineral resources are produced over geologic time, sometimes called "deep time." Although Earth processes are still forming new mineral deposits, the processes are too slow to be of use to us today. Moreover, because mineral deposits are generally in small, hidden areas, they must be discovered. Unfortunately, most of the easy-to-find deposits have already been found and exploited. As a result, if modern civilization were to vanish, our descendants would have a harder time than we did discovering rich mineral deposits. Perhaps they might mine landfills for metals thrown away by our civilization.

Unlike biological resources that are renewable, often on an annual basis, minerals cannot be easily managed to produce a sustained yield. The supply is limited. Recycling and conservation will help, but eventually the supply will be exhausted.

15.2 How Mineral Deposits Are Formed

High concentrations of Earth materials form ore deposits. When metals are concentrated in unusually high amounts by geologic processes, **ore deposits** are formed. Metals in mineral form are generally extracted from such deposits. The discovery of natural ore deposits allowed early peoples to exploit copper, tin, gold, silver, and other metals while slowly developing skills in working with metals.

The origin and distribution of mineral resources are intimately related to the history of the biosphere and to the entire geologic cycle (see Chapter 3). Nearly all aspects and processes of the geologic cycle are involved to some extent in producing local concentrations of useful materials. In this section, we look first at the distribution of mineral resources on Earth and then describe processes that form mineral deposits.

Distribution of Mineral Resources

Earth's outer layer, or crust, is silica-rich, made up mostly of rock-forming minerals containing silica, oxygen, and a few other elements. The elements are not evenly distributed in the crust: Nine elements account

---

☐ TABLE 15.1   MINERAL PRODUCTS IN A TYPICAL U.S. HOME

Electronic equipment (TV, DVD, iPod, telephone, Blackberry, computer): metals, nonmetals, plastic from petroleum
Building materials: sand, gravel, stone, brick (clay), cement, steel, aluminum, asphalt, glass
Plumbing and wiring materials: iron and steel, copper, brass, lead, cement, asbestos, glass, tile, plastic
Insulating materials: fiberglass, gypsum (plaster and wallboard)
Paint and wallpaper: mineral pigments (such as iron, zinc, and titanium) and fillers (such as talc and asbestos)
Floor tiles and carpets: petroleum products
Appliances: iron, copper, aluminum as well as plastics (petroleum products) and many rare metals
Furniture: synthetic fibers made from minerals (principally coal and petroleum products); steel springs
Clothing: natural fibers grown with mineral fertilizers; synthetic fibers made from minerals
    (principally coal and petroleum products)
Food: grown with mineral fertilizers; processed and packaged by machines made of metals
Drugs and cosmetics: mineral chemicals
Other items: windows, screens, lightbulbs, porcelain fixtures, china, utensils, jewelry made from mineral products

Source: Modified after U.S. Geological Survey, Professional Paper 940, 1975.

for about 99% of the crust by weight (oxygen, 45.2%; silicon, 27.2%; aluminum, 8.0%; iron, 5.8%; calcium, 5.1%; magnesium, 2.8%; sodium, 2.3%; potassium, 1.7%; and titanium, 0.9%). In general, remaining elements are found in trace concentrations.

The oceans also have lots of minerals, but not in high concentrations. The oceans, covering over two-thirds of Earth, are a reservoir for many chemicals other than water. Most of these elements have been weathered from crustal rocks on the land and carried to the oceans by rivers. Some are transported to the oceans by wind or glaciers. Concentrations of most minerals in ocean water, with the exception of salt and magnesium, are very low compared with those in the near-surface rocks, and therefore the minerals cannot economically be extracted.

Why do the minerals we mine occur in unusually high concentrations? Earth scientists now believe that Earth, like the other planets in the solar system, formed when gravitational attraction brought together matter dispersed around the forming sun. As the mass that would become Earth increased, the material condensed and was heated by the process. The heat was sufficient to produce a molten liquid core, consisting primarily of irons, nickel, and other heavy metals, which sank toward the center of the mass. The crust formed from generally lighter elements and is a mixture of many different elements. The elements in the crust are not uniformly distributed because geologic processes and some biological processes selectively dissolve, transport, and deposit elements and minerals. We discuss these processes next.

Plate Boundaries

Plate tectonics is responsible for the formation of some mineral deposits. According to the theory of plate tectonics (see Chapter 3), the continents are composed mostly of relatively light crustal rocks. As the tectonic plates slowly move across Earth's surface, so do the continents. Metallic ores are deposited in the crust both where the tectonic plates separate (diverge) and where they come together (converge).

At divergent plate boundaries, cold ocean water comes in contact with hot molten rock. The heated water is lighter and more active chemically. It rises through fractured rocks and leaches metals from them. The metals are carried in solution and deposited as metal sulfides when the water cools.

At convergent plate boundaries, rocks saturated with seawater are forced together, heated, and subjected to intense pressure, which causes partial melting. The combination of heat, pressure, and partial melting mobilizes metals in the molten rocks. Most major mercury deposits, for example, are found near the volcanic regions near convergent plate boundaries. Geologists believe that the mercury is distilled out of the tectonic plate as the plate moves downward. As the plate cools, the mercury migrates upward and is deposited at shallower depths, where the temperature is lower.

Igneous Processes

Igneous processes involve heat deep inside the Earth. The heat and pressure turn rock material into *magma* (molten rock), and as the magma cools, ore deposits may form. Heavier minerals that crystallize (solidify) early may slowly sink or settle toward the bottom of the cooling magma, whereas lighter minerals that crystallize later are left at the top. Deposits of chromite (an ore of chromium) are thought to be formed in this way. When magma containing small amounts of carbon is deeply buried and subjected to very high pressure during slow cooling (crystallization), diamonds—which are pure carbon—may be produced[3] (Figure 15.2).

Hot water moving within the crust is the source of many ore deposits. When circulating groundwater is heated and enriched with minerals from deeply buried rocks, then meets cooler rocks, the dissolved minerals are deposited by the cooled water.[4]

Sedimentary Processes

Sediments carried by water, wind, and glaciers can also form mineral deposits. Sediments are derived from weathering (chemical and mechanical decomposition at

■ **FIGURE 15.2**
Diamond mine near Kimberley, South Africa. This is the largest hand-dug excavation in the world.

or near Earth's surface) and erosion. In transporting the sediments, running water and wind help separate them by size, shape, and density. This sorting is useful to people. For instance, the best sand, or sand and gravel, deposits for construction purposes are those in which the finer particles have been removed by water or wind. Sand dunes, beach deposits, and deposits in stream channels are good examples. The sand and gravel industry amounts to several billion dollars annually and, in terms of the total volume of materials mined, is one of the largest nonfuel mineral industries in the United States.[2]

Streams transport and sort all types of materials according to size and density. Therefore, if the bedrock in a river basin contains heavy metals, such as gold, streams draining the basin may concentrate the metals in areas where the water is less turbulent or slower-moving. These concentrations, called *placer deposits,* are often found in open crevices or fractures at the bottoms of pools, on the inside curves of bends, or where shallow water flows over rocks. Placer mining of gold (known as a poor man's method because it required only a shovel, a pan, and a strong back) played an important role in settling California, Alaska, and other areas of the United States.

Dried-up lakes and basins are often rich in minerals. Rivers and streams that empty into oceans and lakes carry tremendous quantities of dissolved material from the weathering of rocks. Over geologic time, a shallow marine basin may become isolated by tectonic activity that uplifts its boundaries. In other cases, climate variations, such as the ice ages, produce large inland lakes with no outlets. These basins and lakes eventually dry up. As the water evaporates, the dissolved materials precipitate out of solution and form a wide variety of compounds, minerals, and rocks that have important commercial value. These *evaporites* (deposits left by evaporation) are generally of three types:[5]

- *Marine evaporites* (solids): potassium and sodium salts, gypsum, and anhydrite.

- *Nonmarine evaporites* (solids): sodium and calcium carbonate, sulfate, borate, nitrate, and limited iodine and strontium compounds.

- *Brines* (liquids from wells, thermal springs, inland salt lakes, and seawater): bromine, iodine, calcium chloride, and magnesium.

Evaporites are valuable mineral resources. Heavy metals (such as copper, lead, and zinc) associated with brines and sediments in the Red Sea, Salton Sea, and other areas are important resources that may be exploited in the future. Evaporite minerals are widely used in industrial and agricultural activities, and their annual value is several billion dollars.[5] Evaporite and brine resources in the United States are substantial, ensuring an ample supply for many years.

A variety of sedimentary rocks are mined. Examples include limestone (for construction and building), clay (for bricks, paper filler, medicine, and pottery), and phosphorus for fertilizer.

Biological Processes

Some mineral deposits are formed by biological processes, and many are formed under conditions of the biosphere that have been greatly altered by life. Examples include phosphates (discussed in Chapter 3) and iron ore deposits.

The major iron ore deposits are in sedimentary rocks that were formed more than 2 billion years ago.[6] There are several types of iron deposits. Gray beds, an important type, contain unoxidized iron. Red beds contain oxidized iron (red is the color of iron oxide). Gray beds formed when there was little oxygen in the atmosphere, and red beds formed when there was relatively more oxygen. Although we don't fully understand the processes, it appears that major deposits of iron stopped forming when the atmospheric concentration of oxygen reached its present level.[7]

Organisms can form many kinds of minerals, such as the calcium in shells and bones. Thirty-one different biologically produced minerals have been identified. Some of them cannot be formed inorganically in the biosphere. Minerals of biological origin contribute significantly to sedimentary deposits.[8]

Weathering Processes

Weathering can produce mineral deposits. When insoluble ore deposits—such as native gold and other materials that do not dissolve in water—are weathered from rocks, they may accumulate in the soil unless removed by erosion. This is most likely when the parent rock is relatively soluble, as is limestone. Intensive weathering of certain soils formed from aluminum-rich igneous rocks may concentrate oxides of aluminum and iron. (The more-soluble elements, such as silica, calcium, and sodium, are removed by soil and biological processes.) If sufficiently concentrated, aluminum oxide forms an ore of aluminum known as *bauxite.* Important nickel and cobalt deposits are also found in soils from igneous rocks rich in iron and magnesium.

Weathering can improve low-grade ore by *secondary enrichment.* Near the surface, primary ore containing minerals such as iron, copper, and silver sulfides is in contact with slightly acidic soil water in an oxygen-rich environment. As the sulfides oxidize, they dissolve, forming solutions rich in sulfuric acid as well as silver and copper sulfate. These solutions migrate downward, leaving a leached zone that has few ore minerals (Figure 15.3). Below the leached zone and above the ground-

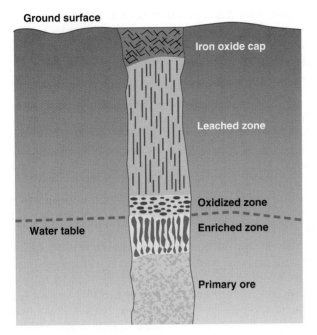

Ground surface

Iron oxide cap

Leached zone

Oxidized zone

Water table

Enriched zone

Primary ore

■ **FIGURE 15.3**

Mineral deposits from secondary enrichment:
Typical zones that form during secondary enrichment
processes. Sulfide ore minerals in the primary ore vein are
oxidized and altered and then are leached from the
oxidized zone by descending groundwater and redeposited
in the enriched zone. The iron oxide cap is generally a
reddish color and may be helpful in locating ore deposits
that have been enriched. [*Source*: R. J. Foster. *General
Geology*, 4th ed. Columbus, OH: Charles E. Merrill, 1983.]

water table, oxidation continues, and sulfate solutions
continue their downward migration. Below the water
table, if oxygen is no longer available, the solutions are

deposited as sulfides, enriching the metal content of the
primary ore by as much as ten times.

Secondary enrichment makes low-grade primary
ore more valuable, and makes high-grade primary ore
even more attractive.[9, 10] For example, secondary enrichment of a copper deposit at Miami, Arizona, increased the grade of the ore from less than 1% copper
in the primary ore to as much as 5% in some areas.[9]

15.3 Resources and Reserves

We can classify minerals as *resources* or *reserves.*
A **resource** is a mineral concentrated in a form that we
can extract to obtain something that can be bought and
sold. It is assumed that we can extract it economically, or
at least that it has the potential for economic extraction.
A **reserve** is the portion of a resource whose location and
amount are known and which can be legally and economically extracted *at the time of evaluation* (Figure 15.4).

Whether a mineral deposit is classified as part of
the resource base or as a reserve is a question of economics. For example, if an important metal becomes
scarce, the price may rise, which would encourage exploration and extraction (mining). As a result of the
price increase, previously uneconomic deposits (part
of the resource base before the scarcity and price rise)
may become profitable, and those deposits would be
reclassified as reserves.

The main point here is that *resources are not reserves.* An analogy from a student's personal finances
may help clarify this point. A student's *reserves* are liquid
assets—funds available for purchasing things right now,
such as money in the bank. The student's *resources,* on

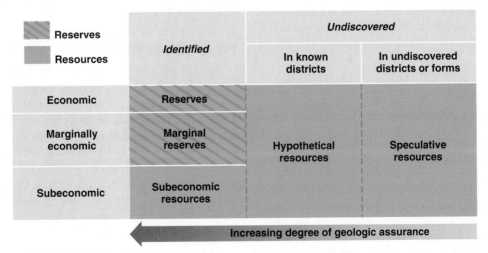

■ **FIGURE 15.4**

Classification of mineral resources used by the U.S. Geological Survey and the U.S. Bureau of Mines. [*Source*: Principles of a resource preserve classification for minerals. U.S.
Geological Survey Circular 831, 1980.]

the other hand, include nonliquid assets (such as an education or perhaps a trust fund available later) as well as income the student can expect to accumulate over his or her lifetime. This distinction is often critical to a student, because resources available in the future cannot be used to pay this month's bills.[11]

Estimating future resources is important for planning purposes. This task requires continual reassessment of all components of a total resource, taking into account new technology, the probability of discovering new deposits, and shifts in economic and political conditions.[1]

Silver illustrates some important points about resources and reserves. Based on geochemical estimates of the concentration of silver in rocks, the amount of silver in Earth's crust (to a depth of 1 kilometer, or 0.6 mile) is much larger than the amount the world uses annually. If this silver existed as pure metal concentrated in one large mine, it would be a supply sufficient for millions of years at current levels of use. However, most of the silver exists in concentrations too low to be extracted economically with current technology. The known reserves of silver, the amount we could obtain immediately with current techniques, is about a 20 year supply at current use levels.

The problem with silver, as with all mineral resources, lies not in its total abundance but in its concentration and ease of extraction. When an atom of silver is used, it is not destroyed but is simply dispersed and may become unavailable. In theory, given enough energy, all mineral resources could be recycled, but this is not possible in practice. Consider lead, which comes from mining minerals in which it is concentrated. The lead that was used in gasoline for many years is now scattered along highways across the world and deposited in low concentrations in forests, fields, and salt marshes close to these highways. Recovery of this lead is virtually impossible.

15.4 Use and Availability of Mineral Resources

We also classify minerals by use and by abundance. Earth's mineral resources can be divided into several broad categories, depending on our use of them: elements for metal production and technology; building materials; minerals for the chemical industry; and minerals for agriculture. Metallic minerals can be further classified according to their abundance. The abundant metals include iron, aluminum, chromium, manganese, titanium, and magnesium. Scarce metals include copper, lead, zinc, tin, gold, silver, platinum, uranium, mercury, and molybdenum.

Some mineral resources, such as salt (sodium chloride), are necessary for life. Primitive peoples traveled long distances to obtain salt when it was not locally available. Other mineral resources are desired or considered necessary to maintain a particular level of technology.

The most-used minerals are not metals. When we think about mineral resources, we usually think of the metals; but except for iron, the predominant mineral resources are not metallic. Consider the annual worldwide consumption of a few selected elements. Each year the world uses several hundred to 1,000 million metric tons of sodium and iron, and up to 100 million metric tons of nitrogen, phosphorus, sulfur, potassium, and calcium (primarily as soil conditioners or fertilizers).

In contrast, metallic elements such as zinc, copper, aluminum, and lead have annual world consumption rates of several to 10 million metric tons, and the consumption rates of gold and silver are a very small fraction of the rates for other metals. Nickel, chromium, cobalt, and manganese are used mainly in alloys of iron (as in stainless steel). If you add up the total amount (millions of tons per year) of all metals we use, you would find that iron would be about 95% of the total.

Availability of Mineral Resources

The cost of mining determines availability. The basic issue with maintaining an adequate stock of mineral resources is not whether we will use them all up but that at some point the costs of mining exceed the worth of material. When the availability of a particular mineral becomes a limitation, there are four possible solutions:

▪ Find more sources.
▪ Recycle and reuse what has already been obtained.
▪ Reduce consumption.
▪ Find a substitute.

Which choice or combination of choices is made depends on social, economic, and environmental factors.

Earth's history determines the availability of a mineral resource in a certain form, in a certain concentration, and in a certain amount. What is considered a resource and at what point a resource becomes limited are ultimately social questions. Before metals were discovered, they could not be considered resources. Before smelting was invented, the only metal ores were those containing metals in their pure form. Gold, for example, was obtained as a pure, or native, metal. We usually get gold by digging deep beneath the surface and reducing tons of rock to ounces of gold.

Mineral resources are limited, which raises important questions. How long will a particular resource last? How much short-term or long-term environmental deterioration are we willing to accept to ensure that resources are developed in a particular area? How can we make the best use of available resources? These questions have no easy answers. We are now struggling with ways to better estimate the quality and quantity of resources.

Mineral Consumption

We can use a mineral resource in three ways: (1) rapid consumption; (2) consumption with conservation; or (3) consumption and conservation with recycling. Which option we select depends in part on economic, political, and social criteria. Figure 15.5 shows the hypothetical depletion curves corresponding to these three options. Historically, with the exception of precious metals, rapid consumption has been the dominant way we use most resources. However, as the supply of resources becomes short, increased conservation and recycling are likely. The trend toward recycling is already well established for metals such as copper, lead, and aluminum.

Rising population and standard of living increase demand. From a global viewpoint, limits on our mineral resources and reserves threaten our affluence. As the world population and the desire for a higher standard of living increase, the demand for mineral resources expands at an increasing pace. Today the more developed and rapidly industrializing countries consume a disproportionate amount of the minerals extracted. For example, the United States, Western Europe, Japan, China, and India together use most of the aluminum, copper, and nickel that is extracted from the Earth.[4]

Production is unlikely to keep up with increasing consumption. Predicted increases in worldwide use of iron, copper, and lead, along with expected population increases, suggest that the rate of production of these

metals will have to increase by several times if the world's per-capita consumption rate is to rise to the level of consumption in developed countries today. Such an increase is unlikely, which means that affluent countries will have to find substitutes for some minerals or use a smaller proportion of the world's annual production. This situation highlights the Malthusian predictions discussed in Chapter 2: It is impossible in the long run to support unlimited population growth on a limited resource base.

U.S. Supply of Minerals

U.S. use exceeds its own supplies of many minerals. As a result, minerals must be imported from other nations. For example, the United States imports many of the minerals needed for its complex military-industrial system. These so-called strategic minerals include bauxite, manganese, graphite, cobalt, strontium, and asbestos. Of particular concern is the possibility that the supply of a much-desired or much-needed mineral may be interrupted by political, economic, or military instability in the supplier nation.

Importing may sometimes simply be more practical than mining. The fact that the United States—along with many other countries—depends on a steady supply of imports to meet the mineral demand of its industries does not necessarily mean that the U.S. doesn't have these minerals in quantities that could be mined. Rather, it suggests that there are economic, political, or environmental reasons that make importing easier, more practical, or more desirable. This situation has resulted in political alliances that otherwise would be unlikely. Industrial countries often need minerals from countries whose policies they do not necessarily agree with. As a result, they may make political concessions that they would not otherwise make on human rights and other issues.

15.5 Impacts of Mineral Development

Mineral exploitation can have both environmental and social effects. The effects on the environment depend on such factors as ore quality, mining procedures, local hydrologic conditions, climate, rock types, the size of the operation, topography, and many other interrelated factors. The impact also varies with the stage of development of the resource. For example, the exploration and testing stages have considerably less environmental impact than the mining and processing stages.

Environmental Impacts

Exploration has minimal impact. Exploration for mineral deposits varies from collection and analysis of remote-sensing data gathered from airplanes or satellites,

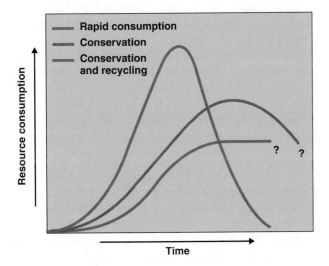

■ **FIGURE 15.5**
How we consume minerals: several hypothetical depletion curves.

to fieldwork involving surface mapping and drilling. Generally, exploration has a minimal impact on the environment if care is taken in sensitive areas, such as arid lands, marshes, and areas underlain by permafrost. Some arid lands are covered by a thin layer of pebbles over fine silt several centimeters thick. The layer of pebbles, called "desert pavement," protects the finer material from wind erosion. When the desert pavement is disturbed by road building or other activities, the fine silt may be eroded, damaging the soil and possibly scarring the land for many years. Similarly, marshes and other wetlands, such as the northern tundra, are sensitive to even seemingly small disturbances, such as vehicular traffic.

Mining and processing minerals generally have a considerable impact on land, water, air, and biological resources. Furthermore, as it becomes necessary to use ores of lower and lower grades, larger mines are necessary and negative effects on the environment tend to become greater problems. For example, there is concern about asbestos fibers in the drinking water (from Lake Superior) of Duluth, Minnesota, as a result of the disposal of waste (tailings) from mining low-grade iron ore.

A major practical issue is whether to use surface or subsurface mines. Surface mining is cheaper but has more direct environmental effects. The trend in recent years has been away from subsurface mining and toward large, open-pit mines, such as the Bingham Canyon copper mine in Utah (Figure 15.6). The Bingham Canyon mine is one of the world's largest human-made excavations, covering nearly several square kilometers to a maximum depth approaching one kilometer (3,300 ft.).

Local disturbances often become broader problems. Surface mines and quarries today cover less than 1% of the total area of the United States. But even though the impact of these operations is a local phenomenon, numerous local occurrences will eventually constitute a larger problem. Environmental degradation tends to extend beyond the land actually being mined. Large mining operations disturb the land by directly removing material in some areas and dumping waste in others, thus changing topography. In addition, dust at mines may affect air quality, even though care is often taken to reduce dust production by sprinkling water on roads and on other sites that generate dust. At the very least, the sites become eyesores.

A potential problem is the possible release of harmful trace elements. Water resources are particularly vulnerable even if drainage is controlled and sediment pollution is reduced. Mining often alters surface drainage, and runoff from precipitation may infiltrate waste material, leaching out trace elements and minerals. Trace elements—such as cadmium, cobalt, copper, lead, molybdenum—when leached from mining wastes and concentrated in water, soil, or plants, may cause illness in people and other animals that drink the water, eat the plants, or use the soil. Specially constructed ponds to collect such runoff help, but cannot eliminate all problems. The white streaks in Figure 15.7 are mineral deposits apparently leached from tailings from a zinc mine in Colorado. Similar-looking deposits may cover rocks in rivers for many kilometers downstream from some mining areas.

Groundwater may also be polluted by mining operations when waste comes into contact with slow-moving subsurface waters. Surface water infiltration or groundwater movement causes leaching of sulfide minerals that may pollute groundwater and eventually seep into streams to pollute surface water. Groundwater problems are particularly troublesome because reclaiming polluted groundwater is difficult and expensive. (See Chapter 10 for a discussion of acid mine drainage.)

Mineral mining directly and indirectly affects the biological environment by causing physical changes in the land, soil, water, and air. Examples of direct impacts include plants and animals that may be killed by mining activity or by contact with toxic soil or water. Indirect impacts include changes in nutrient cycling, total biomass, species diversity, and ecosystem stability owing to alterations in groundwater or surface water availability or quality. Periodic or accidental discharge of low-grade pollutants through failure of barriers, ponds, or water

Royce Bair/ProfFiles West, Inc.

▪ **FIGURE 15.6**
Bingham Canyon Copper Pit, Utah.
The mine is one of the largest artificial excavations in the world.

Ed Keller

Tailings from a lead, zinc, and silver mine in Colorado. White streaks on the slope are mineral deposits apparently leached from the tailings. Such deposits are often the result of past mining practices that are no longer allowed.

diversions or through breach of barriers during floods, earthquakes, or volcanic eruptions also may damage local ecological systems.

Social Impacts

Large-scale mining brings a rapid influx of workers. In areas unprepared for growth, this puts stress on local services, such as water supplies, sewage and solid-waste disposal systems, schools, and housing. Land use shifts from open range, forest, and agriculture to urban patterns. The additional people also increase the stress on nearby recreation and wilderness areas, some of which may be in a fragile ecological balance. Construction activity and urbanization affect local streams through sediment pollution, lower water quality, and increased runoff. Air quality suffers as a result of more vehicles, dust from construction, and generation of power.

Adverse social impacts also occur when mines close down. Towns surrounding large mines come to depend on the income of the mine workers. Mine closures produced the well-known ghost towns of the old American West. Today, the price of coal and other minerals directly affects the livelihood of many small towns. This relationship is especially evident in the Appalachian Mountain region of the United States, where closures of coal mines have taken their toll. Mine closings are the result of depletion of the coal, price of the coal, and mining cost.

One reason costs are rising is greater environmental regulation of the industry. Of course, regulations have also helped make mining safer and have helped restore damaged land. Some miners, however, believe the regulations are not flexible enough, and there is some truth to their arguments. For example, some mined areas could be reclaimed for use as farmland now that the original hills have been leveled. Regulations, however, may require that the land be restored to its original hilly state, even though hills make inferior farmland.

Wars and minerals sometimes go together. Wars and revolutions are often funded by money from resources such as gold, diamonds, copper, and petroleum. People living in resource-rich areas of some developing countries may be attacked and displaced in the rush to exploit mineral resources to support a political agenda. Recently attention has been on the West African country of Sierra Leone and the so called "blood diamonds". During a recent civil war rebels financed their fighting through the sale of illegal diamonds. The war displaced and killed thousands of people as rebel forces fought to control the diamond mining.

15.6 Minimizing Environmental Impacts of Mineral Development

Generation of waste is a major issue. The cycle of mineral resources in Figure 15.8 shows that many components of the cycle involve the generation of waste material. In fact, the major environmental impacts of developing and using mineral resources are related to waste products. Waste produces pollution that may be toxic to living things, may harm natural ecosystems and the biosphere, and may be aesthetically undesirable. Waste not only may degrade air, water, and soil, but also depletes nonrenewable mineral resources and, when improperly disposed of, provides no offsetting benefits for human society.

Minimizing environmental effects of mineral development can take several paths:

Environmental regulations at the federal, state, and local levels. The regulations address such issues as sediment, air, and water pollution resulting from all aspects

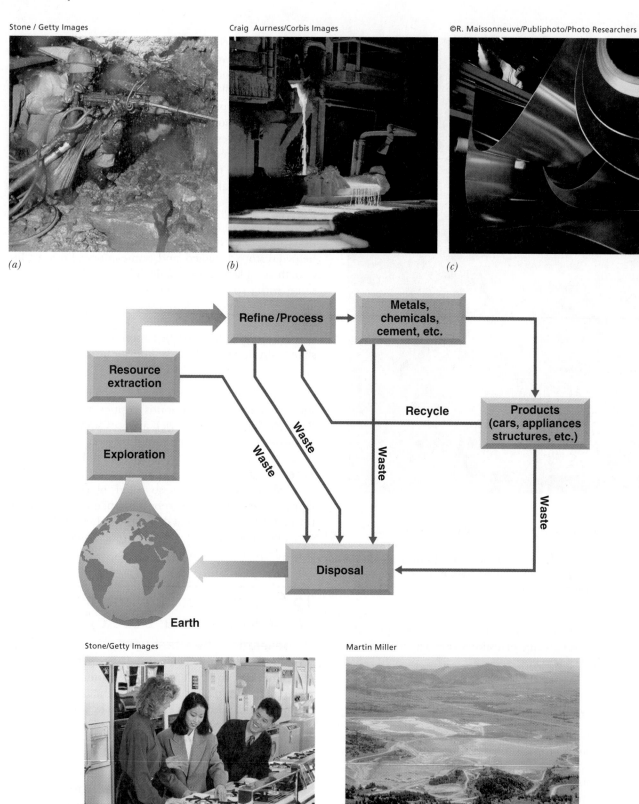

Stone / Getty Images

Craig Aurness/Corbis Images

©R. Maissonneuve/Publiphoto/Photo Researchers

(a)

(b)

(c)

Stone/Getty Images

Martin Miller

(d)

(e)

■ **FIGURE 15.8**

Simplified flowchart of the resource cycle: (a) mining gold in South Africa; (b) a copper smelter in Montana; (c) sheets of copper for industrial use; (d) appliances made in part from metals; and (e) disposal of mining waste into a tailings pond from a Montana gold mine.

of the mineral cycle. Regulations may also address reclamation of land used for mining. Today in the United States, approximately 50% of the land used by the mining industry has been reclaimed.

On-site and off-site treatment of waste. Minimizing on-site and off-site problems by controlling sediment, water, and air pollution through good engineering and conservation practices is an important goal. Of particular interest is the development of biological processes such as biooxidation and the genetic engineering of microbes. These have enormous potential for both extracting metals and minimizing environmental degradation. For example, engineered (constructed) wetlands are being used at several sites; acid-tolerant plants in the wetlands remove metals from mine wastewater and neutralize acids by biological activity.[12] At the Homestake Gold Mine in South Dakota which closed in 2001 after 125 years of mining biooxidation was used to convert contaminated water from the mining operation into substances that are environmentally safe. The process demonstrated that bacteria with a natural ability to oxidize cyanide to harmless nitrates can be used to treat contaminated water.[13]

Practicing the three R's of waste management: **R**educe the amount of waste produced. **R**euse materials in the waste stream as much as possible. And maximize **R**ecycling opportunities.

Let's look at this third option in greater detail.

Recycling

A lot of what our society throws away is still useful. Wastes from some parts of the mineral cycle may themselves be referred to as ores because they contain material that might be recycled and used again to provide energy or useful products.[14, 15] The idea of reusing waste materials is not new. Such metals as iron, aluminum, copper, and lead have been recycled for many years and are still being recycled today. The metal from automobiles, cell phones, computers, and many appliances discarded annually in the United States is being recycled.[15, 16]

The total value of recycled metals is about $50 billion. Of all recycled metals, iron and steel account for approximately 90% by weight and 40% by total value. Iron and steel are recycled in such large volumes for three reasons.[17] First, the market for iron and steel is huge, and as a result there is a large scrap collection and processing industry. Second, an enormous economic burden would result from failing to recycle. Third, disposing of millions of tons of iron and steel, instead of recycling, would have serious environmental impacts. It is estimated that each ton of recycled steel eliminates the need for over a ton of iron ore and one-half ton of coal. In addition, only one-third

as much energy is required to produce steel from recycled scrap as from native ore.[16]

Other metals that are recycled in large quantities include lead (two-thirds), aluminum (one-third), and copper (one-third).[2] Recycling aluminum reduces our need to import raw aluminum ore and saves about 95% of the energy required to produce new aluminum from bauxite.[16]

15.7 Minerals and Sustainability

Exploiting and at the same time sustaining mineral resources is problematic. This is because nonrenewable mineral resources are consumed over time, while sustainability is a long-term concept that includes finding ways to assure future generations a fair share of Earth's resources. Nonetheless, recently it has been argued that with human ingenuity and sufficient lead time, we can find solutions for a broader sustainable development that incorporate nonrenewable mineral resources.

Human ingenuity is important because often it is not the mineral itself we need so much as what we use the mineral for. For example, we mine copper and use it to transmit electricity in wires or transmit electronic pulses in telephone wires. It is not the copper itself we desire but the properties of copper that allow these transmissions. We can use fiberglass cables in telephone wires, eliminating the need for copper. Digital cameras have eliminated the need for film development that uses silver. We are also learning that we can use raw mineral materials more efficiently. For example, when the Eiffel Tower was constructed in the late 1800s, 8,000 metric tons of steel were used. Today the tower could be built with a fourth of that amount of steel.[18] The message is that it is possible to compensate for a nonrenewable mineral by finding new ways to do things.

How long does it take to develop new approaches? Finding substitutes for nonrenewable resources, or more efficient ways to use them, generally requires several decades of research and development. A measure of the time available for finding the solutions to depletion of nonrenewable reserves is the ***R* to *C* ratio,** where *R* is the known reserves (for example, hundreds of thousands of tons of a metal) and *C* is the rate of consumption (for example, thousands of tons per year used by people). The *R* to *C* ratio is often misinterpreted as the time a reserve will last at the present rate of consumption. During the past 50 years, the *R* to *C* ratios for metals such as zinc and copper have fluctuated around 30 years; during that time, consumption of the metals increased by about three times. This was possible because we discovered new deposits of the metals.

Although the R to C ratio is a present analysis of a dynamic system in which both the amount of reserves and rate of consumption may change over time, it does provide a view of how scarce a particular mineral resource may be. Metals with relatively small ratios can be viewed as being in short supply, and those are the resources for which we should find substitutes through technological innovation.[18]

In conclusion, we may approach sustainable development and use of nonrenewable mineral resources by finding ways to more wisely use resources, developing more efficient ways of mining resources, recycling more, and applying human ingenuity to find substitutes to perform the functions that nonrenewable mineral resources are now performing.

Return to the Big Question

Is it possible to use nonrenewable mineral resources sustainably?

Our mineral resources—iron, copper, aluminum, silver, and gold, among others—are basically nonrenewable because we use them much faster than Earth's processes can produce them. For this reason, the long-term use of nonrenewable mineral resources is not sustainable. However we can make our nonrenewable mineral resources last longer through a variety of strategies, such as extensive reuse and recycling, along with finding alternatives to the nonrenewable mineral resources that we are using up the most rapidly. In sum, when we speak of sustainability with respect to nonrenewable mineral resources, we must qualify our statements to refer not to the perpetual availability of these resources but to the goal of maintaining their availability over a relatively long period.

Summary

- Minerals are usually extracted from naturally occurring, unusually high concentrations of Earth materials. Such natural deposits allowed early peoples to exploit minerals while slowly developing technological skills.

- The origin and distribution of mineral resources are closely related to the history of the biosphere and the geologic cycle. Nearly all aspects and processes of the geologic cycle are involved to some extent in producing local concentrations of useful materials.

- Mineral resources are not mineral reserves. Unless discovered and developed, resources cannot be used to address present shortages.

- The availability of mineral resources is one measure of the wealth of a society. Modern technological civilization would not be possible without the exploitation of mineral resources. However, it is important to recognize that mineral deposits are not infinite and that we cannot maintain exponential population growth on a finite resource base.

- The United States and many other affluent nations import many minerals. Sometimes they rely on imports even if they have their own deposits of these minerals, because it's more economical. As other na-

tions industrialize and develop, such imports may be more difficult to obtain, and affluent countries may have to mine their own deposits, find substitutes for some minerals, or use a smaller portion of the world's annual production.

- The environmental impact of mineral exploitation depends on many factors, such as mining procedures, local hydrologic conditions, climate, rock types, size of operation, and topography.

- The mining and processing of mineral resources greatly affect the land, water, air, and biological resources and have social impacts as well due to increased demand for housing and services in mining areas.

- Because the demand for minerals will increase in the future, we must strive to minimize both on-site and off-site problems by controlling sediment, water, and air pollution through good engineering and conservation practices.

- Reducing consumption and reusing, recycling, and finding substitutes for mineral resources are environmentally preferable ways to delay or alleviate possible crises caused by a rapidly rising population and a limited resource base.

Key Terms

ore deposits
reserve

resource
R to C ratio

Getting It Straight

1. What is the difference between a resource and a reserve?

2. Under what circumstances might sewage sludge be considered a mineral resource?

3. If surface mines and quarries cover less than 0.5% of the land surface of the United States, why is there so much environmental concern about them?

4. When is recycling a mineral a viable option?

5. Which biological processes can influence mineral deposits?

6. A deep-sea diver claims that the oceans can provide all our mineral resources with no negative environmental effects. Do you agree or disagree?

7. What factors determine the availability of a mineral resource?

8. Using a mineral resource involves four general phases: (a) exploration, (b) recovery, (c) consumption, and (d) disposal of waste. Which phase do you think has the greatest environmental effect?

9. How can the use of nonrenewable mineral resources be compatible with sustainable development?

What Do you Think?

1. Make an inventory of the items you use in a given day. What minerals are they made from? Do you believe we could do without any of them? Which ones? Why?

2. Assuming wars and terrorism are being financed by money gained from extracting minerals, how might the global community or individual countries avoid violence that depends on minerals for its funding.

3. How could we reduce our dependence on metals such as iron, aluminum, and copper? Could we use significantly less than we use today? What would have to happen to achieve this goal?

Pulling It All Together

1. Does society depend on the availability of mineral sources? What mineral resources do you use on a regular basis and how might your life be altered if those minerals no longer existed?

2. On a global basis how are mineral deposits formed? How does knowing this help us discover new mineral deposits at the regional to local level?

3. What are the long-term social consequences of the so called "blood diamonds" in Africa? What can be done to reduce or eliminate civil wars (and other wars) faught over valuable minerals? Hint: Start with a google search of "blood diamonds".

Further Reading

Brookins, D. G. 1990. *Mineral and energy resources.* Columbus, OH: Charles E. Merrill.—A good summary of mineral resources.

Kesler, S. F. 1994. *Mineral resources, economics and the environment.* Upper Saddle River, NJ: Prentice Hall.—A good book about mineral resources.

Peter Ryan/Photo Researchers, Inc.

16

Waste Management

Big Question

Is Zero Waste Possible?

?

Learning Objectives

The waste-management concept of "dilute and disperse" (for example, dumping waste into a river) is a holdover from our frontier days, when we believed land and water to be limitless resources. We next attempted to "concentrate and contain" waste in disposal sites—a practice that also proved to pollute land, air, and water resources. We are now focusing on managing materials to eliminate waste, a concept known as zero waste. Finally, we are getting it right! After reading this chapter, you should understand . . .

- what "zero waste" means and how we might achieve it;

- what industrial ecology is, and its links to waste management;

- the advantages and disadvantages of the major approaches to integrated waste management;

- that hazardous chemical waste is one of our most serious environmental concerns;

- the various methods of managing hazardous chemical waste;

- the major pathways by which hazardous waste from a disposal site can enter the environment;

- what pollution prevention is and how it can be implemented.

Case Study

New York City's Zero Waste Campaign

The city of New York generates many thousands of tons of municipal solid waste every day—trash, recyclables, and construction/demolition debris (Figure 16.1). Since the closing of the Fresh Kills landfill on New York's Staten Island (Figure 16.2), the city has exported its waste at a cost of more than $100 per ton. Although recycling of bottles, cans, plastic containers, cardboard, and newspapers has reduced the volume of waste, a more innovative waste-management program is needed.

Recently the New York City Zero Waste Campaign, which speaks for more than 40 organizations, released a community-based waste-management plan. The object of the plan is to reduce the export of waste from the city to near zero over the next two decades. The plan calls for a combination of vigorous waste prevention, reuse, recycling, and composting (turning organic waste to mulch). The idea is to eliminate the high costs of shipping waste out of the city. The money that will be saved will circulate within the economy of the city, creating new industries and encouraging economic development. In addition, new jobs will be created for recovering materials of many kinds from what is now waste.[1]

Waste management continues to place a tremendous financial burden on society and reminds us that we have failed in the past 50 years to move from a throwaway, waste-oriented society to a society that sustains natural resources through improved materials management. We are now moving in that direction by producing less waste and recycling more. With this in mind, we introduce in this chapter concepts of waste management applied to urban waste, hazardous chemical waste, and waste in the marine environment.

Newsmankers/Getty Images

■ FIGURE 16.1
New York waste management in action:
New York City sanitation crew cleaning up after Yankees' victory parade in 2000.

Corbis SABA

■ FIGURE 16.2
Fresh Kills landfill, closed in 2001, was one of the largest waste disposal sites in the world.

16.1 What Is This Waste We Are Talking About?

Before discussing concepts of waste disposal and management, it's a good idea to make clear what we are talking about. When we speak of waste, we are talking about what is called the **waste stream**, which is waste we produce from our homes and businesses, manufacturing, agriculture, and numerous other processes. Solid waste from our urban areas is known as **municipal solid waste (MSW)**. Some of the sources of MSW are shown in Figure 16.3a. The composition of that waste before recycling is shown in Figure 16.3b.

Composition of Solid Waste

In a contest for "most abundant solid waste," paper would be the clear winner. Publicity about the long life of fast-food packaging, polystyrene foam coffee cups, and disposable diapers has led many people to assume that these products make up a large percentage of the total waste stream and are responsible for the rapid filling of landfills. However, excavations into modern landfills indicate that fast-food packaging, disposable diapers, and polystyrene cups together account for less than 2% of the average landfill.[2] Paper is the most abundant material found in landfills, accounting for about 35% by weight.

The single largest item is newsprint, which accounts for about one-fifth by volume.[3] Newsprint is one of the major items targeted for recycling because big environmental dividends can be expected. This doesn't mean that we need not reduce our use of other paper, polystyrene, plastic, metals, and other products. In addition to creating a disposal problem, these are resources that might be better managed.

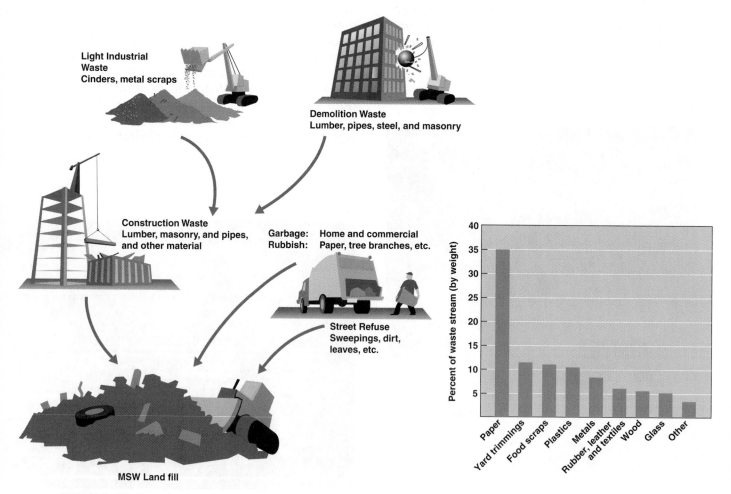

■ **FIGURE 16.3**

Municipal solid waste (MSW).
(a) Sources of municipal solid waste; (b) composition of U.S. urban solid waste (by weight), 2003.
[*Source:* U.S. Environmental Protection Agency, Office of Solid Waste. Available at http://www.epa.gov/. Accessed October 9, 2006.]

16.2 Early Concepts of Waste Disposal

Waste disposal looked easy at first—just let the river dilute and disperse it. During the first century of the Industrial Revolution, the United States produced relatively little waste and managed its waste through the concept of "dilute and disperse." Factories were located near rivers because the water provided a number of benefits, including easy transport of materials by boat, sufficient water for processing and cooling, and easy disposal of waste into the river. The result was polluted rivers.

"Concentrate and contain" came next. As industrial and urban areas expanded and pollution problems were recognized, a new concept, known as "concentrate and contain," came into use. It has become apparent, however, that containment is not always achieved. Containers, whether simple trenches in the ground or metal drums and tanks, may leak or break. Health hazards resulting from such practices in the past have led to a situation in which many people now have little confidence in government or industry to preserve and protect public health.[4]

We produce great quantities of waste, but where to put it? Today, communities in many parts of the world are facing a serious solid-waste disposal problem. The problem is that we are producing an enormous amount of waste and can't find enough acceptable places to put it. It's not that we are actually running out of space for landfills—landfills occupy only a small fraction of the land area of the United States. Rather, existing sites are getting filled up, and it is difficult to find new landfill sites because no one wants to live near one. In fact, we don't want to live near any kind of waste-disposal site, whether it be a sanitary landfill for municipal waste, an incinerator that burns urban waste, or a hazardous-waste disposal operation for chemical materials. This attitude is widely known as NIMBY ("not in my backyard").

Another major limiting factor is the cost of disposal. In the past few decades, the cost of disposing of 1 metric ton of urban waste has increased about tenfold. Today, disposal or treatment of municipal solid waste is one of our most costly environmental expenditures.

16.3 Modern Trends

Viewing waste as resources out of place is the environmentally correct approach to waste management today. Although we may not soon be able to reuse or recycle all waste, it seems apparent that the rising costs of raw materials, energy, transportation, and land will make it more economical to better manage waste. We are moving toward an environmental view that there is no such thing as waste. Waste would not exist because it would not be produced, or, if produced, it would be a resource to be used again. This is the essence of **zero waste**.

Let's think about waste from an ecological point of view (see Chapter 3). The zero-waste movement includes, in part, what is known as **industrial ecology,** which draws an analogy between urban waste management and ecosystem function. What might be thought of as waste in one part of an ecosystem is often a resource for another part or another species. For example, an elephant eats plants and produces waste that becomes a resource for the dung beetle.

As another example, consider a coal-burning power plant that produces electricity for a town. Waste from producing the power includes ash from the coal, exhaust heat, and products from combustion that might go up a smokestack, including carbon dioxide and sulfur dioxide. The waste heat is used to warm homes in the town and provide heat for industrial activities. Sulfur dioxide is removed from the system by scrubbing to produce gypsum, the major component of wallboard used in construction. Carbon dioxide is used in local greenhouses, along with waste heat, to force and prolong the growing cycle of plants. We are left with the ash from the coal, which is used for road surfacing.[5]

The goal is to turn waste into a resource. Zero waste and industrial ecology evolved from a grassroots recognition that waste disposal is inefficient and damaging to the environment. The goal is to eliminate the entire concept of waste and transform waste disposal into a resource program. Pollution of air, water, and land would be reduced as we move toward a sustainable future.

It's not just a matter of science and technology, but also of values. Industrial ecology and zero waste using science and technology alone are not adequate to achieve sustainability of global systems. We also need to clarify our values. Science can offer potential solutions to problems we face, but which solutions we choose will reflect our values. That is one of the key themes of this book.

Zero waste was considered impossible, but is catching on. The city of Canberra, Australia, is one of the first communities to propose a plan to have zero waste, a goal it hopes to reach by 2010. Thousands of kilometers away, the Netherlands has set a goal of reducing its waste by 70–90%. How this is to be accomplished is not entirely clear; but a large part of the planning involves taxation of waste in all its forms, from smokestack emissions to solids delivered to landfills. The Netherlands has already nearly eliminated discharges of heavy metals into waterways by levying pollution taxes. The government is also considering programs—known as "pay as you throw"—that tax people according to the volume of waste they dispose of, including household waste. The idea is that taxing waste will persuade people to produce less of it.[6]

Some "waste management" is really just shuffling, not managing. Many of our waste-management programs involve simply moving waste from one site to another. For example, waste from urban areas may be placed in landfills, but eventually these landfills may cause new problems by producing methane gas or noxious liquids that leak from the site and contaminate the surrounding areas. Managed properly, however, methane produced from landfills is a resource that can be burned as a fuel (an example of industrial ecology).

Old notions of waste disposal are no longer acceptable—we are rethinking how we deal with materials, with the objective of eliminating the concept of "waste" entirely. Viewing waste as a resource will lead to reduced consumption of virgin materials (such as trees and unmined metals) and enable us to live within our environment more sustainably.[6]

16.4 Integrated Waste Management (IWM)

Integrated waste management (IWM) is the dominant approach today in managing waste. It includes *reuse, source reduction, recycling, composting, landfill,* and *incineration.*[4]

Reduce, Reuse, Recycle

The three R's of IWM are *Reduce, Reuse,* and *Recycle.* The ultimate objective of the three R's is to reduce the amount of urban and other waste that must be disposed of in landfills, incinerators, and other waste-management facilities. Study of the **waste stream** (the waste produced) in areas that use IWM suggests that the amount (by weight) of urban waste disposed of in landfills or incinerated can be reduced by at least 50% and perhaps by as much as 70%. A 50% reduction by weight could be achieved by the following:[4]

▪ Better design of packaging with less waste, an element of source reduction (10% reduction).
▪ Large-scale composting programs (10% reduction).
▪ Recycling programs (30% reduction).

As this list indicates, recycling is a major player. Recycling is so common today that we accept it as part of daily life. Recycling bins are common in parks, on university campuses, in apartment houses, and many other places. Most collection is going to bins with mixed recyclable waste, such as paper, bottles, and cans. The materials are separated later at recycling centers. Can re-

cycling in fact reduce the waste stream by 50%? Recent work suggests that it can. In fact, it has done so in some parts of the United States, and the upper limit for waste reduction from recycling could be much higher. It is estimated that as much as 80–90% of the U.S. waste stream might be recovered through intensive recycling.[7] A pilot study involving 100 families in East Hampton, New York, achieved a level of 84%. More realistic for many communities is partial recycling, which targets specified materials, such as glass, aluminum cans, plastic, organic material, and newsprint. Partial recycling can reduce waste significantly—in many places it is approaching or exceeding 50%.[8, 9]

Business and industry are getting with the program. An encouraging sign of public concern for the environment is increased willingness of industry and business to support recycling on a variety of scales. For example, fast-food restaurants are using less packaging and providing on-site bins for recycling used paper and plastic. Grocery stores are encouraging the recycling of plastic containers and paper bags by providing bins for their collection and recycling. Some food stores offer inexpensive canvas shopping bags to people who prefer them to disposable plastic and paper bags. Companies are redesigning products that can be more easily disassembled after use and the various parts recycled. As the idea catches on, small appliances, such as irons and toasters, may be recycled instead of ending up in landfills. The automobile industry is designing automobiles with coded parts so that they can be more easily disassembled by professional recyclers, rather than left to become rusting eyesores in junkyards.

Consumers, too, have become active in recycling. Consumers are now more likely to choose products that can be recycled or that come in containers that can be recycled or composted. Many consumers have purchased small compactors that crush bottles and aluminum cans, reducing their volume and making them easier to recycle. Some of the ways you can reduce your personal waste are listed in Table 16.1. The entire area is rapidly changing, and innovations and opportunities will undoubtedly continue.

Markets for Recycled Products

Designing a successful IWM program can be a complex undertaking. In some communities, enthusiasm for recycling has been so great that recycled products have glutted the market, sometimes requiring temporary stockpiling or suspension of recycling of some items. It is apparent that if recycling is to succeed, markets and processing facilities will also have to be developed to ensure that recycling is a sound financial venture as well as an important part of IWM.

Recycle as much as you can: Make sure your cans, glass, paper get to a recycling bin or center (curbside pickup in most cases or bins at public places such as parks or your university). Deliver hazardous materials—such as batteries, cell phones, computers, paint, used oil, solvents— to a hazardous-waste pickup facility.

Reduce packaging: Buy your staple food items in bulk or concentrated form whenever possible.

Select reusable products: Items such as sturdy, washable utensils, tableware, cloth napkins, and dishcloths can be used over and over again.

Use durable products: Choose automobiles, furniture, sports equipment, toys, and tools that will last a long time and therefore need not be replaced as often.

Reuse products: Reuse newspaper, boxes, and "bubble wrap" to ship your packages.

Purchase products made from recycled material: Many bottles, cans, boxes, cartons, and other containers, as well as books, carpets, floor tiles, some clothing, and other products are made from recycled material.

Manage your e-waste: Be sure cellphones, iPods, computers are recycled safely.

Source: U.S. Environmental Protection Agency. Wastes. Available at www.epa.gov. *Accessed April 21, 2006.*

Recycling Human Waste

Use of human waste on croplands is an ancient practice. In Asia, recycling of human waste ("night soil") has a long history. Chinese agriculture was sustained for thousands of years by collecting human waste and spreading it over the fields. By the early 20th century, land application of treated sewage was one of the primary disposal methods in many metropolitan areas in countries including Mexico, Australia, and the United States.[10]

Using wastewater for agriculture can spread diseases. Today, with the globalization of agriculture, we still are occasionally warned about fruits and vegetables contaminated by bacteria, viruses, and parasites contained in wastewater applied to crops. Another problem is that, along with human waste, thousands of chemicals and metals flow through our modern waste stream. Even garden waste that is composted may contain harmful chemicals such as pesticides. Heavy metals, petroleum products, industrial solvents, and pesticides may end up in our wastewater collection systems and sewage treatment plants, and therefore we must be skeptical of applying sewage sludge (solids remaining after sewage-treatment) to the land.

Of course, the contents of sewage sludge vary from place to place and even from day to day. Nevertheless, studies have shown that high levels of toxic chemicals and metals may be present in the sludge of cities or towns with industries that use toxic materials.[10]

The situation is better but still needs improvement. Fewer toxic materials with our human waste end up at sewage-treatment plants today because many industries are now pretreating their waste to remove materials that used to contaminate wastewaters. Federal, local, and other government agencies, as well as industries, have discussed the question of how much toxic material in the waste stream constitutes a problem. But this is really not the correct question to ask. The question is how to make sure that sewage sludge contains no toxic materials at all.

Sewer lines from urban homes are the same ones used by industry. As a result, conventional waste-disposal technology is unlikely to produce sludge that is safe for living things. A possible solution is to separate urban waste from industrial waste. A second possibility is to pretreat waste from industrial sources to remove hazardous substances before they enter the wastewater stream. As noted above, many industries are already doing this. Some communities are considering smaller wastewater-treatment facilities to treat waste from homes; the recycled waste would be used locally by farms.

What lies ahead? In the future, as the cost of oil, which is necessary to produce fertilizers, continues to rise, the age-old practice of recycling human waste may again be economical and necessary in many more places than today.[10]

16.5 Materials Management

Recycling alone can't do the whole job. Recycling for over two decades has generated entire systems of waste management, produced tens of thousands of jobs in the United States, and reduced the amount of urban waste sent from homes to landfills. Many firms have combined waste reduction with recycling to reduce the waste they deliver to landfills by more than 50%. Despite this success, IWM has been criticized for overemphasizing recycling and failing to promote policies to prevent waste production.

The futuristic goal of zero waste also requires **materials management**—the more sustainable use of materials (such as recycling) combined with resource conservation. The goal might be pursued in the following ways:[11]

▪ Eliminate subsidies for extracting virgin materials, such as minerals, oil, and timber.

▪ Establish "green building" incentives to encourage the use of recycled-content materials and products in new construction.

▪ Assess financial penalties for production that uses poor materials-management practices.

▪ Provide financial incentives for industrial practices and products that benefit the environment by enhancing sustainability (for example, reducing waste production and using recycled materials).

▪ Increase the number of new jobs in the technology of reuse and recycling of resources.

Materials management is starting to influence where industries are located. For example, about half of the steel produced in the U.S. now comes from scrap, so new steel mills are no longer located near steel resources such as coal and iron ore. New steel mills are located in a variety of places, from California to North Carolina and Nebraska; and their resource is the local supply of scrap steel. Because they are starting with scrap metal, the new industrial facilities use far less energy and cause much less pollution than older steel mills that start with virgin iron ore.

Similarly, recycling paper is changing where new paper mills are built. In the past, mills were built near forested areas where timber needed for paper production was being logged. Because enormous amounts of paper are now being recycled, mills are now being built near cities where supplies of recycled paper exist. For example, the state of New Jersey, which has little forested land and no iron mines, has 13 mills producing paper from recycled paper and 8 steel "mini-mills" producing steel from scrap metal, thanks to the power of materials management.[12]

16.6 Solid-Waste Management

Managing solid waste continues to be a problem, both in the United States and in other parts of the world. In many areas, particularly in developing countries, waste management is inadequate, with poorly controlled open dumps and illegal roadside dumping, which can spoil scenic areas, pollute soil and water, and pose health hazards.

Illegal dumping is a social problem as much as a physical one, because many people are simply disposing of their waste as inexpensively and as quickly as possible and may not see dumping their trash as an environmental problem. If nothing else, this is a tremendous waste of resources, since much of what is dumped could be recycled or reused. In areas where illegal dumping has been reduced, the keys have been awareness, education, and alternatives. People are made aware of environmental effects of unsafe, unsanitary dumping through education programs, and funds are provided for cleanup and for inexpensive collection and recycling of trash at sites of origin.

Unintentional hazardous waste ends up in our landfills. Infectious waste from hospitals and clinics sometimes ends up in disposal sites, where it can cause health problems if it has not been sterilized before disposal. Some hospitals have facilities to incinerate infectious waste, and that is probably the surest way to manage it. In urban areas, a large amount of toxic material also may end up at disposal sites. People place all sorts of toxic materials in their trash cans, including cans and jars with small amounts of paint, varnish, cleaning solvents, pesticides, nail polish, and such. As a result, many older urban landfills are now considered hazardous-waste sites that will require costly cleanup. We turn now to specific ways to dispose of solid waste.

On-Site Disposal

Mechanical grinding of kitchen food waste is a common on-site disposal method in urban areas. Garbage-disposal devices are installed in the wastewater pipe system at the kitchen sink, and the garbage is ground and flushed into the sewer system. This reduces the amount of handling and quickly removes food waste. At sewage-treatment plants, solids remaining as sewage sludge still must be disposed of.[13]

Composting

Composting turns organic material into a substance that enriches soil. It is a biochemical process in which organic materials such as lawn clippings and kitchen scraps decompose to a rich, soil-like material. The process involves rapid partial decomposition of moist solid organic waste by aerobic organisms. Although simple backyard compost piles may come to mind, large-scale composting as a waste-management option is generally done in the controlled environment of mechanical "digesters." This technique is popular in Europe and Asia, where intense farming creates a demand for the compost.[13]

Composting has a couple of drawbacks. A major drawback is that organic material has to be separated from other waste, so it is probably economically advantageous only where organic material is collected separately from other waste. Another problem is that composting plant debris that has been treated with herbicides may produce compost that is toxic to some plants. Nevertheless, composting is an important component of IWM, and its contribution will undoubtedly grow in the future.

Incineration

In *incineration*, combustible waste is burned at temperatures high enough (about 1,000°C or 1,830°F) to consume all combustible material, leaving only ash and noncombustibles to dispose of in a landfill.[13] However, besides reducing a large volume of waste to a much smaller volume of ash, incineration has another advantage: Burning waste can supplement other fuels and generate electrical power.

Incineration of urban waste is not necessarily a clean process. It produces air pollution and toxic ash. Incineration in the United States apparently is a significant source of environmental dioxin, a carcinogenic toxin (see Chapter 8).[14] In addition, smokestacks from incinerators may emit oxides of nitrogen and sulfur, which lead to acid rain; heavy metals such as lead, cadmium, and mercury; and carbon dioxide (a greenhouse gas associated with global warming).

Open Dumps

Open dumps are still in use but are becoming obsolete. In the past, solid waste was often disposed of in open dumps, where the refuse was piled up without being covered or otherwise protected. Thousands of open dumps have been closed in recent years, and new open dumps are banned in the United States and many other countries. Nevertheless, many are still being used worldwide (Figure 16.4).

Dumps have been located wherever land is available, often without regard to safety, health hazards, or aesthetic degradation. Common sites are abandoned mines and quarries where gravel and stone have been removed (sometimes by ancient civilizations); natural low areas, such as swamps or floodplains; and hillside areas above or below towns. The waste is often piled as high as equipment allows. In some instances, the refuse is ignited and allowed to burn. In others, the refuse is periodically leveled and compacted.[13]

Open dumps generally create a nuisance by being unsightly, providing breeding grounds for pests, creating a health hazard, polluting the air, and sometimes polluting groundwater and surface water. Fortunately, open dumps are giving way to better-planned and better-managed landfills.

Municipal Solid Waste (MSW)

Covering the waste with soil is what makes this landfill safer. An **MSW landfill** is designed to concentrate and contain refuse without creating a nuisance or hazard to public health or safety. The idea is to confine the waste to the smallest possible area, reduce it to the smallest possible volume, and cover it with a layer of compacted soil at the end of each day or more frequently if necessary. The compacted layer restricts (but does not eliminate) continued access to the waste by insects, rodents, and other animals, such as seagulls. It also isolates the refuse, minimizing the amount of surface water seeping into it and gas escaping from it.[15] A diagram of an MSW landfill and

Alex Quesada/Matrix International, Inc.

■ **FIGURE 16.4**
An open garbage dump in Rio de Janeiro, Brazil.
At this site, people are going through the waste and recycling materials that can be reused or resold.

possible paths for pollution to enter the environment is shown in Figure 16.5.[16]

The most significant hazard from an MSW landfill is **leachate**. *Leachate* is water that has moved through a substance and picked up some of that substance's characteristics. A cup of coffee, for example, is a good-tasting leachate—it is water that has percolated down through coffee grounds, picking up their color and taste. However, when water percolates down through a landfill, or groundwater moves through it, a noxious, organic-rich leachate carrying many toxic materials is produced.[17]

For example, two landfills dating from the 1930s and 1940s on Long Island, New York, have produced subsurface leachate trails (plumes) several hundred meters wide that have migrated far from the disposal site. The nature and strength of such leachate depends on the composition of the waste, the amount of water that moves through it, and how long the water is in contact with the waste.[13]

Where an MSW landfill is situated is very important. In choosing a site, a number of factors must be considered, including topography, location of the groundwater table, amount of precipitation, type of soil and rock, and the surface water and groundwater flow system. Basically, you don't want to put your waste where it can contaminate the ground and the water. A favorable combination of climatic, hydrologic, and geologic conditions helps to ensure reasonable safety in containing the waste and its leachate.[18]

Choosing a site also involves social considerations. Often, planners choose sites where they expect little local resistance or where the land seems to have little value. Waste-disposal facilities are frequently located in areas where residents tend to have low socioeconomic status or belong to a particular racial or ethnic group.

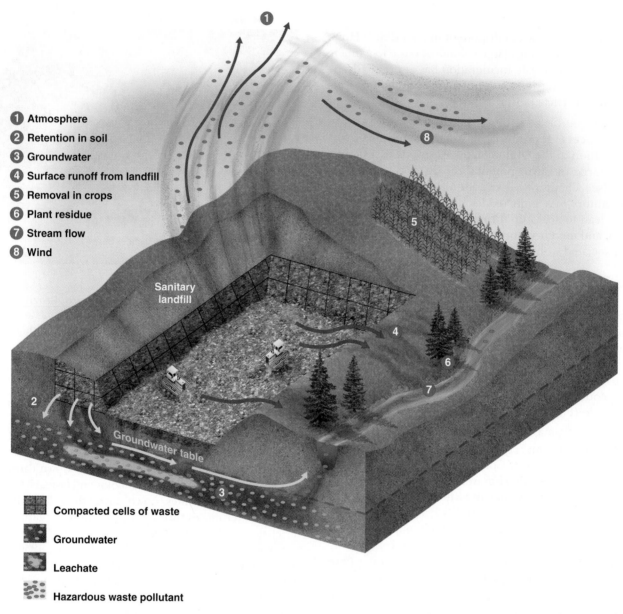

1. Atmosphere
2. Retention in soil
3. Groundwater
4. Surface runoff from landfill
5. Removal in crops
6. Plant residue
7. Stream flow
8. Wind

Sanitary landfill

Groundwater table

Compacted cells of waste

Groundwater

Leachate

Hazardous waste pollutant

■ **FIGURE 16.5**

Municipal solid waste (MSW) landfill.
Idealized diagram showing possible paths that pollutants from a landfill site may follow to enter the environment.

Environmental justice is a new field that focuses on these social issues. Many people object to living near waste-disposal facilities and chemical plants on environmental grounds. Their concerns include health risks from accidental spills, fires, explosions, or illegal discharge of waste or chemicals. Communities near industrial activity are most likely to contain such facilities and generally have a large population of working-class people of color. *Environmental justice* addresses the fact that these people are being placed at greater risk of harm.[19, 20]

Once a site is chosen, pollution must be monitored. Before filling starts at an MSW landfill, monitoring the movement of groundwater should begin. Monitoring is done by periodically taking samples of water and gas from specially designed monitoring wells. Monitoring the move-

ment of leachate and gases should continue as long as there is any possibility of pollution. This is particularly important after the site is completely filled and a final, permanent cover material is in place. Continued monitoring is necessary because a certain amount of settling always occurs after a landfill is completed, and if small depressions form, surface water may collect, infiltrate, and produce leachate. Monitoring and proper maintenance of an abandoned landfill reduce its pollution potential.[15]

Modern MSW landfills are designed to include multiple barriers: clay and plastic liners to limit the movement of leachate; surface and subsurface drainage to collect leachate; systems to collect methane gas produced as waste decomposes; and groundwater monitoring to detect leaks of leachate below and next to the

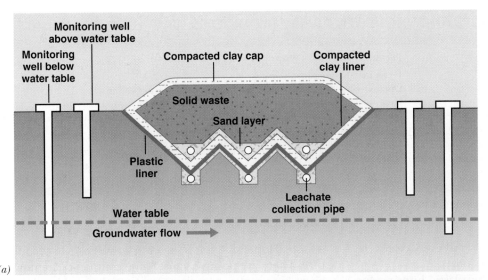

(b)

Courtesy John H. Kramer

■ **FIGURE 16.6**
MSW landfill design:
(a) Idealized diagram of a solid-waste facility (MSW landfill) illustrating multiple-barrier design, monitoring system, and leachate collection system. (b) Rock Creek landfill, Calaveras County, California, under construction. This municipal solid-waste landfill is underlain by a compacted clay liner (exposed light brown slope in the center left portion of the photograph). The green slopes, covered with gravel piles, overlie the compacted clay layer. These form a vapor barrier designed to keep moisture in the clay and help avoid cracking of the clay liner. The sinuous brown trench is lined with blue plastic and is part of the leachate collection system for the landfill. The excavated squared pond (upper part of photograph) is a leachate evaporation pond under construction. The landfill is also equipped with a system to monitor the zone below the leachate collection system.

landfill. It is particularly important to monitor the zone above the water table to spot potential pollution problems before they contaminate groundwater resources, where correction would be very expensive. Figure 16.6 shows an idealized diagram of an MSW landfill that uses the multiple-barrier approach and a photograph of a landfill site under construction.

Federal legislation governs newer MSW landfills. New landfills that open in the United States must comply with requirements of the Resource Conservation and Recovery Act of 1980. The legislation is intended to strengthen and standardize design, operation, and monitoring of MSW landfills. Landfills that cannot comply with regulations face closure.

16.7 Hazardous Waste

So far we have discussed zero-waste, integrated waste management, and materials management for the everyday waste stream from homes and businesses. We now consider the important topic of **hazardous waste**.

Creation of new chemical compounds has increased in recent years. In the United States, approximately 1,000 new chemicals are marketed each year, and about 70,000 chemicals are currently on the market. Although many of these chemicals have been beneficial to people, about 35,000 chemicals used in the United States are classified as definitely or potentially hazardous to the health of people or ecosystems (Table 16.2).

Most hazardous waste is produced by chemical products industries. The United States currently produces about 700 million metric tons of hazardous chemical waste per year, referred to more commonly as *hazardous waste*. About 70% of the total is produced east of the Mississippi River, and about half of the total by weight is produced by chemical products industries, with the electronics industry and petroleum and coal products industries each contributing about 10%.[21, 22]

Toxic "e-waste": Where do our old computers and iPods end up? The average life of a computer is about three years, and the hundreds of millions of computers, printers, cell phones, iPods, televisions, computer games,

☐ TABLE 16.2 PRODUCTS AND THE POTENTIALLY HAZARDOUS WASTE THEY GENERATE

Products We Use	Potentially Hazardous Waste
Leather	Heavy metals, organic solvents
Medicines	Organic solvents and residues, heavy metals (e.g., mercury and zinc)
Metals	Heavy metals, fluorides, cyanides, acid and alkaline cleaners, solvents, pigments
Oil, gasoline, and other petroleum products	Oil, phenols and other organic compounds, heavy metals, ammonia salts, acids
Paints	Heavy metals, pigments, solvents, organic residues
Pesticides	Organic chlorine compounds, organic phosphate compounds
Plastics	Organic chlorine compounds
Textiles	Heavy metals, dyes, organic chlorine compounds, solvents

Source: U.S. Environmental Protection Agency, SW-826, 1980.

and other electronic devices we discard every year were not constructed with recycling in mind.

Electronic waste includes the plastic housing for computers, printers, and electronic devices. When burned, this plastic may produce toxins. Computer parts also contain small amounts of heavy metals—including gold, tin, copper, cadmium, and mercury—that are toxic and may cause cancer if breathed, ingested, or absorbed through the skin. At present, many millions of computers are disposed of by what is billed as "recycling." However, when we take e-waste to a designated disposal location, we should not assume it will be handled in a way that won't cause environmental problems.[23]

The United States has no official process administered by the Environmental Protection Agency (EPA) for proper handling of e-waste, and in the United States computers cannot be recycled profitably without charging the people who dump them a fee. Even with that, many U.S. firms "recycle" e-waste by shipping it to countries such as China and Nigeria. China's largest e-waste facility is in Guiyu, near Hong Kong. People in the Guiyu area process over 1 million tons of e-waste each year with little thought to the potential toxicity of the material they are handling (Figure 16.7). The revenue to the Guiyu area is about $1 million per year, and the central government is resistant to regulating the activity. The problem is that workers at the locations where computers are disassembled, including over 5,000 family-run facilities where people scavenge e-waste for raw materials, may not be aware that they are exposing themselves to a variety of toxins and potential health problems.

To date, the United States has not taken a proactive stance to regulate the computer industry so that less waste is produced. In fact, the United States is the only major nation that did not ratify an international agreement that restricts and bans exports of hazardous e-waste.[23]

Present practices in our handling of e-waste are not sustainable. Attempts to assure a quality environment should include the safe handling and recycling of e-waste. Hopefully, that is the path we will take in the future. There are positive signs. Some companies are now processing e-waste to reclaim metals such as gold and silver. Others are designing computers with materials that are less toxic and easier to recycle. The European Union is taking a leadership role in requiring more responsible management of e-waste.

Natural disasters can also release hazardous chemicals into the environment. Buildings destroyed by events such as fires and hurricanes can release hazardous chemicals, such as stored paints, solvents, and pesticides, when

■ FIGURE 16.7

Person working at one of thousands of shops in China that process e-waste by hand.
Workers at these sites are exposed to toxic materials.

the buildings are burned or buried. For example, when floodwaters in the city of New Orleans from Hurricane Katrina in 2005 inundated about 80% of the city, toxic waste from flooded homes, businesses, industry, and vehicles, along with dead bodies of people and animals, caused water pollution in some locations of the city. Thus, collecting potentially hazardous chemicals after natural disasters and disposing of contaminated soil and water are important in managing hazardous materials.

Until recently, as much as half of U.S. hazardous waste was simply dumped (Figure 16.8).[22, 24–26] Some was illegally dumped on public or private lands, a practice called "midnight dumping." Illegally buried drums full of hazardous waste have been discovered at hundreds of sites by contractors constructing buildings and roads. Cleanup of the waste has been costly and has delayed projects.[21]

Hazardous waste at previously unregulated dumping sites is a serious problem. The United States has an estimated 32,000–50,000 waste-disposal sites, often abandoned, where past dumping was totally unregulated. Of

these, probably 1,200–2,000 contain enough hazardous waste to be a threat to public health and the environment. For this reason, many scientists believe management of hazardous chemical materials is one of the most serious environmental problems in the United States.

Uncontrolled dumping of chemical waste has polluted soil and groundwater in several ways:

- Chemical waste may be stored in barrels, either stacked on the ground or buried. The barrels eventually corrode and leak, polluting surface water, soil, and groundwater.
- When liquid chemical waste is dumped into an unlined lagoon, contaminated water may percolate through soil and rock to the groundwater table.
- Liquid chemical waste may be illegally dumped in deserted fields or even along roads.

Some sites pose particular dangers. The floodplain of a river, for example, is not an acceptable site for storing hazardous chemical waste. Yet that is exactly what occurred on the floodplain of the River Severn near a village in one of the most scenic areas of England. Several fires at the site in 1999 were followed by a large fire on October 30, 2000. About 200 tons of chemicals—including industrial solvents, cleaning solvents, and various insecticides and pesticides—produced a fireball that rose into the night sky (Figure 16.9).

The fire occurred during a rainstorm with wind gusts of hurricane strength. Toxic smoke and ash spread to nearby farmlands and villages, where people had to

Courtesy N.Y. State Dept. of Environmental Conservation

■ **FIGURE 16.8**
Love Canal, in New York State.
Hazardous chemical waste dumped from 1920 to 1952 rose to the surface in 1976, making Love Canal a household name for hazardous-waste problems. In this aerial infrared photograph of the Love Canal area, healthy vegetation is in bright red. The canal, running from the upper left to the lower right, is a scar on the landscape. The old dump became a housing tract with an elementary school. Many homes and the school had to be destroyed. Cleanup costs have exceeded half a billion dollars.

Courtesy of Gloucester Fire Service & Sandhurst Area Action Group

■ **FIGURE 16.9**
Fire at a hazardous-waste site.
On October 30, 2000, fire ravaged a site on the floodplain of the River Severn in England, where hazardous waste was being stored. Approximately 200 tons of chemicals burned.

Courtesy of Gloucester Fire Service & Sandhurst Area Action Group

▪ **FIGURE 16.10**
Flooding after the fire.
Flooding on November 3, 2000, followed the large fire at a hazardous-waste storage site on the floodplain of the River Severn in England (see Figure 16.9).

be evacuated. People exposed to the smoke complained of headaches, stomachaches and vomiting, sore throats, coughs, and difficulty breathing. Then, a few days later, on November 3, the site flooded (Figure 16.10). The floodwaters interfered with cleanup after the fire and increased the risk that water carrying hazardous chemical wastes would contaminate downstream areas. In one small village, contaminated water apparently flooded farm fields, gardens, and even homes.[27] Of course, the solution to this problem is to clean up the site and move waste storage facilities to a safer location.

16.8 Hazardous-Waste Legislation

The recognition in the 1970s that hazardous waste was a danger to people and the environment and that the waste was not being properly managed led to important federal legislation in the United States.

Resource Conservation and Recovery Act (RCRA)

Management of hazardous waste in the U.S. began in 1976 with passage of RCRA. At the heart of the act is identification of industrial hazardous wastes and their life cycles. The idea was to establish guidelines and responsibilities for those who manufactured, transported, and disposed of hazardous waste. This is known as "cradle-to-grave" management. Regulations require stringent record-keeping and reporting to verify that wastes do not present a public nuisance or a health problem.

What qualifies as "hazardous"? A waste is hazardous if its concentration, volume, or infectious nature may contribute to serious disease or death, or if it poses a significant hazard to people and the environment as a result of improper management (storage, transport, or disposal).[21] RCRA applies to solid, semisolid, liquid, and gaseous hazardous wastes and it classifies hazardous wastes in several categories: materials highly toxic to people and other living things; wastes that may ignite when exposed to air; extremely corrosive wastes; and reactive unstable wastes that are explosive or generate toxic gases or fumes when mixed with water.

Comprehensive Environmental Response, Compensation, and Liability Act (CERCLA)

Congress passed CERCLA in 1980 and strengthened it in 1984 and 1986. The act defined policies and procedures for release of hazardous substances into the environment (for example, landfill regulations). CERCLA also (1) mandated development of a list of sites where hazardous substances were likely to cause, or already had caused, the most serious environmental problems; (2) mandated establishment of a revolving fund ("Superfund") to clean up the worst abandoned hazardous-waste sites; and (3) established provisions for government to recover costs of cleaning contaminated sites.[28]

Superfund has had management problems, and cleanup efforts are far behind schedule. Unfortunately, not enough funds are available to decontaminate all targeted sites. Furthermore, there is concern that current technology is not sufficient to treat all abandoned waste-disposal sites, and thus it may be necessary to simply try to confine waste to those sites until better technology is developed. It seems apparent that abandoned disposal sites are likely to remain problems for some time to come.

Other Legislation

Federal laws have changed the way real-estate business is conducted. For example, federal laws have provisions by which property owners may be held liable for costly cleanup of hazardous waste on their property even if they did not directly cause the problem. As a result, banks and other lending institutions might be liable if their tenants release hazardous materials.

The Superfund Amendment and Reauthorization Act (SARA) of 1986 provides a possible defense against such liability if the property owner completes an *environmental audit* before purchasing the property. Such an audit involves studying past land use at the site, usually from analyzing old maps, aerial photographs, and reports. It may also involve drilling and sampling groundwater and soil to determine whether hazardous materials are present. Environmental audits are now routinely done before purchasing property for development.

SARA requires that some industries report all releases of hazardous materials, and a list of companies that release hazardous substances was made public. No property owner or industry wants to be on this list—known as the "Toxic 500"—and this likely placed some pressure on industries identified as polluters to develop safer ways of handling hazardous materials.[29]

In 1990, the U.S. Congress reauthorized hazardous-waste control legislation. Priorities include:

- Establishing who is responsible (liable) for existing hazardous-waste problems.
- When necessary, assisting in or providing funding for cleanup at sites that have a hazardous-waste problem.
- Providing measures to compensate people who suffer damages from the release of hazardous materials.
- Improving the standards for disposal and cleanup of hazardous waste.

16.9 Hazardous-Waste Management: Land Disposal

Managing hazardous chemical waste involves several options, including recycling, on-site processing to recover by-products with commercial value, microbial breakdown, chemical stabilization, high-temperature decomposition, incineration, and disposal by secure landfill or deep-well injection. A number of technological advances have been made in toxic-waste management; and as land disposal becomes more expensive, a recent trend toward on-site treatment is likely to continue. However, on-site treatment will not eliminate all hazardous chemical waste, so it will still be necessary to dispose of some of it.

Secure Landfill

Secure landfills are designed to confine waste and leachate. A secure landfill for hazardous waste is designed to confine the waste to a particular location, control the leachate that drains from the waste, collect and treat the leachate, and detect possible leaks. This type of landfill is similar to the MSW landfill and is an extension of the landfill for urban waste. Because in recent years it has become apparent that urban waste contains much hazardous material, the design of landfills and the design of secure landfills for hazardous waste have converged to some extent.

Some argue that there is no such thing as a really secure landfill—that they all leak to some extent. Plastic liners, filters, and clay layers can fail, even with several backups, and drains can become clogged and cause overflow. Animals—such as gophers, squirrels, woodchucks, and muskrats—can chew through plastic liners and burrow through clay liners, creating or worsening leaks. However, careful siting and engineering can minimize problems. As with sanitary landfills, preferable sites are those with good natural barriers to prevent leachate from migrating, thick clay deposits, an arid climate, or a deep water table. Nevertheless, land disposal should be used only for chemicals that are suitable for this method.

Land Application: Microbial Breakdown

Land application means applying waste materials to near-surface soil. This is also called "land spreading" or "land farming." We discussed land application of human waste earlier, and land application may be an efficient way to treat certain *biodegradable* industrial waste (waste that can be broken down by microorganisms), such as oily petroleum waste and some organic wastes from chemical plants. A good indicator of a waste's suitability for land application is its *biopersistence*—how long it remains in the biosphere. The longer its biopersistence, the less suitable it is for land application. Land application is not an effective way to treat or dispose of inorganic substances, such as salts and heavy metals.[29]

When biodegradable waste is added to the soil, it is attacked by microflora (bacteria, molds, yeasts, and other organisms) that decompose the waste material in a process known as *microbial breakdown*. The soil thus can be thought of as a microbial farm that constantly recycles organic and inorganic matter by breaking it down into more fundamental forms useful to other living things in the soil. Because the upper soil zone contains the largest microbial populations, land application is restricted to the uppermost 15–20 centimeters (6–8 inches).[30]

Surface Impoundment

This is a controversial way to store or dispose of hazardous waste. Both natural topographic depressions and human-made excavations have been used to hold hazardous liquid waste in a method known as *surface impoundment*. The depressions or excavations are primarily formed of soil or other surface materials but may be lined with manufactured materials such as plastic. Examples include aeration pits and lagoons at hazardous-waste facilities.

Surface impoundments are prone to seepage, resulting in pollution of soil and groundwater, and evaporation from surface impoundments can also cause an air-pollution problem. As a result, many sites have been closed.

Deep-Well Disposal

Deep-well disposal is another controversial method. It involves injecting waste into deep wells. The well must go below all freshwater aquifers and be isolated from them, to assure that injected waste will not contaminate existing or potential water supplies. Typically, the waste is injected into rock layers several thousand meters below the surface.[31] Even where geologic conditions are favorable for deep-well disposal, there are a limited number of suitable sites; and within these sites, there is limited space for disposal of waste.

Summary of Land Disposal Methods

Direct land disposal of hazardous waste is often not the best initial alternative. Even with extensive safeguards and state-of-the-art designs, land disposal methods—including landfills, surface impoundments, land application, and injection wells—cannot guarantee that the waste will remain contained and not cause environmental problems later on. Pollution of groundwater is perhaps the most significant risk, because groundwater provides a convenient route for pollutants to reach people and other living things.

Figure 16.11 shows some of the paths that pollutants may take from land disposal sites to contaminate the environment. These paths include leakage and runoff to surface water or groundwater from improperly designed or maintained landfills; seepage, runoff, or air emissions from unlined lagoons; percolation and seepage from land application of waste; leaks in pipes or other equipment used for deep-well injection; and leaks from buried drums, tanks, or other containers.

16.10 Alternatives to Land Disposal of Hazardous Waste

We should be using a combination of methods. In addition to the disposal methods just discussed, chemical-waste management should include source reduction, recycling and resource recovery, treatment, and inciner-

ation.[32] Recently, some people have argued that these alternatives to land disposal are not being used to their full potential—that we could reduce the volume of waste and recycle the rest or treat it and then dispose of the treatment residues.[32] Advantages to source reduction, recycling, treatment, and incineration include the following:

- Useful chemicals can be reclaimed and reused.
- Treatment may make wastes less toxic and therefore less likely to cause problems in landfills.
- The actual waste that must eventually be disposed of would be reduced to a much smaller volume.
- The reduced volume of waste would place less stress on the dwindling capacity of waste-disposal sites.

Although some of the following techniques have been discussed as part of IWM, the techniques have special implications and complications where hazardous wastes are concerned.

Source Reduction: The object of source reduction is to reduce the amount of hazardous waste generated by manufacturing or other processes. For example, changes in the chemical processes, equipment, raw materials, or maintenance measures may reduce the volume of hazardous waste produced and/or its toxicity.[32]

Recycling and Resource Recovery: Hazardous chemical waste may contain material that can be recovered for future use. For example, acids and solvents collect con-

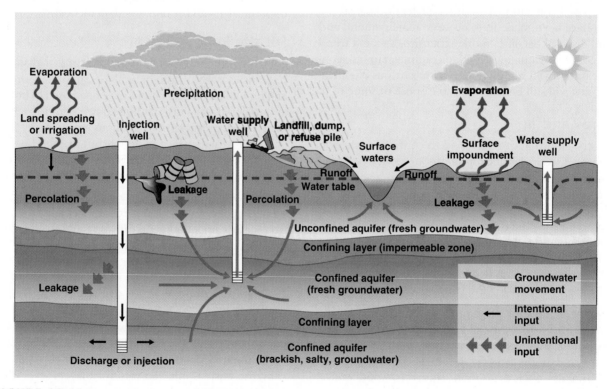

■ **FIGURE 16.11**
Disposal of hazardous waste can cause environmental contamination.
Examples of how land disposal and treatment of hazardous waste may contaminate the environment. [*Source:* Modified from C. B. Cox. The buried threat. Sacramento: California Senate Office of Research, No. 115-5, 1985.]

taminants when they are used in manufacturing processes. These acids and solvents can be processed to remove the contaminants and can then be reused in the same or different manufacturing processes.[32]

Treatment: Hazardous chemical waste can be treated by a variety of processes to change the physical or chemical composition of the waste and thus reduce its toxic or hazardous characteristics. For example, acids can be neutralized, heavy metals can be separated from liquid waste, and hazardous chemical compounds can be broken up through oxidation.[32]

Incineration: Hazardous chemical waste can be destroyed by high-temperature incineration (Figure 16.12). However, incineration is considered a waste treatment rather than a disposal method because the process leaves an ash residue, which must then be disposed of in a landfill. Hazardous waste has also been incinerated offshore on ships, creating potential air pollution and ash disposal problems for the marine environment.

16.11 Pollution Prevention

The early emphasis was on waste disposal. During the first few decades of environmental concern and management (the 1970s and 1980s), the United States approached waste management through government regulations and control of waste disposal. Waste was controlled by chemical, physical, or biological treatment and collection (for eventual disposal), by transformation into another chemical that is not harmful or can be used in other processes, or by destruction of pollutants. This was considered the most cost-effective way to control waste.

With the 1990s came a growing emphasis on *prevention*—**pollution prevention** involves finding ways to stop generating waste pollution rather than ways to dispose of it. This approach, which is part of materials management, reduces the need to manage waste, because less waste is produced. Approaches include the following:[32]

- Purchasing the proper amount of raw materials so that no excess remains to be disposed of.
- Exercising better control of materials used in manufacturing, so that less waste is produced.
- Substituting nontoxic chemicals for hazardous or toxic materials currently in use.
- Improving engineering and design of manufacturing processes so that less waste is produced.

These are often called P-2 approaches, for "pollution prevention." Probably the best way to illustrate the P-2 process is through a case history.[33]

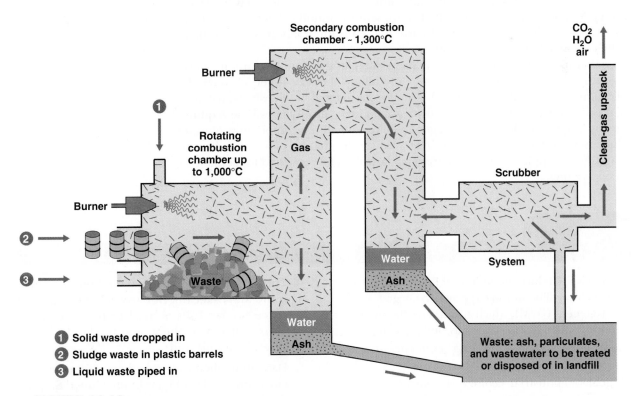

■ **FIGURE 16.12**
Incineration of hazardous waste:
Generalized diagram of a high-temperature incineration system.

Waste Disposal at a Cheese Company

A Wisconsin firm that produced cheese was faced with the need to dispose of about 2,000 gallons a day of a salty solution left over from the cheese-making process. Initially, the firm spread the salty solution on nearby agricultural lands—common practice for firms that could not discharge wastewater into publicly owned treatment plants. This method of waste disposal, if the solution was applied incorrectly, made the soil so salty that crops were damaged. As a result, the Department of Natural Resources in Wisconsin placed limitations on how much salt could be discharged onto the land.

The cheese firm decided to modify its cheese-making processes. It would recover salt from the solution and reuse it in producing more cheese. This required developing a process to recover the salt. The recovery process used an evaporator to reduce the salty waste by about 75%. At the same time, it reduced the amount of the salt the company had to purchase by 50%. The operating and maintenance costs for recovery were about three cents per pound of salt recovered, and it took only two months for the new equipment to pay for itself in savings. The firm saved thousands of dollars a year by purchasing less salt.

The case history of the cheese firm suggests that often small changes can result in large reductions of waste. And this story is not an isolated example—thousands of similar cases exist today as we move from the era of recognizing environmental problems and regulating them at a national level to providing economic incentives and new technology to better manage materials.[33]

Return to the Big Question

Is zero waste possible?

Zero waste is a theoretical possibility in the distant future. We know that zero waste is possible in ecosystems, because there is no waste but only resources. That is, materials produced as waste in some part of an ecosystem are used as resources in another part with energy flow and mineral cycling. Human systems, on the other hand, bring a lot of resources to areas where, as a result of rapid consumption, packaging, and other urban processes, we end up with waste. Much of that waste may be reused and recycled, but today nearly one-half is still sent to landfills. Extensive recycling and reuse can reduce the amount of waste that ends up in landfills to as little as 10–20% of the total waste we generate.

Just because zero waste is not generally possible at this time does not mean we should stop trying to eliminate waste and move toward the zero target. Through the application of industrial ecology, which is analogous to the flow of waste in an ecosystem, it is theoretically possible to reduce waste to near zero. What will be required are many innovative ideas and strategies to better use all the components of waste we currently produce, along with efforts to reduce the amount of waste we have to manage. For example, animal waste from pig farms can be converted to methane or even oil as a fuel.

Summary

- Waste disposal since the Industrial Revolution has progressed from dilution and dispersion to integrated waste management (IWM), which emphasizes the three R's: reducing waste, reusing materials, and recycling.

- Zero waste is an emerging concept of waste management.

- A goal of industrial ecology is a system in which "waste" doesn't exist, because waste from one part of the system would be a resource for another part.

- The most common disposal method for solid waste is the MSW landfill. But around many large cities, space for landfills is hard to find, people don't want to live near them, and the landfills must be carefully monitored.

- Hazardous chemical waste is one of the most serious environmental problems in the United States. Hundreds or even thousands of uncontrolled disposal sites could be time bombs that will eventually cause serious public-health problems. We know that we will continue to produce some hazardous chemical waste. Therefore, it is imperative to develop and use safe disposal methods.

■ Managing hazardous chemical wastes involves several options, including on-site processing to recover by-products that have commercial value, microbial breakdown, chemical stabilization, incineration, and disposal by secure landfill or deep-well injection.

■ Pollution prevention (P-2)—identifying and using ways to prevent generation of waste—is an important emerging area of materials management.

Key Terms

composting

environmental justice

e-waste

hazardous waste

incineration

industrial ecology

integrated waste management (IWM)

leachate

materials management

municipal solid waste (MSW)

MSW landfill

pollution prevention

reduce, reuse, recycle

waste stream

zero waste

Getting It Straight

1. What are links between industrial ecology and zero waste?

2. What are the 3 R's of waste management, and which is most important?

3. Would you approve the siting of a waste-disposal facility in your part of town? If not, why? And where do you think such facilities should be sited?

4. Is government doing enough to clean up abandoned hazardous-waste dumps? Do private citizens have a role in choosing where cleanup funds should be allocated?

5. Considering how much waste has been dumped in the nearshore marine environment, how safe is it to swim in bays and estuaries near large cities?

6. Should companies that dumped hazardous waste years ago, when the problem was not understood or recognized, be held liable today for health problems possibly related to their dumping?

7. What is pollution prevention?

8. What are the components of integrated waste management (IWM)?

9. What methods are available to manage municipal solid waste (MSW)?

What Do You Think?

1. Have you ever contributed to the hazardous-waste problem through disposal methods you use in your home, school laboratory, or other location? How big a problem do you think such actions are? For example, how bad is it to dump paint thinner down a drain?

2. Many jobs will be available in the next few years in the field of hazardous-waste monitoring and dis-posal. Would you take such a job? If not, why? If so, do you feel secure that your health would not be jeopardized?

3. How much garbage do you think you generate in a week? For seven days keep track of your garbage. Identify the total number of bags produced, how much waste is or is not recycled and how much waste per person is produced in your home.

Pulling It All Together

1. Suppose you found that the home you had been living in for 15 years was built over a buried waste-disposal site. What would you do? What kinds of studies could be done to determine whether there is a problem?

2. Do you think we should collect household waste and burn it in special incinerators to make electrical energy? What problems and what advantages do you see for this method compared with other disposal options?

3. Can a large city like New York reach their goal of reducing the export of waste from the city to near zero? Use the information in the case study at the beginning of the chapter to rationalize your answer. Do you think other large cities can achieve a zero goal?

4. Reaching "zero waste" on a global scale will require many things to happen and for the people of Earth to view and value differently what is now waste. List the things, views, and values and prioritize them to work toward zero waste. What is the value of education in the process?

Further Reading

Allenby, B. R. 1999. *Industrial ecology: Policy framework and implementation.* Upper Saddle River, NJ: Prentice Hall.—Primer on industrial ecology.

Ashley, S. 2002 (April). It's not easy being green. *Scientific American,* pp. 32–34.—A look at the economics of developing biodegradable products and a little of the chemistry involved.

Kreith, F., ed. 1994. *Handbook of solid waste management.* New York: McGraw-Hill.—Thorough coverage of municipal waste management, including waste characteristics, federal and state legislation, source reduction, recycling, and landfilling.

Rhyner, C. R., L. J. Schwartz, R. B. Wenger, and M. G. Kohrell. 1995. *Waste management and resource recovery.* Boca Raton, FL: CRC, Lewis.—Discussions of the archaeology of waste, waste generation, source reduction and recycling, wastewater treatment, incineration and energy recovery, hazardous waste, and costs of waste systems and facilities.

Watts, R. J. 1998. *Hazardous wastes.* New York: John Wiley.—A to Z of hazardous wastes.

Adalberto Roque/AFP/Getty Images, Inc.

17

Natural Hazards

Big Question

Why Are More of Them Becoming Disasters and Catastrophes?

Learning Objectives

Natural hazards dominated the news in 2004 and 2005, and with each new event, the toll in lives and damages grew higher. This chapter should help you understand . . .

- that natural hazards are both recognizable and predictable;
- that links exist between different hazards as well as between the physical and biological environment;
- that hazards that used to cause mostly disasters are now producing catastrophes;

- that risks from hazards can be estimated;
- that harmful effects of hazards can be minimized;
- that natural-hazard events also have natural service functions for ecosystems and people.

Case Study

La Conchita Landslide, 2005

Tragedy hit the small beachside community of La Conchita (Spanish for "little shell"), about 80 kilometers (50 mi) northwest of Los Angeles, California, on January 14, 2005. Ten people were killed and about 30 homes were destroyed or damaged when a fast-moving landslide roared through the upper part of the community (Figure 17.1).

The landslide was a partial reactivation of an earlier event, which occurred in 1995 and destroyed several homes but caused no fatalities. The winter of

2004–2005 was a particularly wet one, with high-intensity rainfall at times. However, neither residents nor local officials recognized that another landslide was imminent. What differentiated the 2005 mudslide from the one that occurred in 1995 was that this one moved farther into the community, trapping some people in their homes while others ran for their lives.

While this landslide took far fewer lives than recent hurricanes and earthquakes, which took thousands of lives and caused billions of dollars' worth of damage, there is a major lesson to be learned from La Conchita. The lesson is that what happened was predictable. Geologists say that the question is not *whether* landslides will continue to occur at La Conchita, but *when*. Landslides above La Conchita and to the east and west have been occurring for thousands of years, and a study of La Conchita and the surrounding area suggests that the landslides of 2005, 1995, and earlier are part of a much larger, complex, landslide that probably started moving about 20,000 years ago (Figure 17.2). The entire large, complex landslide is not moving today but parts of it are.

Putting it simply, the 200-meter-high (600-foot) slope directly behind the community is a serious continuing hazard for people living there. La Conchita should never have been constructed at the foot of the slope. Indeed, the community is built on about 15 meters (50 feet) of older landslide deposits.

We have learned from La Conchita that the potential for landslides can be recognized, and where they are likely to occur can be predicted. We have also painfully learned that there may be a very high price for living in areas known to be hazardous. Finally, when it comes to natural hazards that have the potential to produce catastrophes, we need to take steps well before that happens.

In the remainder of this chapter, we will discuss some of the principles of natural processes that are hazardous, and how they produce disasters and catastrophes. Through case histories, we will explore how unwise land uses and changing land uses, coupled with population increase, greatly increase the risk of some hazards.

©AP/Wide World Photos

■ **FIGURE 17.1**

Killer landslide in La Conchita, California (2005).
The landslide shown here was a rapid downslope movement of material mostly reactivated from a slide that occurred in 1995.

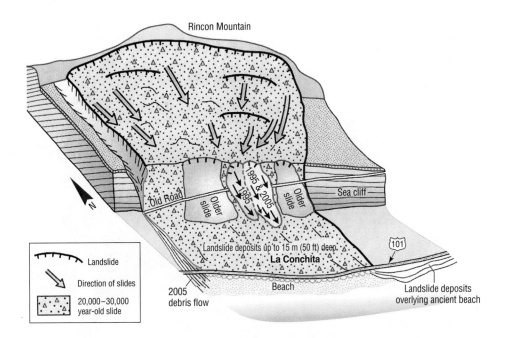

La Conchita, part of a much larger complex landslide.
This simple block diagram shows the slides of 1995 and 2005 at La Conchita,
California. The slides are a small part of a much larger ancient, complex landslide
small parts of which are active today. Such large, complex slides are all too frequent
and may be activated and reactivated periodically over thousands of years.

17.1 Hazards, Disasters, and Catastrophes

A natural process can become a disaster. Natural
processes cause physical, chemical, and biological
changes that modify the landscape. Some processes are
internal, such as earthquakes or volcanic eruptions,
driven by changes deep in the Earth. Other processes
occur much closer to the surface—such as landslides,
flooding, coastal erosion, violent storms, and wildfire.
Natural processes become hazards, disasters, or catastrophes when people interact with them or live and work
where they occur. Some natural processes may occur in
remote places where few people live. Such events, if they
significantly damage ecosystems, rise to the level of an
environmental disaster, if not a human disaster.

What is a natural hazard? We define a **natural
hazard** as any natural process that is a potential threat to
human life and property. A **disaster** is an event that occurs over a limited time in a limited geographic area and
causes significant losses of life and property. A **catastrophe**, in human terms, is a massive disaster, so serious that
recovery from catastrophic events requires very significant expenditures of money and time. For example,
Hurricane Katrina, which flooded the city of New Orleans and damaged part of the coastline of Mississippi in
2005, was one of the most damaging and costly catastrophes in the history of the United States. Recovery from
the event began in the days following the storm but may
take years. In developed countries such as the U.S.,
there are more-advanced warning systems, and buildings
may be designed to withstand earthquakes or high
winds. Economic losses from a natural-hazard event may
be high while loss of life is generally much lower than if
the event occurred in a developing country.

Is anyplace on Earth free of natural hazards? The
occurrence of major hazards in the United States is shown
in a general sense in Figure 17.3. Some major hazards are
not shown—such as blizzards, ice storms, droughts, and
wildfires. In essence, no area of the United States, or the
world, for that matter, is considered hazard-free. During
the past few decades, hurricanes, floods, and earthquakes
have taken the lives of several million people. Globally, the
annual loss of life from natural hazards has averaged about
150,000, and the financial losses have exceeded $50 billion
per year. And bear in mind that the above statistics do not
include social losses, such as unemployment, mental an-
guish, and reduced productivity.

Selected natural hazards in the United States, in terms
of human influence and potential to produce a catastrophe, are summarized in Table 17.1. The hazards that produce the greatest loss of property are not necessarily the

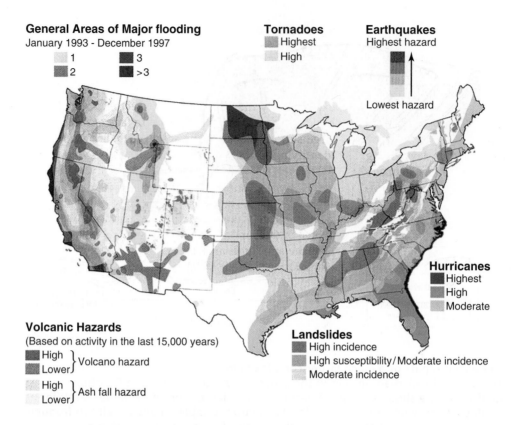

General Areas of Major flooding
January 1993 - December 1997

1   3
2   >3

Tornadoes
Highest
High

Earthquakes
Highest hazard

Lowest hazard

Hurricanes
Highest
High
Moderate

Volcanic Hazards
(Based on activity in the last 15,000 years)

High } Volcano hazard
Lower}

High } Ash fall hazard
Lower}

Landslides
High incidence
High susceptibility / Moderate incidence
Moderate incidence

▪ **FIGURE 17.3**
Selected natural hazards in the United States.
This simplified map of the United States shows areas at risk from hurricanes, earthquakes, landslides, tornadoes, and volcanic eruption.
[*Source:* U.S. Geologic Survey.]

□ **TABLE 17.1 SELECTED U.S. HAZARDS: OCCURRENCE INFLUENCED BY HUMAN ACTIVITY AND POTENTIAL TO PRODUCE A CATASTROPHE**

Hazard	No. of Deaths per Year	Occurrence Influenced by Human Use	Catastrophe Potential
Flood	86	Yes	H
Earthquake	50 + ?	Yes[1]	H
Landslide	25	Yes	M
Volcano	1	No	H
Coastal erosion	0	Yes	L
Expansive Soils	0	No	L
Hurricane	55	Perhaps[2]	H
Tornado and windstorm	220	Perhaps[2]	H
Lightning	120	Perhaps[2]	L
Drought	0	Perhaps[2]	M
Frost and freeze	0	Yes[2]	L
Wildfire	20[3]	Yes[4]	H

[1.] Human activity such as building reservoirs and pumping waste underground has caused small to moderate earthquakes.
[2.] Global warming with warmer sea temperatures increases intensity of hurricanes, storms, tornadoes, droughts, and frost.
[3.] Fire fighter deaths from 1990–1998. Includes: over-run by fire, heart attacks, aircraft accidents, heat stroke, falling, etc.
[4.] Global warming and urbanization in regions with a wildfire hazard increases the incidence of wildfire.
L = Low; M = Medium; H = High

Douglas C. Pizac/©AP/Wide World Photos

■ **FIGURE 17.4**
Urban earthquake in the Los Angeles area.
The Northridge earthquake in 1994 inflicted billions of dollars' worth of property damage and claimed about 60 lives. The damage shown here is to the freeway system in the Los Angeles urban area.

ones that cause the greatest loss of human life. The natural hazards that cause the greatest number of deaths per year are tornadoes, windstorms, lightning, hurricanes, and flooding.

Of course, there is a fair amount of interaction between hazards. For example, hurricanes often cause floods, both coastal and inland. Loss of life from events such as earthquakes in the United States is difficult to put a number on. When a large earthquake occurs, many people may be killed, and damage may amount to billions of dollars. For example, the 1994 Northridge earthquake in Los Angeles killed about 60 people and inflicted as much as $30 billion in property damage (Figure 17.4). Some experts on losses in urban areas suggest that a large urban earthquake in southern California could inflict about $100 billion in damage and kill several thousand people.

How most natural processes occur is moderately well known. Earth and atmospheric scientists have developed the science necessary to be more proactive with respect to natural hazards—that is, to take steps before trouble strikes. A common denominator of natural events that cause the greatest damage—such as earthquakes, volcanic eruptions, tsunamis, landslides, hurricanes, wildfires, and tornadoes—is the transport of material (water, air, and earth) and expenditure of energy. Heat wave and drought are two other natural hazards more related to weather and atmospheric processes. The various hazards are briefly described in Table 17.2.

The cost of natural disasters in the United States is between $10 billion and $50 billion per year, and a single major event may exceed the annual average by several times. Because the U.S. population is expected to increase, and

many people are moving from the interior toward the coasts, where hazards tend to occur, losses of life and property damage are likely to increase in coming decades.

Taking a Historical Point of View

History tells us that if we do not learn from past experiences, we will suffer the same consequences repeatedly. This is certainly true of natural hazards. Since many of them are repetitive events, their history provides us with basic information for any hazard-reduction program. Consider the history of La Conchita, the small town that was built below a slope that has experienced landslides for more than a century—in fact, the town was built on top of old landslides. Or consider flooding, one of the most common of all hazards that people are exposed to.

To assess the flood hazard of a river, look first at that river's history of flooding. We can examine flow records and aerial photos taken during times of floods, and evaluate the deposits and landforms produced by past flooding. In some parts of the world, people have been recording past floods for hundreds, even thousands, of years. In Egypt, early hydrologists kept careful records of the heights of floodwaters and were able to forecast when floods would recede and how high the floods were likely to be. That was critical for predicting crop yields in the Nile Valley. In Great Britain people marked on some cathedral walls the height of floods and the year they occurred. Marking high water from floods of the distant past has helped extend the record of flooding to the modern era, when we regularly measure the elevation and amount of water from floods.

☐ TABLE 17.2 BRIEF DESCRIPTIONS OF SELECTED NATURAL HAZARDS

▪ **Earthquakes**—result when rocks under stress from internal Earth processes (sea plate tectonics, see Chapter 3) rupture along faults. An earthquake can release more energy than a large nuclear explosion.

▪ **Volcanic eruption**—the extrusion of molten rock (magma) onto Earth's surface. Eruptions may be explosive and violent or less energetic lava flows. Volcanoes generally occur at boundaries between tectonic plates (see Chapter 3), where active geologic processes favor the melting of rocks and upward movement of magma. Some volcanoes also occur in more central parts of tectonic plates where "hot spots" deep below heat the rocks above. Examples of hot spots include volcanic activity at Yellowstone National Park and the Hawaiian Islands.

▪ **Landslide**—the downslope movement of soil and rock. Landslides occur when the driving forces that tend to move soil, rock, vegetation, houses, and other materials down a slope are more powerful than the resisting forces that hold the slope material in place. The resisting forces include the strength of the material on the slope (such as natural cementing material in rock and soil, and plant roots that bind the slope materials together). Weak rock, such as shale, on steep slopes that have weak resisting forces favor the development of landslides. The steeper the slope and the heavier the slope materials, the greater the driving forces are. If we make a slope steeper, we increase the driving forces. Resisting forces may be weakened by increasing the amount of water on or in a slope or by removing vegetation whose roots anchor the soil and rock.

▪ **Hurricane**—a tropical storm with circulating winds of more than 120 km (74 ml) per hour that move across warm ocean waters of the tropics. Hurricanes gather and release huge quantities of energy as water is transformed from liquid in the ocean to vapor in the storm.

▪ **Tsunami**—a series of large ocean waves produced when ocean water is suddenly disturbed vertically by processes such as earthquakes, volcanic eruptions, an underwater landslide, or the impact of an asteroid or comet. Over 80% of tsunamis are produced by earthquakes.

▪ **Wildfire**—a rapid, self-sustaining, biochemical oxidation process that releases light, heat, carbon dioxide, and other gases and particulates into the atmosphere. The primary cause of periodic wildfire is the vegetation itself. If there is no vegetation, there is no fire. Plant material and other fuel are rapidly consumed during wildfire (see Chapter 5). Since microbes in the environment usually can't decompose plants fast enough to balance the carbon cycle, fire can help to maintain a balance between plant productivity and decomposition in ecosystems.

▪ **Tornado**—a funnel-shaped cloud of violently rotating wind that extends downward from large cells of thunderstorms to the surface of Earth. Severe thunderstorms may occur when a cold air mass collides with a warmer one. Water vapor in the warmer part of the atmosphere is forced upward, where it cools and produces precipitation. As more warm air is drawn in, the storm clouds grow higher and thunderstorms grow more intense and form lines of storm activity (squall lines hundreds of kilometers long or large cells of updraft called *supercells*). In the United States, tornadoes are concentrated in the Plains states between the Rocky Mountains and Appalachian Mountains, where severe thunderstorms are more common. Parts of this region are often referred to as "Tornado Alley."

▪ **Flood**—the inundation of an area by water. Floods are produced by a variety of processes, including intense rainstorms, melting of snow, storm surge from a hurricane or tsunami, and the rupture of flood-protection structures, such as levees or dams. River flooding, one of our most commonly experienced hazards, shapes the landscape through erosion and deposition. Erosion has produced features as small as gullies and as large as the Grand Canyon of the Colorado River.

▪ **Heat wave**—a period of days or weeks with unusually hot weather. Heat waves are recurring weather phenomena related to heating of the atmosphere and movement of air masses. Human-induced global warming is a factor in the increased number and intensity of heat waves in recent decades.

▪ **Drought**—a period of months or years of unusually dry weather. Droughts are related to natural cycles of wet years that alternate with a series of dry years, but we don't fully understand why these cycles occur. Droughts in California are thought to be due to a shift every ten years or so in zones of high pressure that form in the central Pacific Ocean and the jet stream, allowing winter storms to extend south or remain farther north. Dry years in southern California occur when storm tracks remain north of central California for several years. Prolonged droughts in the midwestern states (such as the Dust Bowl that developed in the 1930s in Kansas and other regions) were associated with gigantic dust storms that commonly occur in desert regions. Droughts in central Africa have been devastating to human populations and are thought to be related in part to global warming associated with changing positions of air masses that move through the region.

Linking the past to the present and the future. Putting all the historical information together can help us make more informed judgments about the future flood hazard at a particular site. That is, we link the historical record with prehistoric and modern measurements to gain increased insight into the flood hazard.

Fundamental Concepts Related to Natural Hazards

Although a variety of natural processes may produce disasters and catastrophes, the following general ideas can help in understanding the nature and extent of hazards and how they might be minimized or eliminated at a particular area or site.

▪ Hazards are predictable.

▪ Links exist between different hazards and between the physical and biological environment.

▪ Hazards that used to cause mostly disasters are now producing catastrophes.

▪ Risk from hazards can be estimated.

▪ Harmful effects of hazards can be minimized.

Nature Can Play a Dual Role, Performing Natural-Service Functions and Posing Hazards

As we have noted throughout this book, nature performs many natural-service functions for people and the biosphere. For example, trees trap dust and other pollutants on their leaf surfaces, helping to clean the air. Wetland plants take up nutrients that otherwise could cause problems in the environment. Coastal wetland plants are also a natural buffer to winds and waves from storms that move inland.

Plate tectonics causes earthquakes but also builds mountains. The mountains offer some of the most spectacular scenery on Earth, and water flowing down hills and mountains moves sediments to river floodplains and other places where nutrient-rich soils may accumulate.

Volcanic eruptions can be catastrophic but also beneficial. Volcanoes emerging from the sea produce new land. In fact, the entire Hawaiian Island chain was created by volcanic processes over tens of millions of years (Figure 17.5). Volcanic ash may be transformed to young, fertile soils.

Floods, landslides, and dust storms are both hazardous and helpful. Huge dust storms in Africa and Asia cause problems and are hazardous to people. However, the dust may be transported thousands of kilometers, to enrich soils on far sides of the planet that would lose fertility without this periodic nourishment. Periodic

flooding brings silt and nutrients necessary for agriculture. For example, floodwaters in the Mississippi River valley nourish the floodplain and make the soils more fertile. Landslides, while often damaging, also perform some service functions, particularly where landslides form dams that produce lakes in mountains where otherwise lakes would be rare.

Crushing force can produce havens of tranquillity. The movement of continental plates creates faults (fractures in Earth's surface), crushing rocks beneath the surface. The crushed rock can block migrating groundwater, often forcing it to the surface. Along some parts of the San Andreas Fault in the very arid Coachella Valley, this produces desert oases, with pools of clear water surrounded by palm trees and sometimes inhabited by rare fish (Figure 17.6).

In sum, periodic disturbances are necessary. Without them, Earth would be much less interesting topographically and biologically: Soils might not be so fertile, water not so available, and the diversity of life not so great. And let us not overlook the important aesthetic aspects of natural disturbances that produce the splendid mountains, valleys, and seascapes that we flock to for leisure and for spiritual renewal.

USGS/Earth Surface Processes Team

Frank Balthis/Courier

■ **FIGURE 17.5**
Volcanic eruption producing new land in Hawaii.
Volcanic activity on the Big Island of Hawaii has added considerable land to the island in recent years. In fact, the entire island was built of volcanic rocks over a relatively short period of geologic time.

■ **FIGURE 17.6**
Earthquake faults dam groundwater, forcing it to the surface as springs or oases.
This photograph shows palm oases along the San Andreas Fault in California's Coachella Valley. These oases are refuges for a variety of plants and animals, some of which are endangered.

17.2 Natural Hazards Are Predictable

Mapping and monitoring are keys to spotting danger. Floods, landslides, volcanic eruptions, and earthquakes are natural processes that can be identified and studied by scientific methods. Most hazardous events and processes can be mapped as to where they have occurred in the past (as, for example, landslide deposits at La Conchita), and their present activity can be monitored. Using records of the location and frequency of past events, patterns in their occurrence, and observations of what happens before these events occur, we can often predict where hazardous events will occur in the future. In addition, once a particular event—such as a hurricane or tsunami (tidal wave)—has been identified, it is possible to forecast when it might arrive at a specific location.

For example, we can predict that the Mississippi River will flood in the spring in response to snowmelt. We can forecast when the river will reach flood stage and how fast the flood is likely to move down the valley toward the Gulf of Mexico. We have also learned that earthquakes often tend to be clustered in time and place. Therefore, we know the most likely place for a large earthquake to occur may well be where one recently happened.

Sometimes it is possible to forecast an event and issue a warning, as for example with the Mississippi River floods. Advance notice could have lessened the tragedy of the catastrophic Indonesian tsunami of 2004. A tsunami is an ocean wave that presents a serious natural hazard because of its potential size. Tsunamis are generated when ocean water is displaced vertically by a large underwater earthquake, volcanic eruption, or landslide. The tsunami of 2004 claimed the lives of nearly 250,000 people. Over three-quarters of the deaths were in Indonesia, which suffered not only from the tsunami but also from the violent shaking of the earthquake that caused the tsunami.

The site of the magnitude 9 earthquake and its generation of the tsunami, as well as its travel over several hours, are shown in Figure 17.7. Notice that the time from generation of the tsunami to its arrival in Somalia was about seven hours. It reached India in just two hours, and arrived

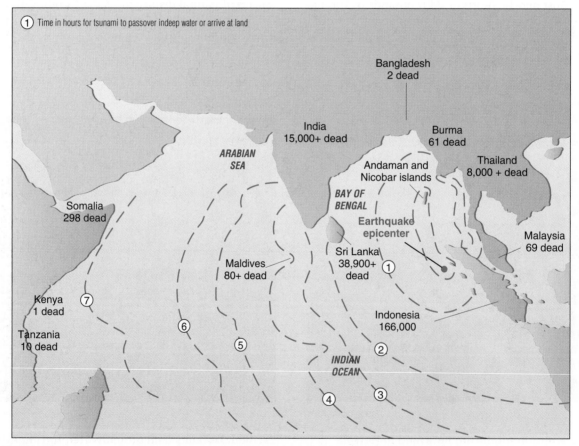

▪ **FIGURE 17.7**

Tsunami in December of 2004 killed several hundred thousand people in Indonesia.
This map shows the epicenter of the magnitude 9 earthquake that produced the Indonesian tsunami. Shown is the movement of tsunami waves that devastated many areas in the Indian Ocean. Notice that the waves took approximately seven hours to reach Somalia, where almost 300 people were killed. Most of the deaths were in Indonesia, where the waves arrived only about one hour after the earthquake. [*Source:* National Oceanic & Atmospheric Administration (NOAA).]

earlier or later in other places depending on their distance from the earthquake. The point is that if there had been a tsunami warning system in the Indian Ocean similar to that of the Pacific, warnings would have been triggered automatically. Even after the earthquake, when it was known that a tsunami was headed toward Africa, there were poor communication lines to warn people in its path.

The 2004 Indonesian tsunami completely destroyed some areas. Figure 17.8 shows Banda Aceh, Indonesia, before and after the tsunami, where almost total destruction occurred. The tsunami hit many places along the coastline where tourists were staying, sometimes in luxury huts constructed over the water, or in shoreline hotels. Without adequate warnings, deaths

Digital Globe/ Getty Images

■ **FIGURE 17.8**
Urban development completely destroyed by Indonesian tsunami of 2004.
Photographs here were taken before and after the tsunami that struck the Indonesian provincial capital of Banda Aceh on the northern end of the island of Sumatra. Essentially all of the development was damaged or destroyed by the tsunami and subsidence of the land as a result of the earthquake. Notice along the top of the photograph the beach with extensive erosion, leaving what appears to be a number of small islands where there had been a more continuous coast.

John Russell/Zuma Press

▪ **FIGURE 17.9**
Tourists running for their lives.
The man in the foreground is looking back at the tsunami rushing toward him that is higher than the building. The location is Phuket, Thailand. Many of the people living there, as well as the tourists, did not initially think the wave would inundate the area where they were, and when the waves arrived they thought they would be able to outrun the rising water. In some cases, people did escape, but all too often people were drowned.

were common. Figure 17.9 shows tourists attempting to run for safety from the rooftop-high tsunami rushing toward them.

What we learned from the Indonesian tsunami. We learned, first of all, that while tsunamis are rare in the Indian Ocean, they can have catastrophic consequences. Thus, a tsunami warning system is clearly needed in the Indian Ocean as well as in the Atlantic and Caribbean. The first part of a warning system was in place by early 2007. In addition, people must be educated about the tsunami hazard. People who had some education in geology or geography were able to recognize that when the sea suddenly receded, going way out from shore, it meant that a dangerously big wave was coming. In some cases they were able to warn others to evacuate before a deadly wall of water surged onto the shore. Some native people were also saved by tribal elders who recognized warning signs from tribal lore about tsunamis in the past and moved people inland.

The toll was made greater by growth in population and tourism. The human population has grown very rapidly in recent decades, as has the tourist industry. As a result, more and more people are living in the coastal zone where tsunamis occur. This makes it even more imperative that we develop better hazard-mitigation plans for areas where tsunamis are likely to recur.

The role of probability in hazard prediction may be significant. This is particularly important for rivers, where we have a long history of flows and floods. Based on that record, it is possible to estimate the probability of a flood recurring at intervals of 10 years to 100 years or more in a particular place. The probability of a 100 year flood is relatively low, about 1% (one chance in 100) per year. However, as we discuss in more detail in Section 17.5, it is like throwing dice, and it is possible for two 50 year floods or even two 100 year floods to occur in the same year.

Location and regional geography are also important. On a global scale, we know where most earthquakes and volcanic eruptions are likely to occur. We regularly map landslide deposits, and based on this we develop hazard maps showing where the risk to people and property is likely to be. We can accurately predict where flooding is likely, based on the landform known as the floodplain, which is flat land adjacent to the river, and by mapping that floodplain and observing the extent of recent floods.

17.3 Linkages Between Hazards and Between the Physical and Biological Environments

Considering potential linkages between hazards and the physical and biological environment is an important part of understanding consequences of natural hazards. First of all, hazards themselves may be linked. For example, volcanic eruptions often cause landslides. Blocking rivers with lava or landslide deposits increases the likelihood of flooding. Volcanic eruptions can profoundly change the landscape and ecosystems. The eruption of Mount St. Helens in 1980 severely disrupted the landscape and rivers. However, recovery since 1980 has been dramatic (Figure 17.10).

Here are some other examples of linked hazards. Hurricanes are huge storms that produce high winds and coastal flooding. When hurricanes move inland, voluminous rainfall may cause flooding and landslides in adjacent mountainous or hilly topography.

Roger Ressmeyer/Corbis Images

(a)

Layne Kennedy/Corbis Images

(b)

■ **FIGURE 17.10**
Recovery following the 1980 eruption of Mount St. Helens in southwestern Washington. (a) After the eruption, much of the region was devastated as trees were blown down and the area was transformed into a desolate looking landscape. (b) In the several decades since the major eruption, considerable natural restoration has taken place and mountain has grown due to addition of new volcanic rock and vegetation has began to regrow. The volcano still poses a threat for future eruptions.

Large submarine earthquakes can cause serious damage on land from seismic shaking, which can cause landslides. And as already discussed, large submarine earthquakes are one of the mechanisms that cause tsunamis.

Natural hazards are linked to earth materials. For example some materials, such as weak soils and rocks, are prone to landsliding, and volcanic eruptions of hot ash can melt snow and ice on the flank of the volcano and cause mudflows and flooding.

Links between hazards and the biological environment are common. Of particular importance are the disruption of ecosystems and the fragmentation of habitat by catastrophic events. On the global scale, the impact of a large extraterrestrial object may cause extinction of many species. On a regional scale, storms, wildfires, floods, and landslides disrupt habitats and disrupt ecosystems over variable periods. Hurricanes erode beaches and uproot vegetation in coastal salt marshes and wetlands, inflicting damage that may take years to recover from. Storm waves may also erode nearshore coral reefs, disrupting habitat for marine organisms. Wildfire kills vegetation, increasing soil erosion and landslides. Eroded sediment enters streams and fills pools, damaging habitat for fish. Severe floods can erode stream banks, widening the channel and removing vegetation that is important habitat for birds and mammals. Landslides may block entire valleys. If a landslide dam fails, downstream flooding may damage aquatic and floodplain environment.

Links to human health are likely after hazardous events. When people and animals are killed in large numbers, a sanitation problem may quickly emerge. Following a large earthquake, hurricane, tsunami, or flood, the water supply may become polluted, increasing the chances of water-borne diseases, such as cholera. Damaged buildings may also release toxic materials, and ruptured gas tanks and lines often cause fires. (Indeed, the Great Earthquake in San Francisco in 1906 is often referred to as the 1906 fire.)

17.4 Hazards That Used to Produce Disasters Now Produce Catastrophes

Human beings are creatures of the Pleistocene. We have evolved in the past few hundred thousand years from a species of small numbers to over 6 billion people today. Early in our history, survival was a day-to-day struggle as we interacted with the natural environment. When our numbers were not large and we had little effect on Earth processes, our losses from natural hazards were probably not as significant.

There are a lot more of us now. During the Holocene, the most recent 10,000 years of Earth's history, our numbers increased dramatically. Especially important was the development of agricultural practices about 7,000 years ago. With a more stable food base, our population increased to about half a billion, with several hundred times the density of people living in the hunter-gatherer period that preceded agriculture. By around the year 1800, we were in the early Industrial Period, and the population had doubled to about 1 billion. With industrialization, our cities grew. As we learned more about sanitation, our numbers continued to rise, until today there are more than 6 billion of us and quickly approaching 7 billion.

As the human population increased, cities grew larger. Today, 15 cities in the world have populations of more than 10 million people. Table 17.3 lists population increases since 1950 and projected growth by the year 2015 for these urban regions. Most are in areas vulnerable to several natural hazards. As the population has grown and the better building spots have been used up, people have been pushed into more hazardous areas, such as the steep hills above Los Angeles, California, which are subject to numerous landslides and an even more serious hazard, wildfire, which occurs every few decades.

Increased population density makes hazards more dangerous. As an example of how population trends can elevate natural hazards from disasters to catastrophes, consider the 1985 eruption of Nevado del Ruiz in Colombia. When Nevado del Ruiz erupted 140 years earlier, in 1845, it produced a disastrous mudflow (a very fluid landslide) that roared down the mountain, killing about 1,000 people in the Lagunilla River valley. The deposits from that flow created level ground in the valley, enticing people to move there, and an agricultural center was established. By 1985 the population of Armero, the center of the agricultural activity, had grown to about 23,000. But on November 13 of that year, the volcano again erupted, and again produced a large mudflow that roared through the river valley. This time, however, the event was a catastrophe: The flow killed 21,000 people, essentially wiping out the town of Armero[1] (Figure 17.11).

Warnings went unheeded. In 1845 there had been no warnings of a volcanic eruption. But before the 1985 eruption, there were a number of indications that it might occur, including increased earthquake activity and hot-spring activity the year before. As early as July 1985, volcanologists had begun to monitor the volcano, and by October 1985 they had completed a hazards map and an accompanying report predicting the events of November 13. The risk assessment stated that there would be a potentially damaging mudflow if the expected eruption occurred. The hazards map was circulated but largely ignored. After the eruption, it took about two hours for the mudflows to reach the town, which was subsequently buried as buildings were swept off their foundations and people were killed in great numbers.

If there had been better communication between civil-defense headquarters and the local towns, and a greater understanding of the hazard, Armero could have been evacuated and thousands of lives saved. Today there is a permanent volcano observatory, and one hopes that the lessons learned from this event will help minimize loss of life in the future.[1]

Land Transformation and Natural Hazards

How we use the land may affect the size and frequency of hazardous events. Landscape transformations—such as changing forestland to agriculture or urban uses or large-scale logging—may turn what were formerly disasters into catastrophes. This is demonstrated by two events in 1998.[2]

□ **TABLE 17.3 POPULATION INCREASE OF SEVERAL CITIES (URBAN REGIONS) FROM 1950 PROJECTED TO 2015. *(POPULATIONS ARE IN MILLIONS OF PEOPLE.)***

City/Urban Region	Population 1950	Population 2000 (estimated)	Population 2015 (projected)
Tokyo, Japan	6.2	27.7	28.7
Mumbai (Bombay), India	2.8	16.9	27.4
Lagos, Nigeria	1.0	12.2	24.4
Shanghai, China	4.3	13.9	23.4
Jakarta, Indonesia	2.8	9.5	21.2
São Paulo, Brazil	2.3	17.3	20.8
Karachi, Pakistan	1.1	11.0	20.6
Beijing, China	1.7	11.7	19.4
Mexico City, Mexico	3.5	17.6	19.0
Dhaka, China	4.0	18.0	19.0
New York City, U.S.	12.0	16.5	17.6
Calcutta, India	4.5	12.5	17.3
Los Angeles, U.S.	4.0	12.9	14.2
Cairo, Egypt	2.1	10.5	14.0
Buenos Aires, Argentina	5.3	12.2	13.9

Source: Data from B. Boyle Torrey. 2000. Forces of change. Smithsonian Institution. Washington DC National Geographic Society 2000, P. 16s.

N. Banks/USGS/Earth Surface Processes Team

Jacques Langevin/Corbis Sygma

(a) *(b)*

■ **FIGURE 17.11**

A volcanic eruption triggered a mudflow that killed 21,000 people.
(a) Nevado del Ruiz as viewed from the northeast on December 10,1985, a month after the destruction of Armero. The white plume is a minor eruption rising from the summit crater, where large mudflows were generated. A previous eruption in 1845 produced a mudflow that killed about 1,000 people. This event shows how natural hazards may become more dangerous in more densely populated places, especially if warnings and lessons are not heeded. (b) The town of Armero, in Colombia, was nearly destroyed by a mudflow when Nevado del Ruiz erupted on November 13, 1985. The rectangular pattern in the upper part of the photograph outlines the building foundations visible through mudflow deposits.

One 1998 event was the flooding of the Yangtze River in China, which took approximately 4,000 lives. The floods were probably made worse by land transformation. In the years prior to the flood, there had been a loss of about 85% of the forest in the upper Yangtze River basin as a result of both timber harvesting and transformation of the land to agricultural uses. As a result of these land-use changes, the amount and speed of runoff increased, and flooding became much more common than it was previously.[2] Recognizing the cause of the flood hazard, China banned timber harvesting in the upper Yangtze River basin and allocated several billion dollars for reforestation.

The other 1998 event, Hurricane Mitch, is a similar story. The hurricane destroyed large areas in Central America, particularly Honduras, causing about 11,000 deaths. In Honduras, approximately one-half of the nation's forest had been removed, and a large fire had occurred prior to the hurricane. The deforestation and fire weakened the materials on the hill slopes, and they washed away when intense rains from the hurricane arrived. Along with the slopes went the homes, roads, farms, bridges, and other infrastructure necessary for human existence.

The lesson in both cases is that care of the landscape and conservation of landforms and ecosystems are necessary if we want to live in harmony with natural processes. Doing otherwise is to invite catastrophe.

Hurricane Katrina: One of the Worst Natural Catastrophes in U.S. History

Hurricane Katrina has been labeled an American tragedy. The hurricane off shore was a Category 5 out of a possible 5. Hurricanes grow fiercer and more dangerous as they scale up from 1 to 5, with greater wind speeds and a higher storm surge (the rise in ocean water pushed ashore by the hurricane). In general, a Category 2 hurricane causes about 10 times as much damage as a Category 1, a Category 3 causes about 100 times as much, a Category 4 about 1,000 times as much, and a Category 5 about 10,000 times as much. Category 4 and 5 hurricanes can produce catastrophes, with great destruction.

Hurricane Katrina made landfall in the early evening of August 29, 2005, just to the east of New Orleans as a category 3 storm (in terms of wind velocity). The storm produced a category 5 storm surge (waves of water pushed in front of and with the storm) 3 to 6 meters (3–20 feet) high. Much of the coastline of Louisiana and Mississippi was devastated, coastal barrier islands and beaches were eroded, and homes were destroyed. At first it was thought that the famous old city of New Orleans had dodged a bullet again, since the hurricane did not make a direct hit. However, the situation changed unexpectedly when water from Lake Pontchartrain, north of the city and connected to the Gulf of Mexico, flowed through failed levees and flooded the city. Approximately 80% of New Orleans was underwater, ranging in depth from knee-deep to rooftop level or higher.

The city had spread out into low-lying areas vulnerable to flooding. And when the worst happened, the people who took the brunt of it were those who could have evacuated but didn't, and those who couldn't evacuate because they lacked transportation. New Orleans, with its population of about 1.3 million people, became a catastrophe of gigantic proportions. About the only part of the city that wasn't flooded was the French Quarter (Old Town), the area of New Orleans famous for music and Mardi Gras. The people who built New Orleans over 200 years ago realized that much of the area was low in elevation and they built on the higher places, which were on natural levees of the Mississippi River. Only as marshes and swamps were drained did the city expand into the lower areas where the flood hazard was much greater. Ultimately, much of the city lay in a natural bowl, with parts of it a meter or so (3 to 9 feet) below sea level (Figure 17.12).

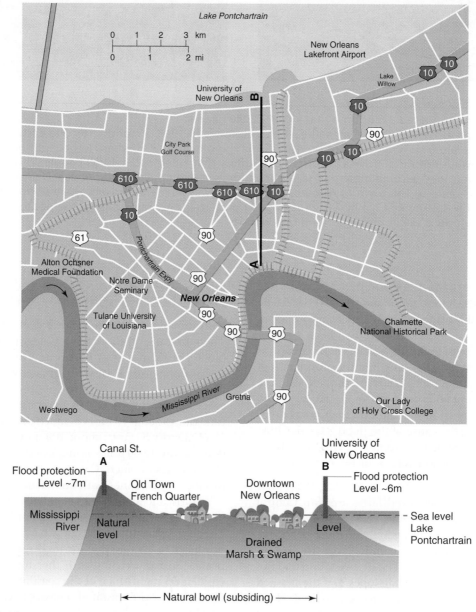

■ **FIGURE 17.12**

New Orleans is between Lake Pontchartrain and the Mississippi River.
The city's location is hazardous because, as a result of loss of wetlands and extraction of oil, much of the area is sinking and has become more vulnerable to flooding in recent years. Floodwalls and levees on the banks of the river and lake were designed for a Category 3 hurricane. Unfortunately, Hurricane Katrina in 2005 was a Category 3 with a category 5 storm surge, and 80% of the city flooded.

There had been warnings, but money was short. It had long been known that if a strong hurricane made a direct or near-direct hit on the city, extensive flooding and losses would result. The warnings were not completely ignored, but insufficient funds were forthcoming to maintain the levees and the system of flood walls to protect low-lying areas of the city from even a Category 3 hurricane.

Another problem was that the whole area is sinking (subsiding)—it has sunk as much as one meter (more than three feet) in the last 100 years, while at the same time sea levels have risen. The subsidence has been caused by a number of things, including extraction of oil and gas and the fact that freshwater wetlands compact and sink when they are drained or when they do not receive sediment from the Mississippi River. Because the Mississippi is artificially leveed (lined with embankments) and no longer delivers sediment to the wetlands, they have stopped building up from sediment accumulation. The wetlands had been a buffer against winds and storm waves. It is well known that one of the natural public services of both saltwater and freshwater coastal wetlands is to protect inland areas from storms. But the freshwater wetlands near New Orleans had been largely removed during past decades.

Damage to homes and belongings was a near-total loss in much of New Orleans (Figure 17.13). The number of deaths by late September was several hundred and still rising as the city was being pumped out to the point where structures could be searched for survivors and bodies. All told, deaths from Hurricane Katrina exceeded 1,800, and property damage and costs to rehabilitate or rebuild exceed $100 billion, making Hurricane Katrina the most costly natural catastrophe in the history of the United States.

There were environmental consequences as well. Think about what is in an average home or building, and remember also the number of vehicles that were left behind. In addition, oil refineries and other facilities were damaged and oil spills were inevitable. All told, a wide variety of chemicals, including gasoline and oil, were released into the stagnant waters that filled the city, and some areas were left covered with thick, oily sludge after the water receded (Figure 17.14). What New Orleans had was a gigantic toxic soup containing a lot of organic material, including bodies of animals and people and everything that goes with humanity.

Afterward, the heavily polluted floodwaters were pumped out of the city and into Lake Pontchartrain, which drains into the Gulf of Mexico. The volume of

■ **FIGURE 17.13**
A city flooded.
Shown here is part of the city of New Orleans that was flooded in September 2005 when Hurricane Katrina caused high waters to overtop the city's flood protection.

▪ **FIGURE 17.14**

New Orleans floodwaters were polluted.
Shown here is a person wading in polluted water surrounded by oil slicks. All sorts of toxic materials entered the floodwaters, which was deepest in the lowest parts of the city. The oil here probably came from automobiles or other vehicles.

water in the lake is large compared with the volume of floodwater pumped into it. As a result, early work suggests that serious water pollution of the Gulf of Mexico did not occur.

New Orleans will likely be rebuilt, though perhaps on a smaller scale. Many people who were evacuated to other states will find work in their new locations and decide not to return. Others, however, remain eager to go home to this famous historic city. Flood-protection levees can be constructed higher and stronger. Homes in low-lying areas can be built on stilts with garages below the living areas, allowing flooding below without damaging homes. Hopefully we and they will have learned something from this event and will take steps to make this kind of catastrophe less likely in the future.

Can it happen again even if stronger flood defenses are built? Of course it can—when a bigger storm strikes, damage is inevitable. But if freshwater marshes are restored and the Mississippi's waters are allowed to flow through them again, the wetlands will help to buffer the city against winds and waves. Making New Orleans more resistant to future storms will be expensive. But in light of the many billions of dollars in dam-

ages caused by catastrophes, it seems prudent to spend the money beforehand to protect people and important resources, particularly in our major cities.

17.5 Risk from Hazards Can Be Estimated

A simple way to define *risk* of a hazardous event is in terms of the probability of its occurring and the consequences (loss of life and property) if it does occur.[3] We will discuss each of these in turn.

First, we try to determine the probability of an event's time and place. Although we don't know exactly when a disaster will occur, we can estimate its frequency and, to some extent, when to expect it. As we mentioned earlier, it's not unlike games of chance. For example, mathematicians have long known that the most likely number to come up from a roll of two dice is seven. So the safest bet (and therefore usually the least profitable bet) is on the number seven. In the same way, although the rules may be a lot more complicated, we make a "bet" as to when a flood, earthquake, hurricane, or other natural disaster will occur. Insurance companies stay in business by trying to make these kinds of bets as accurately as possible. It's often hard, and sometimes the estimates turn out wrong and companies lose a lot of money. This happened when serious wildfires occurred in California more frequently than the historical record suggested. Some of the world's biggest insurance companies had a bad year that time.

We can also estimate the cost of a disaster (something insurance companies also try to do as well as they can). A lot of applied mathematics has been developed to help estimate risks of all kinds. For example, the simplest formula is to multiply the probability of an event by its likely consequences (costs in economic terms) when it does occur.

There are two ways to go about estimating risks. One is to develop a scientific or mathematical theory (as with the role of dice). The other is to make use of a long history of events to see how often severe events occur. As we saw earlier, this is done for river floods. Where a long record of river flow exists, this can be examined to see how often events of a certain severity occurred. Typical planning is often based on the site of the 100 year flood so experts talk about the probability of the 100 year flood, which is one chance in 100 for any given year.

Predicting a hurricane's intensity is difficult. Maps are available that help us estimate the number of hurricanes likely in a particular season and the probability of a hurricane impacting the East and Gulf coasts of the United States. However, less is known about what the intensity of the storm will be. We are all aware that scientists with the National Hurricane Center in Florida monitor hurricanes approaching the U.S. and make predictions as to the intensity of a storm and where landfall is most likely. The closer the storm gets to land, the better the predictions are. However, it is not unusual for a storm to change path and decrease or increase in intensity. Hurricane Katrina went from a Category 1 storm to a Category 5 in a day or so.

We are making progress in forecasting earthquakes. We can estimate the probabilities of future earthquakes based on the prehistoric and historical record of past earthquake activity. However, this area is highly speculative, and the more we learn about earthquakes, the more we realize that they tend to be clustered in time and geographically and thus are not completly-random events, such as most floods.

To estimate the consequences, we estimate property damage and loss of life. For example, before Hurricane Katrina, a number of studies had been published as to what would happen if such a large hurricane ever struck New Orleans. Those studies were right on! They predicted that the low-lying city, located between Lake Pontchartrain and the Mississippi River, would probably flood, and that the consequences of that flood would be catastrophic. As in the case of the volcano Nevado del Ruiz in Colombia, the hazards maps and analysis were also on target, in this case stating that debris and mudflows were almost a certainty and that the consequences would be catastrophic for the river valleys affected. Finally, it is also fairly straightforward to map development on floodplains, such as along the Mississippi River, and estimate what the losses would be if these areas were inundated.

Determining *acceptable risk* is more complicated. With *acceptable risk*, we are talking about the risks that individuals or society are willing to take. For example, we know that driving a car can be dangerous, but most of us accept that risk for the convenience of getting around our cities, getting to work, and going on trips. The risk of being injured or dying in an automobile accident is high compared with the risk of being injured or killed in a nuclear power plant accident. However, the risk from the accident at the nuclear power plant is often unacceptable to people, and as a result people in the United States have turned away from nuclear power. At a personal level, we make choices as to which risks we are willing to take. At the institutional level, what is acceptable risk is an economic decision. A bank decides to loan money for development based on the risk. You may not get a loan for building in a flood-prone area without flood insurance. Your decision might be not to build in hazardous areas, insurance or not.

Everyone knows that hurricanes tend to strike along the coast of the southeastern United States and along the Gulf Coast. Those who choose to live there accept this risk and rely on forecasts, hurricane shutters, official warnings, and evacuation plans before the storm season arrives. Similarly, it is well known that earthquakes are much more likely in California than in other parts of the country, but that is a risk that millions of Californians accept. Their reasons are many—the climate, the beaches, the spectacular scenery, perhaps a good job. The same can be said for the eastern coast of North Carolina, Georgia, and Florida, and into the Gulf states, where hurricanes are common.

In sum, risk analysis is increasingly used in evaluating natural disasters and catastrophes. We know more about how to do risk analysis and are trying to determine and implement the measures that individuals and societies should take in response to the perceived risk. With this in mind, we will now consider how the adverse effects of hazards might be minimized.

17.6 Adverse Effects of Hazards Can Be Minimized

Active vs. Reactive Response

We have often focused mainly on what to do after a disaster happens—search-and-rescue, firefighting, and providing emergency food, water, and shelter. This is all well and good and obviously necessary. It is critical to provide support for people who need to be rescued or who are made homeless by such events.

We need to anticipate hazardous events and be proactive—that is, take steps beforehand to prevent a disastrous event or minimize its effects. Proactive steps include (1) land-use planning to limit construction in hazardous locations; (2) constructing floodwalls, levees, and other hazard-resistant structures; and (3) protecting ecosystems on coastal floodplains and wetlands that provide natural protection from such hazards as flooding and hurricanes.[4]

As we noted, if wetlands surrounding New Orleans and along the coastline had been maintained, they would have buffered the effects of the wind and the storm surge, helping to protect inland areas. If funds for improving the floodwalls and levees had been allocated as requested, the flooding might have been minimized. The floodwalls and levees were designed for a smaller event than Katrina, but it was recognized that a larger hurricane was possible, and that the consequences of such an event would be catastrophic. The response after the disaster struck could have been much better as well if better planning had been done in advance. Since we knew days ahead of time that the storm was coming, it is difficult to understand why it took so long to get aid to the people after the event occurred.

The U.S. is not the only nation that has problems with quick response. In 1995 a large earthquake struck Japan, and it was several days before the government responded there, which engendered a lot of criticism. Governments, like people, may go into shock when a catastrophe happens. However, because other countries are ill prepared does not mean that we should be. Certainly one of the lessons learned from Hurricane Katrina is that we have to be much more aware of what needs to be done before and after an expected catastrophe. This is the essence of proactive response.

Impact and Recovery from Disasters and Catastrophes

Hazardous events may affect society directly and indirectly. The **direct effects** are the people killed, injured, dislocated, made homeless, or otherwise harmed by the event. The **indirect effects** come afterward and often include donating money and goods, providing shelter for people, paying taxes to help finance recovery, and dealing with emotional distress. The direct effects are experienced by far fewer individuals than are the indirect ones, which may affect society as a whole.[5, 6]

A generalized view of recovery after disaster is provided in Figure 17.15. Notice that the first two weeks after a disaster are often spent in a state of emergency where normal activities cease or are changed. This is the period when power may be out; we may be pumping water out of low-lying areas; and people are being rescued and sent to shelters. This period may last much longer than two weeks in some catastrophes. It generally gives way to the restoration period, a somewhat patchy period in which there is some return of functions. This is followed by reconstruction that may take several years or longer.

When should we begin taking proactive measures? The best time to take proactive steps to avoid similar events in the future is during restoration and reconstruction. A case in point is the 1972 flooding of Rapid City, South Dakota. Flash floods downstream of a flood-control reservoir killed over 200 people and destroyed many homes on the floodplain. Unfortunately, the dam gave the community a false sense of security. Below the dam was a tributary to Rapid Creek, and that is where intense rain fell, causing the flood.

In Rapid City, the most important measure was a change in land use. The flood was a catastrophe for the city. But instead of rushing to restore and reconstruct the hazardous area, the people waited a number of weeks while they thought out a solution that would minimize the problem in the future. As a result, today Rapid City uses land on the floodplain in an entirely different way—for greenbelts, golf courses, and other activities appropriate

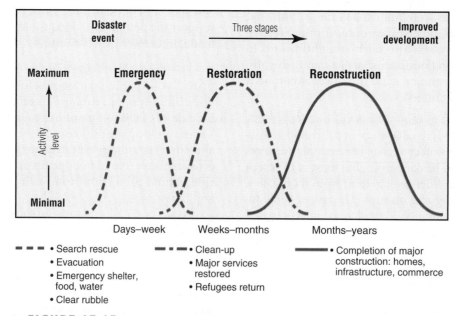

■ **FIGURE 17.15**

Recovery following disasters may take years.
Shown here is a highly generalized idea of what happens after a disaster, from emergency to restoration and reconstruction. The period of emergency lasts from a few days to a few weeks, and is followed by restoration activities, which can take a number of months or longer. The longest part is the reconstruction, which may take several years or more. [*Source:* Modified from R. W. Kates and D. Pijawka. *From rubble to monument: The pace of reconstruction.* Cambridge, MA: MIT Press, 1977.]

for land that is likely to flood again. In short, Rapid City has done what it can to minimize the flood hazard.[7] One Rapid City flood survivor, when asked if she was going to rebuild, said yes and pointed to an area high above the floodplain, far from future floods!

Perceiving, Avoiding, and Adjusting to Hazards

People tend to believe that bad things happen to others, not to themselves. They may believe that some other slope, not the one right behind *their* house, will fail in a landslide, and that their town's peacefully flowing river is unlikely to become a raging torrent.

It is probably not healthy to dwell obsessively on all the dire possibilities. However, many times the threat is very real, and we must carefully consider the probability of danger and do more to be ready for it. It is hard to tell a person who has lived in a house for 50 years on a floodplain and never experienced a flood that they live in a hazardous area and have simply been lucky so far. The same holds for landslides. After the 1995 landslides in La Conchita, California, many people believed that once it was over, the slope was much safer and that they could move back in. Geologists warned that in fact landslides were very likely to occur again. Another landslide occurred, in 2005, and ten people lost their lives.

Laws can protect people who fail to see a hazard. Since people so often fail to perceive the threat from hazards, our cities and states often have laws and regulations that help to ensure public safety by controlling what can be built in particular locations. In California we have setbacks from active faults that may rupture during earthquakes. In many other places, floodplains are regulated so that people will not build homes in the most hazardous places. Finally, because landslides are common in many places, particularly in southern California, a detailed geologic evaluation of slope stability is required prior to construction. Some homes are not allowed on slopes that are deemed unstable, while others require engineering solutions to stabilize the home site.

One of the best adjustments to natural hazards is to avoid them. This means not building our homes on floodplains, on or near active landslides, or directly over active faults. Land-use planning is one of the best tools we have for avoiding some hazards. For example, we may require setbacks from the ocean so that homes built there will not soon erode into the sea. This takes an evaluation by geologists and other scientists, and often laws requiring setbacks estimated to afford adequate protection against, say, 60 or 100 years of expected erosion. Of course, even with setbacks, erosion will continue, and structures will eventually be threatened.

We commonly adjust to hazards in a number of other ways as well. These include attempts to control natural processes (through, for example, dams and levees); insurance; evacuation; disaster preparedness; and doing nothing. We are not going to dwell on the doing-nothing option, but unfortunately it is the option many people choose with regard to natural hazards. People know these events are going to occur, but they are optimistic and assume that they will not be impacted. This can be dangerous thinking (remember La Conchita's landslide). Some people who had ridden out earlier hurricanes along the Mississippi Gulf Coast and declined to evacuate ahead of Hurricane Katrina said later that it was the worst decision they ever made!

Insurance is an important adjustment to natural hazards. If you live in a flood area, you may be required to purchase federally subsidized flood insurance. If you live in an area prone to earthquakes, you have the option of buying earthquake insurance. One problem with flood insurance programs has been that following floods, insured people will rebuild in hazardous areas. Steps are being taken to try to discourage this, but we still build far too many structures on floodplains that are likely to soon be inundated.

For example, terrible floods along the Mississippi River in 1993 caused billions of dollars' worth of damage. Some towns were relocated afterward, but in the St. Louis area, where floodplain regulations are weak, over 20,000 new homes and buildings have been constructed on land inundated by the 1993 flood. The residents are relying on new flood-control structures to protect them, but this strategy cannot be completely successful, because eventually there will be floods larger than the new protective structures are designed to control, and also because after a while the levees and walls may weaken and be more vulnerable to failure. A number of other communities have come to the conclusion that floodplain regulation must be coupled with structural controls if we are to minimize the hazard.[8, 9]

Another important adjustment to catastrophes is evacuation. Hurricanes can often be spotted days or weeks prior to arrival, and their position of landfall, as well as their strength, can be fairly well predicted. For people living in a hurricane's predicted path, evacuation is a wise decision. In fact, before Hurricane Katrina made landfall, much of the population of the Gulf Coast and New Orleans did evacuate. Unfortunately, however, many did not. A very large number of people were unable to evacuate because they had no means of transportation and no place to go, or were too ill or elderly to leave. Some were in hospitals and nursing homes. And some just didn't believe they wouldn't be safe in their homes.

Plans to evacuate people in buses are now being formulated as a result of Katrina. Such plans must take into account people who may not be able to evacuate without help—people in hospitals, nursing homes, and other areas where assistance will be necessary to get people out before the storm hits.

Disaster preparedness is important in minimizing effects of hazards. This is where we have a good chance to be proactive rather than reactive. Water, food, and medical items should be stockpiled, and vehicles, including helicopters and trucks, must be on standby. This is particularly true if we know a large event is likely to occur, such as a hurricane or a downstream flood wave. Good disaster preparedness requires good communication at all levels, from individuals who are being affected by the hazard to city, state, and federal authorities.

It is especially important to have a working chain of command so everyone knows who is in control. Lack of central control was evidently a problem in the aftermath of Hurricane Katrina and greatly complicated efforts to evacuate, rescue, and sustain people affected by the hurricane, particularly in the city of New Orleans. People outside the catastrophe could only stand and watch, horrified by the pain and suffering inflicted upon people in New Orleans and by the length of time it took for rescue and support to arrive. One can only hope that we have learned from this catastrophe and will never experience such a miscommunication in disaster preparedness again.

A final adjustment to hazards is to try to control them—for example, by building dams to hold water back and reduce flooding. Dams are all well and good, unless there are downstream tributaries that are likely to cause flooding below the dams. This was the case with the Rapid City, South Dakota, floods of 1972. The dam had lulled people into believing there was little danger of flooding. The same is true for people who build homes behind levees along rivers. The levees are often nothing more than earthen walls (embankments) that allow the flow in the river to be higher before it actually spills onto the floodplain. Many of these people have been led to believe they are protected by levees, only to find out later that in fact they are living in a very hazardous area.

Many people along the Mississippi and Missouri rivers live behind flood-control levees but it is common throughout the world. One person, when flooded out behind levees, stated that she was moving and that it would be stupid to rebuild on the floodplain even if new levees were constructed. Some people, at least, do emerge wiser from the experience.

Just upstream of Sacramento, the capital of California, the river is prone to periodic flooding[9]. In recent years many homes have been constructed behind levees, and some are between the riverbank and the levee (Figure 17.16), making eventual flooding almost inevitable. It is irresponsible for government to allow such widespread development on active floodplains, and yet it continues to happen almost everywhere.

Courtesy Ed Keller

(a)

(b)

▪ **FIGURE 17.16**

Homeowners are lulled into a false sense of security behind levees of the Sacramento River, in California.
(a) Shown here are homes built behind a large levee in the Sacramento Area, California. These homes are essentially sitting ducks subject to potential future flooding. (b) Also shown is a vulnerable home between the riverbank and levee with some floodproofing because the home is elevated above the flood-prone area.

17.7 What Does the Future Hold with Respect to Disasters and Catastrophes?

Losses from disasters and catastrophes keep setting new records. If we look at it by the decade, we can see that the number of disasters has increased significantly in the last half-century (Figure 17.17). This trend will likely continue as the human population increases and more people are in harm's way.

We may be directly causing the greater severity of some hazards. For example, the oceans of the world are warming, partly in response to our activities. The warmer oceans feed greater energy into storms, so while the number of hurricanes has not increased in past decades, their intensity is apparently increasing. If storms continue to grow more intense, the damage they cause will increase as well. Given the likelihood of more catastrophes in the future, we must pay greater attention to being prepared. Anticipating hazards, rather than simply reacting after the event, is the way to help minimize the losses and reduce pain and suffering.

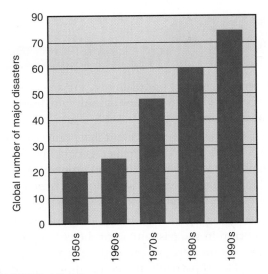

■ FIGURE 17.17
Major disasters are increasing.
The number of major disasters from 1950 to 1990 by decade is shown here. Notice the dramatic increase in the number of events. [*Source:* Data from J. M. Abramovitz. *Averting unnatural disasters*, 2001; in L. R. Brown, *State of the world*, 2001. World Watch Institute, New York: W. W. Norton, pp. 123–142.]

Return to the Big Question

Why are more hazards becoming disasters and catastrophes?

Natural hazards have long produced disasters, but what is new is that the events are now more likely than ever to produce catastrophes. One reason disasters are becoming catastrophes is that there are far more people living on Earth today in areas where hazards occur. A second reason is that we have sometimes made poor land-use choices that tend to increase the possibility of a catastrophe. For example, we built homes in New Orleans in areas below sea level, protected by a flood-management program that proved to be inadequate. We also removed wetlands on the ocean side of New Orleans that had helped to buffer winds and waves. For these reasons, the storm surge from Hurricane Katrina moved inland, contributing to increased flooding.

The tsunami of 2004 in the Indian Ocean was produced by a great earthquake offshore of Indonesia. The loss of life was horrific, at about 250,000 people. The huge number of deaths was in part a result of recent growth in population and tourism in the coastal areas inundated by the tsunami. The size of the tsunami assured that it would produce a catastrophe regardless of other factors, but the great number of people living in harm's way certainly increased the number of deaths.

Global change may also become important in changing what would have been disasters into catastrophes. For example, the ocean is warming, which feeds more energy into storms. A growing body of evidence suggests that while the number of hurricanes has not increased, their intensity has. As hurricanes become more intense, the potential for catastrophe increases. Therefore, global change, and in particular global warming, may be a factor in producing more catastrophes in the future.

Summary

■ Unwise land use, poor planning, and an ever-increasing population have turned hazards into disasters and catastrophes.

■ A disaster is a hazardous event that occurs over a relatively short period in a defined geographic area with significant loss of human life and property. A catastrophe is a massive disaster that requires large expenditures of money and time for recovery to take place.

■ By studying when and where hazardous events have occurred in the past and what the consequences of those events were, we can better prepare for the future. This is particularly true for those hazards that tend to occur repeatedly.

■ Some of the fundamental concepts are that (1) hazards are predictable; (2) linkages exist between different hazards and between the physical and biological environments; (3) the risk from hazards can be estimated; and (4) adverse effects of hazards can be minimized.

■ People are often overly optimistic about natural hazards and the likelihood that they will suffer damages. This is particularly true for those events that occur rarely.

■ It seems apparent that the human population will continue to increase, at least in the 21st century. As a result more stress will be placed on the natural environment as people look for places to live and work, and more people will be in harm's way. We must work harder to lessen losses of human life and property.

Key Terms

acceptable risk

catastrophe

direct effects

disaster

indirect effects

natural hazard

Getting It Straight

1. Why are more disasters becoming catastrophes?
2. What are typical responses to natural hazards?
3. What is the difference between reacting to and anticipating in hazard-reduction programs?
4. Does global warming have any influence on hazards such as hurricanes? Explain.
5. What were the main lessons learned from Hurricane Katrina, which devastated New Orleans in 2005?
6. What are some of the natural-service functions of natural-hazard events?
7. How wise is it to rely on trying to control natural processes? Explain.
8. What are the trends in death and damage caused by natural hazards?
9. What is the role of history in trying to understand natural hazards?

What Do You Think?

1. What are the potential hazards where your home is located? How are the hazards being evaluated, and what are the plans to minimize their effects?
2. If you were in charge of disaster preparation for New Orleans, how would you have prepared for the hurricane and possible flooding?
3. Do you think a natural disaster could strike your home or place where you live? What makes you feel this way about your community's safety or vulnerability?

Pulling It All Together

1. What do you think natural sciences, social sciences, and humanities have to offer in the attempt to reduce the consequences of future natural-hazard events? Which do you think are most important, and why?

2. Is it always possible to forecast natural events of tragedy? What can be done to help improve our system of tracking these events to help notify or portect community members?

3. Is the National Weather Service (NWS) really able to know when natural disaster will strike? What might help the NWS in understanding the occurrence of these natural events? How might they better communicate with the public?

Further Reading

Bolt, B. A. 2004. *Earthquakes*, 5th ed. San Francisco: W. H. Freeman.

Decker, R., and B. Decker. 2005. *Volcanoes*—4th ed. New York: W. H. Freeman.—Primer on Volcanoes.

Keller, E. A. and R. H. Blodgett. 2006. *Natural hazards*. Upper Saddle River, NJ: Prentice Hall. An introduction to natural hazards and their significance to society

Pinter, N. 2005. One step forward, two steps back on U.S. floodplains. *Science* 308:207–208.—Good example of how poor flood plan management is still problem in the U.S.

Pyne, S. J., Andrews, P. L., and Laven, R. D. 1996. *Introduction to wildland fire*, 2nd ed. New York: John Wiley & Sons, Inc.

Yeats, R. S. 2001. *Living with earthquakes in California: A survivor's guide.* Corvallis, OR: Oregon State University Press.—A good primer on the earthquake hazard.

Superstudio/Getty Images, Inc.

18

Environmental Economics

Big Question

Can We Put a Price on Scenic Beauty, Endangered Species, and the Quality of Life?

Learning Objectives

Other chapters in this text have explained the causes of environmental problems and discussed technical solutions. The scientific solutions, however, are only part of the answer. This chapter introduces some basic concepts of environmental economics and shows how these concepts have been applied in the analysis of environmental issues. After reading this chapter, you should understand . . .

■ when and how it is possible to put a dollar value on environment;

■ what the "tragedy of the commons" is, and how it leads to overexploitation of resources;

■ how our perception of the future value of an environmental benefit affects our willingness to pay for it now;

■ what externalities are, and why it is important to evaluate them in determining the costs of actions that affect the environment;

■ what factors may be involved in deciding how much risk to the environment and to human life is acceptable;

■ why it is difficult, yet important, to evaluate environmental intangibles, such as landscape beauty.

Case Study

We can easily find out the price of salmon on the table, but what is the economic value of salmon swimming in a river?

People like to fish for salmon and eat salmon, but many salmon-fishing streams are in trouble, and the causes remain unclear. Some blame overfishing, others blame deforestation, farming, urbanization, or changes in ocean conditions where the salmon mature (see Chapter 12, especially about *El Niño*).

How much is it worth to us to be able to fish for salmon and eat salmon? Can we put a dollar value on it? One way to do this is to figure out how much money people spend for recreation related to salmon in a specific stream and its watershed—how much money they spend to go fishing, or to watch fishermen fish, or to enjoy spending time along a stream filled with salmon.

A group of economists did this for all the rivers in Oregon that flowed into the Pacific Ocean—23 rivers—that are south of the great Columbia River. They considered all the ways that people spent money because there were salmon in a stream—money spent on fishing gear, hotels, buying meals, gasoline, gifts, postcards, and so forth. And they found some interesting things.

It turned out that the money spent within the watersheds of just 5 of the 23 rivers accounted for 62% of the total, and 10 of the 23 accounted for 84%. This suggested a new way to solve the problem of conflicting uses of these rivers and help salmon. In the past, the typical approach had been to try to solve all problems in every stream, to make it possible to fish for salmon, cut timber, grow crops, and build new homes in every watershed. Instead, people could agree that the 10 rivers that accounted for 84% of all the money spent on activities relating to salmon could be designated as salmon-fishing streams and watersheds and used just for that. Forestry, agriculture, urban development, and other activities could take place in the other 13 watersheds.

The state of Oregon has not yet accepted this suggestion, but this case does show that economic analysis can suggest new solutions to environmental problems and may play a useful role as we try to solve environmental problems.

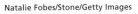
Natalie Fobes/Stone/Getty Images

■ **FIGURE 18.1**

American Indians fishing for salmon on the Columbia River.
As long as people have lived along the Columbia and other rivers that flow to the ocean in the Pacific Northwest, they have fished for salmon, and enjoyed it. Here, at Celilo Falls on the Columbia, Indians told the explorers Lewis and Clark in 1805 that they could take "salmon as fast as they wish."[1]

18.1 Some Environmental Dollar Values

The dollar cost of pollution control is rising. In the mid-1990s, the United States spent about $115 billion a year to deal with pollution, about 2% of the nation's gross national product. The defense budget was only two and a half times larger. Present costs are closer to $170 billion, including amounts spent by consumers, corporations, and government.[2] This total is much greater than the $6 billion budget of the Environmental Protection Agency (EPA).[3]

Although costly, cleaning our environment has economic benefits. Populations in areas with high levels of certain pollutants (people in inner cities, for example) have lower average life expectancies and higher rates of certain diseases. Particulate air pollution in U.S. cities contributes to 60,000 deaths annually,[4] and 2–9% of total mortality in cities is associated with particulate air pollution.[5] It is estimated that by the year 2010, amendments that Congress added to the Clean Air Act in 1990 will prevent 23,000 premature deaths in the United States, 1.7 million asthma attacks, and more than 60,000 hospital admissions due to respiratory problems. The dollar value of the benefits from these amendments alone is estimated to reach $110 billion in 2010, while the costs are estimated at only $27 billion.

It's easier to assess tangible factors than intangibles. As we found in other chapters, making decisions about the environment often leads us to think about both *tangible* and *intangible* factors. A mudslide that results from changing the slope of land is an example of a tangible factor. The beauty of the slope before the mudslide and its ugliness afterward are an example of an intangible factor. Of the two, the intangibles are obviously more difficult to deal with because they are hard to measure and to put a dollar value on. Nonetheless, evaluation of intangibles is becoming more important. One task of environmental economics is to develop ways to evaluate intangibles, ways that provide good guidelines, are easy to understand, and are quantitatively credible. Not an easy goal!

18.2 The Environment as a Commons

One would think that if we benefit from something, we'll make sure we can keep on benefiting. But often people do not use a natural resource in a way that maintains that resource and its environment in a renewable state—they do not seek sustainability. At first glance this seems puzzling. Why would people not act in their own best interest? Here are two reasons: First, because many environmental resources are held in common (shared), and, second, because of the low growth rates, and therefore low productivity, of many renewable resources. These two factors lead to long-term decisions that are environmentally unsound.

The first factor has to do with "the tragedy of the commons," a term coined by the ecologist Garrett Hardin.[6] It means that when a resource is shared, an individual's personal share of the pleasure or profit from exploiting the resource is usually greater than his or her share of the resulting loss.

A **commons** is land (or another resource) owned and used by the public, such as a park. In fact, in New England towns, the parklike town square is called "the commons". The word originally referred, however, to land owned by the public and used by all the farmers of towns in England and New England to graze their cattle. Sharing the grazing area worked as long as the number of cattle wasn't large enough to cause overgrazing, and you might think that people of goodwill would understand the limits of a commons. But take a moment to think about the benefits and costs to each farmer.

Each farmer tries to maximize personal gain, and must periodically consider whether to add more cattle to the herd on the commons. The addition of one cow has both a positive and a negative value. The positive value is the benefit when the farmer sells that cow. The negative value is the additional grazing by the cow. The benefit to the farmer of selling a cow for personal profit is greater than his share of the loss in the degradation of the commons. Each farmer's short-term game plan, therefore, is always to add another cow, and for a while, everyone seems to gain. But eventually the common grazing land is so crowded with cattle that the pasture is destroyed and everyone loses.

The point is, complete freedom in a commons inevitably ruins the commons. The implication seems clear: Without some management or control, all natural resources treated like commons will inevitably be destroyed. That's how ecologist Hardin saw it.

Today's commons include forests, fisheries, the atmosphere, and more. Of forests in the United States, 38% are on publicly owned lands. Resources such as ocean fisheries away from coastlines and the deep-ocean seabed, where valuable mineral deposits lie, are international commons not controlled by any single nation. We saw in Chapter 12 that most marine fisheries are in a mess, and now we see that this must be partly because they are commons. Most of the continent of Antarctica is a commons, although there are some national territorial claims there, and international negotiations have continued for years about conserving Antarctica and about the possible use of its resources.

The atmosphere, too, is both a national and international commons. Consider the issue of global warming and the role that human activity may play (discussed in Chapter 13). Individuals, corporations, public utilities, and nations add carbon dioxide to the air by burning fossil fuels. Just as Garrett Hardin suggested, people tend to focus on the short-term benefit to themselves, rather than on the long-term damage to the commons. The picture here is quite mixed, however, with much ongoing effort to bring cooperation on this issue.

Overuse of a commons may require societal change. In the 19th century, burning wood in fireplaces was the major source of heating in the United States (it is still the major way to provide heat in many nations). A fireplace or woodstove was enjoyed not only for its warmth but also for its beauty—sitting around a fire and watching the flames must have a long history in human societies.

But in the 1980s, with growing populations and widespread building of vacation homes and lodges for enjoying winter sports in states such as Vermont and Colorado, the burning of wood began to pollute the air locally. Especially in valley towns surrounded by mountains, the air became fouled, visibility declined, and there was a potential for effects on human health and the environment. As a result, some communities restrict or prohibit the use of fireplaces and woodstoves. The local air is a commons, and its overuse required a societal change.

Recreation is a problem of the commons, leading to overcrowding of national parks, wilderness areas, and other nature-recreation areas. An example is Voyageurs National Park in northern Minnesota. Located within the boreal forest biome of North America (see Chapter 8), the park has many lakes and islands and is an excellent place for fishing, hiking, canoeing, and viewing wildlife. Before the area became a national park, it was used for motorboating, snowmobiling, and hunting, and many people in the region made their living from tourism based on these kinds of recreation. Some environmental groups argue that Voyageurs National Park is ecologically fragile and should be legally designated a U.S. wilderness area to protect it from overuse and from damage by motorized vehicles. Others argue that the nearby million-acre Boundary Waters Canoe Area provides ample wilderness and that Voyageurs can withstand a moderate amount of hunting and motorized transportation.

At the heart of this conflict is the problem of the commons, which in this case can be phrased as: What is the appropriate public use of public lands? Should all public lands be open to all public uses? Should some public lands be protected from people? At present, the United States has a policy of different uses for different lands. In general, national parks are open to the public for many kinds of recreation, whereas areas designated as wildernesses restrict the number of visitors and the kinds of uses.

18.3 Low Growth Rate and Therefore Low Profit as a Factor in Exploitation

As noted, a resource's low growth rate is the second reason for overexploiting it.[7] For example, one economic view of whales and whaling is to consider them solely in terms of the value of whale oil. Whale oil, a marketable product, can be thought of as the capital investment of the industry. How can whalers get the best return on their investment?

Let us examine two approaches: resource sustainability and maximum profit. Remember, populations increase only if there are more births (biological productivity) than deaths. Thus, if whalers adopt a simple sustainability policy, they will harvest only the net biological productivity each year and thus maintain the whale population at its current level—in short, they will be able to stay in the whaling business indefinitely. In contrast, if they adopt a simple approach to maximizing immediate profit, they will harvest all the whales now, sell the oil, get out of the whaling business, and invest the profits (Figure 18.2).

The New Bedford Whaling Museum

■ **FIGURE 18.2**

Why didn't whalers try to conserve whales? Economics has answers. Nineteenth-century whaling ships like this one sailed from New Bedford, Massachusetts, around South America, then north into the Bering Sea to catch bowhead whales.

If they adopt the first policy, what is the maximum gain they can expect? Whales, like other large, long-lived creatures, reproduce slowly—typically, a female gives birth to one calf every three or four years. The total net growth of a whale population is unlikely to be more than 5% per year. If all the oil in the whales in the oceans today was worth $100 million in the current market, then the most the whalers could expect to take in each year would be 5% of this amount, or $5 million. Meanwhile, they would have to pay the cost of maintaining their ships and other equipment, interest on loans, and salaries of employees—all of which would lower their profit.

If they adopt the second policy, how much profit can they expect? If whalers harvest all the whales, not just 5%, then they can invest $100 million, not just $5 million. Although investment income varies, even a conservative investment of $100 million would very likely yield more than 5%, and this income would come without the annual cost of paying a crew, maintaining the ships, buying fuel, marketing the oil, and so on. In short, in terms of profit, the second approach is the most practical: Harvest all the whales, invest the money, and relax.

Whales simply are not a highly profitable long-term investment. No wonder there are fewer and fewer whaling companies, with many leaving the whaling business when their ships become old and inefficient. Few nations support whaling, and those that do have stayed with whaling for cultural reasons. For example, whaling is important to the Eskimo culture, and some harvest of bowheads takes place in Alaska. Whale meat is a traditional Japanese food, and the whale harvest is maintained by the Japanese for this reason.

Another factor to consider is the relative scarcity of a necessary resource. Scarcity affects the resource's value and therefore its price. For example, if a whaler lived on an isolated island where whales were the only food and he had no trade with other people, then he could not choose to sell off all the whales to maximize profit, since he would have no one to sell them to. Besides, he would need the food himself. He might decide to harvest them in a way that would maintain their population, or he might estimate that his own life expectancy was, say, only ten more years, and that if he timed it right he could harvest a greater number than they reproduced and they would not become extinct until he died. "You can't take it with you" would be his attitude.

If ships began to visit the island, he could trade, and thus begin to benefit from some of the future value of the whales. If ocean property rights existed so that he could "own" the whales that lived within a certain distance of his island, then he might consider the economic value of owning this right to the whales. He could sell rights to future whales and whaling, or mortgage them, and thus reap the benefits during his lifetime from whales caught after his death. Causing the extinction of whales would not be necessary.

We must think beyond immediate economic advantages. Policies that seem ethically good may not be the most profitable for an individual. Economic analysis clarifies how an environmental resource is used, what is perceived to be its intrinsic value, and therefore its price—and this brings us to the question of externalities.

18.4 Externalities: Costs that Don't Show Up in the Price Tag

This is a problem in the economic analysis of whaling. One gap in our thinking about whales, an environmental economist would say, is that we often don't factor in the **externalities**. An externality is a factor that is treated as a side effect of one activity and has consequences for another activity but is not reflected in market prices. In this way it is an indirect cost.[7, 8] The major environmental controversy about externalities is whether they should be treated as internal. If they were, their real costs to the environment would be considered in the market costs. Even where there is an agreement making an external economic factor internal, it's often difficult to figure out how to do this in a free, open market. A variety of ways have been tried, including placing a tax on an activity to cover the cost of its side effects.

Externalities are sometimes easier to understand when we look at specific cases. For example, think about global warming as discussed in Chapter 13. If burning fossil fuels is raising Earth's temperature, who is responsible for the costs to the environment, human health, or specific jobs?

As another example, in the case of whaling, externalities include the loss of revenue to "whale-watching" tourist boats and the loss of an ecological role played by whales in marine ecosystems. Economists agree that the only way for a consumer to make a rational decision is by comparing the costs against the benefits. But if not all of the costs are revealed, then the price tag will be wrong and the comparison won't be helpful.

Air and water pollution provide other good examples of externalities. Consider production of nickel in Sudbury, Ontario. Sudbury is home to the world's largest smelter, and the smelting has serious environmental effects (Figure 18.3). Traditionally, the economic costs of producing commercially usable nickel from an ore are the direct costs—that is, those borne by the producer and passed directly on to the user or purchaser. In this case, direct costs include building and maintaining

Thomas Nilsen/Photo Researchers, Inc.

■ **FIGURE 18.3**

Who should bear the costs of cleaning up air pollution from a nickel smelter?
Emissions from the famous nickel smelter at Sudbury, Ontario, the world's largest, killed trees for miles around, creating a barren landscape. The smelter's smokestack, shown here, is the tallest in North America and the second tallest in the world. Should the polluter pay, or should this be considered an externality, so that the cost is borne by society?

the plant, purchasing the ore, buying energy to run the smelter, and paying employees. The externalities—not taken into consideration in pricing the product—include costs associated with the smelter's degradation of the environment. For example, prior to pollution control, the Sudbury smelter destroyed vegetation over a wide area, which led to an increase in erosion.

Although air emissions from smelters have been substantially reduced and restoration efforts have begun to bring about a slow recovery of the area, pollution remains a problem, and total recovery of the local ecosystem may take a century or more.[9] There are costs associated with the value of trees and soil, and with restoring vegetation and land to a productive state.

Problem number one: What is the true cost of clean air over Sudbury? Economists say there is a lot of disagreement about the price, but all agree that it is larger than zero. Despite this, clean air and water are traded and dealt with in today's world as if their value were zero.

In some cases, the dollar value can be determined. Water resources for power or other uses may be evaluated by the amount of flow of the rivers and the quantity of water storage in rivers and lakes. Forest resources may be evaluated by the number, types, and sizes of trees and the amount of lumber they yield. Mineral resources may be evaluated by estimating how many metric tons of economically valuable mineral material there are at particular locations. It is now standard procedure to evaluate the amount of tangible natural resources—such as air, water, forests, and minerals—in an area before developing or managing that area.

Problem number two: Who should bear the burden of these costs? There are two common answers: (1) Polluters should pay (which would add to the producer's cost and therefore perhaps raise the retail cost), or (2) we should all pay, through general tax revenues used to clean up the environment. Some suggest that companies whose operations result in environmental and ecological costs should be taxed or charged special fees. The company could simply absorb this cost, which would lower its profit from selling the product (nickel in the case of Sudbury), or could pass the cost on to the consumer in higher sales prices.

Stated simply, the question is whether it is better to finance pollution control using tax dollars or a "polluter pays" approach. Today, economists generally agree that the "polluter pays" approach provides much stronger incentives to reduce pollution economically.

18.5 Natural Capital, Environmental Intangibles, and Ecosystem Services

Ecosystems help to maintain clean air and water, but how much can nature do itself? Before the Industrial Revolution, nature did much of the job for us. Forests absorb particulates, salt marshes convert toxic compounds into nontoxic forms, wetlands and organic soils treat sewage. Ecologists call these *public-service functions of nature*. Economists call them *natural capital*.

Bees pollinate an estimated $20 billion worth of U.S. crops. We rarely think of this benefit of bees, but in fact the cost of pollinating these crops by hand would be exorbitant, and a pollutant that killed bees would thus have large economic consequences (Figure 18.4). Today, an outbreak of bee parasites in the United States has reduced the abundance of wild honeybees, and farmers are paying beekeepers to bring their hives to the fields at pollination time. This has turned an intangible factor into a tangible one and has brought this public-service function of bees to public attention.

Another example: Bacteria fix nitrogen in the oceans, lakes, rivers, and soils. The cost of replacing this function by artificially producing and then transporting nitrogen fertilizers would be immense, but, again, we rarely think about this benefit we derive from bacteria. Bacteria also clean water in the soil by decomposing toxic chemicals. And the atmosphere, too, performs a public service by acting as a large disposal site for toxic gases. For instance, carbon monoxide is eventually converted to nontoxic carbon dioxide either by inorganic chemical reactions or by bacteria. Only when our environment loses a public-service function do we usually begin to recognize its economic benefits. Then, what had been accepted as an economic externality (an indirect cost) suddenly may become a direct cost.

The dollar value of nature's public-service functions is hard to measure. Right now, nature's services that benefit people and other living things are roughly estimated at $3 trillion to $33 trillion per year.[10, 11]

Valuing the Beauty of Nature

The beauty of nature ("landscape aesthetics") is another environmental intangible, one that has probably been important to people as long as our species has existed. We know for certain that it has been important since people began producing art and writing, because the beauty of nature is a continuous theme in both. Once again, as with forests cleaning the air, we face the difficult question: How do we arrive at a price for the beauty of nature? The problem is even more complicated because among the kinds of scenery we enjoy are many that have been modified by people.

While it may be difficult to put an exact dollar value on scenery, we know what people pay to enjoy scenery and to travel sometimes long distances to visit scenic places. We also know that people build expensive houses where there are magnificent views, so we can get an idea of the dollar value of scenery by comparing the prices of houses that have views with the prices of similar houses without views. People even build in risky places for the sake of the view (Figure 18.5). Additional evidence of the value of scenery involves the farm fields in Vermont. The open fields improved the view of the mountains and forests in the distance, so when farming declined in the 1960s, the state began to offer tax incentives for farmers to keep their fields open to bolster the tourism economy.

One problem in aesthetic valuation is personal preference. One person may appreciate a high mountain meadow far removed from civilization. Another person prefers visiting with others on a patio at a trailhead lodge. A third person may prefer to visit a city park. A fourth may prefer the austere beauty of a desert or a rocky shore. If we are going to consider aesthetic factors in economic analysis, we must develop a method that allows for individual differences.

Some philosophers suggest that there are specific characteristics of landscape beauty that we can use to

China Photos /Stringer/Getty Images, Inc.

■ **FIGURE 18.4**
Bees are "natural capital."
Bees perform a public service for us by pollinating crops and many flowering plants.

Alamy Images

©AP/Wide World Photos

(a)

(b)

■ FIGURE 18.5
What is the price of scenery?
(a) People build expensive houses in risky places, in part because of the views, as along the coast of Malibu and in the mountains to the east. (b) These houses have beautiful views, but are at risk from earthquakes and mudslides, as well as fires from vegetation that burns readily.

help put a value on intangibles. Some say that the three key elements of landscape beauty are coherence, complexity, and mystery (something not completely seen nor fully explained). Other philosophers think the primary aesthetic qualities are unity, vividness, and variety.[12] *Unity* refers to the wholeness of the landscape—not a group of separate parts but a single harmonious unit. *Vividness* means that the scene is visually striking. People differ in what they believe are the key qualities of landscape beauty, but, once again, almost everyone would agree that the value is greater than zero.

18.6 How Is the Future Valued?

Is a bird in the hand worth two in the bush? The discussion about whaling—explaining why whalers may not find it valuable to conserve whales—reminds us of the old saying "A bird in the hand is worth two in the bush." In economic terms, it means that a profit now is worth much more than a profit in the future. This brings up another economic concept important to environmental issues—the future value compared with the present value of anything.

As an example, suppose you are dying of thirst in a desert and meet two people. One offers to sell you a glass of water now, and the other offers to sell you a glass of water if you can be at the well tomorrow. How much is each glass worth? If you believe you will die today without water, the glass of water today is worth all your money, and the glass tomorrow is worth nothing. If you believe you can live another day without water, but will die in two days, you might place more value on tomorrow's glass than on today's.

We are constantly weighing the present against the future. We know we are mortal, so we tend to value personal wealth and goods more if they are available now than if they are promised in the future. Still, we are accustomed to thinking of the future—planning a nest egg for retirement or for our children. Indeed, many people today argue that we have a debt to future generations and must leave the environment in at least as good a condition as we found it. These people would argue that we must not value the future environment less than the present one.

But if it's hard to assess present value, it's even harder for future value. Since the future existence of whales and other endangered species has value to those interested in biological conservation, the question arises: Can we place a dollar value on the future existence of anything? The future value depends on how long a period you are talking about. For example, when we talk about the future in considering some important global environmental topics, such as stratospheric ozone depletion and global warming, we are talking about a period extending out more than a century. This is because chlorofluorocarbons (CFCs) have such a long residence time in the atmosphere (see Chapter 13) and because it will take a long time to see the benefits of changing our energy policy to offset global climate change.

Is spending on the future always a good investment? The failure of the levees in New Orleans during Hurricane Katrina in August 2005 is a case in point (Figure 18.6). The Army Corps of Engineers had requested $4 billion to strengthen the levees several years before, but the federal government decided against it. The resulting damage greatly exceeded $4 billion, and this is without counting the cost in human suffering and loss of life.

▪ **FIGURE 18.6**

Flooding in New Orleans when the levees gave way. The cost of strengthening the levees would have been less than the dollar value of the flood damage during Hurricane Katrina in 2005.

However, spending on the environment can sometimes be viewed as diverting money from other investments that will be of benefit not just in the future but right now. For example, in the 20th century, asbestos was used as insulation in buildings, around hot-water pipes and heaters. Then we discovered that asbestos can cause cancer, and that asbestos was used as insulation in some schools. As long as the asbestos is well contained, it does not pose an immediate threat. The danger lies in inhaling or ingesting asbestos particles released from old, crumbling insulation or when buildings burn down or are torn down.

In some cases, people decided to spend money now to remove the asbestos in order to reduce future risk. In other cases, people decided to leave the well-protected asbestos intact and spend the money on improving educational facilities in a school—the asbestos could be dealt with if and when it began to be a problem.

The wealthier society becomes, the more it values the environment. This raises another issue: If history is a guide, Americans in the 22nd century will be far more affluent than Americans today, and will place greater value on the environment and be more able and more willing to invest in it. To what extent should we ask the average American today to sacrifice now for much richer great-great grandchildren? And how can we be sure of the future usefulness of today's sacrifices? Put another way, what would you have liked your an-

cestors in 1900 to have sacrificed for our benefit today? Should they have worked to develop electric automobile transportation? Or done research on whale populations and placed restrictions on whaling? Or saved more tall-grass prairie?

The two basic issues have one general answer. The issues are (1) that we are so much richer and better off than our ancestors that their sacrificing for us might have been inappropriate, and (2) that even if they had wanted to sacrifice, how would they have known what sacrifices would be important to us? As a general rule, one answer to these thorny questions about future value is, *Do not throw away or destroy something that cannot be replaced if you are not sure of its future value.* For example, if we do not fully understand the value of the wild relatives of potatoes that grow in Peru but do know that their genetic diversity might be helpful in developing future strains of potatoes, then we ought to preserve those wild strains.

18.7 Risk–Benefit Analysis

What is the value of a human life? Death is the fate of all individuals, and many of our everyday activities involve some risk of injury or death. How, then, do we place a value on saving a life by reducing the level of a pollutant? This question leads us to another important area of environmental economics—**risk–benefit analysis**, in which the riskiness of an action (its possible negative outcomes) is weighed against the potential gain from the action.

Acceptability of Risks and Costs

With some activities, the relative risk is clear. It is much more dangerous to stand in the middle of a busy highway than to stand on the sidewalk, and hang gliding has a much higher mortality rate than hiking. The effects of pollutants are often more subtle, so the risks are harder to pinpoint and quantify. Table 18.1 shows the lifetime risk of death from a variety of activities and from some forms of pollution. In looking at the table, remember that since the ultimate fate of everyone is death, the total lifetime risk of death from all causes must be 100%. If you are going to die of something and you smoke a pack of cigarettes a day, you have 8 chances in 100 of dying as a result of smoking. At the same time, your risk of death from driving a car is 1 in 100. Since risk tells you the chances that an event will occur, but not its timing, you might smoke all you want and die in an automobile accident before the smoking gets you.

Table 18.1 tells us some interesting things about air pollution. One is that death from outdoor air pollution in

□ TABLE 18.1 RISK OF DEATH FROM VARIOUS ACTIVITIES

Activity	Result	Risk of Death	
Cigarette smoking (pack a day)	All causes: cancer, effect on heart, lungs, etc.	8 in 100	
Air in the home containing radon	Cancer	1 in 100	Naturally occurring
Automobile driving		1 in 100	
Death from a fall		4 in 1,000	
Drowning		3 in 1,000	
Fire		3 in 1,000	
Artificial chemicals in the home	Cancer	2 in 1,000	Paints, cleaning agents, pesticides
Sunlight exposure	Melanoma	2 in 1,000	Of those exposed to sunlight
Electrocution		4 in 10,000	
Air outdoors in an industrial area		1 in 10,000	
Artificial chemicals in water		1 in 100,000	
Artificial chemicals in foods		less than 1 in 100,00	
Airplane passenger (commercial airline)		less than 1 in 1,000,000	

Source: From *Guide to Environmental Risk.* 1991. U.S. Environmental Protection Agency. Region 5 publication no. 905/91/017.

most places is comparatively low—even lower than the risk of drowning or of dying in a fire. This suggests that the primary benefit from lowering air pollution is to improve the quality of our lives, rather than to lengthen our lives. Another striking fact in this table is that natural indoor air pollution is much deadlier than most outdoor air pollution—unless, of course, you live near to toxic-waste facility.

Is zero risk possible? If not, how much risk is acceptable? Future discoveries will likely help to lower various risks, perhaps eventually allowing us to come close to a zero-risk environment. However, complete elimination of risk is generally either technologically impossible or prohibitively expensive. How much risk is socially, psychologically, and ethically acceptable varies among societies, but we can make some generalizations about the acceptability of various risks.

One factor is the number of people affected. Risks that affect a small population (such as employees at nuclear power plants) are usually more acceptable than those that involve all members of a society (such as risk from radioactive fallout).

Novel risks seem less acceptable than long-established or natural risks, and society tends to be willing to pay more to reduce such risks. For example, in recent years France spent about $1 million annually to reduce the likelihood of one air-traffic death but only $30,000 for the same reduction in automobile deaths.[13] Some argue that the reason commercial air travel is safer than automobile travel is partly that flying is relatively new, we are less accustomed to it and still more fearful of it, and therefore willing to spend more per life to reduce the risk from flying than to reduce the risk from driving.

Acceptance of risk depends also on an activity's desirability. For example, many people accept much higher risks for athletic or recreational activities than for transportation or employment-related activities. They consider that the risks in playing football, skiing, flying, and hang gliding are just part of the activity. As such, people will accept less risk from pollution than from driving a car or playing a sport and are willing to pay for reducing that risk.

In an ethical sense, it is impossible to put a dollar value on a human life. However, it is possible to determine how much people are willing to pay for the chance of reducing risk, or for increasing the chances of living longer. For example, a study by the Rand Corporation considered measures that would save the lives of more heart-attack victims, including increasing ambulance services and initiating pre-treatment screening programs. According to the study, which identified the likely cost per life saved and the willingness of people to pay, people were willing to pay approximately $32,000 per life saved, or $1,600 per year of longevity.[13]

We can attempt to compare the costs versus the benefits. Although information is incomplete, it is possible to estimate the cost of extending lives in terms of the dollars spent per person per year for various measures (Table 18.1). For example, in terms of direct effects on human health, it costs more to increase longevity by reducing air pollution than to directly reduce deaths by adding a coronary ambulance system.

Such a comparison is useful as a basis for decision-making. Clearly, though, when a society chooses to reduce air pollution, many factors beyond the direct, measurable health benefits are considered. Pollution directly affects more than just our health, and ecological and aesthetic damage can also indirectly affect human health (see Section 18.4). We may choose a slightly higher risk of death in a more pleasant environment (spend money to clean up the air, not to increase ambulance services) rather than increase the chances of living longer in an unpleasant environment (spend the money to reduce deaths from heart attacks).

Decisions like these are not easy. The need to make choices of this kind may make you uncomfortable, but like it or not, we cannot avoid making them. The issue boils down to whether we should improve the quality of life for the living or extend life expectancy regardless of the quality of life.[14]

Risk, especially "unreasonable risk," is an important concept in our legal processes. For example, the U.S. Toxic Substances Control Act states that no one may manufacture a new chemical substance or process a chemical substance for a new use without obtaining a clearance from the Environment Protection Agency (EPA). The act establishes procedures to estimate the hazard to the environment and to human health of any new chemical before it becomes widespread. The EPA examines the data and judges the degree of risk associated with all aspects of producing the new chemical or process, including extraction of raw materials, manufacturing, distribution, processing, use, and disposal. The manufacture or use of the chemical can be banned or restricted if the evidence suggests that it will pose an unreasonable risk of injury to human health or to the environment.

But what is unreasonable?[15] This question brings us back to Table 18.1 and makes us realize that determining whether risk is unreasonable involves judgments about the quality of life as well as the risk of death. Establishing an acceptable level of pollution (and thus risk) requires a social/economic/environmental trade-off. Moreover, the level of risk that a society considers acceptable will change over time, depending on changes in scientific knowledge, changes in the comparison with risks from other causes, changes in the cost of lowering the risk, and changes in the social and psychological acceptability of the risk.

For example, when DDT was first used, no one understood the subtle environmental and ecological effects of this chemical—scientific observations revealed these effects later. At that time, people were not as concerned with environmental issues, and society was not yet willing to pay a lot to protect the environment. Now, people widely agree that the environment is a major concern, and we are less willing to accept indirect environmental effects. What had been considered ex-

ternalities to the use of DDT have become internal cost factors.

How much are we willing to pay to reduce risk? With adequate data, we can take scientific and technological steps to estimate the level of risk from a particular pollutant and the cost of reducing the risk, and then compare the cost with the benefit. The question is, how much is a given reduction in risk from that pollutant worth to us? How much will we, as individuals or collectively as a society, be willing to pay for a given reduction in that risk? The answers will depend partly on societal and personal values. We must also not forget to factor into the equation that costs of cleaning up pollutants and polluted areas and costs of restoration programs can be minimized or even eliminated if we are quick to control a pollutant as soon as we recognize it. The total cost of pollution control need not increase indefinitely.

The average cost of pollution control per family in the U.S. is low, especially compared with other costs. It has been estimated that in the United States the cost of pollution control is $30–60 per year for a family with a median income. In addition to the low cost per family, pollution control has many benefits whose quantitative value can be estimated. For example, federal air-quality standards are estimated to reduce the risk of asthma by 3% and to reduce the risk of chronic bronchitis and emphysema in locally exposed adults by 10–15%.

Compare the costs of pollution control to the costs of pollution. The total cost of direct and indirect effects on human health from stationary sources of air pollution is estimated at $250 per family per year. Air pollution contributes to inflation by reducing the number of productive workdays, reducing work efficiency, adding to direct expenditures for health treatments, and necessitating repair of environmental damage. On this basis, air-pollution control appears to be cost-effective—in fact, it has economic benefits.[16]

18.8 Global Issues: Who Bears the Costs?

Should developing nations pay as much as industrialized nations? Global environmental problems make us more aware of the public-service functions of the environment of our planet and of life around us, as well as raising new economic questions. An important case in point is the possibility of global warming.

The problem is that the carbon dioxide and other greenhouse gases that our technological society is adding to the atmosphere have the potential to warm the climate (see the discussion of global warming in Chapter 13).[17, 18] The direct solution is to release less of these gases, but that would require a worldwide decrease in the burning of fossil fuels. Although most greenhouse gases today come from the industrial nations, in the future the developing nations, especially China and India, will contribute large quantities of these gases.

Developing nations think industrial nations should shoulder the costs. The economist Ralph D'Arge points out an economic problem arising from this global issue: The less-developed countries did not share in the economic benefits of burning fossil fuels during the first two centuries of the Industrial Revolution, but they are sharing the disadvantages.[19] Now the industrialized nations are suggesting that all nations, including the less-developed ones, restrict their use of fossil fuels and therefore endure the future disadvantages without enjoying the benefits of cheap energy. Developing nations tend to think that industrial nations, which enjoyed the past benefits, should accept most of the future costs. At the same time, why shouldn't the developing nations proceed to develop and burn fossil fuels?

Viewing a global issue in terms of what benefits individual nations may be too restricted. To remedy the global environmental effects of our technological civilization, it may be necessary to reduce the total production of greenhouse gases by all nations. If so, then the economic question remains: Who pays, and how? At present, this is an unresolved issue in **environmental economics**. One suggestion is that the developed nations pay for the reduction in greenhouse emissions of the less-developed nations. Another suggestion is that the developed nations share their technology with developing nations, helping them to reduce both local and global pollution. These issues were a major concern at the 1992 Earth Summit in Rio de Janeiro.[20]

18.9 Environmental Policy Instruments

How does a society achieve an environmental goal? Societies use what economists call **policy instruments** to establish policies on social issues, including environmental issues. Policy instruments include (1) moral suasion (which politicians call "jawboning"—persuading people by talk, publicity, and social pressure); (2) direct controls, such as laws and regulations; (3) market processes, which affect the price of goods and include taxes of various kinds, subsidies, li-

censes, and deposits; and (4) government investments, such as research and education.

Pollution Control and the Law of Diminishing Returns

How do we know when to stop? When have we done enough to consider the environment "good" and to feel that we have achieved a reasonable balance between benefits and costs? To decide how far to go, we have to weigh the cost of each additional step we take against the amount that will be gained from it.

At some point, you need to spend more and more to get less and less. That, basically, is the Law of Diminishing Returns. Let's say, for example, that we are cleaning up a contaminated lake. We have removed 20% of the pollutants, and it will cost 5 cents to remove each additional kilogram. We decide to continue. But when 80% of the pollutants have been removed, we find that it will cost 49 cents for each additional kilogram. At this rate of increasing costs, we realize, it would cost an infinite amount to remove all the pollution.

Three common methods of direct control of pollution are (1) setting maximum levels of pollution emission, (2) requiring specific procedures and processes that reduce pollution, and (3) charging fees for pollution emission (as discussed earlier in this chapter). In the first case, a political body could set a maximum allowable level for the amount of sulfur emitted from the smokestack of an industry. In the second, it could restrict the kind of fuel the industry could use. Many areas have chosen this method and prohibited the burning of high-sulfur coal.

The problem with the first approach—controlling emissions—is that careful monitoring is required on an ongoing basis to make certain the allowable levels are not exceeded. Such monitoring may be costly and difficult to carry out. The disadvantages of the second approach—requiring specific procedures—are that the required procedures may place a severe financial burden on the producer of the pollutant, restrict the kinds of production methods an industry may use, and become technologically obsolete. (See the discussion of air pollution and laws regulating it in Chapter 14.)

Many people feel the third approach works best. Although the United States has emphasized the use of direct regulation to control pollution, other countries have succeeded in controlling pollution with the third approach: charging fees for emitting pollutants. For example, charges for emitting polluted water into the Ruhr River in Germany are assessed on the basis of both the concentration of pollutant and the total quantity of polluted water emitted into the river. In response, plants have been redesigned to recirculate polluted water and treat the water.[21]

Return to the Big Question

Can we put a price on scenic beauty, endangered species, and the quality of life?

While we may not all agree about what is beautiful, we can consider what people pay to visit certain environments, such as places to go fishing or to camp out in the wilderness, and what people pay to visit scenic areas, such as the Grand Canyon, Yellowstone National Park, the Alps, the Canadian Rockies, and Niagara Falls. These help us determine a dollar value for what are called "intangibles." Similarly, by looking at the amount of money our society has been willing to invest to reduce risks of death and injury, we can get an idea of the dollar value of a healthful, unpolluted environment. All of these also show that we consider the quality of our lives very important, sometimes as important or even more important than simply living a long time.

Summary

- Two reasons why people sometimes act against their own best long-term interests from an environmental point of view are the commons and the low rate of production of many renewable resources.

- The tragedy of the commons is that when people own things in common, the tendency is to act in one's own short-term interest, which usually means sacrificing the long-term common good.

- Most living renewable resources, such as whales and their oil, increase at a low rate. This makes them poor economic investments and tends to lead people who do invest in them to exploit and destroy the very resource on which their livelihood depends.

- How do we place a future value on nature and the environment? Economists tell us that the comparison between the future value and the present value can

be an important factor in deciding how much to exploit a resource But they also tell us that it is difficult to put a dollar value on the future environment.

- An economic analysis can help us understand why environmental resources have been poorly conserved in the past and how we might more effectively achieve conservation in the future.

- Economic analysis is used in two different kinds of environmental issues: how to use desirable resources (fish in the ocean, oil in the ground, forests on the land) and ways to minimize undesirable pollution.

- Weighing the risks versus the benefits affects our willingness to pay for an environmental good.

- Placing a value on environmental intangibles—such as landscape beauty and nature's public-service functions— is becoming more common in environmental analysis.

Key Terms

commons
environmental economics
externality

policy instruments
risk–benefit analysis

Getting It Straight

1. Which of the following can be thought of as commons in the sense meant by Garrett Hardin's concept of "the tragedy of the commons"? Explain.
   a. tuna fisheries in the open ocean
   b. catfish in artificial freshwater ponds
   c. grizzly bears in Yellowstone National Park
   d. a view of Central Park in New York City
   e. air over Central Park in New York City

2. What are tangible and intangible factors? Give an example of each.

3. What are today's shared environmental resources?

4. Should polluters pay the cost of environmental degradation? Explain.

5. What should you do with items that might have future value to you or someone else?

6. Do more people die from indoor air pollution or exposure to outdoor air pollution?

7. What is someone more likely to die from, cigarette smoking, automobile driving/accidents, or cancer? Use Table 18.1 to formulate your answer.

8. Who is contributing to global pollution issues, more-developed or undeveloped countries? Explain.

9. What are the three common methods of direct control of pollution?

What Do You Think?

1. Is it reasonable to continue to treat the possible effects of global warming as economic externalities? Is it possible to treat these otherwise—as internal to costs and benefits? If so, how would you do this? Pick a particular effect and discuss it.

2. In some cities, a view is considered a real economic factor and land ownership includes the right to the view. What might be the environmental consequences of this kind of internalization of a previously external economic factor? Under what conditions would you be in favor of it?

3. A continuing debate at the International Whaling Commission is whether nations should be allowed to hunt whales if this activity is part of their cultural heritage. From the viewpoint of the environment, if whaling can never be very profitable except in the very short run, but it is important in a human cultural context, how might the activity be made sustainable?

4. Exotic species are often introduced into new environments with very good-willed, innocent motivations, but sometimes with environmentally disastrous effects. A proposal is made that all new introductions of exotic species require either (a) a tax to cover the cost of any unanticipated undesirable consequences or (b) an insurance policy that would serve the same purpose. From an environmental point of view, would you favor the tax? The insurance policy? Neither?

5. Where are most people likely to suffer from exposure to high levels of certain pollutants, inner cities or rural communities? Why would the levels of pollution be higher or lower in these areas?

6. What is the cost of the "beauty of nature" to you? Are you willing to pay the price of its losses or advocate for its protection?

Pulling It All Together

1. Flying over Los Angeles, you see smog below. Your neighbor in the next seat says, "That smog looks bad, but eliminating it would save only a few lives, so it probably isn't worth the cost. We should spend the money on other things, like new hospitals." Do you agree or disagree? Give your reasons.

2. In New York City, a large building project included high-rises that blocked the view of the Hudson River from an older set of residential buildings. Is it fair to consider this loss of a view an economic externality, for which the builder has no responsibility? If not, then who is responsible, who should pay, and how should the amount of payment be decided?

3. One approach to dealing with oil spills along ocean coasts is to maintain a fleet of specially designed boats and ships to clean up the spills.
   a. Is an oil spill a tragedy of the commons?
   b. Should the costs of maintaining this fleet be considered an economic externality?

Further Reading

Daly, H. E., and K. N. Townsend. 2003. *Ecological economics textbook principles and applications.* Washington, DC: Island Press.—Discusses an interdisciplinary approach to the economics of environment.

Goodstein, E. S. 2000. *Economics and the environment,* 3rd ed. New York: Wiley.

Hardin, G. 1968. Tragedy of the commons. *Science* 162: 1243–1248.—One of the most cited papers in both science and social science, this classic work outlines the differences between individual interest and the common good.

Hodge, I. 1995. *Environmental economics: Individual incentives and public choice.* New York: St. Martin's Press.

Kolstad, C. D. 1999. *Environmental economics.* New York: Oxford University Press.—One of the best introductions to the field.

Schnaiberg, A., and K. A. Gould. 1994. *Environment and society: The enduring conflict.* New York: St. Martin's Press.—An examination of several myths related to economics and environmental problems.

John Boykin/NewsCom

19

Planning for a Sustainable Future

Big Question

How Can We Plan, and Achieve, a Sustainable Environment?

Learning Objectives

Ernest Callenbach introduced the vision of ecotopia in 1975 while writing for the University of California Press. He presented a metaphorical ecotopia in a novel that had a far-reaching impact on the incipient environmental movement of the time. His vision of ecotopia was regional, involving northern California, Oregon, and Washington. In the story, these three states separate from the rest of the United States to form a new country, based on what we would today call sustainable development.[1] The basic idea was to structure society around environmental principles while providing a model for an ecologically aware society. At that time, few recognized that environmental problems would become worldwide by the 21st century. Today, when we talk about a sustainable environment, it is understood that we need to consider the global environment, and in particular the urban environment.

This chapter should give you a greater understanding of the following:

■ what constitutes a sustainable environment;

■ environmental planning and why we need it;

■ our need to be connected to nature;

■ environmental law.

19.1 The Ideal Sustainable Environment

Imagine a **sustainable environment** (Figure 19.1). Achieved and maintained by using our environment wisely. What would such an environment be like? Among other things, it would likely be a world in which the human population is within the carrying capacity of Earth; living resources are sustained, and their harvest is sustainable; sources of energy are sustainable, and their use doesn't harm the environment or change the climate; agricultural practices sustain the biodiversity of soils, crops, and animals and do not harm the environment; water resources are managed to ensure sufficient water for people and ecosystems; industrial processes use resources sustainably and do not pollute the environment; and representatives of natural ecosystems in their dynamic ecological state are sustained, as is biodiversity.

Must we tear our whole system down and start over? Some people assume that achieving a sustainable environment will require a crash and restart of the entire system that present-day society is based on. We disagree—we believe that people have the ability to change the ways society works and adapt to the change, which may bring with it new opportunities and prosperity.[2]

How do we begin? Reducing, or even stabilizing, the size of our population and our use of living and nonliving

■ **FIGURE 19.1**

Idealized image of a sustainable enviroment.
(1) We live in sustainable numbers; (2) use mass transportation between urban areas; (3) use renewable energy; (4) practice sustainable agriculture; (5) maintain natural water supply and rivers; (6) maintain sustainable fisheries. We use less total energy and materials per person than today but are just as happy.

resources will not be easy, given how dependent our society is today on fossil fuels, economic growth, and expanding consumption of resources. Two keys to this are education and information. With a clear understanding of why changes are necessary, we can work to reduce our use of living and nonliving renewable and nonrenewable resources. There is fear that we are now in a period of "overshoot" in which an ever-increasing population is using more resources than can be replenished by the planet. As we apply science to come up with solutions based on core values of sustainability, many new opportunities for economic development will naturally arise. We are already seeing this in the remarkably swift growth of alternative energy and in new ways to produce our food and deal with waste. One example, "slow food," has the objective of maintaining biodiversity of the crops and animals we use for food, and we are challenging the very concept of waste by applying industrial ecology to move toward a zero-waste society in which what is now waste would become resources for other sectors of society.

With these comments on what a sustainable environment might look like and what needs to be done to achieve it, we will move to discuss some aspects of environmental planning and environmental law that undoubtedly will play an important role in our future.

19.2 The Process of Planning a Future

Planning a future is a social activity in which we all participate. **Environmental planning** occurs at every level of activity, from a garden to a house, a neighborhood, a city park and its surroundings, a village, town, or city, a county, state, or nation.

All societies have tried to plan their use of land and resources—through custom, or by fiat of a king or emperor, or by democratic processes. For thousands of years, experts have created formal plans for cities and for important buildings and other architectural structures, such as bridges. In some cases, however, land development has "grown like Topsy"—as the saying goes—without a specific plan but as a result of need and custom.

For example, cities develop at important transportation centers and where local resources can support a high density of people. In medieval Europe, bridges and other transportation aids developed in response to local needs. People arriving at a river would pay the farmer whose land lay along the river to row them across. Sometimes this would become more profitable than farming, or an important addition to the farmer's income. Eventually he might build a toll bridge.[3]

Our society has formal planning processes for land use. These processes have two features. The first is a set of rules (laws, regulations, etc.) requiring forms to be filled out and certain procedures to be followed. The second feature is an imaginative attempt to use land and resources in ways that are beautiful, economically beneficial, and sustainable.

Environmental planning and review are closely related to how land is used. Land use in the United States is dominated by agriculture and forestry—only a small portion (about 3%) of U.S. land is urban. However, each year about 9,000 km^2 (about 3,500 mi^2) of rural land are converted to nonagricultural uses. About half of that land is converted to wilderness areas, parks, recreational areas, and wildlife refuges. The other half is for urban development, transportation networks, and other facilities. Nationwide, relatively little rural land is converted to urban uses. Even so, urban areas that are growing rapidly may be viewed as destroying agricultural land and worsening urban environmental problems, and urbanization in remote areas with high scenic and recreational value may be viewed as potentially damaging to important ecosystems.

In a democracy, environmental planning leads to a tug-of-war between individual freedoms and the welfare of society as a whole. On one hand, citizens of a democracy want freedom to do what they wish, wherever they wish, especially on land that, in Western civilizations, is "owned" by the citizens, or where citizens have legal rights to water or other resources. On the other hand, land and resource development and use affect all of society—everyone benefits or suffers, directly or indirectly, from a specific development. Society's concerns about this lead to laws, regulations, bureaucracies, forms to fill out, and limitations on land use.

In planning, we need to achieve a balance. We need to balance individual freedom of action with those actions' effects on society. We need to achieve a sustainable use of resources that keeps the environment beautiful, spiritually fulfilling, open to many kinds of recreation, and supportive of many kinds of employment.

Who speaks for nature? And who legally represents the environment? The landowner? Society at large? At this time, we have no definitive answers, but the history of our laws provides insight into our modern dilemma.

19.3 In Planning a Nation's Landscapes, How Big Should Wildlands Be?

How big an area do we need for wilderness, open space, and undeveloped lands? Two of the first questions that come up in visualizing an environmentally better world is how much of the land should be protected from human activities—set aside as nature preserves—and how big do these areas need to be? Recent thinking about the environment has focused on the big picture—what is necessary on a national scale or on a global scale. But what should *our* planning unit be? An acre? A square mile? A continent? The whole Earth?

Some argue that nature can be saved only by thinking big. A group called the Wildlands Project argues that big predators, referred to as "umbrella species," are keys to ecosystems, and that these predators require large home ranges. The assumption is that large, wide-ranging carnivores offer a wide, protective umbrella under which many species that are more abundant but smaller and less noticeable find safety and resources. Leaders of the Wildlands Project feel that even the largest national parks, such as Yellowstone, are not big enough and that America needs "rewilding."[4]

They propose that large areas of the United States be managed around the needs of big predators, and that we replan our landscapes to provide a combination of core areas, corridors, and inner and outer buffers. No human activities would be allowed in the core areas, and human activity in the corridors and buffers would be restricted. One proposal is to reclaim American prairie, removing towns and cities and re-creating a landscape where bison are once again allowed to roam free. Reed Noss, one of the founders of the Wildlands Project, says that rewilding is not an attempt to re-create Eden, but is "simply scientific realism," with the goal of ensuring "the long-term integrity of the land community." He argues that we have a "moral obligation to protect wilderness . . . animals and plants . . . not only for our human enjoyment, but because of their intrinsic value."

The Wildlands Project has created a major controversy. Some groups see the project as a fundamental threat to American democracy. Others criticize the Wildlands Project's scientific foundation. These critics say that while some ecological research suggests that large predators may be important, we still don't fully understand what controls populations in all ecosystems, and that the idea of "keystone species," a central idea in the Wildlands Project, lacks an adequate scientific base.

The question is, what do you want? Would you like to see a vast area of the United States returned to what might be self-functioning ecosystems? Or would you choose some open system of conservation that integrates people and allows for more freedom of action? The choices lie with your generation and the next, and tests of their validity are also yours. Your answer, as we have pointed out throughout this book, will involve both science and values, and the implications for the world's environment and for the world's people are huge.

19.4 Our Need for Nature in an Increasingly Urban Environment

Modern humans first appeared in the Pleistocene era. At that time, thousands of years before the rise of civilization, our ancestors had to use all of their human senses to survive: their eyes and sense of smell and taste to find edible plants; and acute hearing as well as sharp eyes to search for game and avoid predators. Those who used these senses more successfully were more likely to survive and pass on their genes to future generations.

Fast-forward many thousands of years and we find ourselves in large urban complexes surrounded not by trees and grass but by steel, concrete, asphalt, the steady roar of traffic, subways, construction machinery. It is little wonder that we often feel stressed and yearn for something more in harmony with our basic nature.

Our world is becoming increasingly urban. About 75% of the people in developed countries, and nearly one-half of all people on Earth, live in urban areas. As a result, one of the ironies of our modern times is that although environmentalism is a popular political and social movement, fewer and fewer people have much direct contact with nature. And yet, recent studies verify our need for nature in cities. When parks and gardens and green spaces are available, crime in nearby areas tends to decrease, people are less stressed, and there is a stronger sense of community. As a result, people feel safer when surrounded by green belts and trees than by concrete and asphalt.[5]

Trees also help to protect us from air pollution and heat. Trees not only pull carbon dioxide from the atmosphere in photosynthesis, but they also remove particulates, nitrogen dioxide, carbon monoxide, and ozone. In addition, trees provide shade that can make the air near them cooler than the surrounding area. In a blacktop area, the temperature may be as much as 18°C lower under the tree than over the blacktop. Thus, we are able to stay within our comfort zone more easily when trees are present. Most of all, green areas support our heritage and help fill our need to be in a more natural environment.

How much contact with nature do we need? As a nation, we are not in agreement about how much contact with the outdoors, with nature, is necessary for the good life. At one extreme are those who believe that no one should live in cities. At the other are those who think that only urban life is worthwhile and that nature is something to be enjoyed only on a vacation, much the same way one views a trip to Disneyland or a visit to a foreign city.

However, there is a long tradition of assuring contact with nature in cities, through urban planning and landscape architecture, developed by people who believed that some contact with nature was essential. Henry David Thoreau, one of the greatest American naturalists and conservationists—indeed, he is widely considered the father of conservation in America—thought that life should involve both town and country. To him, the ideal place was what he called a swamp by the edge of town, a place where nature dominated but which was a short walk from people who created, argued, talked, and benefited from each other.[6]

The City Park

Parks have become more and more important in cities, and are a must for a good urban environment in the future. A significant advance for U.S. cities was the 19th-century planning and construction of Central Park in New York City, the first large public park in the United States. The park's designer, Frederick Law Olmsted, was one of the most important modern experts on city planning. For Olmsted, the goal of a city park was to provide relief from city life through access to nature and beauty. Olmsted said that vegetation in a city played social, psychological, and medical roles. He felt that vegetation and the experience of it by people in parks was essential to health and well-being.

Olmsted's parks were naturalistic. He took site and situation into account and attempted to blend improvements to a site with the aesthetic qualities of the city. Vegetation was one of the keys, and he carefully considered the opportunities and limitations of topography, geology, hydrology, and vegetation. In contrast to the approach of a preservationist, who might simply have strived to return the area to its natural, wild state, Olmsted created a naturalistic environment in Central Park, keeping the rugged, rocky terrain but putting ponds where he thought they were desirable. To add variety, he constructed "rambles" that were densely planted and followed winding paths. He created a "sheep meadow" by using explosives, and planned recreational areas in the southern part of the park, where there were flat meadows. To meet the needs of the city, he built transverse roads and depressed roadways that allowed traffic to cross the park without detracting from the vistas enjoyed by park visitors.

Central Park is an example of "design with nature," a term coined much later, and this design influenced other U.S. city parks (Figure 19.2). Olmsted remained a major figure in American city planning throughout the 19th century, and the firm he founded continued to be important in city planning into the 20th century.[7, 8]

A water-control project in Boston further illustrates Olmsted's skill. The original site of the city of Boston, Massachusetts, had certain advantages: It was a

iStockphoto

©AP/Wide World Photos

■ **FIGURE 19.2**
New York City's Central Park is one of the world's most successful urban parks. An ideal city would have parks of many sizes in many parts of town, accessible to all the people, providing relief from the intensity of city life, a place for athletics, and a place to enjoy naturalistic scenic beauty.

narrow peninsula with several hills that could be easily defended, a good harbor, and a good water supply. As Boston grew, however, demand increased for more land for buildings, a larger area for docking ships, and a better water supply. The need to control ocean floods and to dispose of solid and liquid waste grew as well. Much of the original tidal-flats area, which had been too wet to build on and too shallow to navigate, had been filled in. Hills had been leveled and the marshes filled with soil. The largest project had been the filling of Back Bay, which began in 1858 and continued for decades. Once filled, however, the area suffered from flooding and water pollution.[7]

Olmsted's solution to these problems was a water-control project called "the fens." His goal was to keep sewage out of the streams and ponds and build artificial banks for the streams to prevent flooding—and to do this in a natural-looking way. To accomplish this, he created artificial watercourses by digging shallow depressions in the tidal flats, following meandering patterns like natural streams. He used other artificial depressions as holding ponds for tidal flooding, and restored a natural salt marsh planted with vegetation that could tolerate brackish water. He planted the entire area to serve as

a recreational park when not flooded. He put a tidal gate on Boston's major river, the Charles, and diverted two major streams directly through culverts into the Charles so that they flooded the fens only during flood periods. He reconstructed the Muddy River primarily to create new, accessible landscape.

The result not only controlled water but lent beauty to the city. Olmsted's blending of several goals made the development of the fens a landmark in city planning. Although it appears to the casual stroller to be simply a park for recreation, the area serves an important environmental function in flood control. Good urban planning, filled with imagination and concern about both people and nature, would be part of an ecotopia.

An extension of the park idea was the "garden city," a term coined in 1902 by Ebenezer Howard. Howard believed that city and countryside should be planned together. A garden city was a city surrounded by a greenbelt (Figure 19.3). The idea was to locate garden cities in a set connected by greenbelts, forming a system of countryside and urban landscapes. The idea caught on,

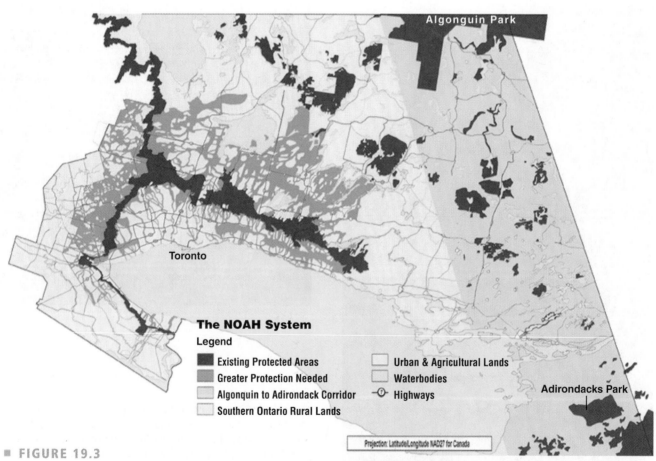

■ **FIGURE 19.3**
Northern Ontario's urban greenbelt provides open space for settled areas along Lake Ontario.
Parks and other open space shown in green in the map. [*Source:* Ontario Greenbelt Alliance Available at http://www.greenbelt.ca/. Accessed October 14, 2005.]

John Maier, Jr./Peter Arnold, Inc.

■ FIGURE 19.4
City streets of Curitiba, Brazil.
The quality of life for the people in this busy city has improved since the 1970s and the city has become a positive model for urban planning and sustainability.

and garden cities were planned and developed in Great Britain and the United States. Greenbelt, Maryland, just outside Washington, D.C., is one of these cities, as is Letchworth, England. Howard's garden-city concept, like Olmsted's use of the natural landscape in designing city parks, continues to influence city planning today.

The Ecological Capital of Brazil: How a City Transformed Itself

In 1950, the city of Curitiba in Brazil had 300,000 inhabitants, but now the population has grown to more than 1.5 million (Figure 19.4), making it the tenth-largest city in Brazil (Figure 19.5). The growth of Cu-

■ FIGURE 19.5
Curitiba is near the Brazilian coast, southwest of Rio de Janeiro and São Paulo.

ritiba resulted primarily from the migration of farm workers displaced by the mechanization of agriculture. At first, the newcomers lived in squatter huts at the edge of the city in conditions of great poverty, with poor sanitation and frequent flooding caused by conversion of rivers and streams into artificial canals. By 1970, Curitiba was well on the way to becoming an example of environmental degradation and social decay.

The story of how Curitiba turned itself from an urban disaster into a model of planning and sustainability by 1995 illustrates that cities can be designed in harmony with people and the environment.[9, 10]

The transportation system was a key factor. Much of the credit for the transformation of Curitiba goes to its three-time mayor, Jaime Lerner, who believed a workable transportation system was the key to making Curitiba an integrated city where people could live as well as work. Rather than constructing an expensive underground rail system, Lerner spearheaded development of a less costly bus system with five major axes, each containing lanes dedicated to express buses (Figure 19.6), and with others carrying local traffic and high-speed automobile traffic. Forty-nine blocks of the historic center of Curitiba were reserved for pedestrians. Passengers pay their fares in tubular bus stations before boarding, which avoids long delays caused by collecting fares after boarding. Circular routes and smaller feeder routes between the major routes maintain vital connections between the central city and outlying areas. As a result, more than 1.3 million passengers ride buses each day. Although Curitiba has the second-highest per-capita number of cars in Brazil, it uses 30% less gas than eight comparable Brazilian cities, and its air pollution is among the lowest in the country.[11, 12]

Recycling solved Curitiba's serious garbage problem. Curitiba required each household to sort recyclables from garbage. As a result, two-thirds of the garbage, more than 100 tons a day, is recycled, with 70% of the population participating. Where streets are too narrow for garbage trucks, residents are encouraged to bring garbage bags to the trucks. They are reimbursed with bus tokens, surplus food, or school notebooks.

The city planned affordable housing and attractive parks. Through a low-cost housing program, 40,000 new homes were built, many placed so that residents have easy access to job sites. The city also embarked on a program to increase the amount of green space. Artificial drainage channels were replaced with natural drainage, reducing the need for expensive flood control. Some areas, including those around the river basins, were set aside for parks. In 1970, Curitiba had only half a square meter of green area per capita, but by 1990 it had 50 square meters for each inhabitant. The accomplishments of Curitiba have led some to call it the "ecological capital of Brazil" and to hope that it is also the "city of the future."[13, 14]

Curitiba transit system

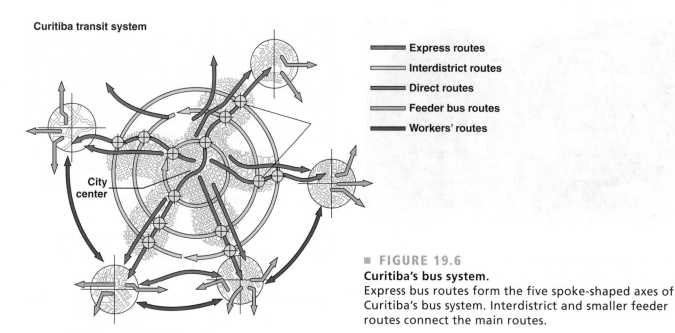

Express routes
Interdistrict routes
Direct routes
Feeder bus routes
Workers' routes

City center

▪ **FIGURE 19.6**
Curitiba's bus system.
Express bus routes form the five spoke-shaped axes of Curitiba's bus system. Interdistrict and smaller feeder routes connect the main routes.

19.5 Regional Planning: The Tennessee Valley Authority

The natural disasters of 2004–2005 highlighted the need for environmental planning. First came the great tsunami in Southeast Asia in 2004, followed in 2005 by two destructive hurricanes, Katrina and Rita, along the Gulf Coast of the United States, and then one of the worst earthquakes in recent times in Pakistan and surrounding countries. The lack of preparation increased the amount of destruction and human misery, and focused people's attention on the need to plan for environmental catastrophes, and the need for environmental planning in general.

Large-scale environmental planning has been unpopular in recent decades, but there was a time in the 20th century when large scale environmental planning by the federal government was seen as a social good. The history of that planning—the reasons for it, and its successes and failures, which are among the largest ever in modern America—is a useful guide to us.

The TVA was a successful experiment in regional environmental planning. During the 20th century, nations tried new approaches to environmental planning. An important experiment was regional planning. In the United States, this meant planning across state boundaries. Seeking new ways to invigorate the economy during the Great Depression of the 1930s, President Franklin D. Roosevelt proposed the establishment of the Tennessee Valley Authority (TVA). He envisioned it as a semi-independent corporation with the power of government but with the flexibility and initiative of a private

enterprise. It would be responsible for promoting economic growth and social well-being throughout parts of seven states, where there had been rampant exploitation of timber and petroleum, and where the people were among the poorest in the country.[15]

Public input is now an important part of managing public lands for recreational use. For example, when management plans are being developed for a national forest, public meetings are often held to inform people about the planning process and to ask for ideas and suggestions. This promotes better communication between those responsible for managing resources and those using them for recreation.

19.6 Environment and Law: A Horse, a Gun, and a Plan

U.S. law has always emphasized individual rights. **Environmental law** refers to laws passed to protect the environment as well as the practice of law that is concerned with the environment. The legal system of the United States has its origins in British common law—laws derived from custom, judgment, and decrees of the courts, rather than from legislation. The U.S. legal system preserved and strengthened British law to protect the individual from society—expressed best, perhaps, in the frontier spirit of "Just give me a little land, a horse, and a gun and leave me alone." Individual freedom—nearly unlimited discretion to use one's own property however one wished—was given high priority, and the powers of the federal government were strictly limited.

But there is a caveat: An individual's behavior can't infringe on the rights of others. When individual behavior infringed on the property or well-being of others, the common law provided protection through doctrines prohibiting trespass and nuisance. For example, if your neighbor's activities cause flood damage on your property, you have recourse under common law. But if the damage was more widespread through the community—for example, polluting air or water, or creating a public nuisance—then only the government has the authority to take action.

Another common-law doctrine is that of *public trust.* Under this doctrine, the common law both grants and limits governmental authority over certain natural areas. Beginning with Roman law, navigable and tidal waters were held in trust by the government for public use. More generally, "the public trust doctrine makes the government the public guardian of those valuable natural resources which are not capable of self-regeneration and for which substitutes cannot be made by man."[16] The government has a responsibility to protect such resources and is not permitted to transfer them to private ownership. This doctrine was considerably weakened by the emphasis on private property rights and by strong development pressures in the United States. More recently, however, it has shown increased vitality, especially concerning the preservation of coastal areas.

Three Stages in the History of Federal Legislation Pertaining to Land and Natural Resources

In the first stage, the goal was to convert public lands to private uses. During this phase, Congress passed laws that were not intended to address environmental issues but did affect land, water, minerals, and living resources—and therefore had large effects on the environment. In 1812 Congress established the General Land Office, whose original purpose was to dispose of federal lands. The government disposed of federal lands through the Homestead Act of 1862 and other laws. Under the Homestead Act, a person received 160 acres free in return for agreeing to improve the land, build a house, and remain on the land for five years.

In the 19th century the U.S. government also granted rights-of-way to railroad companies to promote the development of rapid transportation. In addition to rights-of-way, the government gave the railroads every other square mile along each side of the railway line, creating a checkerboard pattern. The square miles in between were kept as federal land and are administered today by the Bureau of Land Management. Today these lands are difficult to manage for wildlife or vegetation because their artificial boundaries rarely fit the habitat needs of species, especially of large mammals.

The second stage saw the beginning of protection for public lands. In the second half of the 19th century, Congress began to pass laws that conserved public lands for recreation, scenic beauty, and historic preservation. Late in the 19th century, Americans came to believe that the nation's grand scenery should be protected, and that public lands provided benefits, including some direct economic benefits, such as providing rangelands for private ranching. Federal laws created the National Park Service in the second half of the century in response to Americans' growing interest in their scenic resources. Congress made Yosemite Valley a California state park in 1864 (Figure 19.7) and created Yellowstone National Park in 1872 "as a public park or pleasuring-ground for the benefit and enjoyment of the people."[17]

Interest in American Indian ruins led soon after to the establishment in 1906 of Mesa Verde National Park, putting into public lands the prehistoric cliff dwellings of early North Americans, and at the same time creating national monuments. In 1898, President Grover Cleveland appointed Gifford Pinchot to be head of the Division of Forestry, which was soon renamed the U.S. Forest

■ **FIGURE 19.7**
Yosemite Valley's famous half-dome, one of the scenes that helped lead to the creation of the National Park Service.

Service (discussed in Chapter 7). Pinchot believed that the purpose of national forests was "the art of producing from the forest whatever it can yield for the service of man." Although the term *sustainability* had not yet become popular, in 1937 the federal government passed the Oregon and California Act, which required that timberland in western Oregon be managed to give sustained yields.[18] In 1969 Congress created the National Park System, which today consists of 379 areas.

In the third stage, Congress enacted laws about the environment. This began in the 1930s but got going in force in the 1960s and continues today. Legislation was enacted at all levels of government—local, state, and federal—to regulate the use of land and resources. Increasing public concern about deterioration of the environment led the U.S. Congress to pass the National Environmental Protection Act (NEPA) in 1969 and a series of other laws in the 1970s. Well before that, however—as early as the end of World War II—there were already 2,000 laws about managing public lands, often contradicting one another. This led Congress to set up the Bureau of Land Management (BLM) in 1946 to help straighten out the confusion.

Government regulation of land and resources is a subject of controversy: How far should the government be allowed to go to protect what appears to be the public good against what have traditionally been private rights and interests? Today the BLM attempts to balance the traditional uses of public lands—grazing and mining—with the current era's interest in outdoor recreation, scenic beauty, and biological conservation. Part of achieving a sustainable future in the United States will be finding a balance among these uses, as well as a balance between the amount of land that should be public and the amount of land that need not be.

19.7 Skiing at Mineral King Raised a Question: Does Private Enterprise Belong on Public Lands?

Planning for recreational activities on U.S. government lands (including national forests and national parks) continues to be controversial. At the heart of the controversy are two different moral positions, both of which have wide support in the United States. On one side, some argue that if public land is open to public use, this should include profit-making use by individuals and corporations, since they, too, are part of the public. On the other side are those who argue that private enterprise belongs on private land, and that public lands must provide for land uses that the public may not have access to on private lands. The issue becomes even more complicated when the private enterprise in question makes a profit by providing a public service, such as food service or a hotel in a public park.

Private concessions provide food and other services in many public areas today. But permission to operate a business on public land is not granted easily. The size of the enterprise in relation to the size of the area, the pleasure it would provide to the public, the unspoiled natural beauty of an area, and the potential effects of the enterprise on the area's beauty and ecology are all factors that are carefully weighed.

Disney's proposed ski resort on federal land is a classic example. In the 1970s, the Disney Corporation proposed to develop a multimillion-dollar complex of recreational facilities on federal land in Mineral King Valley, an especially beautiful wilderness area in California's Sierra Nevada. The legal battle over this plan is a classic in American environmental history. It is a landmark case, bringing out distinctly different points of view about land use, both viewpoints having some validity in American democracy. The case even created its own terminology, and brought to public attention the idea that nonhuman organisms might have some kind of legal standing in the courts.

Such an idea had not really been brought forward so clearly since the Middle Ages, when, in France, townspeople asked the courts to take legal action against the insects eating their crops. This obviously was a very different point of view than the one we have today about the rights of other living things. The case of Mineral King, therefore, is one that every student of environmental science will benefit from knowing.

The Sierra Club, arguing that this development would mar the natural beauty of the wilderness and also damage its ecological balance, brought a suit against the government to prevent it from allowing this intrusion by private enterprise into public lands.

The case raised a curious question: If a wrong was being done, who was wronged? The California courts decided that the Sierra Club itself could not claim direct harm from the development. Moreover, because the government owned the land but also represented the people, it was difficult to argue that the people in general were wronged. In an article entitled "Should Trees Have Standing? Toward Legal Rights for Natural Objects," Christopher D. Stone, a lawyer, suggested that the Sierra Club's case might be based on the fact that inanimate objects have sometimes been treated as having legal standing—for example, in lawsuits involving ships, the ships have legal standing. Stone suggested that trees should have that legal standing, and that although the Sierra Club could not claim direct damage to itself, it could argue on behalf of the nonhuman wilderness.[19]

The case was taken to the U.S. Supreme Court, which concluded that the Sierra Club itself did not have a sufficient "personal stake in the outcome of the controversy" to bring the case to court. But in a famous dissenting statement, Justice William O. Douglas proposed establishing a new federal rule that would allow "environmental issues to be litigated before federal agencies or federal courts in the name of the inanimate object about to be despoiled, defaced, or invaded by roads and bulldozers and where injury is the subject of public outrage." In other words, trees would have legal standing.

In the end, Disney abandoned the idea, but the case was a landmark. While trees did not achieve legal standing in that case, the case did result in discussion of legal rights and ethical values for wilderness and natural systems, a subject that is still being debated. Should our ethical values be extended to nonhuman, biological communities and even to the life-support system of Earth? The position you take on this issue will depend in part on your understanding of the characteristics of wilderness, natural systems, and other environmental factors and features, and in part on your values.

19.8 How You Can Play a Role in Legal Processes

Environmental groups have been a powerful force since the early 1970s. Working through the courts, groups such as the Sierra Club have helped to shape the direction of environmental quality control. Their influence grew partly because the courts, perhaps in response to the national sense that our environment was in crisis, took a more active role and were less willing to leave the matter to the judgment of government agencies. At the same time, citizens were granted greater access to the courts and, through them, to environmental policy-making.

In the 1980s, a new type of environmentalism arose. These environmentalists (whom some people call radicals) believe that when it comes to defending wilderness, there can be no compromise. Their methods have included sit-ins to block roads into forest areas where mining or timber harvesting was scheduled to take place; sitting in trees or implanting large steel spikes to block or discourage timber harvesting; and sabotaging equipment, such as bulldozers. (This kind of sabotage is known as "ecotage.")

Civil disobedience and ecotage make groups like the Sierra Club look like moderates. The more militant groups have caused millions of dollars' worth of damage to a variety of industries that use natural resources in wilderness areas. However, there is no doubt that civil disobedience has been successful in defending the environment in some instances. For example, members of the group Earth First succeeded in halting construction of a road that was to allow access for timber harvesting in an area of southwestern Oregon. Earth First's tactics included blockading the road by sitting or standing in front of the bulldozers,

which slowed construction considerably. Along with this, the group filed a lawsuit against the U.S. Forest Service.

Environmentalists are now relying more on the law. The Endangered Species Act has been used to halt activities such as timber harvesting and development. Although rarely is the presence of an endangered species responsible for stopping a proposed development, those species are increasingly being used as a weapon in attempts to save remaining portions of relatively undisturbed ecosystems.

Some first seek peaceful ways to avoid the cost and delay of litigation. In environmental disputes, an alternative that has recently received considerable attention is **mediation**, where the two or more sides negotiate a compromise with the help of a neutral person, the mediator. The mediator helps to clarify the issues, makes sure each party understands the position and needs of the other parties, and helps them work out a compromise in which each party gains something and they all avoid the risks and costs of litigation. Often, even the possibility of a citizens' lawsuit gives an environmental group a place at the table in such mediation. Litigation, which may delay a project for years, becomes something that can be bargained away in return for certain concessions, such as reduced environmental impact by a project's developer.

In some states, mediation is required by law as an alternative to litigation or before beginning litigation about the siting of waste-treatment facilities—an issue that is almost guaranteed to arouse strong feelings in a community. For example, in Rhode Island a developer who wishes to construct a hazardous-waste treatment facility must negotiate with representatives of the host community and agree to arbitration of any issues not resolved by negotiation. Costs of the negotiation process are borne by the developer.

Storm King Mountain illustrates mediation's advantages over litigation. In this case—a conflict between a utility company and conservationists—mediation could have saved millions of dollars in legal expenses and years of litigation. It began in 1962, when the Consolidated Edison Company of New York announced plans for a new hydroelectric project in the Hudson River highlands, an area with many unique aesthetic qualities as well as thriving fisheries (Figure 19.8). The utility company argued that it needed the new facility, and the environmentalists fought to preserve the beautiful landscape and the fisheries.

The first lawsuit was filed in 1965, and after 16 years of intense courtroom battles, the litigation ended in 1981. Incredibly, the paper trail exceeded 20,000 pages. After millions of dollars had been spent, the various parties finally managed to get together and settle the case with the help of an outside mediator. If they had been able to sit down and talk about the issues at an early

Stone/Getty Images

▪ FIGURE 19.8

Storm King Mountain and the Hudson River highlands in the state of New York, the focus of a court fight between a utility company and environmentalists for nearly 20 years before an argument about building a power plant was finally resolved by mediation.

stage, mediation might have settled the argument much sooner and at a much lower cost to the individual parties and to society.[20] The Storm King Mountain case is often cited as a major victory for environmentalists, but the cost was great to both sides.

19.9 International Environmental Law and Diplomacy

Legal issues involving the environment are difficult enough within a nation, so it is no surprise that they become extremely difficult in international situations. Some issues of concern to many nations are addressed by a collection of policies, agreements, and treaties that are loosely called **"international environmental law."** But since there is no world government with enforcement authority over nations, international law must depend on nations' agreeing to do, or not do, certain things, even when many of their own citizens oppose such an agreement. Still, there have been encouraging developments in the area of international environmental law, such as agreements to reduce air pollutants that destroy stratospheric ozone (the Montreal Protocol of 1987).

Antarctica is one place where international law protects the environment. Antarctica, a continent of 14 million square kilometers (5.4 million square miles), was first visited by a Russian ship in 1820, and people soon recognized that the continent had unique landscapes and life-forms (Figure 19.9). By 1960, a number of countries had claimed parts of Antarctica to exploit mineral and fossil-fuel resources. Then, in 1961, an international treaty designated Antarctica a scientific sanctuary. Thirty years later, in 1991, a major environmental agreement, the Protocol of Madrid, was reached, protecting Antarctica, including islands and sea south of 60° latitude. The continent was designated "nuclear-free," and access to resources was restricted. This was the first step in conserving Antarctica from territorial claims and establishing the "White Continent" as a heritage for all people on Earth.

Other environmental problems addressed at the international level include persistent organic pollutants (POPs) such as dioxins, DDT, and other pesticides. After several years of negotiations in South Africa and Sweden, 127 nations adopted a treaty in May 2001 to greatly reduce or eliminate the use of toxic chemicals known to contribute to cancer and harm the environment. The Kyoto Protocol of 2005 aims to cut emissions of carbon dioxide and other greenhouse gases. It has been ratified by 166 countries and became a formal international treaty in February 2005. The world's largest emitter of CO_2, the United States, was one of only two major countries that did not sign the treaty (the other being Australia).

19.10 The Challenge to Students of the Environment

To end this book on an optimistic note—and there *are* reasons to be optimistic—we note that the Earth Summit on Sustainable Development, held in the summer of 2002 in Johannesburg, South Africa, had the following objectives:

▪ to continue to work toward environmental and social justice for all the people in the world;

▪ to enhance the development of sustainability;

▪ to minimize local, regional, and global environmental degradation from overpopulation, deforestation, mining, agriculture, and pollution of the land, water, and air;

▪ to develop and support international agreements to control global warming and pollutants, and to foster environmental and social justice.

Solving our environmental problems will help build a more secure and sustainable future. This is becoming your charge and responsibility, as students of the environment and as our future leaders, as you graduate from colleges and universities. This transfer of knowledge and leadership is a major reason why we wrote this book.

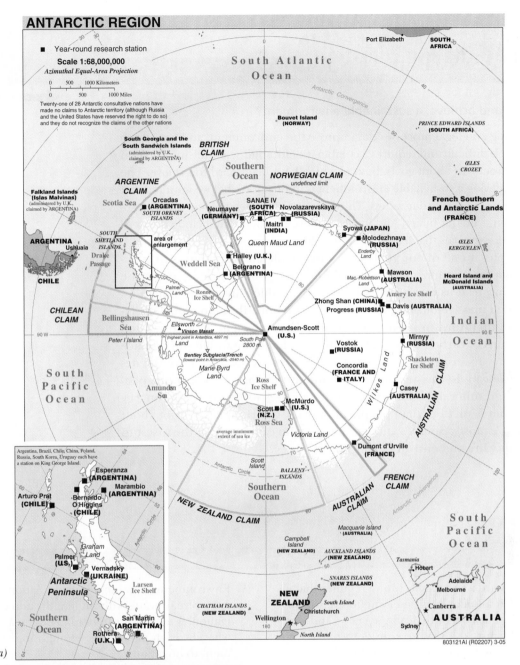

ANTARCTIC REGION

■ Year-round research station

Scale 1:68,000,000
Azimuthal Equal-Area Projection

Twenty-one of 28 Antarctic consultative nations have made no claims to Antarctic territory (although Russia and the United States have reserved the right to do so) and they do not recognize the claims of the other nations

(a)

Argentina, Brazil, Chile, China, Poland, Russia, South Korea, Uruguay each have a station on King George Island.

(b)

Bonne Pioche/Buena Vista/APC/The Kobal Collection, Ltd,

■ **FIGURE 19.9**

International agreements determine environmental practices in Antarctica.

Although different areas of Antarctica are considered to be under the political influence of specific nations, there are international agreements concerning the entire continent. This may be the beginning of a way to deal with regional and global environmental problems. (a) Map of the political divisions of Antartica (b) Emperor penguins in the movie *March of the Penguins*.

Return to the Big Question

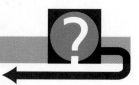

Can we plan, and achieve, a sustainable environment?

Formal planning and legal processes can improve the environment, and have done so, but the task is difficult and requires a combination of solid scientific information and societal agreement about goals, both of which have been hard to achieve. The first step in achieving a sustainable environment is to imagine what we want it to be. The second step is to clarify our values. Third, we apply science to come up with solutions consistent with our values. And finally, we develop plans that include the steps necessary to achieve goals that lead to desired solutions.

For example, if one of our goals for a sustainable environment is to stabilize our climate by minimizing our contribution to global warming, our value clarification may lead us to realize that we value a stable climate more than we value continued massive burning of fossil fuels that contribute to global warming. We would then turn to science to provide alternative energy sources that reduce emissions of carbon dioxide. With those in hand, we would put in place plans and policy for using those alternative energy sources to achieve the desired result. We can do this and still be, productive people. For example, people in Spain and other areas have happy fulfilled lives using much less energy per person from fossil fuels than people in the U.S. do! Happiness is not about who can use the most resources and energy, but who can use them the wisest to sustain a quality life and environment.

Summary

- The path toward a sustainable future requires that we begin to view all people as deserving to live within a framework that is economically, socially, and environmentally just, and that we consider what a desirable environment in the future might be.

- Environmental planning provides a framework for environment management.

- In our increasingly urban environment we have a need to a connection with nature.

- Environmental law including mediation can be a way to resolve disputes without lengthy and expensive litigation. A mediator can help to design a compromise whereby each party gains enough to prefer a settlement to litigation.

- International environmental law is proving useful in addressing several important environmental problems, including preservation of resources and pollution abatement.

Key Terms

environmental law

environmental planning

international environmental law

mediation

sustainable environment

Getting It Straight

1. How can we balance freedom of individual action with the need to sustain our environment?

2. Who are environmentalists?

3. What are three important aspects of ecotopia?

4. Why is environmental planning and review closely related to how land is used?

5. Why does individual freedom and the welfare of society clash when it comes to environmental planning?

6. What three federal legislations are necessary to help keep land and natural resources protected?

7. When was the National Environmental Protection Act passed?

8. What is the National Environmental Protection Act?

9. What is the Bureau of Land Management and what is its responsibility to the land/Earth?

10. How have activists used the Endangered Species Act to help minimize environmental destruction?

11. What are some environmental problems being addressed at international levels?

12. What is environmental justice?

What Do You Think?

1. The famous ecologist Garrett Hardin argued that designated wilderness areas should not have provisions for people with handicaps, even though he himself was handicapped. He believed that wilderness should be truly natural in the ultimate sense— that is, with no trace of civilization. Argue for or against Garrett Hardin's position. In your argument, consider the People and Nature theme of this book.

2. Visit a local natural or naturalistic place, even a city park, and write down what is necessary for that area to be sustainable as it is presently used.

3. It has been suggested that the drier, western states of the U.S. Midwest, such as the Dakotas, ought to be returned to wilderness, where the buffalo would roam again, since the human population is declining there anyway. Argue for or against this.

4. Can people and nature continue to survive together?

5. Can government and activists ever reach a consensus about the Earth and the way we live on Earth?

Pulling It All Together

1. Using maps of your town, design a greenbelt, or, if one already exists, improve on the design. Redraw the map. Take field trips to visit locations.

2. In the 1960s, an infamous riot took place in Watts, a part of Los Angeles where the poorer people lived. Some claim that one of the reasons people were un-happy there was the lack of parks and places for sports. Using a map of Watts, design a park system for the area.

3. What is your vision of ecotopia? How can we begin to use our environment more wisely so that your ideal ecotopia can come to exist?

Further Reading

Beveridge, C. E., and P. Rocheleau. 1995. *Frederick Law Olmsted: Designing the American Landscape.* New York: Rizzoli International.—A beautifully illustrated and beautifully written book about the parks and ideas of America's greatest park planner.

Mumford, L. 1961. *The city in history.* New York: Harvest/HBJ Books.—A classic book by one of the great historians of cities, very readable.

Noss, R. F., A. Y. Cooperrider (contributor), and R. Schlickeisen. 1994. *Saving nature's legacy.* Washington, DC: Island Press.—About the wildlands idea.

Rosenzweig, M. L. 2003. *Win-win ecology: How the Earth's species can survive in the midst of human enterprise.* New York: Oxford University Press.—The author argues that ecological science leads to a belief that people and nature are not in opposition to one another.

Appendix

A

Matter and Energy

The universe as we know it consists of two entities: matter and energy. *Matter* is the material that makes up our physical and biological environments (you are composed of matter). *Energy* is the ability to do work.

Two laws about energy. The first law of thermodynamics—also known as the law of conservation of energy, or the first energy law—states that energy cannot be created or destroyed but can change from one form to another. This law stipulates that *the total amount of energy in the universe does not change.* This means that it is impossible to get something for nothing when dealing with energy; it is impossible to extract more energy from any system than the amount of energy that originally entered the system.

In fact, the second law of thermodynamics states that you cannot break even: *When energy is changed from one form to another, it always moves from a more useful form to a less useful one.* Thus, although energy is conserved as it moves through a system and is changed from one form to another, it becomes less useful.

Matter and the atom. We turn next to a brief discussion of the basic chemistry of matter, which will help you understand biogeochemical cycles. An *atom* is the smallest part of a chemical element that can take part in a chemical reaction with another atom. An *element* is a chemical substance composed of identical atoms that cannot be separated into different substances by ordinary chemical processes. Each element is given a symbol. For example, the symbol for the element carbon is C, and the symbol for phosphorus is P.

Neutrons, protons, and electrons. A model of an atom (Figure A.1) shows three subatomic particles: neutrons, protons, and electrons. The atom is pictured as having a

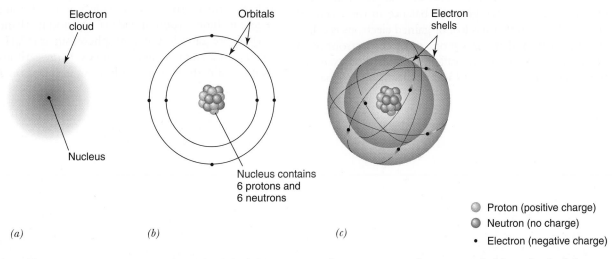

Figure A.1 **Atomic structure.** (a) You can think of the basic structure of an atom as a nucleus surrounded by a cloud of electrons. (b) A model of an atom of carbon with six protons and six neutrons in the nucleus and six orbiting electrons in two energy shells. (c) A three-dimensional view of the carbon atom shown in (b). The size of the nucleus relative to the size of the electron shells is greatly exaggerated. [*Source:* F. Press and R. Siever, *Understanding Earth.* New York: Freeman, 1994.]

405

central nucleus composed of *neutrons, which have no electrical charge, and protons, which have a positive charge.* A cloud of *electrons, which have a negative charge, revolves about the nucleus.* The number of protons in the nucleus is unique for each element and is the atomic number for that element. For example, hydrogen has one proton in its nucleus, and its atomic number is 1. Uranium has 92 protons in its nucleus, and its atomic number is 92. A list of elements with their atomic numbers, called the Periodic Table, is shown in Figure 3.4.

Electrons in our model of the atom are arranged in shells (representing energy levels), and the electrons closest to the nucleus are bound tighter to the atom than those in the outer shells. Electrons have very little mass compared with neutrons or protons; therefore, nearly the entire mass of an atom is in the nucleus.

The sum of the number of neutrons and protons in the nucleus of an atom is known as the atomic weight. Atoms of the same element always have the same atomic number (the same number of protons in the nucleus), but they can have different numbers of neutrons and therefore different atomic weights. Two atoms of the same element with different numbers of neutrons in their nuclei and different atomic weights are known as isotopes of that element. For example, two isotopes of oxygen are ^{16}O and ^{18}O, where 16 and 18 are the atomic weights. Both isotopes have an atomic number of 8, but ^{18}O has two more neutrons than ^{16}O. Study of this sort is proving very useful in learning how the Earth works. For example, the study of oxygen isotopes has given us a better understanding of how the global climate has changed. This topic is beyond the scope of our present discussion, but you can find it in many basic textbooks on oceanography.

An atom is chemically balanced in terms of electric charge when the number of protons in its nucleus is equal to the number of electrons. However, an atom may lose or gain electrons, changing the balance in the electrical charge. An atom that has lost or gained electrons is called an ion. An atom that has lost one or more electrons has a net positive charge and is called a cation. For example, the potassium ion K^+ has lost one electron, and the calcium ion Ca^{+2} has lost two electrons. An atom that has gained electrons has a net negative charge and is called an anion. For example, O^{-2} is an anion of oxygen that has gained two electrons.

What is a compound? A compound is a chemical substance composed of two or more atoms of the same or different elements. The smallest unit of a compound is a molecule. For example, each molecule of water, H_2O, contains two atoms of hydrogen and one atom of oxygen held together by chemical bonds. Minerals that form rocks are compounds, as are most chemical substances found in a solid, liquid, or gaseous state in the environment.

What holds a compound together? The atoms that constitute a compound are held together by chemical bonding. There are several types of chemical bonds. Two examples are covalent and ionic bonds. Covalent bonds result when atoms share electrons. This sharing takes place in the region between the atoms, and the strength of the bond is related to the number of pairs of electrons that are shared. Some important environmental compounds are held together solely by covalent bonds. These include carbon dioxide (CO_2) and water (H_2O). Covalent bonds are stronger than ionic bonds, which form as a result of attraction between positively and negatively charged ions. An example of an environmentally important compound with ionic bonds is table salt (mineral halite), or sodium chloride (NaCl). Compounds with ionic bonds, such as sodium chloride, tend to be soluble in water and thus dissolve easily, as salt does.

In summary, in the study of the environment, we are concerned with matter (chemicals) and energy that moves in and between the major components of the Earth system. An example is the element carbon, which moves through the atmosphere, hydrosphere, lithosphere, and biosphere in a wide variety of compounds. These include carbon dioxide (CO_2) and methane (CH_2), which are gases in the atmosphere; sugar ($O_6H_{12}O_6$) in plants and animals; and complex hydrocarbons (compounds of hydrogen and carbon) in coal and oil deposits.

B Prefixes and Multiplication Factors

Number	10^x, Power of 10	Prefix	Symbol
1,000,000,000,000,000,000	10^{18}	exa	E
1,000,000,000,000,000	10^{15}	peta	P
1,000,000,000,000	10^{12}	tera	T
1,000,000,000	10^9	giga	G
1,000,000	10^6	mega	M
10,000	10^4	myria	
1,000	10^3	kilo	k
100	10^2	hecto	h
10	10^1	deca	da
0.1	10^{-1}	deci	d
0.01	10^{-2}	centi	c
0.001	10^{-3}	milli	m
0.000 001	10^{-6}	micro	μ
0.000 000 001	10^{-9}	nano	n
0.000 000 000 001	10^{-12}	pico	p
0.000 000 000 000 001	10^{-15}	femto	f
0.000 000 000 000 000 001	10^{-18}	atto	a

C Common Conversion Factors

Length

1 yard = 3 ft, 1 fathom = 6 ft

	in.	ft	mi	cm	m	km
1 inch (in.) =	1	0.083	1.58×10^{-5}	2.54	0.0254	2.54×10^{-5}
1 foot (ft) =	12	1	1.89×10^{-4}	30.48	0.3048	—
1 mile (mi) =	63,360	5,280	1	160,934	1,609	1.609
1 centimeter (cm) =	0.394	0.0328	6.2×10^{-6}	1	0.01	1.0×10^{-5}
1 meter (m) =	39.37	3.281	6.2×10^{-4}	100	1	0.001
1 kilometer (km) =	39,370	3,281	0.6214	100,000	1,000	1

Area

1 square mi = 640 acres, 1 acre = 43,560 ft^2 = 4046.86 m^2 = 0.4047 ha
1 ha = 10,000 m^2 = 2.471 acres

	in.^{2}	ft^2	mi^2	cm^2	m^2	km^2
1 in.^{2} =	1	—	—	6.4516	—	—
1 ft^2 =	144	1	—	929	0.0929	—
1 mi^2 =	—	27,878,400	1	—	—	2.590
1 cm^2 =	0.155	—	—	1	—	—
1 m^2 =	1,550	10.764	—	10,000	1	—
1 km^2 =	—	—	0.3861	—	1,000,000	1

Volume

	in.³	ft³	yd³	m³	qt	liter	barrel	gal (U.S.)
1 in.³ =	1	—	—	—	—	0.02	—	—
1 ft³ =	1,728	1	—	0.0283	—	28.3	—	7.480
1 yd³ =	—	27	1	0.76	—	—	—	—
1 m³ =	61,020	35.315	1.307	1	—	1,000	—	—
1 quart (qt) =	—	—	—	—	1	0.95	—	0.25
1 liter (l) =	61.02	—	—	—	1.06	1	—	0.2642
1 barrel (oil) =	—	—	—	—	168	159.6	1	42
1 gallon (U.S.) =	231	0.13	—	—	4	3.785	0.02	1

Mass and Weight

1 pound = 453.6 grams = 0.4536 kilogram = 16 ounces

1 gram = 0.0353 ounce = 0.0022 pound

1 short ton = 2000 pounds = 907.2 kilograms

1 long ton = 2240 pounds = 1008 kilograms

1 metric ton = 2205 pounds = 1000 kilograms

1 kilogram = 2.205 pounds

Energy and Power[a]

1 kilowatt-hour = 3413 Btus = 860,421 calories

2 Btu = 0.000293 kilowatt-hour = 252 calories = 1055 joules

1 watt = 3.413 Btu/hr = 14.34 calorie/min

1 calorie = the amount of heat necessary to raise the temperature of 1 gram (1 cm³) of water 1 degree Celsius

1 quadrillion Btu = (approximately) 1 exajoule

1 horsepower = 7.457×10^2 watts

1 joule = 9.481×10^{-4} Btu = 0.239 cal = 2.778×10^{-7} kilowatt-hour

[a]Values from Lange, N. A., 1967, *Handbook of Chemistry*, McGraw Hill: New York.

Temperature

$F = \frac{9}{5}C + 32$ F is degrees Fahrenheit.
C is degrees Celsius (centigrade).

Fahrenheit		Celsius
32	Freezing of H_2O (Atmospheric Pressure)	0
50		10
68		20
86		30
104		40
122		50
140		60
158		70
176		80
194		90
212	Boiling of H_2O (Atmospheric Pressure)	100

Other Conversion Factors

1 ft³/sec = 0.0283 m³/sec = 7.48 gal/sec = 28.32 liter/sec

1 acre-foot = 43,560 ft³ = 1233 m³ = 325,829 gal

1 m³/sec = 35.32 ft³/sec

1 ft³/sec for one day = 1.98 acre-feet

1 m/sec = 3.6 km/hr = 2.24 mi/hr

1 ft/sec = 0.682 mi/hr = 1.097 km/hr

1 atmosphere = 14.7 lb(in.$^{-2}$) = 2116 lb(ft^{-2}) = 1.013×10^5 N(m^{-2})

Era	Approximate Age in Millions of Years Before Present	Period	Epoch	Life Form
	Less than 0.01		Recent (Holocene)	
	0.01–2	Quaternary	Pleistocene	Humans
	2			
Cenozoic	2–5		Pliocene	
	5–23		Miocene	
	23–35	Tertiary	Oligocene	
	35–56		Eocene	Mammals
	56–65		Paleocene	
	65			
Mesozoic	65–146	Cretaceous		
	146–208	Jurassic		Flying reptiles, birds
	208–245	Triassic		Dinosaurs
	245			
Paleozoic	245–290	Permian		Reptiles
	290–363	Carboniferous		Insects
	363–417	Devonian		Amphibians
	417–443	Silurian		Land plants
	443–495	Ordovician		Fish
	495–545	Cambrian		
	545			
	700			Multicelled organisms
	3,400			One-celled organisms
	4,000	Approximate age of oldest rocks discovered on Earth		
Precambrian				
	4,600	Approximate age of the Earth and meteorites		

Glossary

Acceptable Risk The risk that individuals, society or institutions are willing to take.

Acid mine drainage Acidic water that drains from mining areas (mostly coal but also metal mines). The acidic water may enter surface water resources, causing environmental damage.

Acid rain Rain made acid by pollutants, particularly oxides of sulfur and nitrogen. (Natural rainwater is slightly acid owing to the effect of carbon dioxide dissolved in the water.)

Adaptive radiation The process that occurs when a species enters a new habitat that has unoccupied niches and evolves into a group of new species, each adapted to one of these niches.

Air quality index A numeric description of air quality from good to unhealthy to hazardous.

Air toxins Category of air pollutants that can cause cancer or other serious health problems.

Air-quality standards Levels of air pollutants that delineate acceptable levels of pollution over a particular time period. Valuable because they are often tied to emission standards that attempt to control air pollution.

Alternative energy Renewable and non-renewable energy resources that are alternatives to the fossil fuels.

Aquaculture Production of food from aquatic habitats.

Atmosphere Layer of gases surrounding Earth.

Atmospheric inversion A condition in which warmer air is found above cooler air, restricting air circulation; often associated with a pollution event in urban areas.

Average residence time A measure of the time it takes for a given part of the total pool or reservoir of a particular material in a system to be cycled through the system. When the size of the pool and rate of throughput are constant, average residence time is the ratio of the total size of the pool or reservoir to the average rate of transfer through the pool.

Balance of nature An environmental myth that states that the natural environment, when not influenced by human activity, will reach a constant status, unchanging over time, referred to as an equilibrium state.

Benthos Organisms that live on the bottom of the ocean or other body of water.

Biogeochemical cycle The cycling of a chemical element through the biosphere; its pathways, storage locations, and chemical forms in living things, the atmosphere, oceans, sediments, and lithosphere.

Biological control A set of methods to control pest organisms by using natural ecological interactions, including predation, parasitism, and competition. Part of integrated pest management.

Biological diversity Used loosely to mean the variety of life on Earth, but scientifically typically used to consist of three components: (1) genetic diversity—the total number of genetic characteristics; (2) species diversity; and (3) habitat or ecosystem diversity—the number of kinds of habitats or ecosystems in a given unit area. Species diversity in turn includes three concepts: species richness, evenness, and dominance.

Biological evolution The change in inherited characteristics of a population from generation to generation, which can result in new species.

Biological production The capture of usable energy from the environment to produce organic compounds in which that energy is stored.

Biomagnification Also called *biological concentration*. The tendency for some substances to concentrate with each trophic level. Organisms preferentially store certain chemicals and excrete others. When this occurs consistently among organisms, the stored chemicals increase as a percentage of the body weight as the material is transferred along a food chain or trophic level. For example, the concentration of DDT is greater in herbivores than in plants and greater in plants than in the nonliving environment.

Biosphere Has several meanings. One is that part of a planet where life exists. On Earth it extends from the depths of the oceans to the summit of mountains, but most life exists within a few meters of the surface. A second meaning is: the planetary system that includes and sustains life, and therefore is made up of the atmosphere, oceans, soils, upper bedrock, and all life.

Biota All the organisms of all species living in an area or region up to and including the biosphere, as in "the biota of the Mojave Desert" or "the biota in that aquarium."

Birth rate   The rate at which births occur in a population, measured either as the number of individuals born per unit of time or as the percentage of births per unit of time compared with the total population.

Carbon cycle   Biogeochemical cycle of carbon. Carbon combines with and is chemically and biologically linked with the cycles of oxygen and hydrogen that form the major compounds of life.

Carcinogen   Any material that is known to produce cancer in humans or other animals.

Carrying capacity   The maximum abundance of a population or species that can be maintained by a habitat or ecosystem without degrading the ability of that habitat or ecosystem to maintain that abundance in the future.

Catastrophe   A situation of event that causes sufficient damage to people, property or society from which recovery is a long and involved process. Also is defined as a very serious disaster.

Chemical reaction   The process in which compounds and elements undergo a chemical change to become a new substance or substances.

Chronic patchiness   A situation where ecological succession does not occur. One species may replace another, or an individual of the first species may replace it, but no overall general temporal pattern is established. Characteristic of harsh environments such as deserts.

Climate   The representative or characteristic conditions of the atmosphere at particular places on Earth. Climate refers to the average or expected conditions over long periods; weather refers to the particular conditions at one time in one place.

Climatic change   Change in mean annual temperature and other aspects of climate over periods of time ranging from decades to hundreds of years to several million years.

Commons   Land that belongs to the public, not to individuals. Historically a part of old English and New England towns where all the farmers could graze their cattle.

Community-level effect   When the interatction between two species leads to changes in the presence or absence of other species or in a large change in abundance of other species, then a community effect is said to have occurred.

Competition   The situation that exists when different individuals, populations, or species compete for the same resource(s) and the presence of one has a detrimental effect on the other. Sheep and cows eating grass in the same field are competitors.

Competitive exclusion principle   The idea that two populations of different species with exactly the same requirements cannot persist indefinitely in the same habitat—one will always win out and the other will become extinct.

Composting   Biochemical process in which organic materials, such as lawn clippings and kitchen scraps, are decomposed to a rich, soil-like material.

Contamination   Presence of undesirable material that makes something unfit for a particular use.

Contour plowing   Plowing land along topographic contours, as much in a horizontal plane as possible, thereby decreasing the erosion rate.

Conventional energy source   Sources of energy such as coal and oil that provide most of our energy at the present.

Cultural eutrophication   Human-induced eutrophication that involves nutrients such as nitrates or phosphates that cause a rapid increase in the rate of plant growth in ponds, lakes, rivers or the ocean.

Death rate   The rate at which deaths occur in a population, measured either as the number of individuals dying per unit time or as the percentage of a population dying per unit time.

Demographic transition   The pattern of change in birth and death rates as a country is transformed from undeveloped to developed. There are three stages: (1) in an undeveloped country, birth and death rates are high and the growth rate low; (2) the death rate decreases, but the birth rate remains high and the growth rate is high; (3) the birth rate drops toward the death rate and the growth rate therefore also decreases.

Deoxyribonucleic acid (DNA)   Complex chemical compound that contains the building blocks that carry the genetic code and genes.

Desalination   The removal of salts from seawater or brackish water so that the water can be used for purposes such as agriculture, industrial processes, or human consumption.

Desertification   The process of creating a desert where there was not one before.

Direct effects   With respect to natural hazards refers to the number of people killed, injured, dislocated, made homeless or otherwise damaged by a hazardous event.

Disaster   A hazardous event that occurs over a limited span of time in a defined geographic area. Lose of human life and property damage is significant.

Divergent evolution   Organisms with the same ancestral genetic heritage migrate to different habitats and evolve into species with different external forms and structures, but typically continue to use the same kind of habitats. The ostrich and the emu are believed to be examples of divergent evolution.

Dose response   The principle that the effect of a certain chemical on an individual depends on the dose or concentration of that chemical.

Doubling time   The time necessary for a quantity of whatever is being measured to double.

Dust bowl   Time of drought and extensive soil erosion by wind(dust) storms that in the 1930s occurred in Oklahoma and other midwestern states. It was an environmental disaster.

Dynamic equilibrium   An equilibrium is a condition of constancy for a system, also referred to as the "rest state" for the system, like an ecosystem or a population. A dynamic equilibrium is one in which the "rest state" changes somewhat and slowly over time.

Early-successional species Species that occur only or primarily during early stages of succession.

Ecological community This term has two meanings. (1) A conceptual or functional meaning: a set of interacting species that occur in the same place (sometimes extended to mean a set that interacts in a way to sustain life). (2) An operational meaning: a set of species found in an area, whether or not they are interacting.

Ecological footprint Measure of the total impact a person or society has on the environment. Based on resource use and waste produced.

Ecological gradient A change in the relative abundance of a species or group of species along a line or over an area.

Ecological island An area that is biologically isolated so that a species occurring within the area cannot mix (or only rarely mixes) with any other population of the same species.

Ecological niche The general concept is that the niche is a species' "profession"—what it does to make a living. The term is also used to refer to a set of environmental conditions within which a species is able to persist.

Ecological restoration Restoration of the land and water (rivers, wetlands, beaches, prairie etc.) that utilizes ecologic principals.

Ecological succession The process of the development of an ecological community or ecosystem, usually viewed as a series of stages—early, middle, late, mature (or climax), and sometimes postclimax. Primary succession is an original establishment; secondary succession is a reestablishment.

Ecosystem An ecological community and its local, nonbiological community. An ecosystem is the minimum system that includes and sustains life. It must include at least an autotroph, a decomposer, a liquid medium, a source and sink of energy, and all the chemical elements required by the autotroph and the decomposer.

El Niño Natural perturbation of the physical earth system that affects global climate. Characterized by development of warm oceanic waters in the eastern part of the tropical Pacific Ocean, a weakening or reversal of the trade winds, and a weakening or even reversal of the equatorial ocean currents. Reoccurs periodically and affects the atmosphere and global temperature by pumping heat into the atmosphere.

Electromagnetic fields (EMF) Magnetic and electrical fields produced naturally by our planet and also by appliances such as toasters, electric blankets, and computers. There currently is controversy concerning potential adverse health effects related to exposure to EMF in the workplace and home from such artificial sources as power lines and appliances.

Endangered species A species that faces threats that might lead to its extinction in a short time.

Energy flow The movement of energy through an ecosystem from the external environment through a series of organisms and back to the external environment. It is one of the fundamental processes common to all ecosystems.

Energy An abstract concept referring to the ability or capacity to do work.

Environmental economics Economic effects of the environment and how economic processes affect that environment, including its living resources.

Environmental justice The principle of dealing with environmental problems in such a way as to not discriminate against people based upon socioeconomic status, race, or ethnic group.

Environmental law A field of law concerning the conservation and use of natural resources and the control of pollution.

Environmental planning Planning that emphasizes the environment. Includes environmental impact and land use.

Environmental unity A principle of environmental sciences that states that everything affects everything else, meaning that a particular course of action leads to an entire potential string of events. Another way of stating this idea is that you can't only do one thing.

E-waste Waste generated from computers, televisions, cell phones, and other electronic devices used by people.

Exponential growth Growth in which the rate of increase is a constant percentage of the current size; that is, the growth occurs at a constant rate per time period.

Externality In economics, an effect not normally accounted for in the cost-revenue analysis.

Extinction Disappearance of a life-form from existence; usually applied to a species.

Facilitation During succession, one species prepares the way for the next (and may even be necessary for the occurrence of the next).

Feedback A kind of system response that occurs when output of the system also serves as input leading to changes in the system.

Fission The splitting of an atom into smaller fragments with the release of energy.

Flux Rate of transfer of a chemical from one part of a biogeochemical cycle to another.

Food web A network of who feeds on whom or a diagram showing who feeds on whom. It is synonymous with food chain.

Fossil fuels Forms of stored solar energy created from incomplete biological decomposition of dead organic matter. Includes coal, crude oil, and natural gas.

Fusion Combining of light elements to form heavier elements with the release of energy.

Gaia hypothesis The Gaia hypothesis states that the surface environment of Earth, with respect to such factors as the atmospheric composition of reactive gases (for example, oxygen, carbon dioxide, and methane), the acidity-alkalinity of waters, and the surface temperature, are actively regulated by the sensing, growth, metabolism and other activities of the biota. Interaction between the physical and biological system on Earth's surface has led to a planetwide physiology that began more than 3 billion years ago and the evolution of which can be detected in the fossil record.

Gene A single unit of genetic information comprised of a complex segment of the four DNA base-pair compounds.

Genetic diversity   The diversity of the genetic pool, literally the number of different forms of DNA that are found. The term usually refers to a single species or population, but is also used to refer to the total genetic diversity of the many species that form an ecological community or ecosystem.

Genetic drift   Changes in the frequency of a gene in a population as a result of chance rather than of mutation, selection, or migration.

Geologic cycle   The formation and destruction of earth materials and the processes responsible for these events. The geologic cycle includes the following subcycles: hydrologic, tectonic, rock, and geochemical.

Geothermal energy   The useful conversion of natural heat from the interior of Earth.

Global warming   Natural or human-induced increase in the average global temperature of the atmosphere near Earth's surface.

Greenhouse effect   Process of trapping heat in the atmosphere. Water vapor and several other gases warm the Earth's atmosphere because they absorb and remit radiation; that is, they trap some of the heat radiating from the Earth's atmospheric system.

Greenhouse gases   The suite of gases that have a greenhouse effect, such as carbon dioxide, methane, and water vapor.

Groundwater   Water found beneath the Earth's surface within the zone of saturation, below the water table.

Growth rate   The net increase in some factor per unit time. In ecology, the growth rate of a population, sometimes measured as the increase in numbers of individuals or biomass per unit time and sometimes as a percentage increase in numbers or biomass per unit time.

Gulf Stream   Oceanic surface current carrying warm water from south to north in the Atlantic Ocean.

Habitat   Where an individual, population, or species exists or can exist. For example, the habitat of the Joshua tree is the Mojave Desert of North America.

Habitat diversity   The number of kinds of habitats in an area.

Half-life   The time required for half the amount of a substance to disappear; the average time required for one-half of a radioisotope to be transformed to some other isotope; the time required for one-half of a toxic chemical to be converted to some other form.

Hazardous waste   Waste that is classified as definitely or potentially hazardous to the health of people. Examples include toxic or flammable liquids and a variety of heavy metals, pesticides, and solvents.

Heavy metals   Refers to a number of metals, including lead, mercury, arsenic, and silver (among others) that have a relatively high atomic number (the number of protons in the nucleus of an atom). They are often toxic at relatively low concentrations, causing a variety of environmental problems.

Hormonally active agents (HAAs)   Chemicals in the environment able to cause reproductive and developmental abnormalities in animals, including humans.

Hydrologic cycle   Circulation of water from the oceans to the atmosphere and back to the oceans by way of evaporation, runoff from streams and rivers, and groundwater flow.

Incineration   Combustion of waste at high temperature, consuming materials and leaving only ash and noncombustibles to dispose of in a landfill.

Indirect effects   Effects from a natural hazard or disaster that includes donations of money and goods as well as providing shelter for people and paying taxes that will help finance recovery and relief of emotional distress caused by natural hazardous events.

Indoor air pollution

Industrial ecology   Process of designing industrial systems to behave more like ecosystems where waste from one part of the system is a resource for another part.

In-stream use   A type of water use that includes navigation, generation of hydroelectric power, fish and wildlife habitat, and recreation.

Integrated pest management   Control of agricultural pests using several methods together, including biological and chemical agents. A goal is to minimize the use of artificial chemicals; another goal is to prevent or slow the buildup of resistance by pests to chemical pesticides.

Integrated waste management (IWM)   Set of management alternatives including reuse, source reduction, recycling, composting, landfill, and incineration.

Interference   During succession, one species prevents the entrance of later successional species into an ecosystem. For example, some grasses produce such dense and thick mats that seeds of trees cannot reach the soil to germinate. As long as these grasses persist, the trees that characterize later stages of succession cannot enter the ecosystem.

Interglacial periods   Periods of time lasting thousands of years that are characterized by relatively warm air temperatures and high sea levels. Alternate with glacial periods.

International environmental law   Body of laws consisting of treaties that address environmental problems and solutions.

Intertidal   Coastal zone located between low and high tide.

Island biogeography   The study of geographical principles of islands.

Isotopes   Atoms of an element that have the same atomic number (the number of protons in the nucleus of the atom) but vary in atomic mass number (the number of protons plus neutrons in the nucleus of an atom).

Keystone species   A species, such as the sea otter, that has a large effect on its community or ecosystem so that its removal or addition to the community leads to major changes in the abundances of many or all other species.

Landscape perspective   The concept that effective management and conservation recognizes that ecosystems, populations, and species are interconnected across large geographic areas.

Late-successional species   Species that occur only or primarly in, or are dominant in, late stages in succession.

Leachate   Noxious, mineralized liquid capable of transporting bacterial pollutants. Produced when water infiltrates through waste material and becomes contaminated and polluted.

Life expectancy The estimated average number of years (or other time period used as a measure) that an individual of a specific age can expect to live.

Life history difference

Limiting factor The single requirement for growth available in the least supply in comparison to the need of an organism. Originally applied to crops but now often applied to any species.

Logistic carrying capacity In terms of the logistic curve, the population size at which births equal deaths and there is no net change in the population.

Logistic growth curve The S-shaped growth curve that is generated by the logistic growth equation. In the logistic, a small population grows rapidly, but the growth rate slows down, and the population eventually reaches a constant size.

Longevity How long an organism or species generally lives. Similar to life expectancy.

Mariculture Production of food from marine habitats.

Mass extinction Event that causes the extinction of many forms of life on a global scale.

Materials management Term related to waste management consistent with the ideals of industrial ecology that makes better use of materials, leading to more sustainable use of resources.

Maximum lifetime Genetically determined maximum possible age to which an individual of a species can live.

Maximum sustainable yield The maximum usable production of a biological resource that can be obtained in a specified time period without decreasing the ability of the population to sustain that level of production.

Mediation Negotiation process between adversaries, guided by a neutral facilitator.

Missing carbon sink Substantial amounts of carbon dioxide released into the atmosphere but apparently not reabsorbed and thus remaining unaccounted for.

Municipal solid waste (MSW) Solid waste (paper, glass, plastic, etc.) generated by our urban areas.

MSW landfill Site to dispose of municipal solid waste, also called a sanitary landfill.

Mutation Stated most simply, a chemical change in a DNA molecule. It means that the DNA carries a different message than it did before, and this change can affect the expressed characteristics when cells or individual organisms reproduce.

Natural hazard Any natural process that is a potential threat to human life and property.

Natural selection A process by which organisms whose biological characteristics better fit them to the environment are represented by more descendants in future generations than those whose characteristics are less fit for the environment.

Nitrogen cycle A complex biogeochemical cycle responsible for moving important nitrogen components through the biosphere and other Earth systems. This is an extremely important cycle because nitrogen is required by all living things.

Nitrogen fixation The process by which atmospheric nitrogen is converted to ammonia, nitrate ion, or amino acids. Microorganisms perform most of the conversion, but a small amount is also converted by lightning.

Noise pollution A type of pollution characterized by unwanted or potentially damaging sound.

Non-point sources Sources of pollutants that are diffused and intermittent and are influenced by factors such as land use, climate, hydrology, topography, native vegetation, and geology.

Nonrenewable energy Energy sources, including nuclear and geothermal, that are dependent on fuels, or a resource that may be used up much faster than it is replenished by natural processes.

No-till agriculture Combination of farming practices that includes not plowing the land and using herbicides to keep down weeds.

Nuclear energy The energy of the atomic nucleus that, when released, may be used to do work. Controlled nuclear fission reactions take place within commercial nuclear reactors to produce energy.

Nuclear fuel cycle Processes involved with producing nuclear power from the mining and processing of uranium to control fission, reprocessing of spent nuclear fuel, decommissioning of power plants, and disposal of radioactive waste.

Nuclear reactors Devices that produce controlled nuclear fission, generally for the production of electric energy.

Off-stream use Type of water use where water is removed from its source for a particular use.

Optimum sustainable population The population that is in some way best for the population, its ecological community, its ecosystem, or the biosphere.

Ore deposits Earth materials in which metals are concentrated in high concentrations, sufficient to be mined.

Organic compounds A compound of carbon; originally used to refer to the compounds found in and formed by living things.

Overgrazing When the carrying capacity of land for an herbivore, such as cattle or deer, is exceeded.

Overshoot and collapse Occurs when growth in one part of a system over time exceeds carrying capacity, resulting in sudden decline in one or both parts of the system.

Ozone shield Stratospheric ozone layer that absorbs ultraviolet radiation.

Parasitism An interaction between species that is beneficial to one and generally harmful to the other.

Particulates small particles of solid or liquid substances that are released into the atmosphere by many activities, including farming, volcanic eruption, and burning fossil fuels. Particulates affect human health, ecosystems, and the biosphere.

Pelagic whaling Practice of whalers taking to the open seas and searching for whales from ships that remained at sea for long periods.

Persistent organic pollutants (POPs) Synthetic carbon-based compounds, often containing chlorine, that do not easily break down in the environment. Many were introduced decades before their harmful effects were fully understood and are now banned or restricted.

Phosphorus cycle Major biogeochemical cycle involving the movement of phosphorus throughout the biosphere and lithosphere. This cycle is important because phosphorus is an essential element for life and often is a limiting nutrient for plant growth.

Photovoltaic Technology that converts sunlight directly into electricity using a solid semiconductor material.

Point sources Sources of pollution such as smokestacks, pipes, or accidental spills that are readily identified and stationary. They are often thought to be easier to recognize and control than are area sources. This is true only in a general sense, as some very large point sources emit tremendous amounts of pollutants into the environment.

Policy instruments The means to implement a society's policies. Includes moral suasion (jawboning—persuading people by talk, publicity, and social pressure); direct controls, including regulations; and market processes affecting the price of goods and processes, such as subsidies, licenses, and deposits.

Pollution prevention Identifying ways to avoid the generation of waste rather than finding ways to dispose of it.

Pollution The process by which something becomes impure, defiled, dirty, or otherwise unclean.

Population age structure The number of individuals or proportion of the population in each age class.

Population dynamics The study of changes in population sizes and the causes of these changes.

Population A group of individuals of the same species living in the same area or interbreeding and sharing genetic information.

Precautionary principle The idea that in spite of the fact that full scientific certainty is often not available to prove cause and effect, we should still take cost-effective precautions to solve environmental problems when there exists a threat of potential serious and/or irreversible environmental damage.

Predation-parasitism Interaction between individuals of two species in which the outcome benefits one and is detrimental to the other.

Primary pollutants Air pollutants emitted directly into the atmosphere. Included are particulates, sulfur oxides, carbon monoxide, nitrogen oxides, and hydrocarbons.

Primary production See **Production, primary.**

Primary succession The initial establishment and development of an ecosystem.

R to C ratio A measure of the time available for finding the solutions to depletion of nonrenewable reserves, where R is the known reserves (for example, hundreds of thousands of tons of a metal) and C is the rate of consumption (for example, thousands of tons per year used by people).

Radioactive decay A process of decay of radioisotopes that change from one isotope to another and emit one or more forms of radiation.

Radioactive waste Type of waste produced in the nuclear fuel cycle; generally classified as high level or low level.

Radioisotope A form of a chemical element that spontaneously undergoes radioactive decay.

Reduce, reuse, recycle The three Rs of integrated waste management.

Renewable energy Alternative energy sources, such as solar, water, wind, and biomass, that are more or less continuously made available in a time framework useful to people.

Reserves Known and identified deposits of earth materials from which useful materials can be extracted profitably with existing technology and under present economic and legal conditions.

Resources Reserves plus other deposits of useful earth materials that may eventually become available.

Restoration ecology The field within the science of ecology with the goal to return damaged ecosystems to ones that are functional, sustainable, and more natural in some meaning of this word.

Risk assessment The process of determining potential adverse environmental health effects to people following exposure to pollutants and other toxic materials. Generally includes the four steps of identification of the hazard, dose-response assessment, exposure assessment, and risk characterization.

Risk–benefit analysis In environmental economics, the riskiness of the future that influences the value we place on things in the present.

Scrubbing A process of removing sulfur from gases emitted from power plants burning coal. The gases are treated with a slurry of lime and limestone, and the sulfur oxides react with the calcium to form insoluble calcium sulfides and sulfates that are collected and disposed of.

Secondary pollutants Air pollutants produced through reactions between primary pollutants and normal atmospheric compounds. An example is ozone that forms over urban areas through reactions of primary pollutants, sunlight, and natural atmospheric gases.

Secondary production See **Production, secondary.**

Secondary succession The reestablishment of an ecosystem where there are remnants of a previous biological community.

Sick building syndrome (SBS) Condition associated with a particular indoor environment that appears to be unhealthy to the human occupants.

Sink The location where a chemical that is transferred from one part of a biogeochemical cycle to another ends up being stored.

Smog A term first used in 1905 for a mixture of smoke and fog that produced unhealthy urban air. There are several types of fog, including photochemical smog and sulfurous smog.

Solar energy Collecting and using energy from the sun directly.

Source reduction Process of waste management, the object of which is to reduce the amounts of materials that must be handled in the waste stream.

Species diversity The variety of species in an area or on Earth. Includes factors such as abundance and dominance of species.

Species A group of individuals capable of interbreeding.

Steady state When input equals output in a system, there is no net change and the system is said to be in a steady state. A bathtub with water flowing in and out at the same rate maintains the same water level and is in a steady state. Compare with **equilibrium.**

Strip mining Surface mining in which the overlying layer of rock and soil is stripped off to reach the resource. Large strip mines are some of the largest excavations caused by people in the world.

Successional stages The process of establishment and development of an ecosystem.

Sulfur cycle The biogeochemical cycle of the sources ,flux and sink for sulfur at the, local to blobal scale.

Sustainability Management of natural resources and the environment with the goals of allowing the harvest of resources to remain at or above some specified level, and the ecosystem to retain its functions and structure.

Sustainable environment An environment that is subject to some human use, but at a level that leads to no loss of species or of necessary environmental functions.

Sustainable water use Use of water resources that does not harm the environment and provides for the existence of high-quality water for future generations.

Symbionts Each partner in symbiosis.

Symbiosis An interaction between individuals of two different species that benefits both. For example, lichens contain an alga and a fungus that require each other to persist. Sometimes this term is used broadly, so that domestic corn and people could be said to have a symbiotic relationship—domestic corn cannot reproduce without the aid of people, and some people survive because they have corn to eat.

Synergism Cooperative action of different substances such that the combined effect is greater than the sum of the effects taken separately.

System A set of components that are linked and interact to produce a whole. For example, the river as a system is composed of sediment, water, bank, vegetation, fish, and other living things that all together produce the river.

Terminator gene A genetically modified crop that has a gene to cause the plant to become sterile after the first year.

Thermal pollution A type of pollution that occurs when heat is released into water or air and produces undesirable effects on the environment.

Threatened species Species experiencing a decline in the number of individuals to the degree that a concern is raised about the possibility of extinction of that species.

Tidal power Form of water utilizing ocean tides in places where favorable topography allows for construction of a power plant.

Toxic Harmful, deadly, or poisonous.

Toxicology The science concerned with study of poisons (or toxins) and their effects on living organisms. The subject also includes the clinical, industrial, economic, and legal problems associated with toxic materials.

Trophic level In an ecological community, all the organisms that are the same number of food-chain steps from the primary source of energy. For example, in a grassland the green grasses are on the first trophic level, grasshoppers are on the second, birds that feed on grasshoppers are on the third, and so forth.

Waste stream The waste generated by society.

Wastewater treatment Process of treating wastewater (primarily sewage) in specially designed plants that accept municipal wastewater. Generally divided into three categories: primary treatment, secondary treatment, and advanced wastewater treatment.

Water conservation Practices designed to reduce the amount of water we use.

Wetlands Comprehensive term for landforms such as salt marshes, swamps, bogs, prairie potholes, and vernal pools. Their common feature is that they are wet at least part of the year and as a result have a particular type of vegetation and soil. Wetlands form important habitats for many species of plants and animals, while serving a variety of natural service functions for other ecosystems and people.

Work Force times the distance through which it acts. When work is done we say energy is expended.

Zero population growth A population in which the number of births equals the number of deaths so that there is no net change in the size of the population.

Zero waste The concept of eliminating waste by not generating it or transforming it to resources.

Notes

Chapter 1 Notes

1. Rolett, B., and J. Diamond. 2004. Environmental predictors of pre-European deforestation on Pacific Islands. *Nature* 431:443–446.

2. Stokstad, E. 2004. Heaven or hellhole? Islands' destinies were shaped by geography. *Science* 305:1889.

3. Hunt, T. L. 2006. Rethinking the fall of Easter Island. *American Scientist* 94(5):412–419.

4. Brown, L. R., and C. Flavin. 1999. A new economy for a new century. In L. Star, ed., *State of the world*. New York: W. W. Norton, pp. 3–21.

5. Ehrlich, P. R., A. H. Ehrlich, and P. H. Holdren. 1977. *Ecoscience: Population, resources, environment*, 3rd ed. San Francisco: W. H. Freeman.

6. Margulis, L., and J. E. Lovelock. 1989. Gaia and geognosy. In M. B. Rambler, L. Margulis, and R. Fester, eds., *Global ecology: Towards a science of the biosphere*. Boston: Academic Press, pp. 1–30.

7. Botkin, D. B., M. Caswell, J. E. Estes, and A. Orio, eds. 1989. *Changing the global environment: Perspectives on human involvement*. New York: Academy Press.

8. Lovelock, J. 1995. *The ages of Gaia: A biography of our living earth*, rev. ed. New York: W. W. Norton.

9. Brown, L. R., and J. L. Jacobson. 1987. The future of urbanization: Facing the ecological and economic constraints. In K. Davis, M. S. Bernstam, and H. M. Sellers, eds., *Population and resources in a changing world*. Stanford, CA: The Morrison Institute for Population and Resource Studies, Stanford University.

10. Haub, C., and D. Cornelius. 2000. World population data sheet. Washington, DC: Population Reference Bureau.

11. Bartlett, A. A. 1980. Forgotten fundamentals of the energy crisis. *Journal of Geological Education* 28:4–35.

12. Meadows, D. H., D. L. Meadows, and J. Randers. 1992. *Beyond the limits: Confronting global collapse; envisioning a sustainable future*. Post Mills, VT: Chelsea Green.

13. Western, D., and C. Van Prat. 1973. Cyclical changes in habitat and climate of an East African ecosystem. *Nature* 241(549):104–106.

14. Dunne, T., and L. B. Leopold. 1978. *Water in environmental planning*. San Francisco: W. H. Freeman.

15. Wootton, J. T., M. S. Parker, and M. E. Power. 1996. Effects of disturbances on river food webs. *Science* 273:1558–1561.

16. Leach, M. K., and T. J. Givnich. 1996. Ecological determinants of species loss in remnant prairies. *Science* 273:1555–1558.

17. Gardner, G. T., and P. C. Stern. 2002. Environmental problems and human behavior, 2nd ed. Boston: Pearson Custom Publishing, p. 371.

18. World Wildlife Fund. 2004. Living planet report. Gland, Switzerland, p. 42.

19. Foster, K. R., P. Vecchia, and M. H. Repacholi. 2000. Science and the Precautionary Principle. *Science* 288:979–981.

20. Easton, T. A. and T. D. Goldfarb, eds. 2003. *Taking sides, environmental issues*, 10th ed. Issue 5. Is the Precautionary Principle a sound basis for international policy? Guilford, CT: McGraw-Hill/Dushkin, pp. 76–101.

Chapter 2 Notes

1. Population Reference Bureau, 2002 World Population Data Sheet. The World Bank Data and Statistics, website http://devdata.worldbank.org/externalCP-Profile.asp?SelectedCountry=BGD&CCODE=BGD&CNAME=Bangladesh&PTYPE=CP. Last accessed 4/28/04.

2. EGERTON, F. N. 1975. Aristotle's population biology. *Arethusa* 8: 307–30.

3. Keyfitz, N. 1992. Completing the worldwide demographic transition: The relevance of past experience. Ambio 21:26–30.

4. Population_Reference_Bureau (2006). 2006 World Population DATA SHE E T. Washington, D. C., Population Reference Bureau: **12pp.**

5. World Bank. 1984. World development report 1984. New York: Oxford University Press.

6. Bureau of the Census. 1990. Statistical abstract of the United States 1990. Washington, D.C.

7. Xinhua News Agency, China's cross-border tourism prospers in 2002, December 31, 2002. From the Population Reference Bureau Web site available at http://www.prb.org/Template.cfm?Section=PRB&template=/ContentManagement/ContentDisplay.cfm&ContentID=8661 .U.S. Department of Commerce.

8. U. S. Centers for Disease Control website available at http://www.cdc.gov/ncidod/sars/factsheet.htm and http://www.cdc.gov/ncidod/dvbid/westnile/qa/overview.htm

9. Alan Guttmacher Institute. Sharing responsibility: Women, society and abortion worldwide. New York: AGI, 1999.

10. Source: U S Government Center for Disease Control Website, http://www.cdc.gov/ncidod/dvbid/westnile/wnv_factsheet.ht, accessed 2/6/05.

11. Joint United Nations Programme on HIV/AIDS. 1999. AIDS epidemic update. Geneva: Switzerland.

12. Population Division, United Nations Department of Economic and Social Affairs, 1998. World population growing despite AIDS spread, Publisher. United Nations, N.Y. accessed at www.un.org/esa/population/publications/AID impact. Last accessed 4/28/04.

13. Dumond, D. E. 1975. The limitation of human population: A natural history. Science 187:713–721.

14. World Bank. 2000. World development indicators 2000. Washington, D.C.: World Bank.

15. Central Intelligence Agency. 1999. The world factbook. Washington, D.C.: CIA.

16. World Bank. 1992. World development report. The relevance of past experience. Washington, D.C.: World Bank.

17. Guz, D., and J. Hobcraft. 1991. Breastfeeding and fertility: A comparative analysis. Population Studies 45:91–108.

18. Fathalla, M. F. 1992. Family planning: Future needs. Ambio 21:84–87.

19. Haupt, A., and T. T. Kane. 1978. The Population Reference Bureau's population handbook. Washington, D.C.: Population Reference Bureau.

20. Planned Parenthood Federation of America, Public Policy Division. 1997 (June). International family planning: The need for services. Planned Parenthood Federation of America. N. Y.

21. Xinhua News Agency March 13, 2002 untitled, available at http://www.16da.org.cn/english/archiveen/28691.htm.

Chapter 3 Notes

1. Lehman, J. T. 1986. Control of eutrophication in Lake Washington. In G. H. Orians, ed., *Ecological knowledge and environmental problem solving*. Washington, DC: National Academy of Science, pp. 302–316.

2. Henderson, L. J. [1913] 1966. *The fitness of the environment*. Boston: Beacon.

3. Isacks, B., J. Oliver, and L. Sykes. 1968. Seismology and the new global tectonics. *Journal of Geophysical Research* 73:5855–5899.

4. Dewey, J. F. 1972. Plate tectonics. *Scientific American* 22:56–68.

5. Botkin, D. B. 1990. *Discordant harmonies: A new ecology for the 21st century*. New York: Oxford University Press.

6. Ehrlich, P. R., A. H. Ehrlich, and J. P. Holdren. 1970. *Ecoscience: Population, resources, environment*. San Francisco: W. H. Freeman, p. 1051.

7. Post, W. M., T. Peng, W. R. Emanuel, et al. 1990. The global carbon cycle. *American Scientist* 78:310–326.

8. Keeling, C. D., T. P. Whorf, M. Wahlen, and J. van der Plicht. 1995. Interannual extremes in the rate of rise of atmospheric carbon dioxide since 1980. *Nature* 375:666–670.

9. Hudson, R. J. M., S. A. Gherini, and R. A. Goldstein. 1994. Modeling the global carbon cycle: Nitrogen fertilization of the terrestrial biosphere and the "missing" CO_2 sink. *Global Biogeochemical Cycles* 8:307–333.

10. Woods Hole. 2000. The missing carbon sink. Available at http://www.whrc.org/science/carbon/missingc.htm. Accessed August 18, 2003.

11. Houghton, R. 2003. Why are estimates of the global carbon balance so different? *Global Change Biology* 9:500–509.

12. Houghton, R. 2003. Revised estimates of the annual net flux of carbon to the atmosphere from changes in land use and land management 1850–2000. *Tellus* 55 B:378–390.

13. Herring, D., and R. Kannenberg. 2000. The mystery of the missing carbon. Available at http://earthobservatory.nasa.gov/cgi-bin/printall?/study/BOREAS/missing_carbon.html. Accessed July 5, 2000.

14. Chameides, W. L., and E. M. Perdue. 1997. *Biogeochemical cycles*. New York: Oxford University Press.

15. Agren, G. I., and E. Bosatta. 1996. *Theoretical ecosystem ecology.* New York: Cambridge University Press.

16. Carter, L. J. 1980. Phosphate: Debate over an essential resource. *Science* 209:44–54.

Chapter 4 Notes

1. Line, L. 1996 (April 16). Ticks and moths, not just oaks, linked to acorns, *New York Times.*

2. Ostfield, R. S., C. G. Jones, and J. O. Wolff. 1996 (May). Of mice and mast: Ecological connections in eastern deciduous forests. *BioScience* 46(5):323 330.

3. Leopold, A.. "Deer Irruptions," reprinted in Wisconsin Conservation Department Publication 321(1943):3–11. This is the source usually quoted as initiating the mountain lion-Kaibab deer story. See also A. Leopold, L. K. Sowls, and D. L. Spencer, "A Survey of Over-populated Deer Ranges in the United States," *Journal of Wildlife Management* 11(1947): 162–77.

4. Elton, C. S. 1927. *Animal ecology.* New York: Macmillan.

5. Miller, R. S. 1967. Pattern and process in competition. *Advances in Ecological Research* 4:1 74.

6. Hardin, G. 1960. The competitive exclusion principle. *Science* 131:1292 1297.

7. Park, T., 1954, Experimental studies of interspecific competition. II. Temperature, humidity, and competition in two species of Tribolium, *Physiol. Zool.* 27:177–328.

8. Estes, J. A.., and J. F. Palmisano. 1974. Sea otters: Their role in structuring nearshore communities. *Science* 185: 1058 1060.

9. Kenyon, K. W. 1969. The sea otter in the eastern Pacific Ocean. North American Fauna, no. 68. Washington, D.C.: Bureau of Sports Fisheries and Wildlife, U.S. Department of the Interior.

10. Morowitz, H. J. 1979. *Energy flow in biology.* Woodbridge: CT: Oxbow Press.

11. Lavigne, D. M., W. Barchard, S. Innes, and N. A. Oritsland. 1976. *Pinniped bioenergetics.* ACMRR/MM/SC/12. Rome: United Nations Food and Agriculture Organization.

Chapter 5 Notes

1. http://www.gnofn.org/~swallow/welcome.html 02 MAR 05

2. Bourne, J. 2000. Louisiana's vanishing wetlands: Going, going *Science* 289:1860–1863.

3. http://pubs.usgs.gov/of/2002/of02-206/ Information about Lake Pontchartrain, accessed 02 mar 05.

4. Cicero, *The Nature of the Gods* (44 B.C.).

5. Darwin, C. (1859). *On the Origin of Species by Means of Natural Selection.* London, John Murray. Republished many times. One recent printing is Darwin, C. (2004). *Origin Of Species.* New York, Signet Classics (Penguin).

6. World Health Organization. 1998. *Malaria. Fact Sheet* No. 94.

7. Grant, P. R. 1986. *Ecology and evolution of Darwin's finches.* Princeton, N.J.: Princeton University Press.

8. Cox, C. B., and P. D. Moore. 1993. *Biogeography: An Ecological and Evolutionary Approach.* New York: Blackwell, Scientific Publishers.

9. MacArthur, R. H., and E. O. Wilson. 1967. *The Theory of Island Biogeography.* Princeton, N.J.: Princeton University Press.

10. Morwood, M. J., et al. 2004. Archaeology and age of a new hominid from Flores in eastern Indonesia. *Nature* 431: 1087–1097.

11. The number of species has been estimated by several authors. Here we use estimates by Margulis, L., K. V. Schwartz, and S. J. Gould. 1988. *Five Kingdoms: An Illustrated Guide to the Phyla of Life on Earth,* and Wilson, E. L. 1988. *Biodiversity,* Washington, D. C.: National Academy Press.

12. Principe, P. P. 1989. "The Economic Significance of Plants and Their Constituents as Drugs." In Wagner, H., H. H. Hikino, and N. R. Farnsworth, eds., *Economic and Medicinal Plant Research.* Vol. 3, pp. 1 17. New York: Academic Press.

13. Wallace, a contemporary of Charles Darwin, is also credited with discovering the theory of biological evolution. For some of Wallace's major writings, see Wallace, A. R. (2004). *Natural Selection And Tropical Nature.* Whitefish, MT 59937, Kessinger Publishing.

14. Martin, P. S. (2007). *Twilight of the Mammoths: Ice Age Extinctions and the Rewilding of America.* Berkeley, Univ.of California Press.

15. Information about the Kirtland's warbler and its habitat is from Byelich, J. M.E. DeCapita, G.W. Irvine, R.E. Radtke, N.I. Johnson, W.R. Jones, H. Mayfield, and W.J. Mahalak. 1985. Kirtland's warbler recovery plan. US Fish & Wildlife Service, Rockville, MD, 78pp; and Mayfield, H. 1969. *The Kirtland's Warbler.* Bloomfield Hills, Mich.: Cranbrook Institute of Science.

Chapter 6 Notes

1. The material about the ponderosa pine forest restoration is from: Covington, W. W., et al. 1997. Restoring ecosystem health in ponderosa pine forest of the Southwest. *Journal of Forestry* 95:23–29.

2. Martin, P. S. (2007). *Twilight of the Mammoths: Ice Age Extinctions and the Rewilding of America*. Berkley, University of California Press; and MArtinh, P. S. and D. A. Burney (1999). "Brinf back the elephants." *Wild Earth* (Spring Issue): 57–64.

3. Society of Ecological Restoration International Science and Policy Working Group. 2004. The SER International Primer on Ecological Restoration. Available at www.ser.org and Society for Ecological Restoration International. Tucson, AZ.

4. Houseal, G., and D. Smith. 2000. Source-identified seed: The Iowa roadside experience. *Ecological Restoration* 18(3):173–183.

5. Ingebritsen, S. E., C. McVoy, B. Glaz, and W. Park. 2001. Florida Everglades. In *Land subsidence in the United States*. Washington. DC: U.S. Geological Survey Circular 1182.

6. Roemer, G. W., A. D. Smith, D. K. Garcelon, and R. K. Wayne. 2001. The behavioral ecology of the island fox (*Uroeyon littoralis*). *Journal of Zoology*. 255:1–14.

7. Ibid.

8. Taylor, P. 2000. Nowhere to run, nowhere to hide. *Natural Wildlife*, August/September.

9. Ibid.

10. Walthern, P. 1986. Restoring derelict lands in Great Britain. In G. Orians, ed., *Ecological knowledge and environmental problem-solving: Concepts and case studies*. Washington, DC: National Academy Press, pp. 248–274.

Chapter 7 Notes

1. Information about paper use is from websites http://www.tappi.org/paperu/all_about_paper/paperClips.htm, W ebsite of the technical association for the worldwide pulp, paper, and converting industry. and http://www.epa.gov/epaoswer/osw/conserve/clusters/paper.htm of the EPA, viewed 14 May 2005.

2. United Nations FAO Statistics 14 April 2005.

3. United Nations Food and Agriculture Organization. 1999. State of the world's forests. Rome: UNFAO.

4. Biographical information about Gifford Pinchot is from www.dep.state.pa.us/dep/PA_Env-Her/phichot_bio.thm On 13 April 2005.

5. www.sierraclub.org/john_muir_exhibit/frameindex.html 13 April 2005.

6. Muir, J. 1912. *The Yosemite*, p 256. Quoted by the Sierra Club on its website 13 April 2005.

7. Muir, J. 1901. *Our National Parks*, p. 4. Quoted by the Sierra Club on its website 13 April 2005.

8. From http://./en.thinkexist.com/quotes/gifford_pinchot/ 13 April 2005.

9. Miller, C. 2001. *Gifford Pinchot and the Making of Modern Environmentalism*. Washington, D.C.: Island Press, pp. 138–39.

10. Muir, J. 1912. "The Yosemite," *Century* magazine, pp. 29–262. Quoted in http://nature.berkeley.edu/departments/esmp/env-hist/ 13 April 2005.

11. The Hubbard Brook ecosystem continues to be one of the most active and long-term ecosystem studies in North America. An example of a recent publication is S. W. Bailey, D. C. Buso, and G. E. Likens. 2003. Implications of sodium mass balance for interpreting the calcium cycle of a forested ecosystem. *Ecology* 84(2):471–84.

12. World Resources Institute. 1993. *World Resources 1992–93*. New York: Oxford University Press.

13. Manandhar, A. 1997. Solar cookers as a means for reducing deforestation in Nepal. Nepal: Center for Rural Technology.

14. World Resources Institute. 1993. *World Resources 1992–93*. New York: Oxford University Press.

15. World Resources Institute. 1999. *Deforestation: The Global Assault Continues*. Global trends, resources at risk, world resources 1998–99. Washington, D.C.

16. United Nations Food and Agriculture Organization. 1999. *State of the World's Forests*. Rome: UNFAO.

17. Sedjo, R. A, and D. B. Botkin, 1997, "Using Forest Plantations to Spare the Natural Forest", *Environment* 39(10): 14–20.

18. Botkin, D. B., and L. Simpson. 1990. The first statistically valid estimate of biomass for a large region. *Biogeochemistry* 9:161–74.

19. Quotes about wildlife in Manhattan are from R. H. Boyle. 1969. *The Hudson River: A Natural and Unnatural History*. New York: Norton, pp. 36–37.

20. Haines, F. 1970. *The Buffalo*. New York: Thomas Y. Crowell; and D. B. Botkin, 2004. *Beyond the Stony Mountains: Nature in the American West from Lewis and Clark to Today*. New York: Oxford University Press.

21. Leopold, A.. 1949. *A Sand County Almanac and Sketches Here and There*. New York: Oxford University Press, pp. 130–32.

22. Sheffer, V. B. 1951. The rise and fall of a reindeer herd. *Scientific Monthly* 73: 356–62.

23. Tom Stehn's whooping crane report Aransas National W ildlife Refuge, 10 December 2003. Available at http://www.birdrockport.com/tom_stehn_whooping_crane_report.htm, and Whooping Crane Conservation Association. 2003. Accessed

12/1/2003 Available at http://whoopingcrane.com/wccaflockstatus.htm.

24. The discussion of the Hudson's Bay Company's fur-trading records is from D. B. Botkin, 1990. *Discordant Harmonies: A New Ecology for the 21 Century. st* New York: Oxford University Press. Most long-term records for animal populations come from commercial harvesting—for example, the records of haddock from Icelandic fishing grounds and whales in the Pacific. These records also show variation, rather than constancy, over time. But they are even more likely to be confounded by the effects of variations in effort—the number of boats and the market for the fish—than the Hudson's Bay Company's records of animals caught in traps. This is shown dramatically in the catches for both haddock and whales from 1915 to 1919 and from 1939 to 1945, when fishing was halted by world wars.

25. Braithwaite, W . R.,M. L. Dudzinski, M. G. Ridpath, and B. S. Parker, 1984. The impact of water buffalo on the monsoon forest ecosystem in Kakadu National Park. *Australian Journal of Ecology* 9:309–322.

26. Freeland, W . J. 1990. Large herbivorous mammals: exotic species in northern Australia. *Journal of Biogeography* 17:445–449.

Chapter 8 Notes

1. Committee on Hormonally Active Agents in the Environment, National Research Council, National Academy of Sciences. 1999. *Hormonally active agents in the environment.* Washington, DC: National Academy Press.

2. Krimsky, S. 2001. Hormone disrupters: A clue to understanding the environmental cause of disease. *Environment* 43(5):22–31.

3. Royte, E. 2003. Transsexual frogs. *Discover* 24(2):26–53.

4. Hayes, T., K. Haston, M. Tsui, et al. 2002. Feminization of male frogs in the wild. *Nature* 419:495–496.

5. Hopps, H. C. 1971. Geographic pathology and the medical implications of environmental geochemistry. In H. L. Cannon and H. C. Hopps, eds., Environmental geochemistry in health. *Geological Society of America Memoir* 123:1–11. Boulder, CO: Geological Society of America.

6. Warren, H. V., and R. E. DeLavault. 1967. A geologist looks at pollution: Mineral variety. *Western Mines* 40:23–32.

7. U.S. Geological Survey. 1995. Mercury contamination of aquatic ecosystems. USGS FS 216-95.

8. Ehrlich, P. R., A. H. Ehrlich, and J. P. Holdren. 1970. *Ecoscience: Population, resources, environment.* San Francisco: V. H. Freeman.

9. Waldbott, G. L. 1978. *Health effects of environmental pollutants,* 2nd ed. St. Louis: Mosby.

10. Needleman, H. L., J. A. Riess, M. J. Tobin, et al. 1996. Bone lead levels and delinquent behavior. *Journal of the American Medical Association* 275:363–369.

11. Centers for Disease Control. 1991. *Preventing lead poisoning in young children.* Atlanta: Public Health Service, Centers for Disease Control.

12. Goyer, R. A. 1991. Toxic effects of metals. In M. O. Amdur, J. Doull, and C. D. Klaassen, eds., *Toxicology.* New York: Pergamon, pp. 623–680.

13. Bylinsky, G. 1972. Metallic nemesis. In B. Hafen, ed., *Man, health and environment.* Minneapolis: Burgess, pp. 174–185.

14. Hong, S., J. Candelone, C. C. Patterson, and C. F. Boutron. 1994. Greenland ice evidence of hemispheric lead pollution two millennia ago by Greek and Roman civilizations. *Science* 265:1841–1843.

15. McGinn, A. P. 2000. POPs culture. *World Watch* April 1:26–36.

16. Carlson, E. A. 1983. International symposium on herbicides in the Vietnam War: An appraisal. *BioScience* 33:507–512.

17. Grady, D. 1983. The dioxin dilemma. *Discover* May:78–83.

18. Roberts, L. 1991. Dioxin risks revisited. *Science* 251:624–626.

19. Cleverly, D., J. Schaum, D. Winters, and G. Schweer. 1999. Inventory of sources and releases of dioxin-like compounds in the United States. Paper presented at the 19th International Symposium on Halogenated Environmental Organic Pollutants and POPs, September 12–17, Venice, Italy. Short paper published in *Organohalogen Compounds* 41:467–472.

20. Kaiser, J. 2000. Just how bad is dioxin? *Science* 5473:1941–1944.

21. Ross, M. 1990. Hazards associated with asbestos minerals. In B. R. Doe, ed., Proceedings of a U.S. Geological Survey workshop on environmental geochemistry, pp. 175–176. U.S. Geological Survey Circular 1033.

22. Pool, R. 1990. Is there an EMF–cancer connection? *Science* 249:1096–1098.

23. Linet, M. S., E. E. Hatch, R. A. Kleinerman, et al. 1997. Residential exposure to magnetic fields and acute lymphoblastic leukemia in children. *New England Journal of Medicine* 337(1):1–7.

24. Francis, B. M. 1994. *Toxic substances in the environment.* New York: John Wiley & Sons.

25. Air Risk Information Support Center (Air RISC), U.S. Environmental Protection Agency. 1989. Glossary of terms related to health exposure and risk assessment. EPA/450/3-88/016. Research Triangle Park, NC.

Notes for Chapter 9

1. Kansas Rural Center. 1996 (March). More clean water farm demonstrations selected, p. 5.

2. USDA statistics. Available at http://www.usda.gov/wps/portal/!ut/p/_s.7_0_A/7_0_1OB?navid=DATA_STATISTICS&parentnav=AGRICULTURE&navtype=RT. Accessed May 21, 2005.

3. Raven, P. H., R. F. Evert, and S. E. Eichhorn. 1999. *Biology of plants.* New York: W. H. Freeman/Worth.

4. Field, J. O., ed. 1993. *The challenge of famine: Recent experience, lessons learned.* Hartford, CT: Kumarian Press.

5. United Nations Food and Agriculture Organization. 2000 (April). Food emergencies persist in 34 countries throughout the world. *Food Outlook* 2: p4. Rome: UNFAO.

6. United Nations Food and Agriculture Organization. 1998 (September). Global information and early warning system on food and agriculture. *Global Watch: Food Outlook.* Rome: UNFAO.

7. Pimentel, D. E., C. Terhune, R. Dyson-Hudson, et al. 1976. Land degradation: Effects on food and energy resources. *Science* 194:149–155.

8. Guthrie, Woody. The official Woody Guthrie Web site. Available at http://www.woodyguthrie.org/Lyrics/Lyrics.htm. Accessed January 12, 2007. Copyright 1960 Ludlow Music Inc., NY.

9. Pimentel, D., and E. L. Skidmore. 1999. Rates of soil erosion. *Science* 286:1477–1478.

10. Trimble, S. W., and P. Crosson, 2000. U.S. soil erosion rates—myth and reality. *Science* 289 (July 14): 248–250.

11. Trimble, S. W. 2000. Soil conservation and soil erosion in the upper Midwest. *Environmental Review* 7(1):37–9.

12. Ferry, L. (1995). *The New Ecological Order.* Chicago: University of Chicago Press.

13. Lashof, J. C., ed. 1979. *Pest management strategies in crop protection.* Vol. 1. Washington, DC: Office of Technology Assessment, U. S. Congress.

14. Baldwin, F. L., and P. W. Santelmann. 1980. Weed science in integrated pest management. *BioScience* 30:675–678.

15. Pimentel, D. 2005. Environmental and economic costs of the application of pesticides primarily in the United States. *Environment, Development and Sustainability* 7(2): 229–252.

16. Bear, J. 2005. DDT and human health. In *Science of the total environment* (epub, preprint). New York: Elsevier.

17. USDA. Available at http://www.ers.usda.gov/Briefing/AgChemicals/pestmangement.htm#pesticide. Accessed Aug 17, 2006 May 21, 2005.

18. Michigan State University. 2003. Web site available at http://www.msue.msu.edu/vanburen/ofm.htm.

19. Baldwin, F. L., and P. W. Santelmann. 1980. Weed science in integrated pest management. *BioScience* 30:675–678.

20. Lashof, J. C., ed. 1979. *Pest management strategies in crop protection.* Vol. 1.

21. Pioneer Seed Company news release. May 9, 2005. Available at http://www.pioneer.com/pioneer_news/press_releases/corporate/biotech_acreage.htm.

22. Pew Foundation Initiative on Food and Biotechnology. Available at http://pewagbiotech.org/resources/factsheets/display.php3?FactsheetID=2. Accessed May 22, 2005. From this site comes the information on acreage planted to GMOs except where otherwise noted.

23. According to the Pew Initiative, available at http://pewagbiotech.org/resources/factsheets/display.php3?FactsheetID=2. "Argentina is the next largest producer, with 34.4 million acres, followed by Canada with 10.9 million acres, Brazil with 8.4 million acres, China with 6.9 million acres, and South Africa with 1.0 million acres in 2003. Together, these six countries grew 99 percent of the global GM crop area last year. Australia, Mexico, Romania, Bulgaria, Spain, Germany, Uruguay, Indonesia, the Philippines, India, Columbia, and Honduras also planted significant acreage in GM crops in 2003."

24. Pew Biotechnology Initiative, Pew Charitable Trusts. Genetically modified crops in the United States. Available at http://pewagbiotech.org/resources/factsheets/display.php3?FactsheetID=2. Accessed January 12, 2007.

25. United Nations Food and Agriculture Organization. 2003. Available at www.fao.org. Aug 1, 2006.

26. Botkin, D. B. 1999. *Passage of discovery.* New York: Putnam Books.

27. Buschbacher, R. J. 1986. *Tropical deforestation and pasture development.* Bioscience: 36 (1), 22–88.

28. Jordan, Carl F. ed. 1988, *Amazonian rain forests: Ecosystem disturbance and recovery.* New York: Springer.

29. Grainger, A. 1982. Desertification: *How people make deserts, how people can stop and why they don't,* 2nd ed. London: Russell Press, Ltd.

30. United Nations. 1978. United Nations conference on desertification: Roundup plan of action and resolutions. New York: United Nations.

31. United Nations Food and Agricultural Organization. 1998. The United Nations convention to combat desertification: An explanatory leaflet. New York: United Nations.

32. United Nations Food and Agricultural Organization. 1998. What is desertification? New York: United Nations.

33. Grainger, A. 1982. *Desertification: How people make deserts, how people can stop and why they don't,* 2nd ed.

34. Sheridan, D. 1981. *Desertification of the United States.* Washington, DC: Council on Environmental Quality.

Notes for Chapter 10

1. U. S. Dept. of Energy, International Energy Outlook 2006. Report #:DOE/EIA-0484(2006). Release date: June 2006.

2. Or, if each car owner commuted a total of 2 hours a day, and each had a 100 horsepower car, that would be the energy to run 2,284,117,775 Cars.

3. *Christian Science Monitor.* June 15, 2005. Available at http://www.csmonitor.com/2005/0616/p13s02-stct.html.

4. Youngquist, W. 1998. Spending our great inheritance. Then what? *Geotimes* 43(7):24–27.

5. Edwards, J. D. 1997. Crude oil and alternative energy production forecast for the twenty-first century: The end of the hydrocarbon era. *American Association of Petroleum Geologists Bulletin* 81(8):1292–1305.

6. British Petroleum company 2005. B.P. Statistical Review of world energy. London.

7. Kerr, R. A. 2000. USGS optimistic on world oil prospects. *Science* 289:237.

8. Darmstadter, J., H. H. Landsberg, H. C. Morton, and M. J. Coda. 1983. *Energy today and tomorrow: Living with uncertainty.* Englewood Cliffs, NJ: Prentice-Hall.

9. Peterson, G. 2003. New statute for Canadian Oil Sands. *Geotimes* 48(3):7.

10. Wood, T. 2003. Prosperity's brutal price. Los Angeles Times Magazine, Feb 2, 2003.

11. Nuccio, V. 1997. Coal-bed methane—an untapped energy resource and an environmental concern.

12. Suess, E., G. Bohrmann, J. Greinert, and E. Lauch. 1999. Flammable ice. *Scientific American* 28(5):76–83.

13. Corcoran, E. 1991. Cleaning up coal. *Scientific American* 264:106–116.

14. Caudill, H. 1964. *Night comes to Cumberlands.* Boston: Little, Brown.

15. Miller, E. W. 1993. *Energy and American society, a reference handbook.* Santa Barbara, CA: ABC-CLIO.

16. Energy Information Administration. 1995 (February). Coal data: A reference. Washington, DC: U.S. Department of Energy.

17. M. Holway, M. 1991. Soiled shores. *Scientific American* 265:102–106.

18. National Research Council (NRC) 1985. Oil in the sea: Inputs, fate and effects. Washington, DC: NRC report published by National Academy Press.

19. Peterson, C. H., S. D. Rice, J. W. Short, et al. 2003. Long-term ecosystem response to the *Exxon Valdez* oil spill. *Science* 302:2082–2086.

20. Butti, K., and J. Perlin. 1980. *The golden thread: 2500 years of solar architecture and technology.* New York: Cheshire.

21. Brown, L. R. 2003. *Plan B: Rescuing a planet under stress and a civilization in trouble.* New York: Norton.

22. De Miguel Ichaso, A. 2000 (August). Wind power development in Spain, the model of Navarra. *DEWI* 17:49–54.

23. Flavin, C., and S. Dunn. 1999. Reinventing the energy system. In L. R. Browne, et al., eds. *State of the world 1999: A Worldwatch Institute report on progress toward a sustainable society.* New York: W. W. Norton.

24. Hydropower. Available at http://www.answers.com/topic/hydropower-1. Accessed October 1, 2006.

25. Sierra Club. 2004. Driving up the heat: SUVs and global warming. Available at http://www.sierraclub.org. Accessed March 29, 2004.

26. U.S. Congress, Office of Technology Assessment. 1993. Potential environmental impacts of bioenergy crop production. Washington, DC: Background paper published by U.S. Government Printing Office.

27. Pimentel, D. (2005). Ethanol production using corn, switchgrass, and wood; biodiesel production using soybean and sunflower. *Natural Resources Research* 14(1): 65–76.

28. Jackson, T., and R. Lofstedt. 1998. Royal commission on environmental pollution. Study on energy and the environment. Available at http://www.rcep.org.uk/studies/energy/98-6061/jackson.html. Accessed November 29, 2000.

29. Sterzinger, G. 1995. Making biomass energy a contender. *Technology Review* 98:34–40.

30. U.S. Environmental Protection Agency, Office of Solid Waste and Emergency Response. 2002. Municipal solid waste in the United States: 2000 facts and figures. EPA530-R-02-001.

31. Bisconti, A.S. 2003. The thirds of Americans favor nuclear energy; public divided or building new nuclear plants. Perspectives or Public opinion (July).

32. Duderstadt, J. J. 1978. Nuclear power generation. In L. C. Ruedisili and M. W. Firebaugh, eds., *Perspectives on energy*, 2nd ed., pp. 249–273. New York: Oxford University Press.

33. Till, C. E. 1989. Advanced reactor development. *Ann. Nucl. Energy* 16(6):301–305.

34. Lake, J. A., R. G. Bennet, and J. F. Kotek. 2002 (January). Next-generation nuclear power. *Scientific American*, pp. 73–81.

35. U. S. Department of Energy. 1980. *Magnetic fusion energy*. DOE/ER-0059. Washingtown, D.C.: U.S. Department of Energy.

36. U.S. Department of Energy. 1979. *Enviromental development plan, magnetic fusion*. DOE/EDP-0052. Washingtown, D.C.: U.S. Department of Energy.

37. That Three Mile Island was the most serious event in nuclear power plant operation is stated by the U.S. Nuclear Regulatory Commission. Available at http://www.nrc.gov/reading-rm/doc-collections/fact-sheets/3mile-isle.html. Accessed July 5, 2005.

38. MacLeod, G. K. 1981. Some public health lessons from Three Mile Island: A case study in chaos. *Ambio* 10:18–23.

39. This information is also from the NRC Web site referred to in ref. 25, which stated that "Approximately 43,000 curies of krypton were vented from the reactor building" by 1980. Further information on the TMI-2 accident can be obtained from sources listed below. The documents can be ordered for a fee from the NRC's Public Document Room by calling 301-415-4737 or 1-800-397-4209; or e-mail pdr@nrc.gov. The PDR is located at 11555 Rockville Pike, Rockville, MD; however the mailing address is U.S. Nuclear Regulatory Commission, Public Document Room, Washington, DC 20555.

40. Anspaugh, L. R., R. J. Catlin, and M. Goldman. 1988. The global impact of the Chernobyl reactor accident. *Science* 242:1513–1518.

41. Nuclear Energy Agency. 2002. Chernobyl Assessment of Radiological and Health Impacts: 2002 Update of Chernobyl: Ten Years On.

42. Balter, M. 1995. Chernobyl's thyroid cancer toll. *Science* 270:1758.

43. Skuterud, L., N. I. Goltsova, R. Naeumann, T. Sikkeland, and T. Lindmo. 1994. Histological changes in *Pinus sylvestris L.* in the proximal-zone around the Chernobyl power plant. *The Science of the Total Environment* 157:387–397.

44. Williams, N. 1995. Chernobyl: Life abounds without people. *Science* 269:304.

45. Fletcher, M. 2000 (November 14). The last days of Chernobyl. *Times 2* (London), pp. 3–5.

46. http://www.nbc5i.com/news/2406559/detail.html. Accessed May 27, 2004.

47. Stone, R. 2003. Plutonium fields forever. *Science* 300:1220–1224.

48. Ehrlich, P. R., A. H. Ehrlich, and J. P. Holdren. 1970. *Ecoscience: Population, resources, environment.* San Francisco: Freeman.

49. Van Koevering, T. E., and N. J. Sell. 1986. *Energy: A conceptual approach.* Englewood Cliffs, NJ: Prentice-Hall.

50. Brenner, D. J. 1989. *Radon: Risk and remedy.* New York: Freeman.

51. Office of Industry Relations. 1974. Development, growth and state of the nuclear industry. Washington, DC: U.S. Congress, Joint Committee on Atomic Energy.

52. Weisman, J. 1996. Study inflames Ward Valley controversy. *Science* 271:1488.

53. Roush, W. 1995. Can nuclear waste keep Yucca Mountain dry—and safe? *Science* 270:1761.

54. Hanks, T. C., I. J. Winograd, R. E. Anderson, T. E. Reilly, and E. P. Weeks. 1999. Yucca Mountain as a radioactive-waste repository. U.S. Geological Survey Circular 1184.

55. Bredehoeft, J. D., A. W. England, D. B. Stewart, J. J. Trask, and I. J. Winograd. 1978. Geologic disposal of high-level radioactive wastes—Earth science perspectives. U.S. Geological Survey Circular 779.

56. Nevada Office of Radioactive Waste Management. Information on Web site available at http://www.ocrwm.doe.gov/ymp/index.shtml. Accessed September 5, 2005.

57. Botkin, D. B. *Energy Forever: A Citizens Guide to Energy.* (In preparation.)

Chapter 11 Notes

1. Graf, W. L. 1985. *The Colorado River: Instability and basin management.* Resource Publications in Geography. Washington, DC: Association of American Geographers.

2. Dolan, R., A. Howard, and A. Gallenson. 1974. Man's impact on the Colorado River and the Grand Canyon. *American Scientist* 62:392–401.

3. Hecht, J. 1996. Grand Canyon flood a roaring success. *New Scientist* 151(2045):8.

4. Lucchitta, I., and L. B. Leopold. 1999. Floods and sandbars in the Grand Canyon. *Geology Today* 9:1–7.

5. Henderson, L. J. 1913. *The fitness of the environment: An inquiry into the biological significance of the properties of matter.* New York: Macmillan.

6. Council on Environmental Quality and U.S. Department of State. 1980. *The global 2000 report to the*

President: Entering the twenty-first century. Vol. 2. Washington, DC.

7. Water Resources Council. 1978. *The nation's water resources, 1975–2000.* Vol. 1. Washington, DC.

8. Winter, T. C., J. W. Harvey, O. L. Franke, and W. M. Alley. 1998. Groundwater and surface water: A single resource. U.S. Geological Survey Circular 1139.

9. Conant, E. 2006 (September). Return of the Aral Sea. *Discovery,* pp. 54–58.

10. Morrison, J. How much is clean water worth? *National Wildlife* 43(N.2): 22–28.

11. Solley, W. B., R. R. Pierce, and H. A. Perlman. 1993. Estimated use of water in the United States in 1990. U.S. Geological Survey Circular 1081.

12. Solley, W. B., R. R. Pierce, and H. A. Perlman. 1998. Estimated use of water in the United States in 1995. U.S. Geological Survey Circular 1200.

13. U.S. General Accounting Office. 2003. Freshwater supply: States' view of how federal agencies could help them meet the challenges of expected shortages. Report GAO-03-514.

14. Gleick, P. H., P. Loh, S. V. Gomez, and J. Morrison. 1995. *California water 2020, a sustainable vision.* Oakland, CA: Pacific Institute for Studies in Development, Environment and Security.

15. Holloway, M. 1991. High and dry. *Scientific American* 265:16–20.

16. Levinson, M. 1984 (February/March). Nurseries of life. *National Wildlife,* special report, pp. 18–21.

17. Nichols, F. H., J. E. Cloern, S. N. Luoma, and D. H. Peterson. 1986. The modification of an estuary. *Science* 231:567–573.

18. Day, J. W., Jr., J. M. Rybczyk, L. Carboch, et al. 1998. A review of recent studies of the ecology and economic aspects of the application of secondary treated municipal effluent to wetlands in southern Louisiana. In L. P. Rozas, et al., eds., Symposium on recent research in coastal Louisiana, February 3–5, 1998, Louisiana Sea Grant College Program, pp. 1–12.

19. Hileman, B. 1995. Rewrite of Clean Water Act draws praise, fire. *Chemical & Engineering News* 73:8.

20. Kaiser, J. 2001. Wetlands restoration: Recreated wetlands no match for original. *Science* 293:25a.

21. Gurardo, D., M. L. Fink, T. D. Fontaine, et al. 1995. Large-scale constructed wetlands for nutrient removal from stormwater runoff: An Everglades restoration project. *Environmental Management* 19:879–889.

22. State of Maine. 2001. A brief history of the Edwards Dam. Available at http://janus.state.me.us/spo/edwards/timeline.htm. Accessed January 15, 2000.

23. Pearce, M. 1995 (January). The biggest dam in the world. *New Scientist,* pp. 25–29.

24. Zich, R. 1997. China's three gorges: Before the flood. *National Geographic* 192(3):2–33.

25. Pinter, N. 2005. One step forward, two steps back on U.S. floodplains. *Science* 308:207–208.

26. Brown, L. R. 2003. *Plan B: Rescuing a planet under stress and a civilization in trouble.* New York: W. W. Norton.

27. *Groundwater: Issues and answers.* 1985. Arvada, CO: American Institute of Professional Geologists.

28. Gleick, P. H. 1993. An introduction to global fresh water issues. In P. H. Gleick, ed., *Water in crisis.* New York: Oxford University Press, pp. 3–12.

29. Hileman, B. 1995. Pollution tracked in surface- and groundwater. *Chemical & Engineering News* 73:5.

30. Lewis, S. A. 1995. Trouble on tap. *Sierra* 80:54–58.

31. Kluger, J. 1998. Anatomy of an outbreak. *Time* 152(5):56–62.

32. Mallin, M. A. 2000. Impacts of industrial animal production on rivers and estuaries. *American Scientist* 88(1):26–37.

33. Bowie, P. 2000. No act of God. *Amichs Journal* 21(4):16–21.

34. Hinga, K. R. 1989. Alteration of phosphorus dynamics during experimental eutrophication of enclosed marine ecosystems. *Marine Pollution Bulletin* 20:624–628.

35. Richmond, R. H. 1993. Coral reefs: Present problems and future concerns resulting from anthropogenic disturbance. *American Zoologist* 33:524–536.

36. Bell, P. R. 1991. Status of eutrophication in the Great Barrier Reef Lagoon. *Marine Pollution Bulletin* 23:89–93.

37. Hunter, C. L., and C. W. Evans. 1995. Coral reefs in Kaneohe Bay, Hawaii: Two centuries of Western influence and two decades of data. *Bulletin of Marine Science* 57:499.

38. Mitch, W. J., J. W. Day, Jr., J. W. Gilliam, et al. 2001. The Gulf of Mexico hypoxia—approaches to reducing nitrate in the Mississippi River or reducing a persistent large-scale ecological problem. *BioScience* (in press).

39. Department of Alaska Fish and Game 1918. *Alaska Fish and Game* 21(4). Special issue.

40. Holway, M. 1991. Soiled shores. *Scientific American* 265:102–106.

41. Robinson, A. R. 1973. Sediment, our greatest pollutant? In R. W. Tank, ed., *Focus on environmental geology.* New York: Oxford University Press, pp. 186–192.

42. Poole, W. 1996. Rivers run through them. *Land and People* 8:16–21.

43. Foxworthy, G. L. 1978. Nassau County, Long Island, New York—Water problems in humid county. In G. D. Robinson and A. M. Spieker, eds., Nature to be commanded. U.S. Geological Survey Professional Paper 950. Washington, DC: U.S. Government Printing Office, pp. 55–68.

44. Van der Leeden, F., F. L. Troise, and D. K. Todd. 1990. *The water encyclopedia*, 2nd ed. Chelsea, MI: Lewis Publishers.

45. U.S. Geologic Survey. 1997. Predicting the impact of relocating Boston's sewage outfall. UCGC Fact Sheet, pp. 185–197.

46. Task Force on Water Reuse. 1989. *Water reuse: Manual of practice SM-3*. Alexandria, VA: Water Pollution Control Federation.

47. Kadlec, R. H., and R. L. Knight. 1996. Treatment wetlands. New York: Lewis Publishers.

48. Breaux, A. M., and J. W. Day, Jr. 1994. Policy considerations for wetland wastewater treatment in the coastal zone: A case study for Louisiana. *Coastal Management* (22):285–307.

49. Day, J. W., Jr., et al. 1998. A review of recent studies of the ecology and economic aspects of the application of secondary treated municipal effluent to wetlands in southern Louisiana.

50. Breaux, A., S. Fuber, and J. Day, 1995. Using natural coastal wetland systems: An economic benefit analysis. *Journal of Environmental Management* (44):285–291.

51. Hileman, B. 1995. Rewrite of Clean Water Act draws praise, fire.

Chapter 12 Notes

1. Gordon, B. B. 1993. Pampering our coastlines. *Sea Frontiers* 39(2):5.

2. Mydans, S. 1996. Thai shrimp farmers facing ecologists fury. New York Times, April 28.

3. Pollution wiping out shrimp farms on main Indonesian island of Java. 1996. *Quick Frozen Foods International* 37(3):50.

4. Quarto, A. 1994. Rainforests of the sea: Mangrove forests threatened by prawn aquaculture. E,5(1):16–19.

5. Wickramayanake, S. D. 1995. East Coast shrimp farms face trouble from both nature and protest groups. *Quick Frozen Foods International* 37(1):108–109.

6. Haub, C., and D. Cornelius. 2000. Nine billion world population by 2050. Washington, DC: Population Reference Bureau.

7. World Resources Institute (WRI). 1997. *Water and fisheries. World resources: A guide to the global environment*. Washington, DC: WRI.

8. Cushing, D. 1975. *Fisheries resources of the sea and their management*. London: Oxford University Press.

9. United Nations Food and Agricultural Organization, statistics 2006.

10. National Oceanic and Atmospheric Administration (NOAA). 2003. World fisheries. Available at http://www.st.nmfs.gov/st1/fus/current/04_world 2002.pdf. Accessed January 15, 2007.

11. Cushing, D. 1975. *Fisheries resources of the sea and their management*.

12. Cushing, D. 1975. *Fisheries resources of the sea and their management*.

13. Myers, A., and B. Worm. 2003 (May 15). Rapid worldwide depletion of predatory fish, communities. *Nature*.

14. United Nations Grid, 2004. Overfishing, a major threat to global massive ecology. Enviromental Alert Bulletin. Available at wau.grid.unep.of/ew, statistics.

15. http://www.eubusiness.com/afp/03120412436.eq98k2mi.

16. International Shark Attack File. Available at http://www.flmnh.ufl.edu/fish/sharks/statistics/statsus.htm. Accessed July 23, 2005.

17. United Nations Food and Agriculture Organization, statistics. Available at http://apps.fao.org/lim500/nph-wrap.pl?FishCatch&Domain=FishCatch. Accessed May 3, 2006.

18. Upwelling. Available at http://oceanexplorer.noaa.gov/explorations/02quest/background/upwelling/upwelling.html. Accessed June 4, 2005.

19. World Resources Institute. 1997. Water and fisheries.

20. National Oceanic and Atmospheric Administration (NOAA). 2003. World fisheries.

21. Cushing, D. 1975. *Fisheries resources of the sea and their management*.

22. Friends of the Earth. 1979. *The whaling question: The inquiry by Sir Sidney Frost of Australia*. San Francisco: Friends of the Earth.

23. World Wildlife Fund (WWF). 2000. *Gray whales*. Washington, DC: WWF.

24. Friends of the Earth. 1979. *The whaling question: The inquiry by Sir Sidney Frost of Australia*.

25. Bardach, J. E. 1968. Aquaculture. *Science* 161: 1098–1106.

26. Ibid.

27. Friends of the Earth. 1979. *The whaling question: The inquiry by Sir Sidney Frost of Australia*.

28. Bockstoce, J. R., and D. B. Botkin. 1980. *The historical status and reduction of the western Arctic bowhead whale (Balaena mysticetus) population by the pelagic whaling industry, 1848–1914*. New Bedford, CT: Old Dartmouth Historical Society.

29. United Nations Food and Agriculture Organization. 1978. Mammals in the seas. Report of the FAO Ad-

visory Committee on Marine Resources Research, Working Party on Marine Mammals. FAO Fisheries Series 5, Vol. 1. Rome: UN FAO.

30. Friends of the Earth. 1979. *The whaling question: The inquiry by Sir Sidney Frost of Australia.*

31. International Whaling Commission. 2003. Table of estimates of whale abundances. Available at http://www.iwcoffice.org/Estimate.htm. Accessed May 5, 2006.

32. World Wildlife Fund. 2000. *Gray whales.*

33. Perry, M. 1996. Climate change biggest risk to whales, says IWC. Sydney, Australia: Reuters.

34. United Nations Food and Agriculture Organization, statistics. Available at http://apps.fao.org/lim500/nph-wrap.pl?FishCatch&Domain=FishCatch. Accessed May 10, 2006.

35. Ibid.

Chapter 13 Notes

1. Levy, S. 2000. Wildlife on the hot seat. *National Wildlife* 38(5):20–27.

2. Hartmann, D. L. 1994. Global physical climatology. *International Geophysics Series*, Vol. 56. New York: Academic Press.

3. Hansen, J. 2003. Can we defuse the global warming time bomb? Natural Science www.naturalscience.com.

4. Intergovernmental Panel on Climate Change (IPCC). 2007. Climate Change 2007. Summary for policy makers at www.ipcc.ch. Accessed 2/8/07.

5. Union of Concerned Scientists. 1989. *The greenhouse effect.* Cambridge, MA: Author.

6. Kerr, R. A. 1996. 1995 the warmest year? Yes and no. *Science* 271:137–138.

7. Crowley, T. J. 2000. Causes of climate change over the past 1000 years. *Science* 289:270–277.

8. Union of Concerned Scientists. 1989. *The greenhouse effect.*

9. Kerr, R. A. 1996. 1995 the warmest year? Yes and no.

10. Crowley, T. J. 2000. Causes of climate change over the past 1000 years.

11. Ibid.

12. Ibid.

13. From NOAA's National Geophysical Data Center.

14. Charlson, R. J., S. E. Schwartz, J. M. Hales, et al. 1992. Climate forcing by anthropogenic aerosols. *Science* 255:423–430.

15. Ibid.

16. Crowley, T. J. 2000. Causes of climate change over the past 1000 years.

17. Kerr, R. A. 1995. Study unveils climate cooling caused by pollutant haze. *Science* 268:802.

18. Campbell, I. D., C. Campbell, N. J. Apps, N. W. Rutter, and A. B. G. Bush. 1998. Late Holocene approximately 1500-year climatic periodicities and their implications. *Geology* 26(5):471–473.

19. Broecker, W. 1997. Will our ride into the greenhouse future be a smooth one? *GSA Today* 7(5):2–6.

20. Ibid.

21. Steager, R. 2006. The source of Europe's mild climate. American Scientist 94:334–341.

22. NOAA. What is an El Niño? Available at http://www.pmel.noaa.gov/tao/elnino/el-nino-story.html. Accessed January 17, 2007.

23. Ibid.

24. National Academy of Sciences. El Niño Web site. Available at http://www7.nationalacademies.org/opus/elnino.html. October 1, 2005.

25. Moss, M. E., and H. F. Lins. 1989. Water resources in the twenty-first century: A study of the implications of climate uncertainty. U.S. Geological Survey Circular 1030. Washington, DC: U.S. Department of the Interior.

26. Post, W. M., T. Peng, W. R. Emanuel, et al. 1990. The global carbon cycle. *American Scientist* 78:310–326.

27. Rodhe, H. 1990. A comparison of the contribution of various gases to the greenhouse effect. *Science* 248:1217–1219.

28. Council on Environmental Quality. 1990. *Environmental trends 1989.* Washington, DC: Author.

29. Moss, M. E., and H. F. Lins. 1989. Water resources in the twenty-first century: A study of the implications of climate uncertainty. U.S. Geological Survey Circular 1030. Washington, DC: U.S. Department of the Interior.

30. Titus, J. G., S. P. Leatherman, C. H. Everts, et al. 1985. Potential impacts of sea level rise on the beach at Ocean City, Maryland. Washington, DC: U.S. Environmental Protection Agency, Office of Policy Planning and Evaluation.

31. PhysicalGeography.net. Introduction to the atmosphere. Available at http://www.physicalgeography.net/fundamentals/7h.html. Accessed January 17, 2007.

32. Titus, J. G., and V. K. Narayanan. 1995. *The probability of sea level rise.* Washington, DC: U.S. Environmental Protection Agency.

33. Davis, C. H., L. Yonghong, J. R. McConnell, et al. 2005. Snowfall-driven growth in East Antarctic ice sheet mitigates recent sea-level rise. Available at: DOI:10.1126/science.1110662. *Science Express Reports.* Accessed May 19, 2005.

34. Hartmann, D. L. 1994. Global physical climatology. *International Geophysics Series*, vol. 56.

35. Kerr, R. A. 1995. U.S. climate tilts toward the greenhouse. *Science* 268:363–364.

36. Levy, S. 2000. Wildlife on the hot seat. *National Wildlife* 38(5):20–27.

37. Montaigne, F. 2004. Eco signs. In *The heat is on. National Geographic* 206(3):34–35.

38. Root, T. L., D. P. Mac Mynowsky, M. D. Mastsandren, and S. H. Schneider. 2005. Human-modified temperature-induced species changes. Proceedings of the National Academy of Sciences. 102(21):7465–7469.

39. Holmes, N. 2000. Has anyone checked the weather (map)? *Amicus Journal* 21(4):50–51.

40. Epstein, P. R. 2000. Is global warming harmful to health? *Scientific American* 283(2):50–57.

41. Botkin, D. B., D. A. Woodby, and R. A. Nisbet, 1991, Kirtland's warbler habitats: A possible early indicator of climatic warming, *Biological Conservation* 56(1):63–78.

42. Botkin, D. B., 1993, *JABOWA-II: A computer model of forest growth* (software and manual). New York: Oxford University Press.

43. Botkin, D.B., 1993. *Forest dynamics: An ecological model.* New York: Oxford University Press.

44. Thomas, C. D., et al. 2004. Extinction risk of climate change. *Nature* 427:145–148.

Chapter 14 Notes

1. Friedman, M. S., K. E. Powell, L. Hutwagner, L. M. Graham, and W. G. Teague. 2001. Impact of changes in transportation and commuting behaviors during the 1996 Summer Olympic Games in Atlanta on air quality and childhood asthma. *Journal of the American Medical Association*, 285(7):897–905.

2. Simons, L. M. 1998. Plague of fire. *National Geographic* 194(2):100–119.

3. National Park Service (NPS). 1984. *Air resources management manual.* Washington, DC: NPS.

4. American Lung Association. 1998. American Lung Association outdoor fact sheet. Available at http://www.lungusa.org. Accessed September 18, 1998.

5. Godish, T. 1991. *Air quality*, 2nd ed. Chelsea, MI: Lewis Publishers.

6. Seitz, F., and C. Plepys. 1995. Monitoring air quality in healthy people 2000. Healthy people 2000: Statistical notes no. 9. Atlanta: Centers for Disease Control and Prevention, National Center for Health Statistics.

7. American Lung Association (ALA). 2001. State of the Air 2000. Washington, DC: ALA.

8. Moore, C. 1995. Poisons in the air. *International Wildlife* 25:38–45.

9. Pope, C. A., III, D. V. Bates, and M. E. Raizenne. 1995. Health effects of particulate air pollution: Time for reassessment? *Environmental Health Perspectives* 103:472–480.

10. Stern, A. C., R. T. Boubel, D. B. Turner, and D. L. Fox. 1984. *Fundamentals of air pollution*, 2nd ed. Orlando, FL: Academic Press.

11. Pountain, D. 1993 (May). Complexity on wheels. *Byte*, pp. 213–220.

12. Molnia, B. F. 1991. Washington report. *GSA Today* 1:33.

13. Office of Technology Assessment. 1984. Balancing the risks. *Weatherwise* 37:241–249.

14. Canadian Department of the Environment. 1984. The acid rain story. Ottawa: Minister of Supply and Services.

15. How many more lakes have to die? 1981. *Canada Today* 12(2).

16. Molina, M. J., and F. S. Rowland. 1974. Stratospheric sink for chlorofluoromethanes: Chlorine-atom catalyzed distribution of ozone. *Nature* 249:810–812.

17. Brouder, P. 1986 (June). Annals of chemistry in the face of doubt. *New Yorker*, pp. 20–87.

18. Rowland, F. S. 1990. Stratospheric ozone depletion of chlorofluorocarbons. *AMBIO* 19:281–292.

19. Khalil, M. A. K., and R. A. Rasmussen. 1989. The potential of soils as a sink of chlorofluorocarbons and other man-made chlorocarbons. *Geophysical Research Letters* 16:679–682.

20. Shea, C. P. 1989. Mending the Earth's shield. *World Watch* 2:28–34.

21. Kerr, J. B., and C. T. McElroy. 1993. Evidence for large upward trends of ultraviolet-B radiation linked to ozone depletion. *Science* 262:1032–1034.

22. Showstack, R. 1998. Ozone layer is on slow road to recovery, new science assessment indicates. *Eos* 79(27):317–318.

23. Cutter Information Corp. 1996. Reports discuss present and future state of ozone layer. Global Environmental Change Report V, VIII, 21, no. 22, pp. 1–3. Dunster, B.C., Canada: Cutter Information Corp.

24. Spurgeon, D. 1998. Surprising success of the Montreal protocol. *Nature* 389(6648):219.

25. Makhijani, A., and A. Bickel, 1990. Still working on the ozone hole. *Technology Review* 93:52–59.

26. U.S. Environmental Protection Agency. 2003. Ozone depletion. Available at http://www.epa.gov/ozone/index.html. Accessed October 8, 2003.

27. Brown, L. R., N. Lenssen, and H. Kane. 1995. CFC production plummeting. In *Worldwatch Institute, vital signs 1995*. New York: W. W. Norton.

28. Zimmerman, M. R. 1985. Pathology in Alaskan mummies. *American Scientist* 73:20–25.

29. Ehrlich, P. R., A. H. Ehrlich, and J. P. Holdren. 1970. *Ecoscience: Population, resources, environment.* San Francisco: W. H. Freeman.

30. Conlin, M. 2000 (June 5). Is your office killing you? *Business Week*, pp. 114–124.

31. U.S. Environmental Protection Agency. 1991.Building air quality: A guide for building owners and facility managers. EPA/400/1-91/033, DHHS (NIOSH) Pub. No. 91–114. Washington, DC: Environmental Protection Agency.

32. Zummo, S. M., and M. H. Karol. 1996. Indoor air pollution: Acute adverse health effects and host susceptibility. *Environmental Health* 58:25–29.

33. Committee on Indoor Air Pollution. 1981. *Indoor pollutants.* Washington, DC: National Academy Press.

34. Godish, T. 1997. *Air quality*, 3rd. ed. Boca Raton, FL: Lewis Publishers.

35. O'Reilly, J. T., P. Hagan, R. Gots, and A. Hedge. 1998. *Keeping buildings healthy.* New York: Wiley.

36. Brenner, D. J. 1989. *Radon: Risk and remedy.* New York: W. H. Freeman.

37. U.S. Environmental Protection Agency (EPA). 1992. *A citizen's guide to radon: The guide to protecting yourself and your family from radon*, 2nd ed. ANR-464. Washington, DC: EPA.

38. U.S. Environmental Protection Agency (EPA). 1986. Radon reduction techniques for detached houses: Technical guidance. EPA 625/5-86-019. Research Triangle Park, NC: Air and Energy Engineering Research Laboratory, Office of Research and Development, U.S. EPA.

39. Osborne, M. C. 1988. Radon-resistant residential new construction. EPA 600/8-88/087. Research Triangle Park, NC: Air and Energy Engineering Research Laboratory, Office of Research and Development, U.S. EPA.

40. Kolstad, C. D. 2000. *Environmental economics.* New York: Oxford University Press.

41. Crandall, R. W. 1983. *Controlling industrial pollution: The economics and politics of clean air.* Washington, DC: Brookings Institution.

42. Hall, J. V., A. M. Winer, M. T. Kleinman, et al. 1992. Valuing the health benefits of clean air. *Science* 255:812–816.

43. Krupnick, A. J., and P. R. Portney. 1991.Controlling urban air pollution: A benefits-cost assessment. *Science* 252:522–528.

44. Lipfert, F. W., S. C. Morris, R. M. Friedman, and J. M. Lents. 1991. Air pollution benefit-cost assessment. *Science* 253:606.

45. Tyson, P. 1990. Hazing the Arctic. *Earthwatch* 10:23–29.

46. Brown, L. R., ed. 1991. *The WorldWatch reader on global environmental issues.* New York: W. W. Norton.

47. Lents, J. M., and W. J. Kelly. 1993. Clearing the air in Los Angeles. *Scientific American* 269:32–39.

Chapter 15 Notes

1. McKelvey, V. E. 1973. Mineral resource estimates and public policy. In D. A. Brobst and W. P. Pratt, eds., United States mineral resources, pp. 9–19. U.S. Geological Survey Professional Paper 820.

2. U.S. Geological Survey 1997. The role of nonfuel minerals in the U.S. Economy. Available at minerals.usgs.gov. Accessed 1/15/07.

3. Meyer, H. O. A. 1985. Genesis of diamond: A mantle saga. *American Mineralogist* 70:344–355.

4. Kesler, S. F. 1994. *Mineral resources, economics, and the environment.* New York: Macmillan.

5. Smith, G. I., C. L. Jones, W. C. Culbertson, G. E. Erickson, and J. R. Dyni. 1973. Evaporites and brines. In D. A. Brobst and W. P. Pratt, eds., United States mineral resources, pp. 197–216. U.S. Geological Survey Professional Paper 820.

6. Awramik, S. A. 1981. The pre-Phanerozoic biosphere—three billion years of crises and opportunities. In M. H. Nitecki, ed., Biotic crises in ecological and evolutionary time, pp. 83–102. *Spring Systematics Symposium.* New York: Academic Press.

7. Margulis, L., and J. E. Lovelock. 1974. Biological modulation of the Earth's atmosphere. *Icarus* 21:471–489.

8. Lowenstam, H. A. 1981. Minerals formed by organisms. *Science* 211:1126–1130.

9. Bateman, A. M. 1950. *Economic mineral deposits*, 2nd ed. New York: Wiley.

10. Park, C. F., Jr., and R. A. MacDiarmid. 1970. *Ore deposits*, 2nd ed. San Francisco: W. H. Freeman.

11. Brobst, D. A., W. P. Pratt, and V. E. McKelvey. 1973. Summary of United States mineral resources. U.S. Geological Survey Circular 682.

12. Jeffers, T. H. 1991 (June). Using microorganisms to recover metals. *Minerals Today.* Washington, DC: U.S. Department of Interior, Bureau of Mines, pp. 14–18.

13. Haynes, B. W. 1990 (May). Environmental technology research. *Minerals Today.* Washington, DC: U.S. Bureau of Mines, pp. 13–17.

14. Sullivan, P. M., M. H. Stanczyk, and M. J. Spendbue. 1973. Resource recovery from raw urban refuse. U.S. Bureau of Mines Report of Investigations 7760.

15. Davis, F. F. 1972 (May). Urban ore. *California Geology*, pp. 99–112.

16. U.S. Geological Survey. 2005. Minerals yearbook 2005—Recycling metals. Available at http://minerals.usgs.gov. Accessed 1/15/07.

17. Brown, L., N. Lenssen, and H. Kane. 1995. Steel recycling rising. In *Vital Signs 1995*. Washington, DC: Worldwatch Institute.

18. Wellmer, F. W., and M. Kosinowski. 2003. Sustainable development and the use of non-renewable sources. *Geotimes* 48(12): 14–17.

Chapter 16 Notes

1. Consumers Union. 2005. NYC zero waste campaign announces release of community-based plan for zero waste in New York City. Available at consumersunion.com. Accessed May 7, 2005.

2. Rathje, W. L., and C. Murphy. 1992. Five major myths about garbage, and why they're wrong. *Smithsonian* 23:113–122.

3. Rathje, W. L. 1991. Once and future landfills. *National Geographic* 179(5):116–134.

4. Relis, P., and A. Dominski. 1987. *Beyond the crisis: Integrated waste management.* Santa Barbara, CA: Community Environmental Council.

5. Allenby, B. R. 1999. *Industrial ecology: Policy framework and implementation.* Upper Saddle River, NJ: Prentice Hall.

6. Garner, G., and P. Sampat. 1999 (May). Making things last: Reinventing of material culture. *The Futurist*, pp. 24–28.

7. Relis, P., and H. Levenson. 1998. *Discarding solid waste as we know it: Managing materials in the 21st century.* Santa Barbara, CA: Community Environmental Council.

8. Young, J. E. 1991. Reducing waste-saving materials. In L. R. Brown, ed., *State of the world.* New York: W. W. Norton, pp. 39–55.

9. Steuteville, R. 1995. The state of garbage in America: Part I. *BioCycle* 36:54.

10. Gardner, G. 1998 (January–February). Fertile ground or toxic legacy? *World Watch*, pp. 28–34.

11. McGreery, P. 1995. Going for the goals: Will states hit the wall? *Waste Age* 26:68–76.

12. Brown, L. R. 1999 (March–April). Crossing the threshold. *World Watch*, pp. 12–22.

13. Schneider, W. J. 1970. Hydrologic implications of solid-waste disposal. U.S. Geological Survey Circular 601F. Washington, DC: U.S. Geological Survey.

14. Thomas, V. M., and T. G. Spiro. 1996. The U.S. dioxin inventory: Are there missing sources? *Environmental Science & Technology* 30:82A–85A.

15. Turk, L. J. 1970. Disposal of solid wastes—acceptable practice or geological nightmare? In *Environmental geology*. Washington, DC: American Geological Institute Short Course. American Geological Institute, pp. 1–42.

16. Walker, W. H. 1974 Monitoring toxic chemical pollution from land disposal sites in humid regions. *Ground Water* 12: 213–218.

17. Hughes, G. M. 1972. Hydrologic considerations in the siting and design of landfills. *Environmental Geology Notes*, no. 51. Urbana: Illinois State Geological Survey.

18. Rahn, P. H. 1996. *Engineering geology*, 2nd ed. Upper Saddle River, NJ: Prentice Hall.

19. Bullard, R. D. 1990. *Dumping in Dixie: Race, class and environmental quality.* Boulder, CO: Westview Press.

20. Sadd, J. L., J. T. Boer, M. Foster, Jr., and L. D. Snyder. 1997. Addressing environmental justice: Demographics of hazardous waste in Los Angeles County. *Geology Today* 7(8):18–19.

21. Watts, R. J. 1998. *Hazardous wastes.* New York: John Wiley & Sons.

22. Wilkes, A. S. 1980. Everybody's problem: Hazardous waste. SW-826. Washington, DC: U.S. Environmental Protection Agency, Office of Water and Waste Management.

23. Harder, B. 2005 (November 8). Toxic e-waste is couched in poor nations. *National Geographic News.*

24. Elliot, J. 1980. Lessons from Love Canal. *Journal of the American Medical Association* 240:2033–2034, 2040.

25. Kirschner, E. 1994. Love Canal settlement: OxyChem to pay New York State $98 million. *Chemical & Engineering News* 72:4–5.

26. Westervelt, R. 1996. Love Canal: OxyChem settles federal claims. *Chemical Week* 158:9.

27. Whittell, G. 2000 (November 29). Poison in paradise. *(London) Times* 2, p. 4.

28. U.S. Environmental Protection Agency. 2003. Key Dates in Superfund. Available at http://www.epa.gov/superfund/action/law/keydates.htm. Accessed November 6, 2003.

29. Bedient, P. B., H. S. Rifai, and C. J. Newell. 1994. *Ground water contamination.* Englewood Cliffs, NJ: Prentice Hall.

30. Huddleston, R. L. 1979. Solid-waste disposal: Land farming. *Chemical Engineering* 86:119–124.

31. National Research Council, Committee on Geological Sciences. 1972. *The Earth and human affairs.* San Francisco: Canfield Press.

32. Cox, C. 1985. The buried threat: Getting away from land disposal of hazardous waste. Report no. 115-5. Sacramento: California Senate Office of Research.

33. U.S. Environmental Protection Agency. 2000. Forward pollution protection: The future look of environmental protection. Available at http://www.epa.gov. Accessed August 12, 2000.

Chapter 17 Notes

1. Herd, D. G. 1986. The 1985 Ruiz volcano disaster. EOS, Transactions, American Geophysical Union, 67(19):457–460.

2. Abramovitz, J. N., and S. Dunn. 1998. Record year for weather-related disasters. *Vital Signs Brief* 98-5. Washington, DC: World Watch Institute.

3. Crowe, B. W. 1986. Volcanic hazard assessment for disposal of high-level radioactive waste. In: *Active tectonics.* Geophysics Study Committee. National Research Council. Washington, DC: National Academy Press, pp. 247–260.

4. Advisory Committee on the International Decade for Natural Hazard Reduction. 1989. *Reducing disaster's toll.* National Research Council. Washington, DC: National Academy Press.

5. Kates, R. W., and D. Pijawka. 1977. From rubble to monument: The pace of reconstruction. In J. E. Haas, R. W. Kates, and M. J. Bowden, eds., *Disaster and reconstruction.* Cambridge, MA: MIT Press, pp 1–23.

6. Costa, J. E., and V. R. Baker. 1981. *Surficial geology: Building with the Earth.* New York: Wiley.

7. Rahn, P. H. 1984. Flood-plain management program in Rapid City, South Dakota. *Geological Society of America Bulletin* 95:838–843.

8. Pinter, N. 2005. One step forward, two steps back on U.S. floodplains. *Science* 308:207–208.

9. Mount, J. F. 1997. *California rivers and streams.* Berkeley, CA: University of California Press.

Chapter 18 Notes

1. Moulton, G. E. (1986). *The Journals of the Lewis and Clark Expedition: August 30, 1803 to August 24, 1804.* Lincoln, NE, University of Nebraska Press. Vol V. p. 338.

2. Roberts, L. 1991. Costs of a clean environment. *Science* 251:1182.

3. Fairley, P. 1995. Compromise limits EPA budget cut, removes House riders. *Chemical Week* 157:17.

4. Moore, C. E. 1995. Poisons in the air. *International Wildlife* 25:38–45.

5. Pope, C. A., III, D. V. Bates, and M. E. Raizenne. 1995. Health effects of particulate air pollution: Time for reassessment: *Environmental Health Perspectives* 103:472–480.

6. Hardin, G. 1968. The tragedy of the commons. *Science* 162:1243–1248.

7. Clark, C. W. 1973. The economics of overexploitation. *Science* 181:630–634.

8. Freudenburg, W. R. 2004. Personal communication.

9. Gunn, J. M., ed. 1995. *Restoration and recovery of an industrial region: Progress in restoring the smelter-damaged landscape near Sudbury, Canada.* New York: Springer-Verlag.

10. Costanza, R., et al. 1997. The value of the world's ecosystem services and natural capital. *Nature* 387:253–260.

11. James, A., K., T. Gaston, and A. Blamford. 2001. Can we afford to conserve biodiversity? *BioScience* 51(1):43–52.

12. Litton, R. B. 1972. Aesthetic dimensions of the landscape. In J. V. Krutilla, ed., *Natural environments.* Baltimore. *John Hopkins University Press.*

13. Schwing, R. C. 1979. Longevity and benefits and costs of reducing various risks. *Technological Forecasting and Social Change* 13:333–345.

14. Gori, G. B. 1980. The regulation of carcinogenic hazards. *Science* 208:256–261.

15. Cairns, J., Jr., 1980. Estimating hazard. *BioScience* 20:101–107.

16. Ostro, B. D. 1980. Air pollution, public health, and inflation. *Environmental Health Perspectives* 345:185–189.

17. James, A., et al. 2001. Can we afford to conserve biodiversity?

18. Office of Technology Assessment. 1991. Changing by degrees: Steps to reduce greenhouse gases. Washington, DC: U.S. Superintendent of Documents.

19. D'Arge, R. 1989. Ethical and economic systems for managing the global commons. In D. B. Botkin, M. Caswell, J. E. Estes, and A. Orio, eds., *Changing the global environment: Perspectives on human involvement.* Boston: Academic Press, pp. 327–337.

20. Rogers, A. 1993. *The Earth summit: A planetary reckoning.* NY: Global View Press.

21. Baumol, W. J., and W. E. Oates, 1979. *Economics, environmental policy, and the quality of life.* Englewood Cliffs, NJ: Prentice Hall.

Chapter 19 Notes

1. Callenbach, E. 1975. *Ecotopia.* Berkeley, CA: Bantam Books.

2. Odum, H. T., and E. C. Odum. 2001. *The prosperous way down.* Boulder, CO: The University Press of Colorado.

3. Jusserand, J. 1897. *English wayfaring life in the Middle Ages (XIVth Century).* London: T. Fisher Unwin.

4. Noss, R. F., and A. Y. Cooperrider. 1994. *Saving nature's legacy: Protecting and restoring biodiversity.* Washington, DC: Island Press.

5. Akerman, J. 2006. Space for the soul. *National Geographic* 210(4).

6. Botkin, D. B. 2000 *No man's garden: Thoreau and a new vision for civilization and nature in the 21st century.* Washington, DC: Island Press.

7. McLaughlin, C. C., ed. 1977. *The formative years: 1822–1852.* Vol. 1. *The papers of Frederick Law Olmsted.* Baltimore: Johns Hopkins University Press.

8. Miller, L. B. 1987. Miracle on 104th Street. *American Horticulturalist* 66:14–17.

9. Dobbs, F. 1995. *Curitiba: City of the future?* [Video]. World Bank.

10. Rabinovitch, J. 1997. *Integrated transportation and land use planning channel Curitiba's growth.* Washington, DC: World Resources Institute.

11. Ibid.

12. Hunt, J. 1994 (April). Curitiba. *Metropolis.*

13. Rabinovitch, J. 1997. *Integrated transportation and land use planning channel Curitiba's growth.*

14. Rabinovitch, J., and J. Leitman. 1996 (March). Urban planning in Curitiba. *Scientific American,* pp. 45–53.

15. Steiner, F. 1983. Regional planning: Historic and contemporary examples. *Landscape Planning* 10:297–315.

16. Cohen, B. S. 1970. The constitution, the public trust doctrine and the environment. *Utah Law Review* 388.

17. A brief history of the National Park Service. Available at http://www.cr.nps.gov/history/hisnps/NPSHistory/briefhistory.htm. Accessed January 24, 2007.

18. Bureau of Land Management facts. Available at http://www.blm.gov. Accessed January 24, 2007.

19. Stone, C. D. 1996. Should trees have standing? Dobbs Ferry, NY: Oxford University Press, p.181.

20. Bacow, L. S., and M. Wheeler. 1984. *Environmental dispute resolution.* New York: Plenum Press.

Index

Page numbers followed by f denote material in a figure. Page numbers followed by t denote material in a table.